This book offers the first complete study of the origins of American intelligence testing. It follows the life and work of Henry Herbert Goddard, America's first intelligence tester and author of the most popular American eugenics tract, *The Kallikak Family*. Leila Zenderland traces the controversies surrounding Goddard's efforts to bring Alfred Binet's tests of intelligence from France to America and to introduce them into the basic institutions of American life – from hospitals to classrooms to courtrooms. She shows how testers used their findings to address the most pressing social and political questions of their day, including poverty, crime, prostitution, alcoholism, immigration restriction, and military preparedness. The book also explores the broader legacies of the testing movement by showing how Goddard's ideas helped to reshape the very meaning of mental retardation, special education, clinical psychology, and the "normal" mind in ways that would be felt for the rest of the century.

Measuring Minds

Cambridge Studies in the History of Psychology

GENERAL EDITORS: MITCHELL G. ASH AND
WILLIAM R. WOODWARD

This series provides a publishing forum for outstanding scholarly work in the history of psychology. The creation of the series reflects a growing concentration in this area by historians and philosophers of science, intellectual and cultural historians, and psychologists interested in historical and theoretical issues.

The series is open both to manuscripts dealing with the history of psychological theory and research and to work focusing on the varied social, cultural, and institutional contexts and impacts of psychology. Writing about psychological thinking and research of any period will be considered. In addition to innovative treatments of traditional topics in the field, the editors particularly welcome work that breaks new ground by offering historical considerations of issues such as the linkages of academic and applied psychology with other fields, for example, psychiatry, anthropology, sociology, and psychoanalysis; international, intercultural, or gender-specific differences in psychological theory and research; or the history of psychological research practices. The series will include both single-authored monographs and occasional coherently defined, rigorously edited essay collections.

Also in the series

Measuring Minds

Henry Herbert Goddard and the
Origins of American Intelligence Testing

Leila Zenderland

PUBLISHED BY THE PRESS SYNDICATE OF THE UNIVERSITY OF CAMBRIDGE
The Pitt Building, Trumpington Street, Cambridge, United Kingdom

CAMBRIDGE UNIVERSITY PRESS
The Edinburgh Building, Cambridge CB2 2RU, United Kingdom   http://www.cup.cam.ac.uk
40 West 20th Street, New York, NY 10011-4211, USA   http://www.cup.org
10 Stamford Road, Oakleigh, Melbourne 3166, Australia

First published 1998
Reprinted 1999

Printed in the United States of America

Typeset in Times Roman

Library of Congress Cataloging-in-Publication Data
Zenderland, Leila.
Measuring minds : Henry Herbert Goddard and the origins of
American intelligence testing / Leila Zenderland.
p.   cm. – (Cambridge studies in the history of psychology)
1. Intelligence tests – United States – History.   2. Goddard, Henry
Herbert, 1866–1957.   3. Psychologists – United States – Biography.
I. Title.   II. Series.
BF431.Z46   1997
153.9′3′097309041 – dc21   97-6101

*A catalog record for this book is available from
the British Library*

ISBN 0 521 44373 3 hardback

Contents

v

Preface

Writing about the life of Henry Herbert Goddard has made me appreciate the ironies of history. On a more personal level, nothing seems more ironic in hindsight than my reason for choosing to study Goddard in the first place. As a graduate student, I had become fascinated with the history of intelligence testing after reading Leon Kamin's study, *The Science and Politics of I.Q.* However, since I was already juggling my coursework with a job as an editor, I thought it unwise to undertake such a massive project for a dissertation. So instead I chose to focus on Goddard. After all, I asked myself, how long could it take to explicate the ideas of the author of *The Kallikak Family?* And besides, the broader framework for understanding Goddard's science, the heredity-environment debate, had largely been worked out. At the time, I had only one serious reservation: was it fair, I asked my advisor Charles Rosenberg, to write a biography of a subject whom my reading had already led me to dislike? "Don't worry," I remember him replying. "In studying Goddard, you'll come to see the world as he saw it, and then you'll understand him."

Far too many years later, I have finally completed what was supposed to be a short study. The life of Goddard led me into the history of intelligence testing. Both subjects proved far more complex, more surprising, and more intellectually challenging and rewarding than I had ever imagined. Trying to see the world through Goddard's eyes reoriented my perspective. Moreover, the heredity-environment framework, which seemed so clear when I began, soon proved of relatively little explanatory value. Instead, I had to find new frameworks to explain both Goddard and his science. The result is this book.

Working on such a long project also means incurring a long list of debts. This is especially true in my case, for in order to understand the history of testing, I also had to understand the histories of psychology, medicine, and biology – fields far removed from my own background in American cultural history. Fortunately, I was able to find both institutional support and generous individuals willing to share their expertise.

My experiences as a graduate student in the University of Pennsylvania's Department of American Civilization proved invaluable in preparing me for this study, for its chairman, Murray Murphey, strongly believed that the social sciences and the humanities could be, and should be, interconnected. Since this department encouraged interdisciplinary scholarship, I was able to work closely with faculty from other departments including those of History and the History and Sociology of Science. Bruce Kuklick, Henrika Kuklick, and Charles Rosenberg all profoundly influenced my approach to the history of ideas. Influential as well were graduate students from all three departments doing related research, among them Janet Tighe, Lou Zanine, John O'Donnell, Nancy Tomes, Jim Capshew, and Jack Pressman.

I owe a large debt to the members of Cheiron, the International Society for the Study of the Behavioral and Social Sciences, who willingly shared their expertise in the history of psychology. Although members who helped me are too numerous to name, I am especially grateful to John Burnham, Don Dewsbury, Ben Harris, Horace Marchant, and Kathy Milar, as well as to the supportive community of scholars with special expertise in the history of testing: John Carson, Steven Gelb, Henry Minton, Jim Reed, Peter van Drunen, and Richard von Mayrhauser. Mike Sokal proved a wonderful resource in more ways than I can mention. Historians of biology were generous as well; I thank Hamilton Cravens, Dan Kevles, Phil Pauly, and especially Diane Paul. My efforts to balance the history of science with American social history were also helped by the Los Angeles Social History Study Group; among its members who commented on several chapters of this book are Hal Barron, Phil Ethington, Doug Flaming, Nancy Fitch, Darryl Holter, Sandy Jacoby, John Laslett, Margo McBane, Jan Reiff, Steve Ross, Bob Slayton, and Frank Stricker. Finally, I am especially grateful to Ray Fancher and Franz Samelson, who read this manuscript for Cambridge, and to the series editors Mitchell Ash and Bill Woodward; all four made astute suggestions for revisions that challenged my thinking in the best of ways.

Financial support proved crucial in allowing me to conduct the archival research necessary for this study. I began my work with a summer grant from the Archives of the History of American Psychology at the University of Akron. Summer stipends from both the National Endowment for the Humanities and the History and Philosophy of Science Division of the National Science Foundation, as well as several research grants from California State University, supported additional visits to archives. Numerous archivists, librarians, psychologists, and historians helped me locate and interpret materials used in this study. Marion White McPherson and John Popplestone were unfailingly supportive and extremely resourceful in retrieving archival records during my many trips to Akron. I also thank John Miller and Sharon Ochsenhirt, as well as Bob Zangrando for his hospitality. John Rose of the Vineland Training School helped me locate historic photographs. Pnina and Z'ev Kronish made it possible for me to find numerous materials in New

York City. Mrs. H. Weiss of the Department of Records, Israelitsche Kultusgemeinde of Vienna, tracked down the fate of Goddard's friends, the Krenbergers. Clinical psychologists Richard Flaten and Chris Milar helped me understand contemporary diagnostic methods, while Marie Skodak Crissey shared her memories of Goddard's methods. And I am especially grateful to the people of East Vassalboro, Maine, among them Betty Taylor, Margaret Cates, the relatives of Rufus Jones, and above all Esther Holt, who introduced me to their community and to their Quaker meeting, and who helped me to reconstruct the world Goddard would have known as a child.

Support I received from my own institution, Cal State Fullerton, proved invaluable as well. Barbara Campbell was a wonderful research assistant; other graduate students also helped me track down library materials. Staff members Jo Ann Robinson, Giulii Kraemer, and Doug Temple went out of their way to help me on countless occasions; I also thank Nancy Caudill from Interlibrary Loan and Mike Riley from Photographic Services. Colleagues from other departments generously shared their expertise, among them Chris Cozby and Richard Lindley in psychology, Nancy Fitch in History, Jill Rosenbaum in Criminal Justice, and Brad Starr in Religous Studies. Above all, I know I am very fortunate in working with an exceptionally talented group of colleagues in the Department of American Studies: Allan Axelrad, Jesse Battan, Wayne Hobson, John Ibson, Karen Lystra, David Pivar, Terri Snyder, Michael Steiner, Pam Steinle, and Jim Weaver. I thank them all for their friendship, their thoughtful responses to my work, and their unwavering faith that my often-arcane studies in psychology, medicine, and biology did indeed form an important part of American cultural history. Wayne, David, Allan, and Karen deserve special thanks, for they valiantly read this entire manuscript at different points; I am deeply grateful for their incisive suggestions, many of which are now a part of my text.

I owe a particular debt to family and friends who repeatedly offered me support and advice over the many years it took to write this book, especially Debbie Forczek and Leo Ribuffo. Finally, I thank my parents, Harry and Sylvia Zenderland, who taught me to love history and to know that it matters.

Introduction: Motives, Meanings, and Contexts

"I have been asked to write my autobiography," psychologist Henry Herbert Goddard, America's first intelligence tester, told a colleague in 1947. Such a task would be difficult, he knew, for by then Goddard was over eighty years old. "I have no intention of doing it," he continued, "but if I were to write it, I should call it: 'As Luck Would Have It.' Not one single thing of importance in my career," Goddard confessed, "was the result of my planning, foresight or wisdom."[1]

Whether true or not, such a confession serves as a chastening reminder to historians tempted to interpret the past largely as the product of conscious actions, intentions, or motives. It is especially ironic coming from a psychologist who became world famous for his belief that an individual's future is determined by his intelligence. A few years later, however, Goddard repeated these sentiments. Answering some poignant questions posed by a former student, psychologist Robert Fischer, Goddard again reflected on his life – and reached the same conclusion.

"So many of us," Fischer had written him in 1950, lived lives with "almost no discernible pattern or purpose." Disparaging those with "no thought of the future," Fischer begged Goddard to "tell us of your experiences" and thereby "point the way that we may escape the frustrations that came to you because you were a pioneer in our field." "*What would you do,*" this psychologist asked his mentor, "*if you could relive your professional life?*" Of course, answering such a question required caution, Fischer warned, for the "temptation to retrospectively assign motives to our lives is always with us."[2]

Goddard, however, felt no such temptation. Instead, his reply suggested the opposite: he wished to disavow all motives. "I have not *Made* my life," he responded emphatically. "I have taken things as they came and tried to make the best of them." Goddard conceded that he had been writing up some of his reminiscences for family and friends; these memories, however, had only confirmed his deeper feeling that his life had lacked direction. "Do you know what I have decided to call this composition that the folks refer to as my autobiography?" he asked. "I was not a bit interested to write it until I thought of this TITLE . . . AS LUCK WOULD HAVE IT." The title was

1

appropriate, for "everything of importance in my life," he explained, "has been LUCK and NOT the result of thought and planning." On this point, Goddard was adamant. "I never planned to be a teacher, psychologist or loafer," he concluded. "It just happened."[3]

Such a claim is intriguing, for Henry Herbert Goddard had indeed been a pioneer in his field, and what happened during his lifetime had profoundly altered American social science. In 1908, Goddard became the first American to recognize the potential of the new "intelligence tests" invented three years earlier by French psychologist Alfred Binet. Over the next decade, he became world famous as the leading spokesman of a movement which introduced these new mental measuring devices into the basic institutions of American life.

In less than ten years, Goddard had won legitimacy for intelligence testing in ways only dreamed of by Binet himself. By 1910, he had convinced American physicians to try intelligence testing. By 1911, he had used the same tests in public schools. By 1913, he had tried out his tests at Ellis Island. By 1914, he had become the first psychologist to present evidence from Binet tests in a court of law. By 1918, he had even helped to introduce intelligence testing into the United States Army. Although the leadership of the testing movement had by this time passed to his younger successor, psychologist Lewis Terman, Goddard's place in history was secure, for his actions had led to the institutionalization of intelligence testing, thus transforming both his profession and his society. Intelligence tests, one of his contemporaries bragged, had finally "put psychology on the map of the United States."[4] They had also instigated what would become one of the most intense, long-lasting, and bitterly contested scientific and social controversies of the twentieth century.

The very idea that any single psychologist could introduce so many controversial innovations into so many different American institutions is in itself surprising. Far more surprising is the fact that this psychologist was Henry Herbert Goddard. A former Quaker schoolteacher who had earned a psychology degree from Clark University in 1899, Goddard was hardly a prominent star in his field. To the contrary, his early career had been rather undistinguished, for following his graduation he taught pedagogy at the State Normal School in West Chester, Pennsylvania, before moving in 1906 to the Training School for Feeble-Minded Girls and Boys, an obscure little institution in Vineland, New Jersey. A genial New Englander with a personal passion for mountain climbing, he hardly seemed the sort of scientist capable of transforming American psychology. Yet in introducing intelligence testing into American society, Goddard did just that.

Nearly nine decades later, Binet's approach toward mental measurement and Goddard's actions in institutionalizing it remain both extremely controversial and extraordinarily influential. Intelligence tests have played, and still continue to play, highly publicized roles in numerous American political controversies, from debates over immigration restriction in the 1920s through clashes concerning antipoverty programs in the 1960s to battles over affirmative action in the 1990s. At the same time, they have also served as a model for thousands of other types of mental tests quietly adopted by physicians, educators, and psychologists with far less controversy. "The general mental test," psychologist Lee Cronbach argued the 1960s, was "the most important single contribution of psychology to the practical guidance of human affairs." And while such a claim is itself controversial, it again highlights Goddard's historical importance, for it was he who first showed his contemporaries how to move psychological testing out of the laboratory and into society.[5]

Henry Herbert Goddard's life and the early American intelligence testing movement he led form the subjects of this book. Both subjects have generated much polemical writing in recent decades but relatively little social history. Most histories of testing mention Goddard; yet while several studies examine the life and work of Lewis Terman, Goddard's life has yet to inspire a biography. There are no studies tracing all of Goddard's efforts to institutionalize intelligence testing in the decade before World War I. As a result, much of the early history of this very public controversy remains surprisingly unknown.[6]

Instead, most historical attention has focused on the broader controversy within which intelligence testing ultimately became inextricably enmeshed: the heredity–environment controversy. Binet's ideas about intelligence were soon absorbed into an older, wider, and still ongoing debate – a debate over the origins of human differences, their responsiveness to environmental improvement, and their significance for social change. In its most polarized form, this debate pitted those who blamed most of humanity's failings on poor "nature," or biological inheritance, against those blaming poor "nurture," or social environment. The role played by Goddard's writings within this broader debate is by now well documented, for his most famous works stressed the power of heredity and strongly supported the emerging eugenics movement. Even in an age when social scientists often made bold claims and sweeping generalizations, his statements are striking, for as Stephen Jay Gould concluded in *The Mismeasure of Man,* Goddard "may have been the most unsubtle hereditarian of all."[7]

Ironically, Goddard's writings about heredity have received increasing attention in recent years – not because they merit support, but because they have been so thoroughly and resoundingly rejected. Few works by any American scientist have suffered as dramatic a rise and fall as Goddard's once famous and now notorious 1912 monograph, *The Kallikak Family.* To Goddard, this family study offered convincing proof of the devastating social consequences caused by a feeble inheritance. Its publication led contemporaries to praise his research as a major scientific accomplishment. Yet in the decades that followed, this book was subjected not only to criticism but also to ridicule. By 1940, psychologist Knight Dunlap wrote in *Scientific Monthly,* Goddard's Kallikak study had been "laughed out of psychology."[8]

In the decades since, scholars have continued to expose the deeply flawed nature of this once respected work of science. Yet exposing this book's errors has proven to be a far easier task than explaining its earlier scientific acceptance. By 1969, Goddard's research shortcomings had become "so obvious," psychologists Seymour Sarason and John Doris conceded, "that pointing them out by themselves is without much interest. It only becomes of interest," they emphasized, "when we ask why these studies . . . were so widely accepted . . . not just by unsophisticated and gullible segments of the public but by professionally trained individuals of outstanding intellect."[9]

Even during Goddard's lifetime, a number of critics had begun to ask precisely the same question. Their answers usually focused on the matter of motives – which might in part account for Goddard's insistent denial of *any* intentions in his life or work. This theory received its most elaborate expression in a 1948 dissertation written by educational psychologist Nicholas Pastore. For his thesis, Pastore examined the writings of twenty-four English and American scientists involved in the heredity–environment debate. Published a year later as *The Nature–Nurture Controversy,* it too focused explicitly on motives.

According to Pastore, the stances taken by these scientists were closely linked to their political predispositions – predispositions he classified as either "liberal" or "conservative." Hereditarians, Goddard among them, were political conservatives, he argued, while environmentalists were political liberals. "The philosophy underlying social reform is environmentalism," Pastore explained, whereas "the position of the hereditarian would be to favor the *status quo.*" Pastore's conclusion was clear: in this instance, scientific conflict was really political conflict, thinly disguised.[10]

This thesis has largely shaped most subsequent research into the history of testing. It has proven especially useful in explaining the blind spots, lapses in logic, and overheated rhetoric which often pervade these polemical writings. In the decades since, numerous studies have added supporting evidence to this argument linking science to politics by exposing the nativist, racist, and class-biased assumptions of many testers, and by showing how psychologists used the lower scores earned by the poor, immigrants, and nonwhites to justify the growing gap between capital and labor, to argue for immigration restriction, or to defend racial segregation. They have also shown how these cultural biases were frequently built into the tests themselves, thus invalidating their ability to tell us anything definitive about the relative influence of nature or nurture.[11]

Yet in 1948, Pastore's hypothesis about the political ideas motivating hereditarians drew bitter protest from the two American scientists most active in the early American intelligence testing movement – Henry Herbert Goddard and Lewis Terman. After reading Pastore's summaries of their scientific and political beliefs, Goddard expressed anxiety and alarm, and Terman, anger.

Writing in reply, Terman attacked Pastore's portrayal – not for distorting his science but for misrepresenting his politics. Above all, this psychologist bristled at being called "conservative," and responded with a detailed political statement. Though a Republican, he conceded, he had voted for Woodrow Wilson and Franklin Roosevelt, and among 1948 contenders, he liked the conservative Robert Taft least. "As for political ideology," he added, "I hate every form of national totalitarianism, whether of the Stalin, Hitler, Mussolini, Franco, Peron, or Japanese variety." Although "not a socialist," Terman explained, "I am not afraid of the partial socialization now operating in Britain or Sweden, or Norway." After all, the United States had "socialized education," he argued, "and for 35 years I have believed that every argument for socialized education is valid also for socialized medicine." In sum, Terman believed in "stiff inheritance taxes, old-age pensions, social security measures, unemployment insurance, minimum wage laws," and "the necessity of labor unions." "Most of all," he added, "I believe in civil liberties. . . . " America's "widespread racial and religious discrimination" was disgraceful, while the "witch-hunting and character-smearing activities" of groups like the Dies Committee were "the most un-American thing in the USA. I feel so strongly about such threats," Terman claimed, that "if called . . . I would go to jail rather than answer any questions about my political beliefs. . . . "[12]

Terman's protest was so vehement that Pastore had to recategorize him as an exception to his argument. His earlier writings notwithstanding, Pastore concluded, intelligence tester Lewis Terman was a liberal; more precisely, as Terman defined himself, he was a "New Dealer."[13]

While Terman focused on Pastore's political classification, Goddard challenged his scientific dichotomy. "I think perhaps you have been misled," Goddard noted in drafting a response to Pastore, "having the answer, before you had the problem." His scientific ideas, Goddard maintained, were being taken out of context. "You forget that this work was done about 40 years ago," he reminded the younger scholar, "when the problem of 'nature-nurture' had not been formulated. Accordingly you read into the language that I used in 1912 and 1914, meanings and thoughts that you have acquired in the nineteen-forties." While Goddard's reply was tactful, it also focused on motives. "I do not mean that you have *intentionally* done this," he went on. "Not at all. It is perfectly natural that you should interpret my language in terms of to-day's experiences; but that does not state the facts," he added, "as much as it states what you judge were the facts, or even what you *wish* the facts were." After all, Goddard concluded self-consciously, we all "do a lot of wishful thinking."[14]

These protests have usually been overlooked or dismissed as the defensive reactions of persons whose views, while once popular, were rapidly losing favor in the post–World War II world. To a historian, however, both are provocative. Terman's retort is intriguing, for it suggests that his many nativist, elitist, or racist statements of the 1910s should not be used to predict his politics in the 1940s, by which time the terms "liberal" and "conservative" had acquired very different political meanings. Goddard's reply is more problematic, for it argues that the scientific meaning of the "problem of 'nature-nurture'" had also changed since 1912. For anyone interested in understanding the origins of the American intelligence testing movement, both claims are important, for each implies a protest against presentism and a plea for more complex explanations – explanations which would allow for a wider range of motives, and which would pay much closer attention to changes in meanings and contexts.[15]

The question of meanings is especially difficult in assessing the early history of this controversy, for during the pre-World War I decade in which Goddard's writings proved most influential, both the scientific and the popular connotations of the words most commonly invoked in this debate – words such as "heredity," "environment," and "intelligence" – were themselves undergoing a process of redefinition. Also in flux were the meanings of many other crucial terms used by Goddard and his contemporaries. As a consequence, historians must be especially cautious, for the same words or phrases might connote something subtly different from one decade to the next.

These linguistic problems will be most obvious if we examine the words used to identify the subject that Goddard was studying – the condition then called "feeblemindedness," later called "mental retardation," and more recently called "developmental disability." When Goddard began his research in 1906, the word "feeblemindedness" had itself replaced an older generic term which had since developed a much harsher connotation: "idiocy." Goddard was especially proud of his own semantic contribution in inventing the word "moron" in 1910 as a technical term that as yet carried no popular connotation. Within a decade, this word would be widely used by the public. "In the clever talk of the twenties," one psychologist has observed, "'moron' claimed as large a place as 'Babbitt.'" Scientists soon had to abandon this term, for its popular meaning had come to suggest derision. In the following decades, the terminology used by practitioners in this field has continued to change, leaving each generation to invent its own labels and to criticize the stigmatization evident in the labels used by its predecessors.[16]

The writings of sociologist Stanley Davies illustrate this process especially starkly. In 1923, Davies published his dissertation under the title *Social Control of the Feebleminded*. Reissuing this book in 1930, he changed its name to *Social Control of the Mentally Deficient;* the word "feebleminded" was no longer acceptable. By 1959, the phrase "social control" had been equally stigmatized, and Davies named his next book on the same topic *The Mentally Retarded in Society*. Were it being reissued today, its name would probably change again. "Under no circumstances," states the current editorial policy of the American Association on Mental Retardation (an organization whose own name has changed four times since 1876), "should *retarded* be used as a noun," for such usage suggests an entire identity rather than a characteristic. (Today's title might instead be *People with Mental Retardation in Society*). In none of these works did Davies intend his language to be pejorative. Such changes in terminology may signify an increasing refinement of ideas in this field, as well as a growing self-consciousness concerning the power of words; even so, they still complicate the problem of recapturing past meanings.[17]

Other words pose even more problems. Perhaps none has acquired as negative a connotation as "eugenics," a term coined in 1883 by Francis Galton from a Greek root meaning "good in birth" or "wellborn." To late-twentieth-century readers, this word usually brings to mind the biological policies put into practice in Nazi Germany. Recent historical studies, however, suggest that "eugenics," too, is a term requiring precise historical contextualization, for a wide variety of eugenic ideas proved influential in at least thirty countries during the first four decades of the twentieth century. The broad array of persons who considered themselves eugenicists shared a belief that scientific control over the processes governing human heredity would provide benefits to society. What such advocates meant when they spoke of "control," or of "benefits," however, differed markedly from decade to decade, from country to country, and often from individual to individual.[18]

The pronouncements of its leaders notwithstanding, eugenics in the pre–World War I era was hardly a unified movement with a single objective; instead, it was a broad coalition of persons or groups promoting overlapping yet diverse scientific, social, or political agendas. Thus, while its most prominent American leader, Charles Davenport, might have used his eugenic science to express, as historian Daniel Kevles put it, "the native white Protestant's hostility to immigrants and the conservative's bile over taxes and welfare," the same movement also attracted socialists like Karl Pearson, Fabians like George Bernard Shaw, sexual reformers like Havelock Ellis, and a few immigrant radicals like Emma Goldman, as well as a diverse group of physicians, agriculturalists, researchers, and popularizers. Simply labeling any individual a "eugenicist" thus tells us little in itself, and might even be misleading.[19]

It is not only the scientific vocabulary of this era that requires contextualization; equally troublesome is the political vocabulary. Perhaps no word has proven more slippery to American historians than the most commonly invoked political adjective of Goddard's day, "progressive." In the first two decades of the twentieth century, both Republican reformers who followed

Theodore Roosevelt and Democratic admirers of Woodrow Wilson often used this label to describe their own activities. The reforms they advocated ranged from antitrust laws to women's suffrage, from the city manager movement to the formation of regulatory agencies, from prohibition to segregation. Such a movement defies easy translation into today's political vocabulary, for "progressivism" was neither "liberalism" nor "conservatism."[20]

In recent decades, historians seeking to understand the common currents of this movement have focused less on specific reforms and more on broader changes in the very processes of reform. These included the formation of new issue-oriented lobbying groups, new means of molding public opinion, new strategies to empower the middle class. Especially important was the connection between progressive political reform and the reforming of professions. The process now called "professionalization" brought new political and social power not only to older occupational groups like doctors or engineers but also to newer "social scientists," including psychologists. In this sense, progressivism constituted an organizational revolution, a revolution aimed at restructuring the bureaucratic foundations of American government.[21]

Progressives, too, used language in their own ways. During these decades, activists all along the political spectrum pleaded their cases in metaphors suggesting engineering skill, scientific control, and managerial expertise. Perhaps nothing expresses the progressive ethos better than the many positive connotations these reformers attached to their favorite adjective, "efficient." "Efficient and good came closer to meaning the same thing in these years," historian Samuel Haber has concluded, "than in any other period of American history." Yet interpreting this language also requires contextualization. Most progressive educators, for instance, repeatedly touted the need for increased "school efficiency"; such a phrase, however, might mean extending the school's outreach or curtailing its funding, expecting more from students or resigning oneself to expect less.[22]

Even if one could fully explain the ideological contours of the "eugenics" movement or the "progressive" movement or the "school efficiency" movement in all their complexity, these movements in themselves would not do justice to the range of motives shaping the life of a social scientist such as Henry Herbert Goddard. After all, Goddard's lifetime spanned nearly a century of American social, intellectual, and scientific transformation. He was born in 1866, one year after the end of the Civil War, in a New England village deeply influenced by the reformist, temperance, and abolitionist movements of the previous half-century. At the time of his death nearly ninety-one years later, in 1957, he was living a quiet life in Santa Barbara, California, having completed his last book, *How to Rear Children in the Atomic Age*. Such a life defies all easy assumptions about the influence of any single explanatory "zeitgeist." Instead, both Goddard and the movement he led need to be explained within a framework emphasizing diverse and overlapping contexts – contexts which themselves changed over time.

The main purpose of this study is to provide such a framework. To explore more fully the range of motives influencing the earliest American intelligence testers, this book is really two studies intertwined – the first a biography of Henry Herbert Goddard, and the second an examination of institutional responses to testing in the years between 1908 and 1918. I also pay special attention to what "intelligence" meant within a variety of different contexts – medical, educational, biological, social, and political – and to the ways in which its meaning changed. To find answers to these questions concerning motives, meanings, and contexts, I have approached this study with three broad objectives in mind.

The first of these has been the most crucial: to expand the paradigm within which the history of testing is usually discussed – that is, the heredity-environment controversy. While emphasis on the nature–nurture debate has shed much-needed light on the deeper connections between science and politics, it has also overshadowed other controversies provoked by Binet's innovation. Most obscured by these broader accounts are the more subtle scientific and social debates of the day. To understand the multiple ways in which intelligence tests first entered into American society, we need to examine a wider range of controversies.

This is especially true concerning the years between 1908 and 1918, for during this decade, the testing debate had yet to crystallize fully. By the 1920s, the most vocal participants in this controversy had taken their stances as either "hereditarians" or "environmentalists"; such self-conscious polarization, however, is not nearly so evident a decade earlier. Despite Goddard's own hereditarian pronouncements, his earliest defenders and attackers did not fall neatly along the sides of a single heredity-environment divide. Instead, participants debated a wide range of issues, many of which are still being debated at the end of the century. In order to connect testing's past with its present, historians need to examine not only the controversies that captured the most publicity, but also those that more quietly reshaped professional practices.

For instance, whereas the debate over the causes of mental subnormality has received a great deal of historical attention, an even older medical debate – the debate over how to diagnose just what it was that one either inherited or acquired – has been largely overlooked. Yet this debate was crucial, for Binet's invention offered doctors something they had been seeking for nearly a century: a set of uniform criteria for diagnosing different degrees of mental impairment. It was testing's role in this debate, far more than any concern with heredity, that best explains its medical acceptance. Almost immediately, institutional physicians began to use intelligence tests to help diagnose and classify cases, whatever their cause – a practice that has continued, albeit with many modifications, from Goddard's day to our own.[23]

Testing also entered the schools within multiple contexts. Historians of education have shown how intelligence tests were eventually used to stratify students into different school "tracks," thus delimiting both their academic and their social mobility – a process that paralleled the broader class stratification then under way in the surrounding society. Yet many early testers were also avid participants in the public health movement and promoted efforts to offer free medical care to all schoolchildren, a fact that again challenges any simple dichotomy pitting "hereditarian" against "environmental" reform in this decade. Even more important was testing's close connection to the special education movement. Without exploring this context, it would be difficult to explain why it was Goddard who in 1911 helped to draft this nation's first law requiring a state to provide special education for blind and deaf as well as "feebleminded" children within public schools.[24]

Our understanding of the role played by intelligence testing in shaping social policies also needs to be broadened. Much attention has focused on Goddard's now discredited theory positing hereditary feeblemindedness as a major cause of crime. Far more controversial in his own day, however, were his ideas about criminal responsibility, for Goddard also argued that murderers diagnosed as having subnormal intelligence be spared the death penalty. More than seventy years later, this issue was still being hotly contested in a death penalty case which finally reached the United States Supreme Court – a case in which evidence from intelligence testing once again played a key role.[25]

To understand the range of roles that psychological testing came to play within American society for the rest of the century, we need to regain more of the questions being asked and answered in this crucial first decade. For while Goddard and his contemporaries did indeed debate the relative influence of nature and nurture, so too did they debate other issues. And among these were issues concerning the mind of the child, the mission of the school, the role of the state, and the interrelationships among all three.

The second objective of this study is more specific: to trace the social processes that actually permitted the dissemination of intelligence tests into American institutions. To do this, I have tried to eschew the easy assumption, too often made in histories of testing, that these mental measuring devices were automatically accepted simply because they "fit" well with the larger forces reshaping American society. These forces are usually described in broad terms – the bureaucratic tendencies overtaking public schools, for instance, or the desire to stratify the labor market to meet the needs of industrial capitalism, or the growing hostility toward immigrants and nonwhites. All of these forces did indeed play important roles in both shaping and disseminating tests of intelligence. Even so, institutional adoption was never as automatic as it now appears in retrospect. Instead, it was a complex and contingent historical process – a process that sometimes succeeded and sometimes failed.

At its heart, this process meant that psychologists such as Henry Herbert Goddard were trying to exercise power in ways previously unknown to members of their profession. They were in a position to do so precisely because Binet's invention had offered them a new claim to expertise in an area with important social ramifications: diagnosing mental retardation. With tests in hand, psychologists argued that they were best able to "measure minds" and therefore ought to be allowed to make decisions. Yet such arguments were hardly accepted simply because their claimants asserted them. Among those who challenged them were doctors, teachers, lawyers, reformers, bureaucrats, and politicians. As a consequence, the institutional history of testing is a story of cooperation and conflict, of rivalries and compromises among various groups exercising significant social power.

To trace this process, I have followed Goddard as he tried to convince institutional superintendents, school administrators, courtroom lawyers, social workers, Ellis Island officials, army officers, and many others of both the scientific legitimacy and the social efficacy of Binet's new invention, thus instigating one controversy after another. I have focused not only on Goddard's intentions but also on his interactions with a wide variety of persons in positions to support or resist his ideas. These included medical inspectors as well as institutional doctors, classroom teachers as well as school principals, juries as well as judges.

This study also has a third objective: to reconstruct, as fully as possible, the mental perspective of participants in these early testing debates. To do so, I have paid particular attention to the language used by Goddard and his contemporaries to describe what they saw and did. This includes the rhetorical flourishes they employed to embellish their evangelical pronouncements as well as the technical vocabulary they invented to describe their scientific discoveries. I have also retained the terms they used to describe the persons they were studying – terms such as "idiot," "imbecile," "feebleminded," "moron," "degenerate," "laggard," "mongolian," "cripple," and "defective" – words that now make the modern reader wince.

This problem is by far the most difficult, for a wide gulf separates the consciousness of late-twentieth-century Americans from that of their contemporaries in the opening decades of the same century. Two historical events in particular have profoundly altered the ways that readers respond to these writings of an earlier era: World War II, with its legacy of genocide, and the civil rights movement of the 1960s. Both of these events fostered a painful new awareness of the pervasiveness of ethnic, religious, and racial prejudices, and of their devastating consequences. To many historians, both also suggested a pressing moral imperative: to expose the close connections between popular prejudices and faulty scientific theories which worked to legitimate and bolster them.

This imperative has shaped much recent writing on the history of the testing movement. To readers aware of the modern meaning of culture (a concept that was itself only beginning to be explored in the decade in which testing was introduced), the biases that infuse the writings of these supposedly objective scientists will be transparent. In Goddard's case, the problem of cultural bias will be especially obvious. For although Goddard shared neither the xenophobia nor the extreme racism common to social thinkers of this pre-war decade – a decade in which Madison Grant's anti-immigrant diatribe, *The Passing of the Great Race,* as well as D. W. Griffith's film glorifying the Ku Klux Klan, *Birth of a Nation,* found immense and approving audiences – his writings do suggest a clear and consistent pattern of class prejudice. (In this sense, psychologist Leon Kamin's wry assessment of Lewis Terman – that he "should not be thought of as a racist" because his "stern eugenical judgment was applied even-handedly to poor people of all colors" – fits Goddard even more aptly.) Such ideas have made Goddard's writings relevant to a new generation, but hardly in the ways that he intended: they are now used largely to show how social prejudices can distort scientific judgment.[26]

This very relevance, however, has made it especially difficult to assess Goddard's writings within the context of his own times. To do so, it is necessary to avoid judging these past actors against present-day standards of social consciousness. Such a stance is especially crucial in Goddard's case, for although works such as *The Kallikak Family* have by now become all too easy to attack, they are still hard to explain. Even the concept of prejudice does not fully answer the important question raised by critics – why Goddard's writings appealed so strongly to some of the best scientific minds of his day. To see why Goddard's solutions proved so persuasive to his contemporaries, we need to compare them with available alternatives. This means paying less attention to the ways in which Goddard's science differs from our own, and more to the ways in which it differed from the science that preceded it.

Modern audiences may also be disturbed by the attitudes toward welfare policies in general, and institutions in particular, embraced by reformers of Goddard's generation. To readers sensitized to the potential for institutional abuse and the need to defend individual civil rights (including the rights of disabled citizens to be fully integrated into society), the policies advocated by Goddard and his contemporaries may seem hard to fathom. To understand them, it is necessary to recapture the mentality of a different world, a world in which even secular reform still retained a large dose of missionary fervor, a world that still believed in both the possibility and the desirability of benevolent paternalism.

To Goddard (who had spent much of his own youth in boarding schools), the modern skepticism toward all institutions would be incomprehensible. Institutions, Goddard believed, could be either good or bad places; at their best, they could simultaneously protect society while providing its most desperate, most needy, or most unwanted members with a refuge, a home, a "haven in a heartless world." Perhaps our own recent experiences with widespread homelessness partly traced to deinstitutionalization will engender more historical empathy for these beliefs of a past era. (At a minimum, they ought to illustrate how even well-intentioned reform efforts can often have unintended consequences.) Without understanding these efforts in both their secular and their religious contexts, contemporary readers will miss much of Goddard's motivation, for they will fail to see the intense idealism that emanated from the little Training School at Vineland – an idealism described in poignant terms by countless visitors of Goddard's day, and encapsulated in the simple but powerful phrase they used to express it: "Vineland Spirit."[27]

In writing this study, I have sought to recapture both the spirit and the language of Goddard's life and times. I have also endeavored to keep my own language and late-twentieth-century sensibility out of this story as much as possible. I have chosen not to interject comments by today's critics pointing out when or why Goddard's ideas were right or wrong; instead, I have left the task of criticizing Goddard to his contemporaries. In short, my goal has been to recover the history of intelligence testing by allowing members of an earlier generation to debate their own issues in their own terms – to frame their own questions, to find their own answers, and to make sense of their findings in their own ways. Toward this end, I have made extensive use of a broad range of primary sources.

Of critical importance are Goddard's own words, recorded in his ten books, dozens of articles, and large cache of unpublished papers. Equally significant are materials left by those he worked with, including other psychologists, psychiatrists, superintendents, educators, biologists, and social workers. The annual meetings of medical, educational, and welfare organizations proved invaluable in allowing me to trace the influence of Goddard's ideas; so too did numerous journals in a variety of disciplines which published the earliest controversies over intelligence testing. Government records proved equally crucial in tracing policymaking; among the most interesting were the special census reports on institutionalized populations, the records of the New York City Board of Education, the publications of the U.S. Public Health Service, and the correspondence of the Children's Bureau. Finally, buried within these records was a most uncommon source, and one especially treasured by the historian: the words, often recorded verbatim, of persons who answered questions put to them by psychologists attempting to measure their minds. These included the responses of schoolchildren, delinquents, prostitutes, criminals, immigrants, soldiers, and, of course, the children of the Vineland Training School. When woven together, all of these sources offer a rich body of testimony about the ways that tests of intelligence first made their presence felt in American society.

My study begins with Henry Herbert Goddard's life and work before his discovery of intelligence testing. In exploring Goddard's background, I focus in particular on the intertwining of two powerful ideologies motivating reformers of the late nineteenth century: evangelical Christianity and Darwinian science. It is within this context that I analyze Goddard's involvement in the Child-Study movement, his early efforts to study "feeblemindedness," and his discovery of Binet's ideas about measuring intelligence. I next follow Goddard's efforts to introduce intelligence testing into American society by structuring collaborative working relationships with doctors and teachers – processes that led to the institutionalization of the new profession already calling itself "clinical psychology." Finally, I follow Goddard's search for the causes of mental pathology and what he saw as its corollary, social pathology. In particular, I trace the ever-widening uses of intelligence testing to address social questions of the day, including crime, poverty, prostitution, alcoholism, unemployment, immigration restriction, and military preparedness. These issues allowed Goddard to link psychological means to political ends most explicitly, and finally led him to expound a broad new political philosophy by 1918.

My detailed study of the institutional dissemination of intelligence testing ends where most histories of this movement begin – with the use of intelligence tests during World War I, for here starts a different episode in testing history and one in which Henry Herbert Goddard no longer played a pivotal role. In surveying the decades that followed, I focus on this scientist's declining reputation and his changing understanding of his own role in history. I end by examining Henry Herbert Goddard's complex and multifaceted "legacy," both to psychology and to American society.

In explicating Goddard's science, each chapter explores a different set of questions being debated in the early years of the twentieth century, a different community of debate participants, and a different set of social circumstances which either allowed for or restricted psychological intervention. Each also pays close attention to the ways that participants in these debates understood their own behavior at the time. For despite the role played by luck in Goddard's life, and notwithstanding his late-in-life denials of any conscious intentions, his career still offers striking evidence of purpose, pattern, and motives. These motives, however, cannot easily be reduced to fit within our more modern political categories, for they must be viewed from the perspective of the past, and not the present.

The American intelligence testing movement owes its origins to Henry Herbert Goddard. Its deepest historical roots, like Goddard's, lie buried in the nineteenth century. They can be exposed only by examining the explanatory languages of Goddard's day: the language of evangelical faith, and the language of scientific enlightenment. Neither of these lacked purpose. It is to these languages that I turn first.

1

Spirit and Science:
Faith, Healing, and Mission

Strictly New England

Throughout a lifetime that lasted for nearly a century, Henry Herbert Goddard's words, deeds, and demeanor always reflected the powerful influence of his New England origins. Especially strong were his ties to the small village in central Maine where he was born on August 14, 1866. Located south of Waterville and east of the Kennebec River along the shores of China Lake, East Vassalboro was a traditional New England community still governed by town meetings. Goddard would eventually leave village life behind him; even so, with his sober manner, understated but wry humor, and slightly accented Down East speech, he would always remain, in the words of one who knew him, "strictly New England."[1]

Like other famous Goddards with whom he shared a common lineage, among them anthropologist Pliny Earle Goddard and rocket scientist Robert Goddard, Henry Herbert Goddard could trace his New England roots back at least two centuries. His father, Henry Clay Goddard, was a descendant of William Goddard, who had left Wiltshire, England, for Watertown, Massachusetts, in 1666. By that time, his mother's ancestors had already arrived, for Sarah Winslow Goddard was related to colonial governor Edward Winslow, who had come to America on the *Mayflower*. Henry Herbert was this couple's fifth-born child. Their first and fourth daughters died in infancy. Two middle daughters, named Lucy Maria and Mary Ellen but called Ria and Nellie, were sixteen and fifteen years old respectively when their only brother was born. "I think both my parents loved me uncommonly," Goddard would recall. "I was the child of their old age."[2]

By the time of his birth in the year following the end of the Civil War, this rural frontier area had begun to experience some significant economic transformations. The Maine Central Railroad had reached the region, and a woolens factory had opened in the nearby village of North Vassalboro. Yet the forces of industrialization and urbanization that had already revolutionized the economies of other New England regions seemed largely to have bypassed the small villages that made up Vassalboro Township. As a result, many younger residents would begin to seek opportunities elsewhere.[3]

Like most inhabitants who remained, the Goddards were farmers. Unlike their neighbors, however, this family demonstrated a striking pattern of downward social mobility. Census takers in 1850 had found Mr. Goddard a relatively prosperous twenty-six-year-old landowner. A decade later, Goddard's property, while still adequate, had been reduced by nearly two-thirds. By the time the 1870 census taker recorded the addition of the family's new three-year-old son, called (as was the custom) by his middle name, Herbert, the Goddards no longer owned any land; Mr. Goddard had lost the farm. For the first time, Henry Clay Goddard listed his occupation not as "farmer" but as "day laborer," the poorest category of agricultural worker.[4]

Herbert Goddard later explained his family's financial collapse as the result of a farm injury. His father, he reported, had been left an invalid after being gored by a bull "thought to be quiet and safe"; never fully recovering, he died when his son was nine. "When my Father died, we were left without our breadwinner," he noted. Goddard's memories of his father were warmly positive, for he recalled him as a gentle man who had never punished him. His parents had also been ardent advocates of temperance who taught their son to avoid drinking anything that even "looks like cider." Above all, however, Goddard's boyhood memories were tinged with painful recollections of intense poverty.[5]

"Were it not for the fact that we were all members of the Society of Friends (Quakers) it would probably have gone hard with us," he recalled. "The Friends always take care of their poor," he declared. Yet his mother was reluctant to ask for charity, and "church authorities did not always see the need as soon as might have been desirable." Although Sarah Goddard depended for a time upon her married daughters, she would soon find new ways of sustaining herself and her young son. While her husband was ill, Mrs. Goddard became ever more deeply involved in church activities.[6]

All of these early experiences – New England village life, strong temperance sentiments, intense but still respectable poverty – would leave their mark on Goddard's later ideas. None would prove more crucial, however, than his mother's growing commitment to her Quaker faith. Nor was Sarah Goddard alone, for during the years that young Herbert was growing up, the entire local community would experience a major religious revival. It was this movement which would shape Goddard's own life and work most profoundly ever after.

A Quaker Awakening

The Goddards had long been orthodox followers of the teachings of George Fox, the charismatic shoemaker from Leicestershire, England, who began preaching in 1647 and founded the Society of Friends. Like other religious leaders of the English Reformation, Fox had advocated a "plain" form of

Christianity that would eschew church ritual and instead grant authority
only to "the light of Christ within." [7]

Among Fox's most striking practices was his insistence upon addressing
all persons with the familiar pronouns "thee" and "thou," instead of the
more formal term, "you." In the seventeenth century, such behavior had signified a radical egalitarianism and a defiance of traditional social hierarchies
and had led many Quakers to martyrdom. When nineteenth-century families
like the Goddards continued to use these by then archaic pronouns in their
daily lives, the practice acquired a new symbolism: it kept Quakers a distinct
people who shared a legacy of suffering for the sake of conscience. [8]

New England Friends had also inherited a tradition of social "concern,"
which had included an intense opposition to slavery. The preacher credited
with founding the Vassalboro Quaker community in the eighteenth century,
David Sands, had felt so passionately about this cause that his memoirs had
been bound in fabric solely "the produce of free labour." By the nineteenth
century, nearby Portland had become a Quaker station along the Under-
ground Railroad. Long after the Civil War ended, inhabitants of the country-
side along the Kennebec River continued to boast of their most famous former neighbor, Harriet Beecher Stowe, who had secluded herself in her Maine
home while writing *Uncle Tom's Cabin.* [9]

While Quaker social activism had flourished in the antebellum era, the
faith's spiritualism had languished amidst divisive doctrinal schisms. Traditionally wary of the practices of outsiders, most Quakers had remained skeptical of the evangelical awakenings transforming other denominations. Between 1860 and 1880, however, the Society of Friends began to feel the flames
of revivalism. Inspired by itinerant preachers addressing tent audiences in
styles popularized by Charles Grandison Finney and Dwight Moody, large-
scale revivals began first in midwestern frontier communities and spread
rapidly from meeting to meeting, until the "Quaker Great Awakening" had
reached both coasts. [10]

This revivalistic spirit permeated religious life in and around East Vassalboro when Goddard was a boy. By the 1870s, local Quakers had joined
their neighbors in sponsoring large tent meetings open to Christians from all
denominations. The "old partition walls that have always stood so distinctly
between that worthy class of Christians and other evangelical sects," a local
newspaper wrote of the Quakers' new practices, were "not only crumbling
but falling to pieces." The impact of such events was evident in the changed
behavior of area residents, among them the community's most prominent
businessman, Charles M. Bailey. A director of the Maine Central Railroad,
president of the Maine Steamship Company, and founder of the local Quaker
seminary, Bailey had become religious after hearing Moody speak in Lewiston. By the 1880s, he had joined Moody in supporting Bible institutes and

was spreading the evangelical message musically by performing with his "Bailey Praying Band."[11]

The most famous of local religious leaders were Eli and Sybil Jones, Quaker missionaries from the adjacent village of South China who had worked with freed American slaves in Liberia and started a Quaker school in the Holy Land. Their nephew, Rufus Matthew Jones, although three years older than Herbert, would soon become his schoolmate. Jones would later become the most important Quaker mystical metaphysician of the twentieth century, as well as a founder of the American Friends Service. He would also write many memoirs chronicling religious life within this community. In such an intensely spiritual environment, Jones concluded, "religion was as important an element for life as the air we breathed."[12]

Local awakenings exerted an equally strong influence on Goddard's mother. A "great change" had been wrought in Sarah Goddard, Rufus Jones reported, when she received "a message which Benjamin Jones, a minister among Friends, felt called to deliver to her personally." By the time Herbert was six, the community had recognized his mother's "gift in the ministry." Mrs. Goddard "preached at the Congregational Church last Sunday afternoon," the *Waterville Mail* noted in 1874. The "emphatic utterance of her prayer," it reported, was "'Lord Save the Children!'"[13]

In the following years, Sarah Goddard increasingly found herself, in her son's words, "called to visit Friends Meetings in distant communities." She thus became an avid participant in the unusual system of lay ministerial interchange practiced by Quakers. As Jones explained this system, if a woman such as Sarah felt called to travel and minister to other congregations, she would simply walk up the aisle of the local meeting, remove her bonnet, and "say that for a long time the Lord had been calling her to a service in a distant yearly meeting . . . and now she had come to ask Friends to release her." Those present would then "concur in this concern." Such occasions, Jones recalled, were often "of a heavenly sort, and the voices of strong men choked in tears as a beloved brother or sister was equipped and set free. From this little meeting," he recounted, "heralds went out to almost every part of the world. . . ."[14]

When Herbert was seven, his mother felt called to visit all the Friends meetings in Canada. After her husband's death, she spent nine months in Iowa, "holding meetings, visiting families, jails, prisons, and reformatory institutions," and, according to Jones, "accomplishing great results." Eventually, her ministry led her to visit all the yearly meetings on the American continent. By 1880, she listed her occupation in the census not as "keeping house," the most common designation for local women, but as "Friends preacher." Sarah Goddard, Rufus Jones reported, had become "a faithful messenger of the Gospel."[15]

For young Herbert Goddard, however, the loss of the family's farm, his father's death, and his mother's frequent traveling meant a lonely boyhood. He spent his early years living with his married sister until he was old enough for boarding schools. In the absence of parents, it was the Quaker community that provided Herbert with a proper religious education.

A Guarded Education

Goddard began his schooling under country teachers – the type of teacher for whom, as Jones put it, the very word "pedagogy" would have "meant no more . . . than did 'that blessed word Mesopotamia.'" More formal education started at age eleven when he entered the local Vassalboro academy, Oak Grove Seminary, where he and Jones became roommates. Beside the school stood a Quaker meetinghouse, which Jones remembered for its "occasions of profound silence and . . . old-fashioned rhythmic preaching of too great length."[16]

As a student, Herbert was "just a normal, average boy – without peculiarities or any unusual qualities," his principal later recalled. Nevertheless, both he and Jones soon received scholarships to the Friends School in Providence – the finest Quaker academy in the region. "New England friends gave me a scholarship that covered my tuition and board," Goddard explained, "in appreciation of my mother's services and our poverty." At twelve, he recalled, he entered this prestigious boarding school as its youngest and smallest pupil.[17]

While Jones' memoirs describe this school's "immense buildings with high ceilings, extensive grounds beautifully kept, and the splendid vistas out into a world of novelty and wonder" that had so impressed a country boy, Goddard's memories suggest more ambivalence. His late-in-life sketches describe a large building surrounded by a six-foot fence. "Quaker jail," he scratched in his notes. Yet he was "not unhappy," he conceded. He remained at the school until, at age seventeen, he received a scholarship to enter the sophomore class of Haverford College, another Quaker institution, where Jones had enrolled the previous year.[18]

In many ways, Haverford simply extended the education of the Providence school, for both institutions, Jones wrote, formed a "single system of culture." Friends, Haverford's founders had argued, required a "guarded education" – one that would instruct children while "carefully preserving them from the influence of corrupt principles and evil communications." Of nonreligious subjects, mathematics was considered the "safest," and Quaker schools excelled in it. "Nobody was likely to be swept off his feet and be given a propensity to sin by studying trigonometry!" Jones explained. "No one was in danger of imbibing any subtle poison to infect his soul by whetting his appetite for sines, co-sines and tangents!"[19]

Henry Herbert Goddard as a student in Philadelphia, c. 1885. Courtesy of the Archives of the History of American Psychology, University of Akron.

Goddard's mathematics instructor, Isaac Sharpless, described Haverford's system of "guarding" students in these years. "The students rose at the call of the bell, collected in a room on the first floor and marched down two and two to breakfast in the basement." Other meals were administered in the same manner. There were daily recitations at nine, eleven, and three o'clock. "At 10 in the morning, 2 in the afternoon and 7 in the evening an hour's Collection for all students was held. . . . " Thus, "through six hours a day," Sharpless reported, "all students were on duty." The day ended with a Bible reading and lights turned out by ten o'clock.[20]

Such "guarding" notwithstanding, Goddard's undergraduate career suggests a wide range of activities. He graduated sixth in a class of about twenty, he remembered. Goddard also won the Athenaeum Prize for Declamation as well as the Alumni Prize for Oratory and Composition for a speech extolling New Englander John Quincy Adams' Congressional career as an abolitionist. He was business manager of the school newspaper and vice-president of the Y.M.C.A., an organization that appealed largely to boys from rural areas. He also played on Haverford's football team. "I was very enthusiastic during my undergraduate days," Goddard later recalled.[21]

Yet while Jones' school memories remained enthusiastic, Goddard's grew increasingly bitter. The two disagreed most sharply over the merits of the classical curriculum in general, and their classics professor, Seth K. Gifford, in particular. Gifford, who taught both boys Greek and Latin first at the Providence school and then at Haverford, was a rigid teacher with a work system, in Jones' words, "as unyielding as the procession of the equinoxes." Yet while Jones maintained that studying the classics had led to the "unfolding of *capacities*," Goddard strongly disagreed. Although his training in Greek was evident in words he later coined, he came to resent deeply such teaching. Years later he still recalled his painful attempts to memorize Latin. Gifford "if he were as much as *half*-witted ought to have seen before the end of the first week, that I needed help," Goddard insisted. But no help was offered. Moreover, he vividly remembered this teacher's reaction to finding the lonely schoolboy reading *David Copperfield.* "He said: YOU TAKE THAT BOOK BACK TO THE LIBRARY; AND DONT YOU EVER TAKE ANOTHER BOOK OUT OF THE LIBRARY UNTIL YOU HAVE LEARNED YOUR LATIN. And I was fool enough to obey him," Goddard lamented. "And all those years when I should have been getting acquainted with good literature, I was mulling over a dead language with no one to help me."[22]

Goddard's school years were marred not only by the study of the classics, but also by poverty and insecurity. He resented Haverford's reputation as "a convenient way to keep the sons of rich Philadelphia Quakers out of mischief" while he was struggling to support himself with scholarships and summer jobs. "I came to Haverford," he recounted, "perhaps the most poverty-stricken child that ever entered those 'sacred precincts.'"[23]

He was poor, Goddard later admitted, "not only financially but morally" – not because he was immoral, but because he was morally unprepared for life outside Quaker institutions. "And so I received that 'guarded' education that Friends have always been so mistakenly proud of," he concluded. "I was so well 'guarded' that . . . I was totally *ignorant* of the customs and habits of the world I had to live in."[24]

Above all, Goddard resented his unpreparedness for choosing a career. "There was always more or less talk about the College fitting the youth for life," he remembered. "It was all talk." Instead, he attended school with no idea "of what I *wanted* to do – to say nothing of what I was fitted for." In fact, "nobody even asked me what I was preparing to do in life," he complained. "I was not consciously preparing for anything. I was working on Greek, Latin and Math."[25]

This feeling of unpreparedness haunted him in his later years. "I MUST HAVE BEEN STUPID NOT TO HAVE FOUND OUT THINGS FOR MYSELF!" he insisted. "I admit that I must have had some abilities but I didn't know it; and nobody told me. And so I concluded that I had none." Haverford's influence had been "about as near nil as it could possibly be." Boys "of good intelligence, good inheritance and well brought up, were not injured any." Those "not so well furnished, who needed some real help on the problems of life, and not merely Greek and Latin," had apparently "met life about as well as they would have done if they had never seen Haverford." A third group, however – those that "needed to be made over" – had received no help at all, for they "simply 'didn't belong there'" and soon dropped out. "I claim to understand that third group," Goddard reported when he was in his eighties, "both because they have been the objects of my study for many years and also because I have first-hand knowledge. *I was one of them.*"[26]

Goddard had indeed been one of the failures, at least briefly, for after one year he left college and came home to teach in the district schools of Winthrop, Maine. And although he returned to Haverford a year later and graduated in 1887, this missing school year came to symbolize his emotional immaturity and his inability to find help. "In all my adult life," Goddard confessed to a friend, "I have felt keenly the defects of my early training. Schools in those days, a travesty upon education," he added, "were of little value in preparing the youth for future usefulness."[27]

Goddard's bitter, late-in-life criticisms of the schooling of an earlier era may simply be hindsight acquired after a long career in psychological testing and educational counseling. In the 1940s, when he was in his eighties, he took a vocational aptitude test, which identified his strongest interest as medicine; other suitable occupations, listed by strength of interest, included physicist, chemist, mathematician, engineer, dentist, and psychologist. Goddard's complaints, however, suggest that his education had failed to supply him not only with career guidance but also with social and emotional guidance – the kinds of guidance traditionally supplied by parents.[28]

During his college years, Goddard's mother moved even farther away, for in 1884 she married Jehu Newlin, a poor itinerant missionary from a distinguished Quaker family. Thereafter, the couple left to preach the gospel in England, Ireland, Scotland, Wales, Norway, Sweden, Denmark, Germany, France, and the Holy Land – travels which kept them away for years at a time. Yet even in his old age, when he lashed out angrily against his schools, his teachers, and most of all himself, Goddard never voiced any complaints about his mother. Criticizing Sarah Goddard Newlin would mean questioning the genuineness of a religious calling. Goddard never rebelled this openly against his heritage. Even so, his insistence that teachers accept more responsibility for guiding students is rebellious in a covert sense, for it sug-

gests a loss of faith in an older belief system with which he had been raised –
the belief in divine guidance.[29]

Unfailing Guidance

Had Mrs. Newlin lived to hear her son's complaints about the lack of guidance in his life, she would have been sorely disappointed. By contrast, her
own experiences proved repeatedly that life's apparent purposelessness was
actually a manifestation of God's divine plan. This is illustrated in a Friends
tract entitled *The Unfailing Guide,* which describes a visit of missionaries
Sarah and Jehu Newlin to a London meeting in 1886.

The Newlins had arrived "expressing their belief that the Lord had definitely called them to this service." As the meeting "settled down to its usual
silence," Sarah rose to deliver a message from God. In "tones of love, vibrant
with intense feeling and pleading earnestness," she delivered "the Sword of
the Spirit to one heart in that little Meeting House" by asking, "Woman,
where are thine accusers?" A secret sin would be discovered, she claimed; a
more terrible sin was about to be committed. "I cannot think why I had such
a message," she told her hosts; yet it was "an act of faith . . . taken in simple
reliance upon the unfailing Guide of those who put their trust in Him." The
incident remained a mystery until "the One who brings hidden things to the
light of day, saw fit to vindicate His child," for it was later learned that a
woman in attendance had stolen some money and, while contemplating suicide, had met a friend who brought her to the meeting. After hearing Sarah's
message, she "dedicated her life henceforth to belong to Him, entirely, Who
had so loved her and saved her."[30]

The incident exemplified Sarah's faith that God was guiding her steps. In
letters home, she urged her children to trust God in the same way. Though
she suffered from "a yearning heart and a desire to soon be ready to return
to them who are so closely entwined into our very selves," she also insisted
that "this does not signify that we are not in our places and doing *all,* for
Him, and the gospel's sake." Her advice largely concerned the state of her
son's soul. "I hope to see *all* my precious children walking with Jesus and
praising Him," she wrote.[31]

Such sentiments were of little help to a youth struggling to establish himself financially. His stepfather's advice was equally otherworldly. "As thy
needs are made known," he wrote, and "thou asks Heavenly Father he will
provide." Newlin hoped that "ere long you will outstrip your father and
Mother in preaching the Gospel of Christ to a world lying in wickedness." If
a "way should open in business that will be conducive to thy growth in grace
though it may not be so remunerative," he advised his stepson, "close in with
it," since "a morsel with peace will be more to thee than that which will fill

the purse and trouble the conscience." Newlin may have had few troubles with his conscience, but he did suffer occasionally from an empty purse. Were it "not for the promise that the last shall receive equal wages with the first," he confessed, "we should be discouraged." Within a few years, Goddard would be lending his stepfather money.[32]

Such amorphous career guidance was hardly unusual; in fact, it formed a concomitant part of the community's evangelical faith. "Everybody at home," Jones explained, ". . . believed implicitly in immediate divine guidance." While Jones, like Goddard, left school with no plans, he nonetheless believed that "God's hand was surely guiding." Goddard felt no such heavenly direction. With no parents, prospects, or resources, he borrowed enough money to get to California, where one of his married sisters had settled.[33]

Arriving in the fall of 1887, Goddard stopped first in Los Angeles to drop off some letters of introduction at the recently opened University of Southern California; then he headed north to Oakland. "I crossed the bay every work day for nearly three months, answering 'Help Wanted' Ads," he recalled. "Nobody wanted me."[34]

To his surprise, the University of Southern California offered him a position for the spring, and Goddard began teaching Latin, history, and botany, while also coaching the football team – a position which earned him a small place in sports history as the first coach of the team that later became the USC Trojans. Unfortunately, however, the position was only temporary. "There was nothing for me to do but borrow more money, go back to Haverford and work for my MA. degree," Goddard explained. In 1889, he graduated with a master's degree in mathematics and no plans.[35]

Goddard's need for a steady income became more serious when, on August 7, 1889, one week before his twenty-third birthday, he wed Emma Florence Robbins, a twenty-four-year-old Maine woman whom he had met while teaching in Winthrop. A strong-willed schoolteacher who had been raised as a Universalist, Emma Goddard soon came to share her husband's Quaker faith, love of outdoor life, and dry sense of humor; she also set the rules which kept Goddard's home life both private and satisfying. Years later, Goddard would recount to a friend how he had jokingly placed a picture of Michelangelo's Moses under the glass top of his wife's desk, and the conversation which ensued when she discovered it: "'What's this picture of Moses doing over here?' she asked. And he said, "'Well, he was *another* great law giver.'" Although this couple never had any children of their own, their long union was evidently a close and happy one. More than twenty-five years later, the Goddards were still, according to a friend, "like boy and girl together." Even after forty years of marriage, Herbert could still receive as many as three letters a day when separated from "Emmie," his "sweetheart" and "true friend," and could still close his own letters with "love from your lover."[36]

Goddard receives "A Letter from Her," his future wife Emma, while working for his Haverford M.A. degree, 1889. Courtesy of the Archives of the History of American Psychology, University of Akron.

With a new wife, Goddard took no more chances in the non-Quaker world and quickly accepted a post as principal of a Quaker school in Damascus, Ohio. With Herbert teaching mathematics and mental and moral science (and conducting daily chapel services and weekly prayer meetings) and Emma teaching in the primary department, the newlyweds constituted two-thirds of Damascus Academy's faculty. They remained there for two years, until Goddard got one of those "lucky" breaks he would later make so much of in explaining his life.[37]

"I had not the slightest idea what I was going to do in the world," he wrote, until he came across a book written by his old friend, Rufus Jones. Jones had published an account of the missionary careers of his aunt and uncle, Eli and Sybil Jones, and Goddard wrote him a note of congratulations. By this time, Jones had returned to Vassalboro to become principal of their old alma mater, Oak Grove Seminary, and he offered Goddard a teaching position in the school. Goddard accepted the offer and returned home in 1891. The two men would teach together until 1893, when Jones received an offer to teach

Faculty of Oak Grove Seminary, Vassalboro, Maine, 1895. Goddard is fourth from left, and Emma Goddard third from left. Courtesy of the Archives of the History of American Psychology, University of Akron.

philosophy at Haverford, and Goddard succeeded him as Oak Grove's principal.[38]

In the 1890s, education at Oak Grove was still intensely religious, with daily services, twice-weekly meetings, and Sabbath prayer meetings. Goddard's remarks in the school's catalogue suggest his secular concerns as well. Commercial course graduates *"will be assisted in obtaining a situation,"* he emphasized. Every student, moreover, should be "taught to know the powers he possesses and how to use his talents." It was "the teacher's place," Goddard insisted, "to point out a proper goal of development and to help the young on the road."[39]

The road that Goddard's own career was taking certainly satisfied the community. "Herbert makes an excellent success of it according to accounts, and sits at the head of the meeting . . . and is a very good minister," a neighbor wrote Goddard's mother. "I always feel so glad when I think about it," she continued, "and indeed what cannot the Lord do with willing and obedient instruments listening to, and obeying his divine commands." Goddard's own words suggest expedience more than obedience. Even so, he might have remained within the Quaker school system, as did Jones, were it not for the outreach efforts of a different kind of teacher. While employed at Oak Grove, Goddard first heard psychologist G. Stanley Hall address the Maine State Teachers Association in Lewiston. It was Hall who finally offered Goddard guidance, purpose, and mission.[40]

An Open Education

Goddard was not the only schoolteacher to find his life changed by the speechmaking of G. Stanley Hall. By the 1890s, Hall had become one of the most influential educational lecturers in America. At an 1891 National Education Association meeting, he had drawn 150 teachers to an unofficial lecture simply by pinning a notice to a bulletin board; at the 1893 Chicago Columbian Exposition, he had upstaged traditional educators by leading his own congress on "Experimental Psychology and Education," the most widely attended educational event at the fair.[41]

What teachers heard in Hall's speeches was a passionate call for reform. American education needed to be changed, Hall argued, from the kindergarten to the graduate school. Above all, he insisted, the schools needed to pay more attention to the child. A simple statement summed up his philosophy: "the school is for the child and not the child for the school." And while such romantic sentiments had been proposed since Rousseau's day, Hall spoke with the imprimatur of science.[42]

Teachers must have been surprised to find a scientist of Hall's stature addressing them so frequently. After all, G. Stanley Hall was one of the country's most accomplished scientific psychologists. After earning his Ph.D. under William James at Harvard, Hall had been the first American to work in Wilhelm Wundt's psychological laboratory in Leipzig. He had helped to introduce experimental methods in America by opening one of the nation's first psychological laboratories at Johns Hopkins University. In 1887, Hall had started the *American Journal of Psychology;* five years later, he had founded the American Psychological Association. Since 1889, moreover, Hall had been president of Clark University. It was certainly rare to find such a scholar spending so much of his time with schoolteachers.[43]

Hall's educational interests, like Goddard's, were at least partly expedient, for while his new discipline of experimental psychology struggled to survive within American universities, education courses drew appreciative audiences. Hall repeatedly used pedagogy to bail out psychology in times of trouble. To generate public support for his financially strapped graduate school, for instance, in 1891 he opened a summer school for teachers and started an educational journal called *Pedagogical Seminary.* These educational concerns, however, were more than simply mercenary, for Hall saw a "new pedagogy" as an integral component of what he and his contemporaries called the "new psychology."[44]

Hall's new psychology was based on a Darwinian reconceptualization of "mind" that challenged an older system known as "faculty psychology." Instead of a mind made up of distinct "faculties" such as intellect, emotion, and will, each of which required training and control, new psychologists posited a fluid, complex entity that evolved over time in response to its environment. The best way to understand such an entity was to study its simpler, less evolved manifestations first. The "phenomena necessary for such a study," one of Hall's followers explained, "can be found in the animal world; others in the history of mankind; still others in the primitive races, or in imbeciles or in other undeveloped groups, – all can be found in children." In fact, Charles Darwin himself had published the first evolutionary study of childhood in his 1877 essay, "A Biographical Sketch of an Infant." Such research had profound implications not only for psychology but also for education. To Hall, it meant that teachers ought to emphasize not "mental discipline" but "mental development" as the cornerstone of learning.[45]

Since the 1880s, Hall had been imploring teachers to join him in promoting both the scientific study of the child and the reform of the school. By the 1890s, he had assumed national leadership of a movement he alternatively called "paidology" or "Child Study." His efforts won him the enthusiastic support of classroom teachers throughout the country. By 1895, Child Study was the nation's fastest growing educational movement. Hall's message had even begun to affect small religious institutions like Oak Grove Seminary, for Goddard's students who wished to "fit for teaching" were required to learn School Management, Discipline, Psychology, and Child Study.[46]

What intrigued Goddard most, however, was not Child Study but the psychology behind it. By 1896, he had decided to spend one year at Clark University studying psychology with Hall. Once there, he borrowed money for two more years and earned his doctorate.[47]

After so many years of guarded schooling, what Goddard found in Worcester must have stunned him. The reforms Hall advocated for public education paled in comparison to those already initiated on his own campus, for Clark University in the 1890s was an American institution unlike any other. Hall showed little interest in monitoring Christian character on his campus. In fact, he did not care to educate undergraduates at all, for his university accepted only graduate students. Instead of emphasizing student self-discipline, he embraced the newer German ideals of freedom to teach and learn. The German university, Hall believed, was "the freest spot on earth." He replicated this freedom on his campus by minimizing bureaucracy and maximizing intellectual independence: seminars replaced recitations; there was no required curriculum; few professors even took attendance. Above all, he encouraged a respectful, research-oriented collaboration between faculty and graduate students. Research, Hall believed, ought to constitute "the vital spirit of teaching."[48]

When Goddard arrived in 1896, Hall's experiment in graduate education had already begun to flounder from a shortage of funds. "It was most unfortunate for the world, when Jonas Clark, the donor of Clark University got 'cold feet' and withdrew his support," Goddard observed. Hall's poor handling of the crises that ensued led many of his most promising faculty to accept better offers from the new University of Chicago. Yet the most important element of a Clark education had always been Hall himself, and his charismatic, fatherly presence ensured the loyalty of all who remained. Hall's own inspired teaching style seemed to prove the genuineness of his belief in a new type of pedagogy. "I am not one of the brilliant disciples who reflect great credit to the teacher," Goddard wrote Hall a few years after his graduation, "yet as far as one life goes, I know that mine is incomparably larger, higher and broader than it would have been had I not come under your instruction and inspiration."[49]

Most inspirational was Hall's indefatigable spirit of inquiry. What Hall liked best about German scholarship – its "burning and all-sided curiosity" – also described his own. The questions asked by German intellectuals became his as well, for the new evolutionary science suggested the need to rethink all the older verities. "What is the essence of life, love, freedom, duty, law, state, religion?" Hall asked. "All the old forms of laws and belief that men had lived by," he declared, now merited a reexamination "in quest of deeper foundations."[50]

Under Hall's leadership, no subject was exempt from the probings of the new evolutionary psychology. This questioning spirit was embodied especially well in his favorite research technique, the questionnaire. In the course of his career, Hall would ask many new questions about childhood, adolescence, senescence, sexuality, religion, psychoanalysis, and countless other issues.[51]

In most cases, Hall's questions have outlasted his answers. Even so, the "new psychology" that Hall promoted meant something much more than laboratory skills; it also meant an open-ended inquisitiveness about human nature, grounded in evolutionary theory, that profoundly influenced both his discipline and his disciples. Safely ensconced within the world that Hall had created, Goddard asked and answered a few questions of his own.

A Science of Facts and Laws

Despite the stark differences distinguishing Quaker from Clark pedagogy, Goddard's education in science remained surprisingly consistent. Like other Protestants, Goddard's Quaker teachers had taught a version of natural theology, in which the order found in the physical world illuminated God's orderly mind. Science, Goddard learned, meant discovering the laws of nature.

And the way to achieve such discoveries was through fact gathering. Throughout his lifetime, no phrase in Goddard's lexicon remained as laden with meaning as "the facts." "*I reverence the facts,*" he wrote to a colleague when he was over eighty.[52]

The most influential of nineteenth-century fact gatherers, of course, had been naturalist Charles Darwin, whose discoveries had led to the most controversial law of nature yet proposed: evolution. Goddard had learned about evolution as a schoolboy in Providence, for his geology teacher, his schoolmate Jones recalled, had taught the class "the astonishing fact that the world was not made in six days, six thousand odd years ago, but had a history of uncounted and uncountable years." Because he "made the *fact* as clear and plain as morning sunrise," Jones added, students believed him. "He laid before us the marvelous story of the evolution of the horse. He showed us the array of fossils. He pointed out how the stages of the embryo child run in a parallel order to the stages of the order of evolving life." Such preparation, Jones asserted, had allowed Quakers to follow scientific debates of the day without suffering "any wreckage of faith." It had surely prepared Goddard, for he evidently perceived no open warfare between his Christian heritage and his new career as a scientist studying evolutionary theory.[53]

Neither did his new mentor, G. Stanley Hall. Although he had earned a degree from Union Theological Seminary, Hall's growing scientific interests had led him to abandon the strict Congregationalism of his own childhood. Even as a youth, Hall wrote, he had been "almost hypnotized by the word 'evolution,' which was music to my ear and seemed to fit my mouth better than any other." By the 1890s, however, personal tragedies had brought Hall back to Christianity, albeit in a much less traditional form. In the decades that followed, this psychologist would become increasingly intrigued with religious questions, and would eventually publish his own study of *Jesus, the Christ, in the Light of Psychology.*[54]

The scientific education he offered at Clark, Hall insisted, required no religious renunciation; instead, what he was promoting was the psychological reinterpretation of Christian ideals. "Rightly interpreted," Hall told a Unitarian Club audience in Boston in 1896, the year Goddard began his graduate studies, the new psychology reaffirmed the "five points of the new orthodoxy" – God, Christ, sin, regeneration, and the Bible, the "greatest book of psychology in the world." All he was supplying, Hall argued, were "modern methods of studying the soul."[55]

In addition, Hall promoted these "modern methods" with a revivalistic passion that must have reminded Goddard of home. "It seems to me that the sentiment which we really want to cultivate," he reassured his Boston audience, "is not unlike that of the old Church Father, Tertulian, who said, in substance, that the end, even of Christianity, was that it might be able to say,

'Stand forth, O soul of man, naked, genuine, real, just as thou dost come into the world from the hand of God, and, having stood forth, grow to thy full perfection!'" "I thought perhaps I might end with a little sermon," a minister noted after Hall's address, "but it would not have been a quarter as good as the one you have heard."[56]

If Hall was careful to avoid driving a wedge between psychology and religion, he did attempt to widen the gap between psychology and philosophy. Philosophical beliefs, Hall argued, could themselves be conceptualized as products of human evolution. Thus, the history of logic, ethics, philosophy, and aesthetics ought to be examined "in the same objective way that we study the delusions of the insane or the instincts of animals." Even the "highest conceptions of the great heroes of speculative philosophy" could be studied "in the same objective way as the myths of savages." This distinction between psychology as science and philosophy as something else remained with Goddard throughout his lifetime. Philosophy, Goddard wrote years later, "is a convenient way of arriving at some sort of conclusion when one has no facts" – a process "more injurious to the science of psychology than helpful."[57]

Instead of philosophy, Goddard learned physiology. Especially exciting was the psychological laboratory of E. C. Sanford, which was filled with special equipment designed to measure minute sensory or motor responses to stimuli, and thus to gather the facts of "psychophysics." "Those were the days when experimental psychology was coming in," Goddard recalled, "and those who got interested in it were dubbed 'brass instrument' psychologists" by the old liners who "tried to keep us out of the fold by declaring we were merely physiologists." Such charges seemed founded at the time, for Goddard's earliest psychological work could barely be distinguished from that of Clark biologist C. F. Hodge, who taught him neurology. In the years that followed, Goddard maintained a strong interest in psychological instrumentation. His first contribution to his field was to build an improved new microtome to dissect brains – an instrument he described in the *Journal of Comparative Neurology*. His invention was put to use at Westboro State Hospital and was later copied and reproduced by the Mico Instrument Company.[58]

Yet neurology for Goddard was never an end in itself; his work on the brain microtome was actually an attempt to develop more precise tools with which to study the connections between brain and behavior. The teacher that Goddard came to admire even more than Hodge was Adolf Meyer, a psychiatrist from the Worcester State Hospital for the Insane whom Hall brought to lecture at Clark. "I shall never forget the impression made upon me," Goddard noted a decade later, "when Dr. Adolf Meyer took us into his laboratory at Worcester and taking a brain from a jar said, 'this is the brain of the patient you saw last spring. You recall the symptoms. We shall now examine this, and probably shall find such and such conditions' – describing very minutely hemorrhage, degenerate fibres and cells. His prediction was verified at every point!"[59]

In the following years, Goddard too came to focus on the mind-body relationship. His earliest scholarly endeavors, however, suggest not only intellectual curiosity but also a deeper struggle to reconcile the scientific climate of Clark with the spiritual world of his childhood. This was especially evident in his choice of a research project. For his dissertation, Goddard chose to study "The Effects of Mind on Body as Evidenced in Faith Cures."[60]

Psychological Cures for New England Ills

In choosing to study "faith curing," Goddard placed himself at the contemporary crossroads between the worlds of science and spiritualism. Of course, faith cures were nothing new. Within the Christian tradition, they dated back to Christ himself. Seventeenth-century Quakers had credited George Fox with over 150 such cures. Yet Goddard's topic had special significance in his own day, for nineteenth-century New Englanders had been especially quick to embrace new combinations of optimistic theology, idealistic philosophy, and mystical metaphysics. For intellectuals, these took such forms as Ralph Waldo Emerson's Transcendentalism or the elder Henry James' Swedenborgianism. For plainer folk, they emerged as new forms of divine healing or "mind cures."[61]

The new healing theologies had also attracted medical attention, for they arose alongside a new ailment proving resistant to traditional therapies. This malady seemed to defy somatic classification. Sufferers spoke of "nerve exhaustion" or "nervelessness"; their symptoms ranged from sweaty palms to paralysis. The most detailed description of this condition came from neurologist George Beard, who labeled it "neurasthenia" and explained it as a response to the stresses of urban industrial civilization. If Beard was correct, then neurasthenia, and its largely female cousin, "hysteria," were approaching epidemic proportions, especially in the Northeast.[62]

Hall, a friend of Beard's, was familiar with the literature on neurasthenic ailments. What brought neurasthenia within the sphere of the new psychology, however, were reports of its responsiveness to new hypnotic techniques being tried by European psychiatrists. In France, J. M. Charcot and Hippolyte Bernheim had used hypnosis to treat hysteria. It was not only physicians who had found hypnosis useful, however. In America, theological healers had adopted hypnotic techniques of their own and were claiming equally impressive results on a wide range of illnesses that had defied treatment by more traditional methods. All of these claims suggested a need to reassess the relationship between mental states and physical health. In the name of the new psychology, Goddard accepted this challenge.[63]

What was needed most, Goddard argued, was a new authority. Psychologists ought to arbitrate the competing claims of medicine and theology, he proposed, for faith curing had confounded them both. "The medical man has it to contend with," he reported, while the minister "meets it as a more or less persistent theological doctrine, which he must either uphold or denounce." Lawyers too were perplexed when called upon to decide whether relatives were "guilty of 'culpable neglect' because they trusted to some form of mental therapeutics and did not consult the recognized doctors of medicine." Even more confused was the public, for "no person can see a friend enduring a lingering illness, unbenefited by the arts of the physician, without having this new method urged upon him, and without having at least the beginnings of a query in his own mind as to whether there is 'anything in it' or not." Most desperately in need of a new authority was "the sick one himself," who, "watching the weeks grow into months and the months into years, with no improvement," would also wonder if "it may not be worth while to try the prayer cure or hypnotism or Christian Science."[64]

In elaborating and seeking answers to such questions, Goddard's dissertation may have suggested not only scientific but also personal curiosity. After all, this scientist studying faith curing had spent his earliest childhood years living with a seriously injured father and an intensely religious mother. If anyone had been left with "a query in his own mind as to whether there is 'anything in it' or not," it may have been Goddard himself, for he approached his topic with a subtle mixture of skepticism and empathy.

Adding to his interest were events in the surrounding community. It is easy to see why the Maine-born Goddard would be drawn to study mental curing, for Maine had also been home to Phineas Parkhurst Quimby, the most influential of the new mental healers and the acknowledged father of the American "mind cure" movement. Although Quimby had died in 1866, the year Goddard was born, his impact could still be felt locally. In writing his dissertation, Goddard had a chance to assess Quimby's theories, as well as those of other mind cure advocates, scientifically.[65]

P. P. Quimby's fascination with the cure of diseases (and skepticism toward the medical establishment) had begun when he ignored a doctor's advice to treat consumption with calomel and instead cured himself with horseback riding. A traveling hypnotist further piqued his interests in the power of the mind to cure the body. In the following years, Quimby began to travel around the state, giving performances in which hypnotic techniques were used to cure illnesses. He soon opened offices in the Maine communities of Portland and Belfast.[66]

Quimby himself never claimed any special healing powers; it was his patients' belief in cure, he insisted, that cured them. He soon dispensed with hypnosis and substituted an even more subtle form of suggestion. "I simply sit by the patient's side and explain to him what he thinks is his disease," he reported, "and my explanation is the cure. And, if I succeed . . . I change the fluids of the system and establish the truth, or health." Disease, Quimby argued, was merely a deception in which people "live a lie, and their senses are in it." By establishing "truth" in its place, the patient would be healed. Cures therefore lay "not in medicine, but in the confidence of the doctor or medium."[67]

Goddard's dissertation offered a sympathetic summary of Quimby's ideas. Quimby deserved credit, Goddard claimed, for providing "a valuable addition to our methods of coping with human infirmities." He even excused the excesses of many of Quimby's followers, who called themselves "Mental Scientists."[68]

Goddard also summarized the claims of a variety of other healers offering new versions of the age-old belief that "God heals disease in answer to the prayer of faith." These included prominent figures such as Chicago's Reverend John Alexander Dowie, proprietor of a "Divine Healing Home" and editor of *Leaves of Healing,* as well as "tramp healers" like Francis Schlatter, who claimed that God, or "Father," wanted him to "go forth from Denver on foot," and who eventually starved to death in the deserts of the Southwest.[69]

Goddard was least sympathetic toward the most powerful of the new therapeutic theologies to emerge out of New England, the "Church of Christ, Scientist," or Christian Science. This faith owed its start to Mrs. Mary Baker Eddy, a neurasthenic first successfully treated by Quimby who later used similar methods to cure herself. By the time Goddard began his dissertation, Eddy had become pastor of her own "mind healing church" in Concord, New Hampshire, and had published her testament, *Science and Health.*[70]

Mrs. Eddy's theology, Goddard summarized, "is a sort of absolute idealism. Mind is divine; mind is all. Sin and sickness are delusions of 'mortal mind.' The 'treatment' consists in the assertion that sickness is not a reality but only a 'belief.'" What drew Goddard's ire, he wrote, was this church's dogmatism. "Mental Science is far more 'scientific' than Christian Science," he maintained, for its practitioners manifested a "far greater readiness to accept the facts." Facts, not faith, Goddard asserted, would determine the value of the new methods of healing.[71]

The search for the facts – facts from doctors, healers, and patients, facts about successes and failures – dominated Goddard's research strategy. These facts, he argued, "may eventually be so numerous and so complete that they will fit into each other, and exhibit a more or less perfect picture." Amassing the facts, in other words, would lead to the discovery of laws. And the best way to amass the facts was by using Hall's favorite technique, the questionnaire.[72]

Goddard designed his questionnaire, entitled "The Psychology of Health and Disease," to gather "the actual facts, as experienced by a large number of individuals, in the various departments of healing, such as: Faith Cure, Mind Cure, Christian Science, Hypnotism, Materia Medica, etc." Some questions were directed to physicians. "To what extent do you use suggestion in your practice?" he asked doctors. "Did you, when studying medicine, have any of the symptoms about which you studied?" Most of his questions, however, were designed for patients, and suggested a scientific inquiry into religious experience. "How did the idea come to you that you could be healed?" he asked. "Did you seem to have any 'revelation,' or was there any 'manifestation,' as of 'angels' or 'flames' or 'voices,' or any such thing? Was it comparable to any of the cures wrought by Jesus . . . ?"[73]

To secure responses from individuals involved in the mental healing movement, Goddard published an article in *The Hypnotic Magazine* in 1897. Entitled "Are Drugs Unnecessary to the Cure of Disease?" it outlined his research strategy and encouraged readers to send him information. He who "denies the fact of positive cures by hypnotism, by 'mental science,' by christian science, by divine healing, or at shrines and by charms and relics," Goddard argued, "simply declares his ignorance or his prejudice." Nevertheless, in the name of science, Goddard warned that the "mysticism connected with such practice is so great that much harm is likely to result unless the practice can be rationalized," so that "these new methods may stand on as firm a basis as do now the regular schools of medicine." Since most of the new practitioners assumed that the cure of disease was "a question of mind," this task fell to psychologists, the scientists of mind. "For the mind acts in accordance with fixed laws," Goddard stated, and psychologists needed to find the laws that would "bring order out of chaos." "We want thousands of cases, and of every known kind," he pleaded. Endorsing Goddard's plea was the magazine's editor, who declared his project "scientific and wholesomely free from bias," and who supplied him with his subscription list.[74]

Goddard's research brought him thousands of individual accounts of cures and failures. In addition, he studied medical folklore, learned to administer hypnosis, and even garnered the blessings of at least one tramp healer. He also visited several institutions offering new therapies. In one "Mental Science" home, for instance, Goddard heard from patients who reported varying degrees of success and failure in curing a wide range of complaints, including pneumonia ("cured"), depression ("little improvement"), cancer ("unsuccessful"), ovarian trouble ("cured"), and "over study" ("left well and strong").[75]

Traditional physicians, of course, had remained skeptical of such "cures" and frequently suggested misdiagnosis. Not all claims could be so easily dismissed, however, for advocates of mental healing included many well-educated persons who had exhausted medical remedies before turning to less respectable healers. Moreover, mind curists articulated an equivalent skepticism about medical authority.

In assessing his data, Goddard cautiously steered a course between both positions by trying to establish a rapprochement between medical practice and mental science. Physicians, he argued, had had similar experiences to those of the healers. Doctors using placebos had produced similar "cures." One doctor, for instance, told of curing a nervous condition with drops of water. Others had witnessed "spontaneous cures" as surprising "as any reported by divine healing." Thus, one physician described an athlete whose knee "had given out . . . six months previously." Within fifteen minutes, he reported, this young man was walking without crutches. "Never had any trouble afterwards," the doctor added, "and played on the Harvard foot-ball team Thanksgiving, 1887." [76]

The two approaches intersected most closely in the practices of traditional physicians using hypnosis. Goddard collected over a thousand such cases from doctors with "no theory to defend, no religious dogma to support." And although physicians tended to use hypnosis on a narrower range of diseases than the mental healers, their results were surprisingly similar. One doctor, for instance, claimed that in eighteen months he had "not seen a case of nervous prostration which has not been cured in a few weeks, when suggestion was properly used." Medicine and mental science, Goddard concluded, were responding to similar phenomena. [77]

Hypnosis, Goddard argued, held the key to mental healing. Ironically, however, each of the new healers had charged their competitors with practicing "mere hypnotism." "The Divine Healer thinks Christian Science is hypnotism," Goddard reported; "the Christian Scientist says Mental Science is hypnotism, and so through the whole list of rival schools." Such charges were not literally true, he added, for none put patients to sleep. In a "scientific sense," however, Goddard maintained, "*all* mental therapeutics is hypnotism, *i.e.,* it is suggestion." The "law of suggestion," Goddard concluded, "is the fundamental truth underlying all of them, and that upon which each has built its own superstructure of ignorance, superstition, or fanaticism." The new therapeutic faiths thus rested upon a single psychological principle, the law of suggestion, padded with a "superstructure" of religious trappings. [78]

Goddard's next task was to ask why suggestion worked. His answer owes much to the evolutionary psychology of both Darwin and Herbert Spencer, and to the theories of psychiatrist Hippolyte Bernheim, author of *Suggestive Therapeutics.* Suggestibility, Goddard argued, was actually a survival mechanism that had become increasingly sophisticated through natural selection. In its most primitive form, it referred to an organism's ability to connect an "idea in the mind" with some motor response – a connection critical for any organism's survival. "The very existence of animal life is dependent upon the sequence of idea by its motor equivalent. A sensation, a stimulus, gives rise to the idea – and this idea in turn discharges itself in motor form. Thus the needs of the animal are satisfied." As one ascended the evolutionary scale, however, one could see evidence of a more sophisticated response to ideas. In higher animals, "an aroused idea frees itself by arousing another idea." This was evident in man, where "the idea is commonly followed by another idea and that one by another, and so on until such time as the nervous tension becomes so strong as to discharge into the motor areas, then the motor response appears."[79]

This evolutionary development, Goddard noted, was equally evident in human history. "The differences in men consist in the differences in the suggestions that they have received and the ways in which they have reacted to them." Thus, an "ideo-motor man," who quickly translates every idea into a motor response, is "but little above the brute." Such reactions "will suffice for his vegetative needs, if he is not made the victim of a designing reasoner." Much higher on the evolutionary scale, however, was the "ideo-idea man – the one who inhibits the motor response, and follows the idea by another idea in rapid succession until such time as it is wise to follow the motor expression."[80]

This evolutionary model also explained the efficacy of hypnosis. In man, Goddard argued, harmful ideas might block healthy motor responses, for the idea of movement might also be "met by an idea of rest, or of movement in another direction, or what not; the result is, no movement is made." The hypnotist, he maintained, was able to return the organism to a more primitive state of consciousness, thus allowing ideas to become actions. Since the time of least idea resistance would be while the organism was asleep, he reasoned, hypnosis provided "the condition *par excellence* for suggestion."[81]

The facts, Goddard believed, could now be explained. Mental healers were using hypnotic methods, all of which were forms of suggestion. And the law of suggestion was explained by the evolutionary logic of natural selection. Goddard thus ended his dissertation satisfied that he had found a scientific explanation for faith cures.

Goddard's evolutionary solution to religious questions of his day apparently had little impact on the practices of his contemporaries. His research, however, did attract the attention of the most important American thinker then studying the psychology of religion, William James. In a series of lectures later published as *Varieties of Religious Experience,* James analyzed the American mind cure movement as part of what he labeled "the religion of healthy-mindedness." Like Goddard, he too saw such practices as efficacious for many individuals. "The plain fact remains that the spread of the movement has been due to practical fruits . . . ," James argued. Persons influenced by such phenomena "form a psychic type to be studied with respect," he concluded, citing Goddard's findings as support.[82]

Even more significant was what Goddard's thesis sugggested about the fledging discipline of scientific psychology, especially as promoted by Hall.

Goddard's dissertation is a model illustration of all that was good and bad in Hall's research methods – methods that Hall believed he had modeled on Darwin. The first step in such research was always to raise a provocative question that in itself broadened the boundaries of scientific inquiry. In seeking an answer, one then gathered as many facts as possible. The final step was to explain these facts as illustrating yet again the law of evolution. Such procedures may not have led Hall and his followers to many new discoveries; nonetheless, they still served to expand the intellectual parameters of their new discipline.

Of equal importance was the expanded social role being sought for psychologists. In his carefully worded conclusions, Goddard had proposed a broader role for psychological therapeutics without challenging somatic medicine. Thus, he argued that although he had found "no power that is able adequately to take the place of a thorough knowledge of anatomy and pathology or the skill of the surgeon," his evidence did suggest that "the proper reform in mental attitude would relieve many a sufferer of ills that the ordinary physician cannot touch" and might even "delay the approach of death to many a victim."[83]

Scientific psychology, Goddard believed, could contribute to the "practical side of ameliorating human ills." "*Mind* cure suggests psychology," he insisted, "and the psychologist is appealed to for the laws of mind which may explain the phenomena and give the rationale of the question." Amidst the chaotic contemporary debates over faith curing, Goddard concluded, psychologists could interject the voice of scientific reason. Through his own research, he had sought a means to do so.[84]

Scientific Faith

But if psychological laws of suggestibility explained much of the cure in faith cures, what about the faith? Could Henry Herbert Goddard, the son of a missionary who believed in divine intervention in daily life, explain the power of prayer to heal through scientific method alone? The very fact that Goddard wrote so much about the power of the new psychology, and so little about God's power, is itself indicative of the degree to which he had rejected his mother's version of religion.

This rejection, however, was neither overt nor complete, for it did not mean that Goddard was rejecting the deeper values acquired from his intensely Quaker upbringing. For years to come, Goddard would maintain close ties to the Quaker community. In Worcester, the Goddards joined the local Quaker meeting. A rare glimpse of this couple's home life can be gained from the captions of photographs Emma sent to relatives – "Domestic Bliss," "Sabbath Afternoon," and "Writing Up 'Faith Cure'" – the last suggesting that family and friends back home were aware of Herbert's research interests. In fact, there is nothing to suggest that these Quakers found anything objectionable in Goddard's scientific assessment of questions involving faith. Apparently, Sarah Goddard Newlin recognized no direct challenge to her own belief in God's immanent guidance in the Shakespearean epigram that her son quoted in opening his dissertation: "Our remedies oft in ourselves do lie / Which we ascribe to heaven."[85]

Perhaps Goddard's mother assumed that her son's analysis dealt mainly with the new sects springing up around them. It was certainly true that Goddard had saved his sharpest words for Christian Science. Still, Goddard's conclusions emphasized the similarities rather than the differences between old and new theologies. In fact, he ultimately pronounced the new faiths "healthy and safe." Both Christian Science and Mental Science offered their followers "the best idealism of heathen philosophy and the Christian religion." After all, he argued, it "certainly can do the world no harm to have a body of people devoting themselves to emphasizing the mental side of life in these days of materialism." Even Divine Healing was a "positive perversion of religion" – positive because it fixed the patient's attention on "things higher and beyond himself." Mental healing, Goddard concluded, could be summed up simply: "Do not dwell on your ills, is the key note of it all."[86]

This, however, was also a keynote of older Christian faiths. According to Goddard, traditional Christianity contained within it "all there is in mental therapeutics, and has it in its best form. It teaches temperance in the broadest sense, high ideals and dependence upon the Highest alone" – beliefs that could preserve men from "most of the ills that make up the list of those curable by mental methods." Such beliefs, along with "a wise submission to the inevitable," he concluded, "will do everything for us that can be done."[87]

Christianity, in other words, was itself therapeutic. "Lofty thoughts, high ideals, and hopeful disposition," Goddard noted, "are able to cure many diseases, to assist recovery in all curable cases, and retard dissolution in all others." The most therapeutic of all religious doctrines, moreover, was "altruism," for it took the sufferer's mind off selfish concerns and focused it on others. Altruism thus constituted both "the gist of all mental healing" as well as "the very essence of Christianity."[88]

Yet altruism too could be explained as a product of evolution, as Herbert Spencer had so forcefully shown. After all, it was easy to believe that man had once been as selfish a creature as the less evolved animals. "It is because people will act upon suggestion without thinking," Goddard concluded, "that evil has entered into the world." Fortunately, evolution had slowly allowed man to gain increasing control over his own selfish ideas and violent tendencies.[89]

But if even altruism could be explained scientifically, then what role was left for religious explanations? Unfortunately, Goddard surmised, much of the religion of his day was merely superstition. Theological healers succeeded not because of God's intervention but because of man's superstitious nature. Faith healing worked because man had always identified with something mysterious, a "force or power that was just beyond his understanding." As understanding evolved, man pinned his faith "further back into the unknown." "So that whether the treatment be with the idea that the gods are appeased by the swallowing of nasty compounds, or that certain objects in themselves possess the healing power, or God answers the prayer, or obedience to some transcendental law of mind brings health," Goddard wrote, "the principle is the same. The unknown is powerful; mystery makes the suggestion all potent."[90]

This "appeal to mystery" would vary, Goddard argued, according to the background and sophistication of the subject. Thus, one might believe in a crass form of mysticism that saw hypnosis as related to magnetic forces. By contrast, the man who "understands suggestion, and voluntarily accepts the suggestions of the operator and is cured of his disease" was appealing to a higher mystery suitable even for scientists: "that ultimate mystery of the relation of body and mind."[91]

Thus, in Goddard's science, the universe operated according to natural laws, while remaining a fundamentally mysterious place. Such careful conclusions allowed Goddard to affirm both science and religion, for he could denounce superstition without denying the supernatural. Even so, Goddard's research still evidences a marked transfer of allegiances. In seeking to understand the world, it was science and not theology that supplied him with answers. As his explanations strongly suggest, for Goddard the scientific language of facts and laws had clearly gained an upper hand over the spiritual language of faith.

This shift in allegiances, first evident in Goddard's dissertation, was starkly illustrated in a brief autobiographical essay he wrote several decades later. Entitled "Mysterious Influences," it is one of the few works in which Goddard mentions either his mother or his Maine childhood. The essay explores the source of Goddard's mysterious desire to climb the Alps – a childhood dream for as long as he could remember.

"My first consciousness of the matter," he remarked, "was a strange and weird feeling connected with the word 'Alps.' . . ." This feeling, he wrote, could be called "a child's religion," since it stood for "something as unattainable as heaven itself. Had I ever seen a man who had climbed the Alps," he added, "I think I would have fallen down and worshipped him as, – if not God, at least almost."[92]

To explain the origin of such a feeling, however, Goddard turned not to theology but to psychology – in this case, to free association techniques learned from other scientists of the mind, psychoanalysts. "I would sit down by myself and repeat: 'Climbing the Alps.'. . . Finally I noticed that when I said 'Alps,' I thought of a dog. . . . Then 'snow,'" all of which convinced Goddard that he must be describing a picture from a book. "One day, I asked mother if I had ever had such a book. . . . Just then my older sister came in. . . . She said at once: "Yes. Doesn't thee remember Aunt Phoebe gave him a picture book for Christmas? It was about the St. Bernard monks and their dogs going out to rescue the travelers lost in the snow of the mountains." Psychological methods had thus solved the mystery, Goddard declared, and he ended his essay satisfied that he had "found the origin of my life-long fascination for 'Climbing the Alps.'"[93]

Of course, other twentieth-century psychologists might see such a story as merely a starting point in explaining a lonely boy's fascination with a book about lost travelers, or his lifelong desire to master mountains – a desire that eventually led Goddard to climb the Matterhorn as well as many other dangerous peaks.[94] Goddard himself, however, entertained no such speculations. Even in the decades in which Freudian ideas reached the height of their popularity, this psychologist never tried to analyze his own complex feelings of respect and resentment toward his mother, or his apparently powerful sense of childhood abandonment. Instead, for Goddard, this story from his past held a far more straightforward moral. By eschewing childish religious explanations and adopting psychological methods, he argued, another of life's "mysterious influences" would lose its mystery.

While Goddard may have moved beyond such "childish" explanations, he never fully abandoned the modes of thinking acquired in his childhood. In his later years, both his negative and his positive responses to his early Quaker experiences would be evident in his actions. For the rest of his life, this scientist would always despise rigid pedagogy and strict theological dogma. In a deeper sense, however, his religious background would become intertwined with his very understanding of what a psychologist was. Like his Quaker predecessors, Goddard too would see himself as disseminating an important message of community concern. In the years to come, his career would repeatedly reflect his efforts to advance older Christian ideals through new psychological science.

Through his studies at Clark, he had found a way to fuse the two. In his thesis, Goddard had explained faith curing in terms of mental therapeutics, and mental therapeutics in terms of evolutionary law. At the same time, however, he had affirmed both the efficacy of and the need for such curing. In other words, Goddard had shown that scientific means – the search for facts and laws – could be put to religious ends, in this instance the charge to heal the sick and comfort the suffering. In so doing, he had made both scientific and secular one key aspect of the missionary calling. It was psychology, Goddard believed, that would have to assume such tasks in the modern world. It was psychology that would have to respond to the most pressing needs of the community, for only it could do so on a solid foundation of science.

By the time he graduated, Goddard had found his vocation. He left Clark in 1899 a disciple less of the church's version of the Gospels than of G. Stanley Hall's. Moreover, Goddard embraced his new psychological calling with an evangelical zeal which matched his mother's. He now believed in an evolutionary version of the faith of his fathers. Like its predecessors, this faith too would use as its touchstone the suppression of human selfishness and the promotion of altruism.

2

"A Little Child Shall Lead Them": Educational Evangelism and Child Study

Psychological Science and Pedagogical Practices

"With my PhD. for comfort," Henry Herbert Goddard bragged years later in describing the start of his psychological career, "I outstripped six other Clark men in the race for the headship of Psychology in the biggest and best Normal School in Pa." As the competition proved, Goddard's 1899 appointment as Professor of Psychology and Pedagogy at the State Normal School in West Chester, Pennsylvania, was clearly a desirable one. Yet while his success was not surprising, for G. Stanley Hall saw him as "a thorough scholar, of most charming disposition" – in short, "almost an ideal man" – it was nonetheless ironic, for it meant that three years of studying scientific psychology had returned Goddard to the field he had previously left: pedagogy.[1]

As Goddard's career demonstrates, in turn-of-the-century America, psychology and pedagogy were still closely connected. Their relationship owed much to Hall, who had recruited many of his best students from pedagogical institutions, and who saw the need to find jobs for his graduates in an era when laboratory positions were scarce and public schooling expanding. Hall too had previously earned his living by lecturing on pedagogy while awaiting a psychological appointment. Ever since, he had nurtured the intellectual and institutional connections between the two fields in his NEA speeches, his summer school for teachers, and his journal, *Pedagogical Seminary*. By 1895, Hall had placed half a dozen Clark graduates in joint appointments teaching psychology and pedagogy in state universities or as chairs of psychology in the teacher training institutes then called normal schools.[2]

Goddard was clearly grateful to secure a similar appointment at the West Chester Normal School. Yet such a position was less than ideal in his case, for as a graduate student he had shown no interest in studying education. Following his graduation, Goddard had worked briefly as an instructor in Hall's summer school for schoolteachers. Here too, however, his real interests were apparent, for he had been hired as Adolf Meyer's teaching assistant for a course entitled "Laboratory Histology and Neurology."[3]

Moreover, after moving to Pennsylvania, Goddard quickly came to see the immense difficulties facing anyone who tried to introduce modern scientific methods into old-fashioned pedagogical institutions. In the years that followed he would become increasingly frustrated, for even Pennsylvania's "biggest and best Normal School" contained no psychological laboratory, nor was it interested in opening one. Most troubling of all was the deep hostility toward scientific research expressed by this school's principal, George Morris Philips. "I get pretty much discouraged here, at times," Goddard would soon confide to Hall, for he was tired of being "constantly told by the head of the Institute that Psychology is of no practical value to the teacher" and that students would be much better served by having a textbook "well pounded" into their heads.[4]

It was precisely such attitudes that had led Hall to campaign so vigorously for the reform of American normal schools. For Hall, school reform and scientific research ought to be inseparable. Pedagogical practices should be grounded not upon past traditions, he repeatedly told teachers, but upon a new evolutionary understanding of child development. Educational progress, he insisted, was thus contingent upon "a genetic knowledge of childhood" – that is, "a knowledge of the growth of the body, brain and soul of the young of the human species." It was this knowledge, Hall believed, that would at last produce a science of education.[5]

Yet if by the 1890s, evolutionary science had found a welcome home within many American universities, the same was far from true within other types of educational institutions. While many teachers longed to transform their profession into a science, and were thus wildly enthusiastic about Hall's ideas, they also recognized the difficulties of blending scientific psychology with contemporary pedagogy. Could laboratory methods really be used within public schools? Would American parents allow Darwinian science, with all its anti-Christian implications, to shape the education of their children? Even worse, would they permit schoolchildren to be subjected to psychological experimentation?

For psychologists who hoped to use science to reshape schooling, such questions posed serious challenges which had to be surmounted. In dealing with these issues, the followers of Hall would have to steer an especially delicate course – one which would have to take into account the religious sensibilities of the American public as well as the methodological requirements of modern science. For the next seven years, it was precisely this task that would absorb Goddard's attention.

Science and Revivalism in Teacher Training

As a Clark graduate, Goddard was very familiar with the methods Hall had used to promote Child Study. To expand knowledge and infuse teachers with a scientific spirit, Hall had suggested a means of linking pedagogical with psychological research. Teachers, he had proposed, ought to gather as much data as possible by simply observing their students; their findings would then be interpreted by psychologists. Thus, working together, educators and psychologists would establish a new science of child nature of benefit to them both.

This ingenious strategy, designed to produce maximum results with minimal institutional support, is best exemplified in Hall's own collaborative research. In 1880, Hall began to supply teachers with "syllabi" – lists of topics and questions to be asked to children. His first and most famous syllabus, entitled "The Contents of Children's Minds on Entering School," was modeled on a similar study that had been conducted in Germany. It consisted of 123 questions administered to Boston schoolchildren with the help of four kindergarten teachers and a school superintendent.[6]

Equally influential was Hall's close working relationship with the principal of the nearby Worcester Normal School, Elias Harlow Russell. At Hall's suggestion, in 1885 Russell asked his students to "observe the conduct of children in all circumstances" and record their results. By 1896, this assignment had generated more than 37,000 observations on the subject of "imitation" alone. "Helen has one of these little cloth caps with a visor, and she always lifts it just as her father does his hat," one typical report began. Although such reports might appear individually trivial, Hall argued that when combined, they would allow psychologists to understand the natural evolution of the child's mind, and thus to answer questions of importance to schoolteachers. The new psychology, Hall promised, "is coming to be a court of appeal in all educational questions."[7]

Teachers evidently believed him, for they enthusiastically supported the joint research endeavors he promoted through his Child-Study movement. In state after state, Hall's followers organized Child-Study societies, either as separate divisions of teacher associations or as independent organizations open to scientists, schoolteachers, and parents. In Illinois, for instance, the Society for Child Study devoted itself "honestly and earnestly to investigate the nature and growth of the child"; by 1895, this Society's *Transactions* reached five hundred members. By the time Goddard returned to pedagogy, he could report that Child Study had gained a "recognized position in at least two-thirds of the States of the Union, as well as in Europe, Hawaii, India, Japan and South America," and had generated more than nine hundred publications.[8]

The Child-Study enthusiast who most influenced Goddard was Earl Barnes, one of this movement's most effective organizers. A graduate of Oswego Normal School and Indiana University, Barnes had earned a Master of Science degree from Cornell. "My own approach to the scientific spirit in psychology was typical," he reported. At Oswego, a science-minded instructor had set him to work dissecting a frog, and at Cornell, he had learned of Wilhelm Wundt's findings from psychologist Edward Titchener. Barnes, however, could as yet see no connection between animal physiology and child psychology. "It was a great day when I first heard G. Stanley Hall speak," he recalled, for Hall "saw that the nervous system and the functions of the nervous system belonged together." In 1891, Barnes became Stanford University's first professor of education. Under his leadership, the Child-Study movement in California generated more research than that of any other state except Massachusetts. By the early 1900s, Barnes was working as an educational lecturer in London and Philadelphia and had published two volumes of school research entitled *Studies in Education*.[9]

Barnes' research strategy made sense to Goddard. His own experience with questionnaires had convinced him that after collecting the facts, psychologists would be able to discern laws governing human behavior – in this case the laws of child nature. In thus beginning the scientific study of pedagogy, Goddard's first task was to convince schoolteachers to join him in forming a Child-Study society.

Establishing such a society, however, was not as easy a task as it had once been, for by the time that Goddard returned to teaching, the Child-Study movement had become defensive. By the end of the nineteenth century, a growing number of educators and psychologists had begun to raise serious questions about this movement's goals, methods, and potential. After 1897, the number of Child-Study publications had begun to decline, thus apparently confirming the predictions of critics who had long dismissed this supposed new science as another of "the crazes by which teachers are periodically stampeded." "I shall not even stop to notice," Goddard stated indignantly to Pennsylvania teachers, "the supposedly impregnable position of those who are wont to utter in omniscient tones and with an air of superior intelligence that scathing rebuke, 'fad.'" Yet such rebukes were hard to ignore after two decades of grand promises, stirring rhetoric, and few evident accomplishments.[10]

Even in its heyday, the Child-Study movement had attracted a variety of critics. To some educators, studying children in this way suggested a kind of human "vivisection." Most critics, however, had simply dismissed the work of these "sweet-tempered souls who took themselves seriously." Child-Study research, they argued, had led to little more than commonsense conclusions pronounced "with an air of oracular profundity."[11]

Criticisms by psychologists were usually tempered, for the belief that scientific psychology could contribute something to pedagogy almost constituted a truism. Even the sharpest of Child-Study's detractors, among them Harvard University psychologist Hugo Munsterberg, had no desire to denigrate the ideal of a more scientific pedagogy. "If I regret that something has become the fad of dilettants [*sic*]," Munsterberg maintained, ". . . do I imply by that the belief that we do not need a modern science of education?"[12]

Yet calling for such a "modern science" was far easier than establishing one. While psychology had become more scientific by allying itself with physiology, education, in Barnes' words, had largely remained "theological and philosophical." University psychologists had transformed their discipline by emphasizing laboratory expertise; such methods, however, could not be easily adapted to the schoolroom. Even more difficult was the challenge of convincing both teachers and parents to apply Darwinian concepts to the training of children.[13]

Child-Study advocates tried to meet these challenges by allying new concepts from evolutionary science with older ideas about moral regeneration already familiar to educators. American schoolteachers had long rested their shaky claims for professional respectability less on any specialized expertise than on their "moral influence." Perhaps no other profession experienced so large a gap between the rhetoric glorifying its significance to society and the actual social status of its practitioners. As compensation for their low salaries and equivalently low regard, teachers usually emphasized the "spiritual" aims and rewards of their work. Education was society's bulwark, they insisted, against materialism, idleness, crime, and unchristian conduct in general. Schoolteachers, according to Barnes, had become the "historical descendants of the clergy."[14]

This tenuous professional situation was mirrored in the confused mission of the normal school. If the teacher's job was not merely to disseminate information but, more importantly, to inculcate Christian moralism, then how should one train teachers? While the first convention of normal school faculty had passed a resolution in 1859 declaring pedagogy a science, many of these institutions had instead evolved into what one historian has called "revival agencies." Poor teachers required regeneration, educators maintained, spiritually as well as intellectually. Even the secularizing forces transforming education by the century's end had failed to alter the primacy of religious rhetoric in inspiring teachers in their work.[15]

A similar inspirational rhetoric infused Child Study, especially as promoted by Hall. In talking to teachers, Hall grafted the new promises of science onto older messages, for like the Gospel, he argued, Child Study "makes old things new." This was especially striking in his speech inaugurating the NEA's new Child-Study Department in 1894. So many teachers had come to hear Hall that this meeting had to be moved to a nearby church. A church, however, was especially suited for Hall's message, for "unto you is born this day," he told teachers, "a new Department of Child Study."[16]

Studying "the little child now standing in our midst," Hall preached, would lead to the "regeneration of education, to moralize it, to make it religious." Members of his audience clearly recognized and understood his message. "It has taken us a long time, many centuries, to begin to realize what was in the mind of Jesus Christ when he said that the little child should lead them," one teacher commented in response to Hall. "At last the meaning of childhood begins to dawn upon us. . . . "[17]

Similar messages were often heard at other Child-Study meetings, where scientific presentations were frequently interspersed with what Hall's student William Bryan disparagingly called "brief evangelizing addresses." Barnes, for instance, was almost as adept as Hall in mixing the metaphors of modern science and old-time religion into single sentences. Studying the "laws of child nature," he proclaimed, would raise teachers from "the position of a patcher of personalities to a co-partnership with the Divine spirit in the development of a law-abiding soul in a law-abiding universe." In the past, Barnes lamented, teachers had functioned as a "lesser clergy" because they had failed to understand "the continuous revelations of God in the successive generations of men." Evolutionary science, he believed, would make such understanding possible.[18]

These arguments also made sense to Goddard, for he too saw scientific means and religious ends as compatible. Goddard's connections with the Quaker community still remained strong; a year after graduating from Clark, he taught in the Friends' Summer School of Religious History at Haverford, alongside his old friend Rufus Jones as well as his old nemesis, Seth K. Gifford. And while all that remains of his lecture is its title – "Psychology in its Relation to Revelation and Religion" – its contents evidently satisfied his Quaker audience, for during the 1900–1901 school year, while Jones went to Harvard to earn a philosophy doctorate, Goddard taught his classes at Haverford (in addition to his own at the normal school).[19]

He also spent the year organizing Child Study in Pennsylvania. Like Hall and Barnes, Goddard too used religious enthusiasm to promote science. Thus, at a State Teachers' Association meeting, he quoted the comments of a colleague who had seen in Child Study not just "a scientific awakening" but also "a great educational revival, in which the hearts of parents and teachers are being purified through the holy fires of a regenerated love and newly consecrated devotion to the rights of childhood." If educators had any fears concerning the atheistic underpinnings of evolutionary science, such speeches reassured them that science and Christianity had been safely intertwined in advancing the interests of children.[20]

But was Child Study really a science? Hall had consistently tried to make pedagogy scientific by linking it to psychology. Other psychologists, however, took a much dimmer view of schoolteachers infringing upon their profession. At issue, they argued, was the crucial question of scientific expertise. The study of the child, Hugo Munsterberg stated bluntly, "must be done by trained specialists or not at all," and teachers possessed no special training in science. Not only did Munsterberg distinguish true psychologists from these amateurs, but he further subdivided the amateurs into "people who know that they do not know psychology" and "people who don't know even that."[21]

Acutely cognizant of their limitations, most teachers had no intentions of comparing their own scientific research to that of physicists or chemists. Still, many Child-Study activists insisted that they were capable of emulating Charles Darwin when he claimed he had "without any theory collected facts on a wholesale scale" and thus transformed his field. Schoolteachers could collect facts about children, they maintained, in the same way that other amateur scientists had previously collected facts about flora and fauna. By objectively recording their observations, they could thus function as psychology's "naturalists."[22]

With this goal in mind, Child-Study organizers promoted a mass effort to gather as many facts as possible about childhood. Barnes, for instance, reported that by 1895, California teachers had "gathered and worked up 15,000 children's drawings, 7,000 papers on the historical sense, 37,500 definitions . . . , 3,000 papers on children's rights, 1,200 compositions on heaven and hell, 4,000 papers describing punishments, 2,000 studies on observation, 3,000 comparisons of the horse and cow, 5,000 papers on inference, 3,000 papers on children's ambitions, 1,200 tests on poor spellers, 1,200 color tests, and 2,000 compositions on fear." The most prodigious of fact gatherers was, of course, Hall himself, who continued to distribute syllabi and publish his results in articles filled with an endless array of statistics. In his 1896 "Study of Dolls," for example, Hall had found that of "845 children, with 989 preferences, between the ages of three and twelve, 191 preferred wax dolls; 163, paper dolls; 153, china dolls; 144, rag dolls; 116, bisque dolls; 83, china and cloth dolls; 69, rubber dolls; 12, china and kid dolls; 11, pasteboard dolls; 7, plaster of Paris dolls; 6, wood dolls; 3, knit dolls; while a few each preferred papier-mache, clay, glass, cotton, tin, celluloid, French, Japanese, brownie, Chinese, sailor, negro, Eskimo dolls, etc."[23]

Such research, however, impressed few scientists outside Clark. By the end of the century, even those who had previously been sympathetic, among them Princeton psychologist J. Mark Baldwin, had become skeptical. "Now, what can be said of indiscriminate observation of every conceivable thing which the children do . . . ?" Baldwin inquired. Unfortunately, he could see little merit in using descriptions gathered by "the uninstructed teacher or parent, worked up to a fine pitch of enthusiasm for his new scientific calling." Such research, he now believed, was largely worthless. "Child Study is a fad, a harmless one for the most part," Baldwin concluded. "But it is an insult to the teaching profession . . . to hoodwink them into thinking that they are making contributions to science."[24]

Harvard psychologist William James was equally critical. Although he claimed no intention of impeding those who "take a spontaneous delight in filling syllabuses, inscribing observations, compiling statistics, and computing the per cent," he hoped that educators would feel no guilt being "passive readers." All the science they needed, James told teachers, "might almost be written on the palm of one's hand." For James, teaching could not be reduced to a science. "I cannot too strongly agree with my colleague, Professor Munsterberg," he maintained, "when he says that the teacher's attitude towards the child, being concrete and ethical, is positively opposed to the psychological observer's, which is abstract and analytic. . . . "[25]

Munsterberg's criticisms – aimed, he insisted, not at the teacher but at "those who make the teacher believe that his observations nevertheless have value for psychology" – were even more pointed. Child Study's "untrained and untechnical gathering of cheap and vulgar material," he surmised, had about as much value for psychology as hunting stories had for biology. "It is not scientific botany," Munsterberg complained, "to find out in whose yard in the town cherries, in whose yard apples grow."[26]

Such criticisms angered Child-Study activists, who defended their work as a legitimate variant of psychological science. "The professors who look askance upon child study," insisted Sara Wiltse, a secretary of the NEA's Child Study Department, "are of two kinds: those whose work has shut them within the four walls of a psychophysical laboratory, and the teachers of the old fashioned philosophy." Barnes seconded this. "Physiologists and psychologists who look out from their laboratories and laugh at our clumsy attempts to use their tools," he complained, "make the mistake of thinking that we are trying to do their work." This was not so, Barnes insisted, for Child Study was not a pure but an applied science. Its purpose was to supply useful knowledge to those raising children. "The farmer, the gardener, the schoolmaster, and the parent," he cautioned, "must make their scientific study secondary to their immediate duties to cattle, plants, and children." Yet applied sciences still held promise. "Child-study has today," he concluded, "the same relation to psychology that horticulture has to botany."[27]

Surprisingly, teachers received support for their position from psychologist Edward Thorndike, the most sophisticated American student of animal learning. Thorndike too had been unable to find laboratory employment after graduating from Columbia in 1898. Instead, he taught education at Western Reserve's College for Women before securing a position at Teachers College in New York. And while he had as little patience for Hall's evangelical rhetoric and imprecise research methods as he had for the outmoded normal school curriculum of his day, he was far more tolerant of Child Study than his university peers. Teachers, Thorndike maintained, "can do work as good for the purpose of mental science as much of the work of naturalists has been for biology." Those "even in normal schools and child-study societies" ought to "go ahead making judgments about mental facts, testing their mental judgments by experience, widening their acquaintance with human nature as much as they can," he advised, "in full faith that they are making psychology." Like Hall, Thorndike too hoped to nurture the scientific curiosity of schoolteachers.[28]

Goddard's defense of Child Study also emphasized the effects of science on the schoolteacher. Especially in rural districts, he argued, teachers needed "a greater professional spirit, more originality and sense of power and culture" which would come from "having done something that contributes to the sum of human knowledge." Many teachers needed to be "waked up and pulled out of ruts." The Child-Study movement, Goddard admitted, was, as Hall put it, "directly for the sake of the teacher, indirectly for the sake of the child, and incidentally for the sake of science." "I care not," he conceded in an unusual defense of a science, "if all the material is worthless." A valuable result had been obtained anyway, Goddard asserted, by inducing the teacher "to think of her individual pupils more than she has been accustomed to think." Even the most old-fashioned teacher would modify her practices, he predicted, "as new ideas *which she has discovered* come to her."[29]

Despite such disclaimers, Goddard still felt that Child Study could be a genuine science – a science of facts and laws. In dealing with children, he told teachers, "we are within the realm of law." Behind individual differences, which Child Study "brings home to us as nothing else does," there remained "a law of child nature which we can bank upon when once we have comprehended it." By building upon such laws, pedagogy would become predictable.[30]

Such knowledge, Goddard insisted, was sorely needed. "There is scarcely a thing that we do in school work," he lamented, "no method of teaching or of discipline – that we are absolutely certain is correct, the best method; nothing in practice that we may not doubt or question." Only scientific study could "put us on a foundation that cannot be shaken." Teachers owed it to themselves to produce results. "Let us in Pennsylvania take our place in this new renaissance," he proposed in September 1900, "and throw our weight on the side of a higher appreciation of childhood and a determined stand for a better pedagogy."[31]

Such appeals proved effective, for two months later Goddard announced the formation of a Child-Study Department within the Pennsylvania State Teachers' Association. Remaining before him was the task of designing a research project that would benefit teachers and psychologists alike. "It is our desire to organize the work of Child Study among the teachers of Pennsylvania," Goddard stated, "for mutual help."[32]

Studying the Child

In starting his Child-Study organization, Goddard turned first for guidance to G. Stanley Hall. The evangelism underlying the two men's common mission to the schools, as well as Goddard's growing frustration within his own institution, was apparent in their correspondence. "Will you 'Come over to Macedonia and help us'?" Goddard wrote Hall, echoing the biblical call given to Saint Paul. "I pray that you may come," he added, for it would be "a godsend to these people!" The "humiliating part," he conceded, was that he could pay Hall nothing but expenses, "so if you come it will be a labor of love and a missionary enterprise." Pennsylvania schoolteachers, Goddard wrote at the time of the Boer War, "do need the Gospel of the new education, worse than South Africa needs English Civilization. Pennsylvania Normal Schools are veritable sweat shops," he reported, "and Pennsylvania County Superintendents are – well, curious." His letter ended with a religious plea often used by Quakers to compel a response. As "the good old Friends say," Goddard wrote Hall, he hoped "that 'Thee will feel a Concern' to come."[33]

There is no evidence that Hall felt concerned enough to come to Pennsylvania. Instead, Goddard proceeded on his own by copying Hall's methods. Hall had long maintained that the first task in studying children was to assess their physical health. In fact, one of the most influential of all Child-Study innovations had been Hall's simple suggestion that teachers seat children with poor vision in the front of the room. To allow teachers to make such diagnoses, Goddard's Child-Study organization offered to supply free eye charts to Pennsylvania teachers. Goddard also urged teachers to look for other physical impediments, and to inform parents if children needed medical attention. "If parents are too poor," he told teachers, "you yourself find some physician who will examine the child for the love of childhood."[34]

Again following Hall, Goddard also wrote a series of questions designed to assess children's minds. He thus asked children to identify their favorite books, to describe what made them happy or afraid, or to tell of just punishments they had received or cowardly acts they had witnessed. "Have you ever had any habit that you have tried to break?" asked one question. "Tell all about it – why you wanted to break it and how you tried." Such questions used child introspection to assess self-control, conscience, and the formation of character.[35]

This emphasis on character formation was especially evident in Child-Study research concerned with children's "ideals." Since Barnes was studying the types of individuals that children idealized, Goddard decided to investigate "Negative Ideals." "What person whom you have known or of whom you have heard or read," he asked students, "would you not wish to be like? Why?" His article interpreting his results illustrates the working relationship linking fact gathering, psychological theorizing, and Christian didacticism that defined the new science of Child Study.[36]

"If sin could be swept out of the world at one stroke," Goddard's article began, "would the world be better off?" The reply might seem apparent, he responded, but the "thoughtful man hesitates to answer." If man could not learn to avoid sin by observing the consequences of the sinfulness of others, he could only learn by experimentation – a more destructive process in the long run. Negative ideals, or examples of sin in the world, Goddard concluded, thus served a positive educational function.[37]

The answers to his questions, however, proved disappointing. Only a few students, for instance, had mentioned laziness, "the soil in which all vice grows," in identifying their negative ideals. Moreover, only 2 percent were "repelled by weak will power," while 10 percent viewed negatively "unpopular people." Children, Goddard concluded, were apparently learning the wrong lessons.[38]

His results also suggested changes in the religious awareness of students. "What is the significance of the fact that Satan – the very personification of all that we should wish not to be like," he wondered, "is mentioned by only two boys and one girl? Is it because he has become so vague and abstract that as a warning he cannot compete with a bad man?"[39]

Such scientific questions remained unanswered. Two years later, however, Goddard had a chance to ask them again in a very different setting. In 1903, he was finally able to take a year off from teaching to study in Europe – an essential component of a proper scientific education for American psychologists of his day. While in Zurich, Goddard spent several months working in the laboratory of Ernst Meumann, a psychologist known for his interests in educational questions. Goddard also saw for the first time his own childhood "ideal," the Alps. On a visit to Göttingen, Germany, he tried a comparative approach to Child Study by asking local children some of the same questions that he and Barnes had asked in America and England. His results, mixed with his impressions of German society, were published a few years later as "Ideals of a Group of German Children" in *Pedagogical Seminary.*[40]

"Göttingen has a University of 1,200 students," Goddard reported. "It appears to have not the slightest influence upon the ideals of these children." Other monuments to German learning were apparently equally ineffectual, for none of the children mentioned any of the scientists, poets, or kings whose statues surrounded them in their daily lives. Moreover, in Germany as in America, few children chose ideals from literature, leading Goddard to conclude that "much time is wasted on the modern novel – especially by the young."[41]

Goddard did find some national differences in identifying ideals, however. The "choice of father and mother," for example, was "age for age, from two to four times higher in Göttingen than in New Jersey," thus leading Goddard to conclude that "Germany is, indeed, the Father- (and mother-)land." This deference to national authority, evident as well in the choice of the emperor, he argued, was part of "a condition of things with which we in America have no sympathy," since it created "the narrowest sort of life, limited horizon, in short a personality wholly unfit for life in a Republic."[42]

Goddard was equally unimpressed by the class stratification of the German school system. The "only inference" one could make from the marked differences in school programs was that "they are getting what they want." As a teacher told Goddard, "You see these girls will all be market women or shop girls and they only need enough arithmetic to enable them to make change." Goddard strongly disagreed. "The danger of a narrow utilitarianism is not that it does not provide for the livelihood of the individual," he argued, "but that it does not provide for the expansion of his personality, the growth of his ideals, the complete development of his soul." Such limited learning was dangerous to child and country. "Surely we have not done our duty by the child," Goddard insisted, "or by the state of which he will be a citizen, until we have brought him into contact with the greatest souls that have appeared among men. Not every child will be quickened by such contact," he added, "but as long as we cannot predict which one will be quickened, it is wise as well as just to give every one the opportunity."[43]

In their religious awareness, however, German students seemed superior to their American counterparts, at least according to Goddard's criteria, for they idealized God, Jesus, and biblical heroes far more frequently. Goddard attributed this to the Prussian system of studying religion in school. In America, by contrast, public schools "taboo the subject," leaving it to "Sunday School and the home – both of which seem to be inadequate." The consequences of each system could now be verified statistically. Asking German children the question he had previously asked Pennsylvanians, "What person do you most wish not to be like?" Goddard was pleased to find 20 percent of the boys and 15 percent of the girls citing Satan or another biblical villain. "Taken together," he concluded, "these two studies surely indicate that the time given to study of religion is not without effect upon the ideals of German children."[44]

Still, Goddard preferred American to German idealism. Whereas Germans were more likely to cite "material possessions" and "physical condition or appearance" as traits of persons they idealized, Americans more often chose those with "intellectual or artistic" talents or those manifesting "power to do things." This group of poor German peasant children, Goddard concluded, was "inferior to the average American group, as, indeed, we should expect it to be." Yet since he was assessing "a group of children of the crudest life and narrowest environment and the least efficiency as citizens," he believed he had identified "a lower limit which we must regard as the danger line for American ideals."[45]

In describing such "danger lines," Goddard's research also suggested the cultural limitations of Hall's new science of childhood. By comparing results from America, Germany, and England, Barnes and Goddard did discern some patterns in child development. In all three countries, for example, the percentage of children idealizing acquaintances decreased as age increased.[46] Yet their interpretations actually placed less emphasis on discovering such "laws of child nature" than on assessing contemporary behavior in light of older and still unquestioned moral standards. The ideals unselfconsciously advocated by these scientists were those of their own childhoods – the ideals of rural Protestant republicanism. As their studies repeatedly showed, however, contemporary schoolchildren were moving in the wrong directions. Child-Study research often blended scientific optimism about a controllable future with moral concern about a way of life that was passing.

Such studies may have done little to change the school curriculum or to create a "science of pedagogy." Even so, Hall's vision of Child Study as a movement allowing scientists and schoolteachers to work together did leave its mark, both on American education and on Goddard's career. Goddard's involvement with Child-Study organizations proved crucial, not because of the content of his research but because of the contacts he made. By meeting frequently with teachers from Pennsylvania, New Jersey, and New York, Goddard gradually developed a new understanding of the physical and mental problems afflicting children – problems that could be alleviated neither by moving students to the front of the room nor by moral exhortation.

The "Indiscriminating Agency of Compulsory Education"

In the decades between 1880 and 1900, while the research agenda of the Child-Study movement had remained relatively constant, the American school population had begun to change in dramatic ways. Central to these changes were massive foreign immigration and rural-to-urban migration. In Pennsylvania, there were more urban than rural inhabitants for the first time in the 1900 census; New York and New Jersey had reported this change by 1880. This trend, moreover, was accelerating. In New Jersey, for instance, in 1870 rural residents outnumbered urban by 100,000; three decades later, urban outnumbered rural by 800,000.[47]

Equally crucial was the passage of compulsory education legislation in state after state. By 1880, fifteen states required children to attend school; by 1900 this number had doubled. Ten more states passed legislation over the next decade, and in 1918 Mississippi became the last state to adopt compulsory education. Of course, enforcement of the new laws was still lax. Even so, their very existence at a time of explosive population growth, especially in industrial cities, left teachers facing a heterogeneous school population different from anything Goddard had known in the small academies of rural New England or the Midwest.

The sheer increase in numbers overwhelmed educators. In the last two decades of the nineteenth century, public elementary school enrollments increased by nearly 54 percent. High schools experienced even more growth. Nationwide, between 1890 and 1900, the number of high schools increased from 2,526 to 6,005 to match an enrollment increase of nearly 250 percent. Urban enrollments continued rising the following decade. Between 1900 and 1910, for example, New York City's population increased by 39 percent while its day school enrollments increased by 53 percent and its high school enrollments by 175 percent.[48]

While urban life had yet to alter Child-Study rhetoric, it obviously affected the children being studied. The results of Hall's 1880 investigation into "the contents of children's minds," for instance, had startled him, for he discovered that "60 per cent of the six-year old children entering Boston schools had never seen a robin." Many children had never seen a cow, "some thinking it as big as their thumb or the picture." Most had never seen growing corn, while "71 per cent did not know beans – even in Boston." Especially revealing of these trends toward urbanization and secularization were the children who believed "that good people, when they die, go into the country."[49]

Of course, compulsory education was in itself a responsive measure designed to influence as well as educate these new urban masses. "The state, to provide for its own protection," declared educational researcher Leonard Ayres, "has decreed that all children must attend school, and has put in motion the all powerful but indiscriminating agency of compulsory education which gathers in the rich and the poor, the bright and the dull, the healthy and the sick." Such a gathering, however, also brought unforeseen consequences that changed the relationship between American pedagogy and other fields of expertise.[50]

Gathering together "the healthy and the sick" brought the first such consequence: epidemics. In Boston, enforcing compulsory education had spread waves of contagious diseases, leaving education with "a record of suffering and death." To counter charges that "compulsory education under modern city conditions meant compulsory disease," educators sought help. Physicians thus became the first outside professionals hired to work within American schools. Boston began mandatory medical inspections of schoolchildren

in 1894. By the end of the century, eight other cities had joined the "school
doctor movement," and by 1911, over four hundred school systems employed
medical inspectors.[51]

"When medical inspection began in the New York schools," Goddard
noted, "several thousand children were sent home the first week because they
were unfit to be in school. The parents rebelled," he reported, "and the medical inspector was told that the work must be given up. He replied, 'We can
not stop; we have gone too far. If we stop now, the parents of healthy children
will create a much greater disturbance.'" In other ways, compulsory education had also "gone too far," for even physicians could not alleviate many of
the problems now visible in public schools.[52]

In overcrowded schoolrooms which frequently contained more than fifty
pupils, teachers now found themselves facing children with a wide range of
physical and mental problems. While doctors could spot contagious diseases,
prescribe eyeglasses, and remove tonsils and adenoids, they could do little
for more seriously disabled children, such as those partially blind or deaf,
motor- or speech-impaired, or suffering from epilepsy, tuberculosis, or car-
diac conditions. Even more perplexing were those with no obvious physical
disabilities who nonetheless failed to make school progress. What, for instance, were school authorities to do with a child like "Clara," a twelve-year-
old whom teachers claimed had "no mental power" and was "a source of
disturbance to other children" and who, after four years, had yet to pass the
first grade? The presence of even one or two such children, one administrator
noted, "so absorbs the energies of the teacher and makes so imperative a
claim upon her attention that she cannot under these circumstances properly
instruct the number commonly enrolled in a class." In the past, such children
had simply left school (and many still did, new laws notwithstanding). Compulsory education laws, however, raised new questions about the state's obligation to keep such children in school.[53]

Of course, awareness of the plight of physically or mentally handicapped
children requiring special educational attention was not new. An earlier generation of reformers, among them Dorothea Dix and Samuel Gridley Howe,
had lobbied intensively to raise public consciousness of this very issue. Their
work had led several states to establish special institutions to educate the
blind, deaf, and in the terminology of the day, the idiotic or feebleminded.
Yet by the end of the century, many states had at best only a few such institutions, while others still had none. As compulsory education made painfully
evident, existing institutions could not educate all the needy children.[54]

By the 1890s, new efforts to bridge the gap between specialized institutions
and public schools were under way. In 1896, the first "special class" was
opened in a public school in Providence, Rhode Island. This apparently successful innovation spread quickly: "special classes" were opened in Spring-

field, Mass., in 1897; Chicago in 1898; Boston in 1899; New York in 1900; Philadelphia in 1901; and Los Angeles in 1902.[55]

The merging interests of public school educators with those working in institutions for the handicapped was also evident in the formation of a new "Department of Education for the Deaf and Dumb, the Blind, and the Feeble-Minded" as a part of the NEA in 1896. Sensitive to its stigmatization as the "department for the education of defectives," in 1902 teachers renamed it the "Department of Special Education – Relating to Children Demanding Special Means of Instruction." The department's purpose, according to its organizers, was to bring teachers educating "children requiring special methods of instruction" into regular contact with "teachers in general."[56]

In seeking some means of understanding their "special" students, both institutional and public school teachers turned expectantly to Child-Study organizations. At a Trenton Child-Study meeting in 1900, Goddard met one such "special educator," Edward Ransom Johnstone.

Psychological Expertise and "Special Education"

Like Goddard, Edward Johnstone had been a schoolteacher in Ohio; on the advice of his brother-in-law, social reformer Alexander Johnson, however, he had begun institutional teaching, first at a reform school and then at the Indiana School for Feeble-Minded Youth. In 1898, he became assistant superintendent of what was then called the New Jersey Home for the Education and Care of Feeble-Minded Children, and its superintendent in 1900. Painfully aware of the dearth of information on educating such children, Johnstone sought expertise.[57]

"Many years ago," Johnstone later reminisced, "when this kind of research was called 'child study,' a green young man attended a meeting of the National Education Association" and learned that "it was possible to study a child and perhaps discover why he did some of the things he did." Johnstone also became an enthusiastic participant in his state's Child-Study society, the New Jersey Association for the Study of Children and Youth. In 1900, he presented a paper to this association on "The Training of Normal and Feeble Minds." Over thirty years later, Goddard still remembered his fateful first meeting with Johnstone at the same conference. Since Goddard was clearly intrigued by his ideas, Johnstone invited him to be a guest at his institutional school.[58]

Visiting this school for the first time in 1900, Goddard saw a small and largely unknown institution in the little town of Vineland, surrounded by the farmland of southern New Jersey. Although a private institution, the Training School also accepted many state-sponsored poor children. Yet whereas state institutions frequently housed more than 1,000 residents in large build-

The Training School at Vineland in 1899, located amidst the farmland of southern New Jersey. Courtesy of The Training School, Vineland, New Jersey.

ings, this school was unimposing, for it housed 233 children in a set of small, "homelike" cottages grouped around a small administration building. At Johnstone's school, Goddard also saw for the first time a radically different pedagogy.[59]

Like most institutions of its day, the Vineland school trained its children according to the principles of "physiological education" developed by Edouard Seguin, a physician famous for his success in teaching the mentally handicapped. Seguin, according to Goddard, had found "that in a being where the nervous system and brain was undeveloped," one had to "appeal to that brain through the sense organs by special methods." He had thus used "photographs, cards, patterns, figures, wax, clay, scissors, compasses, glasses, pencils, colors, even books." In the process, Goddard reported, he had "accomplished little short of miracles."[60]

Even more striking than Vineland's pedagogy was its spirit. While most institutions had suffered a dampening, if not a deadening, of enthusiasm by the twentieth century, Vineland retained the religious zeal of its original founder, Reverend S. Olin Garrison. The author of *Forty Witnesses, Covering the Whole Range of Christian Experience,* Garrison had seen both a spiritual and a social mission in opening this school, first in his home in Millville, New Jersey, in 1887 and then in larger quarters at Vineland the following year. Garrison, Goddard wrote, had taken seriously "the words 'Inasmuch as ye have done it unto one of the least of these.'"[61]

In hiring Johnstone to assist and then succeed him, Garrison had found a kindred soul to help him fight the "under-spirit" he saw as endemic in other institutions. How many "*enthusiasts* – medical or other – are engaged in this work in the U.S. today?" Garrison had written Johnstone. "Perfunctory!! Perfunctory! Rut! Mechanical and Formal! In it for a living. . . . Mediocrity exalted!" Johnstone's main mission would be to retain his school's spirit. His enthusiasm was matched by the school's main teacher, Alice Morrison, a graduate of the Literary and Biblical Institute of New Hampton, New Hampshire. By 1900, the Vineland school still resembled the optimistic institutions of the early nineteenth century far more than its contemporary counterparts. This school exuded a missionary atmosphere – one in which Goddard was immediately at home.[62]

To his surprise, he was equally at home talking to Johnstone's unusual pupils. "After it was over," he remarked in recounting his first visit to Vineland, "Johnstone said, by way of reassuring me, 'You talked as though you were accustomed to talking to the feeble-minded.'" The compliment was comforting, Goddard joked, "but when I went back to the Normal School and in a burst of egotism told my class what Johnstone said, I nearly lost my job – too much 'freedom of speech!'"[63] Yet Goddard did demonstrate an immediate rapport with Johnstone's "special" pupils, perhaps because he still identified with insecure and largely abandoned children.

Goddard and Johnstone met again in Newark in December 1901 at another Child-Study meeting. This time, both Goddard and Barnes presented papers, and it was Johnstone's turn to be impressed. The three men continued conversing on the train back to Philadelphia. "On that train," Goddard recalled, "was born another of Johnstone's big ideas – the Feeble-Minded Club. . . ."[64]

The "Feeble-Minded Club," as participants self-mockingly referred to themselves, was Johnstone's suggestion for an informal meeting at his institution of men interested in special education. The "Club," or the "Gang," as they were also called, or the "Paidological [Child-Study] Staff," as Vineland's Board of Trustees called them, first met in March 1902. Besides Johnstone, Goddard, and Barnes, its original participants included Smith Burnham, a West Chester history professor; Maurice Fels, a Philadelphia lawyer; and Edward Allen, principal of Pennsylvania's Institution for the Instruction of the Blind. "I have never seen any one more glad to have visitors," Goddard noted of their host, Edward Johnstone.[65]

Also visiting Vineland were public school administrators seeking guidance on issues related to special education. Since most normal schools offered no instruction on teaching the handicapped, teachers often turned to institutions. When school officials came to Vineland, Johnstone often included Goddard in their deliberations. "As an institution man," Johnstone explained, "I think I am doing my work by making these men who conduct public schools go out and talk to their teachers and present these ideas."[66]

Johnstone's success in bringing schoolteachers and scientists together was evident in the new additions to the twice-yearly informal meetings of the men in the "Feeble-Minded Club." By 1906, this Club's membership had more than doubled, for it now included principals and school superintendents from Trenton, Camden, and Philadelphia, professors of pedagogy, medical inspectors, and Dr. Henry H. Donaldson, Director of Neurology at the University of Pennsylvania's Wistar Institute (and former Hall student), as well as the original six members.[67]

Such gatherings, however, were far from adequate to meet the growing demand for teacher training in special education. In 1903, Johnstone began more formal training of his own staff during the summer. The following year, he opened a small summer school for teachers interested in special education. Since "the school men of the Feeble-minded Club seemed to see in our children the characteristics found in many children in the special classes," he explained, his summer school would offer teachers an opportunity to learn about such children firsthand by spending six weeks living in an institution. Johnstone also tried to influence the attitudes of teachers toward special education by involving himself in Child-Study activities. When train delays kept him from attending a Pennsylvania Child-Study meeting in 1905, members used the opportunity to express their gratitude. Considering "the unattractive and, not unfrequently, repulsive children with whom he deals," one teacher noted bluntly, his success had to be due to "love for his work and love of the children. . . . If only all teachers of normal children could feel that to him or to her the command has come, 'Go thou and do likewise!' what happy places the schoolrooms all over the land would become!" Another teacher credited Johnstone with removing the stigma of shame from special class teaching. "Teachers the first year are not anxious to have their friends know that they are teaching at a school for the feeble-minded," he admitted, "but soon they become enthusiastic and proud of it. They are learning to exercise an infinite patience and have pride and pleasure in their work," he added. "All schools and all teachers would be happier if they felt as these teachers do."[68]

Johnstone influenced Goddard's thinking most of all. Special education, Goddard came to believe, exemplified not only the true Christian spirit but also the scientific spirit of trial and error. Teachers of the feebleminded had been "compelled to invent methods," he reported, "since by the ordinary methods of education they got no results." Of course, the teacher of normal children often got "results of no value whatever," but since there was usually "a response of some sort, he easily lets it pass as good." In contrast, special teachers were "always growing," he remarked, "always watching for results, discarding methods that are not effective, trying new ones that give promise. I would recommend every teacher to visit such an institution," Goddard advised. "It will repay everyone by its suggestiveness and inspiration."[69]

Goddard's experiences at this institution were suggestive in another sense, for the Vineland children had piqued his psychological curiosity. These children, he believed, were passing through the same developmental stages as normal children, albeit more slowly. By watching these minds in slower motion, psychologists might be able to understand the evolution of the child's mind. The children of Vineland suggested another research agenda for psychologists.

Studying the "Special" Child

Goddard's contacts with the Vineland children led him to several preliminary conclusions. First of all, they reinforced his evolutionary views about mental development. Goddard disagreed with those who claimed that such children were "so utterly different from ours that we cannot argue from one to the other." Children, he stated, "all start pretty nearly alike." Differences developed later, for "the normal goes on to higher and higher stages of development, while the feeble-minded has forever stopped somewhere along the line."[70]

Vineland's pedagogy also suggested new directions for educating normal children. Like many of his contemporaries studying handicapped children, among them the Italian special educator Maria Montessori, Goddard was just beginning to see the broader implications of these pedagogical innovations. The "methods which are permanent with these children," he told teachers, "may very well be the methods with which our children must *begin*."[71]

To Goddard, this suggested that teachers of young children had placed too much emphasis on memorization. The verbal memory of institutional children, he noted, was remarkable. "I have heard low-grade idiots repeat Scripture by chapters until I was tired of listening," he reported. Yet such feats did not signify intellectual growth. A child's mind ought to be trained, Goddard argued, along the lines in which it will ultimately function. "Of these, memory has some small place," he conceded, "but thinking, reasoning, judging, are more important and must have the preponderance of attention in our school work." To Goddard, this suggested the need for changes. "Develop the thought by exercising the perception and the action," he counselled teachers, "and memory will take care of itself."[72]

If memory work was overemphasized in early child learning, Goddard surmised, manual work was underemphasized. Not only do children "learn to do by doing," he claimed; more significantly, they "learn to think and to feel by doing." Through manual training, the Vineland children had learned a variety of tasks; in mastering them, the child's "soul expands enormously." Such observations corresponded to broader principles. "History and anthropology unite in showing . . . that man has become what he is by what he has done," Goddard argued. "His mind has developed in accordance with his movements." Such training would benefit even the brightest of children. "All this has made me realize," he concluded, "that we have not begun to exhaust the possibilities of manual training as an educational force in our public schools." After all, Goddard remarked, the "test of a man in actual life is not what he remembers, but what he can do."[73]

Goddard's new ideas suggested for the first time a movement away from the type of Child Study advocated by Hall. Hall, of course, had always welcomed special educators into his movement. Moreover, like other evolutionary social theorists of his day, he had encouraged psychologists to make use of "nature's cruel experiments" by observing those considered to be less evolved – "the insane, blind, deaf, idiots, paupers, and criminals." Yet Hall himself disapproved of focusing too much educational attention on the handicapped. While special education might appeal to the "method-enthusiast" who "prides himself on results gained from stupid children," he conceded, he preferred to "let the bright children set the pace." "I would rather have a teacher who knew nothing of methods for defective children," Hall told one Child-Study audience, "if he but knew the childhood of distinguished men, to put in the model school that I should like to see established."[74]

Goddard, however, was increasingly drawn to Johnstone's views. "Working with special children," Johnstone remarked in summarizing his own philosophy, "makes us realize and see in a new light the statement of the Master, 'A little child shall lead them.' I firmly believe," he predicted, "that our most advanced ideas on educational procedure will come from the study of 'special' children and their mental processes."[75]

Yet studying these children meant abandoning many of the methods most widely used by the Child-Study movement. Neither written questionnaires nor children's introspections worked well with this population. More appropriate was the extension of physical testing to diagnose abnormal conditions. The same was true, moreover, for the children in the special classes.

Schoolteachers seemed especially eager for information about diagnosing and thus perhaps preventing physical and mental problems. "I would like to suggest that a paper be prepared for this section," a normal school principal stated at the Pennsylvania Child Study Department's 1905 meeting, "showing what common school teachers can do in the way of discovering abnormal conditions among young children, that these may be remedied so far as possible by having the children understood. . . . " Through early diagnosis, another teacher maintained, "we may do something to prevent defects becoming so prominent as to seriously cripple the child." Those present asked Goddard "to prepare a paper in accordance with the following resolution: 'That attention be given by teachers to the defects in children. That all teachers should be capable of examining physical defects in children, such as hearing, seeing, etc.'"[76]

To Hall, this emerging concern with diagnosing the special problems of individual children was of far less importance than the search for generalizable natural laws. "'Study the Child' is becoming 'Study *this* Child,'" he complained. Earl Barnes, however, saw this growing specialization as a positive sign of professionalization. "With the creation of a body of knowledge concerning the subjective growth of human beings," he predicted in 1902, the old "Manuals of Education" would be "shipped to the less intelligent parts of the country to be sold with old copies of 'Every Man His Own Lawyer,' and 'The Universal Family Physician.'"[77]

Barnes' prediction implied a change not only in pedagogical literature but in professional behavior as well. "To-day we have lost our old clerical connection and tradition," he explained. And although education showed possibilities of becoming a bureaucracy "like the post-office," Barnes hoped it might instead become a profession "like medicine or engineering." In typical pedagogical hyperbole, Barnes spelled out the consequences of such a choice: "The future of civilization hangs in the balance."[78]

Hope, however, was in sight. Among the newer generation of schoolmen Barnes saw "a tendency to abandon the methods of exhortation and assertion . . . and to seek a basis for theory and practice in definite knowledge." Included in this group were psychologists such as Goddard. "We are slowly, but surely," Barnes declared, "abandoning the methods of the clergy for those of the physicians."[79]

The Institution as a Laboratory

If Pennsylvania teachers believed that Goddard could enlighten them on the proper means of diagnosing children's disabilities, they were mistaken. Even the Feeble-Minded Club felt the limits of its expertise – a fact reflected in the name this group had chosen for itself. "Time after time we found questions raised by the teachers and also by ourselves which we did not seem to be able to answer," Johnstone reported. In order to advance the field scientifically, Barnes argued, Johnstone ought to open a psychological laboratory on institution grounds.[80]

Convincing Johnstone of the need for scientific research was not difficult; convincing the Board of Trustees was another matter. Barnes, however, broached the subject in an address before the school's Annual Association Meeting in 1903 by immediately discounting the specter of atheistic scientists overtaking this school's humanitarian mission. "All who remain here long must feel the deep missionary instinct that pervades the work of the place," he reassured his audience. Then, in what was to become one of the most frequently quoted phrases used at Vineland for the next few decades, Barnes ended his address dramatically. "To me," he exclaimed, "Vineland is . . . a human laboratory and a garden where unfortunate children are to be cared for, protected and loved while they unconsciously whisper to us syllable by syllable the secrets of the soul's growth." Barnes' words linked scientific research on the feebleminded to the Christian belief that the last shall be first. "It may very well be," he predicted, "that the most ignorant shall teach us the most."[81]

Three years were to pass before Johnstone was able to make Vineland a "human laboratory" in more than metaphor. By 1906, however, he was ready to hire a psychologist. What he wanted was "a man with the requisite scientific knowledge, a humanitarian spirit, with the necessary high social sense to enable him to live under the inevitable restrictions of institutional life and yet to have that wide degree of academic freedom that is the right of all good scientists." He turned for advice to G. Stanley Hall, who promptly recommended Goddard. Years later, Johnstone recalled that he had asked Goddard, "I do not suppose there is any chance to get you to give up what you are doing and take up this work?" and that Goddard had replied, "Well, I do not know." According to Johnstone, he then knew that he had his answer. "If *you* do not know, *I do*," he had told Goddard. "You will come."[82]

Especially appealing to Goddard in Johnstone's offer was the chance to use once again the laboratory skills he had acquired from Hodge at Clark and Meumann in Zurich. Such an opportunity would have been not only welcome but also timely, for despite Hall's heroic efforts to broaden the scope of normal school training, research opportunities within Goddard's own institution had remained severely constricted. Although he had offered to build his own equipment, Goddard could not even persuade his principal to supply him with space for a laboratory. "I am glad to know that Prof. M. E. Meumann applies psychology to pedagogy," Philips had written him in 1904, while Goddard was in Zurich. Yet this principal still felt "more strongly than ever," he added, "that, so far as Normal Schools such as ours are concerned, the time spent on psychology is largely wasted unless it is constantly, directly and practically applied to teaching." Philips' conception of teacher training did not include research. "I am sure that for us it is necessary that a simple text book should be used," he admonished, and "that students must study required lessons of moderate length in it every day, and must be required to recite those lessons regularly and constantly as the most important part of their work." Moreover, he pressed Goddard to agree with his philosophy or to seek employment elsewhere. "I do not feel that anyone ought to undertake this work here," he wrote pointedly, "who does not realize the truth of the suggestions I have made or is unwilling to carry them out." After Goddard's return from Europe, their relationship deteriorated even further. In 1905, Goddard drew his principal's wrath after he complained to the State Superintendent of Schools about cheating among his students. After this incident, the two reached a secret agreement: Goddard would look for another job.[83]

Yet even for a psychologist seeking a new position, opening a psychological laboratory within an institution was an extremely risky career move. It had few if any precedents; in 1906, there were barely any psychologists earning a living without some sort of academic support. Goddard's laboratory would be funded solely by philanthropic contributions – an unpredictable income source at best.[84]

In raising these funds, Johnstone carefully outlined the kind of science he had in mind. It was as though "we were constructing a temple," he told contributors. This temple's first story consisting of "custody and care" had already been built, as had its second of "occupation and entertainment" and its third of "training and education." The fourth story of "investigation and research" now needed construction. Johnstone then challenged his patrons. "Shall we not rise to the top and train our sight upon the sky of truth laid out before us by the Great Master of all knowledge, and search the stars that He has placed within our vision?" he asked. "The truth will be hard to find," he warned. "Our eyes are dim, clouds are in the sky, sometimes it is night, but we must be patient, faithful, confident," he added, "and sometime we – you – will know that we have found our task and have performed it as well as we know how."[85]

For Johnstone, the scientific orderliness of God's universe illuminated his own Christian mission. "Truly nothing is in vain," he told parents. "When you ask why these children are in the world, when the heart-broken parent asks why this affliction is placed upon him, let him realize that God makes no mistakes," he pleaded, "and that this child may be the means of uplifting the world."[86]

Johnstone's message again reflected the Child-Study movement's emotionally powerful mixture of science and spirit. Scientific study posed no threat to Christian ideals, he argued; to the contrary, it would advance them. "We must study these children," he insisted, " – not with the knife of the surgeon, not with the scalpel of the vivisectionist, but with a loving heart and mind, and with the eye and ear of the trained observer who bends to catch the slightest whisper, the least movement that leads to truth – truth which is eternal, truth which is indeed scientific." Johnstone ended with a poem which again suggested God's plan of allowing a little child to lead scientists forward:

> O Thou who gavest me this little child,
> Teach me Thy way, let me be reconciled.
> Lift up mine eyes that I may see the light
> Of strength and progress, which Thou in Thy might
> Shall beckon forth e'en from this feeble soul
> To guide his stronger brother to his goal.[87]

Johnstone's evangelical appeals secured him enough contributions to start his laboratory without touching the school's operating funds. They also secured Goddard. In September 1906, at the age of forty, Goddard moved to Vineland to accept his new mission as a psychologist living among and studying feebleminded children.

"When he accepted this call to leave his position as psychologist in a famous normal school to go to an idiot asylum," Goddard later joked in a 1933 speech describing the decision that became the most significant of his life, "many of his friends wondered why – and some did not."[88] By the late 1940s, however, Goddard's self-deprecating humor on this subject had largely disappeared. Desperately trying to convince his detractors that his decision to go to Vineland had nothing to do with any preconceptions concerning the heredity–environment debate, Goddard described this period of his life:

> I was a teacher of psychology in a Normel [*sic*] School – as they were then called. I became acquainted with the Supt. of a school for the feeble minded. He asked me to come to his institution and make a psychological study of the feeble minded children. It sounded interesting and I went. I knew nothing about the Feeble-minded: I had no idea of how to go about making a "psychological study" of the F.M. You may very well say: Why did you accept such a position? I don't know. I liked the Supt. – as I knew him – and I guess I thought that if he wanted me, I must do as he wanted. At Clark Univ. I had had courses in experimental psych. and I think I must have expected to use some of those experiments, though there were not many of them that the feeble-minded could do.
>
> I had visited the institution several times and was greatly interested in the children. I did have a strong desire to talk with them and to see what kind of creatures they were. I suspect that my acceptance of the job was determined more by that desire than by any thought of how I was going to do what was expected of me.
>
> Not only did I have no idea of how to proceed; I had no idea of

Goddard at the time he began his Vineland research, c. 1906. Courtesy of the Archives of the History of American Psychology, University of Akron.

what I should find. I knew nothing of sociology or anthropology. I had nothing in mind that I wanted to prove. *And I never did have anything that I wanted to prove.* I wanted only the facts.[89]

In his later years, Goddard may have regarded his reasons for accepting his new position as unsure and tenuous. In 1906, however, a far more confident, optimistic, and enthusiastic psychologist happily left his teaching job at the West Chester State Normal School to begin his new work as a researcher at the New Jersey Training School for Feeble-Minded Girls and Boys. Before him lay the task of conducting a psychological study of institutionalized children.

Perhaps the best indicator of Goddard's state of mind at the time is the unusual opening paragraph of his very first scientific paper dealing with his new research area, mental deficiency. Entitled "Psychological Work Among the Feeble-Minded," Goddard had written this address for a Philadelphia

conference consisting largely of physicians. He started it by recounting an
episode from religious history. "Carlyle says," his speech began, "that perhaps the most remarkable incident in modern history is George Fox's making
for himself a pair of leather breeches. He finds in this outward act the expression of a mental resolution which was the beginning of religious freedom in
England and in the world."[90]

The meaning of these opening lines has largely been lost on later generations. In Goddard's day, however, this allusion would have been familiar to
any Quaker schoolchild (and thus many Philadelphians). Goddard was para-
phrasing British historian Thomas Carlyle's stirring paean to Quakerism's
founder, George Fox. In particular, Carlyle had lauded this shoemaker's courage in defying convention by sewing himself a practical but unorthodox article of clothing to wear while preaching outdoors – a pair of pants made
of leather.

"Perhaps the most remarkable incident in Modern History," Carlyle had
written, was not "the Battle of Austerlitz, Waterloo, Peterloo, or any other
Battle; but an incident passed carelessly over by most Historians . . . namely
George Fox's making to himself a suit of Leather." Fox, Carlyle concluded,
was "God-possessed," and his "shoe-shop, had men but known it, was a ho-
lier place than any Vatican." To Carlyle, this small act of free thinking illustrated the "outflashing of man's Free will, to lighten more and more . . . the
Chaotic Night that threatens to engulf him." Such incidents often seemed
"the only grandeur there is in History." With this act, Fox had become a
prophet. "Stitch away, thou noble Fox," Carlyle had exclaimed, for "every
prick of that little instrument is pricking into the heart of Slavery, and World
worship . . . were the work done, there is in broad Europe one Free Man,
and thou art he."[91]

Despite his scientific subject, his professional demeanor, and his subdued
tone, Goddard clearly identified his own obscure work with feebleminded
children with Carlyle's impassioned account of Fox's stitching in freeing the
mind of man. "Some one more familiar with the history of institutions for
the feeble-minded, must locate the event which marked the beginning of the
newer view of defectives," he argued. "The leather breeches may be just completed, and not much preaching done and no quakings apparent," Goddard
exclaimed, "but the movement is on."[92]

3

"Psychological Work among the Feeble-Minded": The Medical Meaning of "Mental Deficiency"

A Medical Specialty

"Monday Sept. 17. Arrived in Vineland at about 5:30 p.m.," Goddard jotted as the first entry in the small composition book recording his ideas and activities for the 1906–1907 year. Since his new "laboratory" contained only a desk, tables and chairs, and filing cases, his first priority was equipment. "Worked on ergograph," Goddard wrote a week later; "Fixed an electrical attachment to metronome. . . . This is used to test will power."[1]

Despite his sparse furnishings and financial constrictions, Goddard's writing clearly suggests his excitement at once again undertaking laboratory research. Yet working in a laboratory funded to alleviate the suffering of handicapped children was a far different experience from working in a university, or even a normal school. In such a situation, experimentation in the name of science alone would hardly suffice. Goddard now faced a more difficult and more fundamental challenge: to prove the social worth of scientific psychology.

Such a challenge must have been appealing, for even in his dissertation, Goddard had maintained that the study of psychology could help to alleviate many human ailments. Feeblemindedness, he believed, was obviously an ailment deserving study by the new scientists of mind. Yet how exactly should a laboratory psychologist go about studying institutionalized children? What questions should he ask? What methods should he use to answer them? At the time he began his new work, neither Goddard nor his contemporaries knew much about how to begin such a project.

Even before he arrived at Vineland, Goddard had begun searching for a strategy. This was especially evident in the "S.O.S. calls" he sent to his former Clark professors. Goddard was entering a "rich but uncultivated field," G. Stanley Hall responded. Adolf Meyer was delighted to see Goddard "taking up this work." "It is risky to prophesy science," E. C. Sanford concluded in words that would indeed prove prophetic, "but I think that the next ten or fifteen years will see that subject receiving very much more attention than it receives at the present time."[2]

Yet while Goddard's professors shared a diffuse enthusiasm concerning

71

his new work, they parted company in advising a specific research agenda. Characteristically, Hall suggested gathering the facts and asking lots of questions. There was, for instance, "the whole question of dress, putting it on and off, care for it, interest in it. . . . What can be done in their motor training, their musical abilities, their play?" he inquired. Sanford was more discriminating in suggesting research areas that might illuminate evolutionary development. The two "brightest jewels in my casket," he told Goddard, were "the expression of emotion" and "the fundamental human instincts." In normal children, these were usually "covered up by training and convention"; in the feebleminded, they might be "less obscured." Even so, Sanford recommended that Goddard at first avoid "minute experimental studies carried on with all the enginery of the laboratory."[3]

Meyer seconded this. "I should advise the most direct common-sense method that has as little appearance of fussy test-work as possible," he answered, "and to allow the form to grow out of what one encounters." The "chief thing to guard against," Meyer warned, "is to pile up a lot of apparently very scientific tests which in the eyes of the teacher and any common-sense individual would appear to be topheavy, and therefore bring discredit to the movement."[4]

Meyer's advice was the most sensitive to outside pressures affecting scientists who chose to work within institutional settings. By 1906, Meyer had become director of the Pathological Institute of the New York State Commission in Lunacy; thus he was acutely aware of the need to balance scientific curiosity against social needs. He was also the most cognizant of the fact that schoolteachers, and not scientists, still constituted the primary audience for Goddard's new psychological research. After all, it was teachers like Edward Johnstone, Earl Barnes, and the members of the Feeble-Minded Club who had established his laboratory; it was they who were also the most expectant about its promise.

However, by the time Goddard began his work at Vineland, the power of teachers within schools for the feebleminded was itself becoming increasingly circumscribed. Over the previous decades, control of such institutions had gradually shifted from educators to physicians. By 1906, Edward Johnstone stood out as one of the only institutional superintendents who held no medical degree. If Goddard hoped to prove the value of scientific psychology, his work would have to convince not only teachers but also doctors.[5]

Like other psychologists of his day, Goddard was most familiar with the work of doctors treating "neurasthenics," those fashionable sufferers from "weak nerves." He now began to meet regularly with institutional superintendents working in a far less glamorous field. Among these were Dr. Martin Barr of the Elwyn institution in suburban Philadelphia, Dr. Walter Fernald

A meeting of institutional superintendents, n.d. (c. 1909). l. to r.: Dr. Walter Fernald, superintendent of the Waverley institution near Boston; Edward R. Johnstone, superintendent of the Vineland Training School; H. H. Goddard; and Dr. A. C. Rogers, superintendent of the Faribault institution near Minneapolis. Courtesy of the Archives of the History of American Psychology, University of Akron.

of the Waverley institution outside Boston, and Dr. A. C. Rogers of the Faribault institution near Minneapolis. Since the 1890s, these doctors had called their field "psycho-asthenics," the study of "weak minds."[6]

In the years that followed, Goddard began to attend the same meetings, read the same literature, and most significantly, observe the same patients as did institutional superintendents. As a result, in many ways he began to think less like a psychologist and more like a physician. He also came to understand mental deficiency as the medical community conceptualized it.

Yet this conceptualization suggested more questions than answers. What, Goddard began to wonder, did mental deficiency really mean? Could the diverse cases that he now saw in institutions be consistently diagnosed? Could they be classified in ways that would aid in their treatment? Most important, could this mental condition really be studied in a laboratory setting? Such questions would need to be answered if Goddard hoped to succeed in conducting "psychological work among the feeble-minded."[7]

Diagnosing the Feebleminded

The medical field that Goddard was now entering had a long popular and a short scientific history. For centuries, European literature and art had contained graphic portrayals of persons often called "fools." Yet portraying this condition had proven far easier than explaining it. Since the sixteenth century, English common law had tried to define this well-known mental state. An "idiote, or a naturall foole," it argued, was a person of lawful age and yet "so witlesse that he can not number to twentie, nor can he tell what age he is of, nor knoweth who is his father, or mother, nor is he able to answer to any such easie question."[8]

While insanity had long captured the medical imagination, idiocy had not. Before the nineteenth century, few physicians had even tried to offer medical explanations for mental conditions long regarded as unalterable consequences of Divine Providence. Sustained medical interest in such cases had begun only with the French Enlightenment, when physicians of the Age of Reason accepted the unreasoning mind as a problem worthy of and susceptible to scientific amelioration.[9]

The American medical community was linked to this French tradition through the work of Edouard Seguin. A student of Jean Itard, the physician who had treated the famous "Wild Boy of Aveyron" captured in 1798, Seguin had moved to the United States after 1848. His admonitions to American doctors had emphasized the empiricism and optimism of enlightened thinking. "*PROGRESS,*" he had prophesied, would be "*in proportion to the thoroughness of observation.*"[10]

By the end of the nineteenth century, American physicians had had ample opportunity to observe the expanding population of mentally handicapped individuals living in institutions. Following the example of their colleagues working in institutions for the insane, in 1876 they had formed their own professional organization, the Association of Medical Officers of American Institutions for Idiotic and Feeble-Minded Persons, with Seguin as first president. Thereafter, they met annually, recorded their proceedings, and in 1896 began to publish the *Journal of Psycho-Asthenics.*[11]

Yet despite these outward trappings of professional progress, Goddard found medical authorities as confused as their lay counterparts in answering questions basic to their field. Physicians had trouble defining the condition they were treating. They used different diagnostic criteria and arrived at inconsistent prognoses. Most frustrating of all, they could not agree on a means of classifying the cases which increasingly found their way into institutions.

Such populations required some sort of categorization. Although most physicians crudely categorized their most severely impaired patients as "idiots," those less impaired as "imbeciles," and those only mildly impaired by a variety of names, including the generic term "feebleminded," these categories had no accepted boundaries; an "idiot" in one institution could be an "imbecile" in another. Medical attention had produced a proliferation of case descriptions; yet instead of a single system of diagnosis and classification, these descriptions suggested an ever-increasing heterogeneity.[12]

These problems were strikingly apparent in the best American medical treatise on this subject available to Goddard, Martin Barr's 1904 text, *Mental Defectives: Their History, Treatment, and Training.* "To the student of mental defect," Dr. Barr lamented, the "very first requisite" was a classification "at once simple and comprehensive, definite and clear." None yet existed. Barr could report only that "the conditions, incident upon diversity of times and nationalities, as well as the differences of bases constituting premises," had prevented the adoption of "one common order of classification." Instead, the best that Barr could offer was a summary of over a dozen conflicting systems.[13]

Perhaps it was their frustration with this lack of progress, or, more positively, their curiosity concerning the work of psychologists and educators newly interested in feeblemindedness, that convinced physicians to open their association to nonmedical members. Whatever their reasoning, in 1906 the Association of Medical Officers changed its name to the American Association for the Study of the Feeble-Minded, and Goddard joined it. In the following years, he too came to focus on the problems that had perplexed doctors. "I was early impressed," he later told his physician colleagues, "with the fact that we are not all using the same classification. This results in much confusion."[14]

Yet if Goddard was "early impressed" with the problems facing physicians, he also understood the reasons for their confusion. In his new position, he had grown ever more distant from academic psychology and increasingly close to the medical world. While his psychological contemporaries were conducting studies on university students, Goddard now had a tougher group of subjects on which to examine the mind–body relationship – subjects whose minds and bodies often both appeared to be seriously impaired.

Like his medical colleagues, Goddard turned his attention to the heterogeneous group of individuals that doctors called the feebleminded. The best statistics on this population had been gathered in a census report of 1904; it stated that while "competent authorities" (meaning medical superintendents) believed that at least 150,000 Americans were "so pronouncedly feebleminded as to stand in need of institutional treatment," the number actually living in institutions was 14,347. In 1906, the Vineland school housed more than 300 such children. Classifying these cases, however, proved an extremely complex task. Both in his own and in other institutions, Goddard quickly

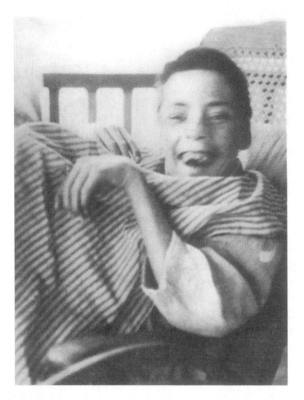

A Vineland resident described as a "microcephalic idiot boy, 22 years old." From *Psychology of the Normal and Subnormal* (New York: Dodd, Mead, 1919), facing p. 78.

learned, the feebleminded varied markedly in their physical appearance, medical histories, social behavior, and aptitudes for learning, all of which complicated the question of classification.[15]

Easiest to describe were persons suffering from obviously serious physical as well as mental handicaps. This included about a third of those confined to American institutions. Of these, over half were epileptics. The remainder, in the medical terminology of the day, were blind, deaf-mute, maimed, crippled, deformed, or paralyzed. Doctors also identified numerous less serious physical "stigmata" frequently found among this population, such as cleft palates, asymmetrical facial features, speech impediments, or unusual gaits.[16]

Even individuals with analogous symptoms, however, often had very disparate medical histories. Some had been ill since birth; others had been born healthy but failed to recover from spinal meningitis, "brain fevers," accidents, or other traumatic events. For those who still bore forceps marks on their skulls, birth itself had proven harmful. Goddard found twenty Vineland children who gave positive readings on the Wasserman test, indicating they suffered from congenital syphilis. In fact, by the time a child reached an institution, he often had a complex medical history. Six-year-old "Upton," for example, had arrived at Vineland with the following history: "Instruments used at birth; child had convulsions at five weeks; spasms from three years on; measles at six months; meningitis at seventeen months; has had whooping-cough and paralysis."[17]

Despite these diverse histories, physicians were able to recognize some distinct physical types. The "cretins," for example, could be identified by their growth deficiencies, while "hydrocephalics" and "microcephalics" could be categorized respectively by their large or small heads. Some recognized the children with Asian-looking eyes that Dr. J. Langdon Down had christened "Mongolians." Many institutionalized children, however, manifested no such obvious physical stigmata; physicians accordingly classified them by other criteria.[18]

Some children could be distinguished by their bizarre behavior. Behavior, however, actually varied as widely as physical appearance. While Goddard described one child brought to Vineland at the age of six as "a cheerful, affectionate little girl, quiet and obedient, very willing, tries, and is making considerable progress," another six-year old had been described by a staff member as "excitable and nervous, cried and laughed without cause, was gluttonous, destroyed clothing and furniture, was dangerous with fire, not truthful, nor trustworthy; active, obstinate, sly and passionate." How could physicians categorize the boy who enjoyed "stealing even from himself," according to Dr. Barr, since he frequently reported his possessions stolen just to watch the staff search for them? In such children, doctors surmised, the physical senses might be operating, but the "moral sense" was surely damaged. Some children even posed a danger to parents, teachers, or other children. Barr described a child removed from school because he "had a habit of kicking children in the stomach and struck a blow like a sledge-hammer." Others had set homes on fire. Parents brought one child to Elwyn after he shot his younger sister. In such cases, Barr lamented, there was "no moral sense to appeal to."[19]

Even more puzzling were children who hurt only themselves. There were, for example, the "rocking" children found in institutions, such as "M.W.," an eleven-year-old girl who "rocks constantly back and forth," or "F.C.," a nine-year-old boy who pounded his head "like a trip-hammer upon the wrists" for hours if left undisturbed. Goddard saw similar children at Vineland, such as "Mattie," a ten-year-old who "has what seem like insane spells; has pulled out two teeth and pulled out her earrings while in one of them; beats her head; digs herself with her nails, screaming all the time."[20]

Still other children both looked and behaved properly and yet seemed un-

" Study of a child. Samuel here I am for I was called."

Drawn by FRANK H., CASE 314, AGE 19. MENTALLY 8.

Pencil drawing of a child produced by "Frank H.," a Vineland resident noted for his artistic talents. Originally diagnosed as an "idiot savant," he was later declared insane. He was also "very strongly religious," as evidenced by the biblical caption used for this drawing. From H. H. Goddard, *Feeble-Mindedness: Its Causes and Consequences* (New York: Macmillan, 1914), facing p. 426.

able to master elementary lessons. The strangest were the "idiot savants" – individuals whose idiocy was accompanied by genius in the areas of music, art, or feats of memory. Seguin, for example, had described the "historical cook," a man who could give "an intimate account of the Peloponnesian War, or the history of Tallyrand," but who was in all other ways a "real simpleton, utterly without judgment." Goddard saw one such exceptionally talented child at Vineland, for notwithstanding his other limitations, "Frank H." was a gifted artist.[21]

These cases, however, were rare. More common, and more frustrating to institutional teachers, were the children who, despite years of instruction, made no school progress. "Keith," a sixteen-year-old who had lived at Vineland for six years, was such a case. Goddard described him as "a handsome boy with no marks of his defect on his body; quite active and pleasant spoken, just the kind of boy to tempt any teacher to believe that with a little special training he could be made thoroughly normal; yet every effort put upon him meets with failure." Despite the absence of physical or behavioral abnormalities, Keith had "never been able to do much with his reading, writing and counting" and Goddard found him "one of the most disappointing cases in the whole School."[22]

It is not surprising that institutional physicians in Goddard's day were having trouble classifying cases that in later years would be labeled as retarded, epileptic, emotionally disturbed, autistic, learning disabled, or by numerous other designations. Even so, doctors continued to search for characteristics commonly shared by the institutionalized population. They never doubted that they were dealing with genuine pathology – pathology somehow connected to feeble "mindedness," or mental weakness. Yet what was feeble-mindedness? What exactly did mental weakness mean?

By 1877, physicians had agreed on a definition broad enough to cover all contingencies. "Idiocy and imbecility," the Association of Medical Officers stated, were "conditions" in which there was "a want of natural and harmonious development of the mental, active and moral powers." These usually were associated with "some visible defect or infirmity of the physical organization, or with functional anomalies" and could be "expressed in various forms and degrees of disordered vital action." Moreover, there was "frequently defect or absence of one or more of the special senses, always irregular or uncertain volition and dullness or absence of sensibility and perception." This did indeed cover all cases. However, while such a definition offered an illusory medical unity, its vagueness rendered it practically useless.[23]

Moreover, such open-ended definitions made it especially difficult to instruct noninstitutional physicians on the means of diagnosing feeblemindedness in children. Seguin had recognized this problem years earlier; in response, he had tried to specify "the minimum of *what a general practitioner must know about idiocy.*" In making such diagnoses, he had recommended, doctors should overlook intellectual deficiencies (e.g., incomprehension) unless these were accompanied by apparent physiological disorders. They should watch for a swinging walk, "automatically busy" hands, saliva dripping from a "meaningless mouth," a "lustrous and empty" look, and "limited" or "repetitive" speech. Based on such observations, physicians could offer prognoses to anxious parents. They could predict a favorable outcome, he advised, if the walk was steady, the hand firm, the look "easily called to action," the words connected in meaning, and the child "active without restlessness, is pleased to obey, sensible to eulogy, quite as capable of giving as of receiving caresses." They should offer negative prognoses, however, if conditions like epilepsy or paralysis were present, or if parental affection did not bring "corresponding intellectual progress." The "criterion of idiocy," Seguin insisted, "is found more in the physiological than in the psychological symptoms"; even so, his medical admonitions actually conflated physiological and psychological observations. Doctors received no guidance, for instance, in how to gauge physiologically the meaning of "corresponding intellectual progress."[24]

While general practitioners had grown more confused, however, institutional physicians had continued to gain confidence in their diagnostic abilities, despite an inability to agree on the criteria they were using. By the twentieth century, most maintained that, after years of experience treating feeblemindedness, they simply knew it when they saw it. Thus, Fernald told Goddard of an incident in which two of his attendants had returned from town with the news that a new boy was coming. Asked how they knew, they had replied, "We saw him." That was all they needed, Goddard explained. "They 'saw him' and knowing the type, that was enough."[25]

Working with more than three hundred mentally handicapped children, Goddard too came to believe that one could indeed learn to diagnose mental deficiency intuitively. "Most people of long experience in institutions for the feeble-minded come to have the power of guessing rather accurately the grade of a child," he argued. Yet this was at best "an unsatisfactory method," he concluded. Moreover, such subjective understandings formed a shaky foundation for laboratory work. Scientists needed a more objective measure of the problem they intended to study.[26]

Physicians too were frustrated, for although they believed they knew who belonged within institutional walls, they still sought a common classification system. Administering a residential, educational, and training facility for hundreds of heterogeneous, disabled individuals required superintendents to separate their charges into manageable subgroups. Children exhibiting different physical, behavioral, and learning characteristics required different amounts of medical care, daily supervision, and educational instruction. Moreover, doctors hoped to distinguish cases that might improve from those that would deteriorate. They longed for a prognostic classification that would tell parents what to expect. The rising expenses of maintaining their expanding institutions led philanthropists and bureaucrats to press them for answers to the same questions.

Institutional physicians had yet to agree upon a classification system that met any, much less all, of these needs. In the absence of consensus, each institution adopted its own methods of subdividing its charges. Meanwhile, doctors continued to debate the merits and drawbacks of a variety of systems.

In beginning his work at Vineland, Goddard also quickly came to focus on the need to understand the different types of institutionalized children. In the process, he found himself in the middle of the debate over classification. His involvement in this debate would soon prove crucial, for it would allow him to make his first important contribution to his new field. Of even more significance, Goddard's ideas about classifying institutional cases would ultimately lead American physicians to accept a fundamental redefinition of the medical meaning of mental deficiency.

Medical Systems of Classification

In examining the crucial question of classification, Goddard was entering into a medical debate whose very contours appeared confused and unfocused. Perhaps for this reason, it would be largely overlooked by later generations, for its many sides and numerous participants make it difficult to discern even the broad outlines of this medical controversy. Yet notwithstanding their diversity and the absence of a common vocabulary, medical writings of the late nineteenth and early twentieth century still evidence the emergence of three distinctly different approaches toward the important problem of classifying "feeble minds." In order to analyze the underlying assumptions guiding medical thinking in Goddard's day, these can be loosely labeled the pathological, psychiatric, and sociological approaches.[27]

The most important advance in organizing the growing body of empirical data being acquired largely by institutional physicians was the pathological classification system developed by Dr. William Ireland in the late nineteenth century. As chief physician to the Scottish National Institution, Ireland had sought "some arrangement in order to say clearly what I wanted to say" and thus approached idiocy "from the standpoint of pathology." Mental handicaps, he suggested, were manifestations of distinct disease processes, and should therefore be classified according to their causes. By 1877, Ireland had elaborated a classification system based on what he assumed were twelve known causes of idiocy: epilepsy, hydrocephalia, microcephalia, paralysis, inflammation, trauma, cretinism, deprivation, eclampsia, syphilis, sclerosis, and "genetous" or congenital conditions.[28]

While many Americans lauded Ireland's pathology as a major advance, others argued against its adoption. Dr. Isaac Kerlin voiced a common objection: Ireland's categories were not mutually exclusive. An "individual in any single class," Kerlin complained, might also "be found in two or more of the other classes." Even more problematic were the many persons that fit no class. Ireland had grouped these cases together in his largest single category, "genetous idiocy," his classification for cases whose cause was still "shrouded in the obscurity of intra-uterine existence." He hoped to reclassify them in the future, when more might be known. Contemporaries, however, pointed out the impracticality of classifying cases by cause when, in most instances, cause could not be determined with any reliability.[29]

Besides impracticality, other physicians raised a more theoretical objection to this system: Ireland had emphasized only the physical differences between patients, while failing to note their mental differences. His system "omits from consideration entirely," Kerlin remarked, "the essential features of idiocy – the mental deficiency," for there was "nothing in the mental incompetency of a hydrocephalic idiot, or a paralytic idiot, that is peculiar, one from the other." Dr. Hervey Wilbur, superintendent of the State Institution of New York at Syracuse, was equally critical. Such systems, he claimed, regarded mental conditions as "mere symptoms of abnormal and pathological physical condition." Before adopting it, Wilbur wanted more proof of this connection between mind and body. He thus insisted that its proponents prove first that there was a constant relationship between "certain physiological and pathological conditions" and "corresponding manifestations of defective intelligence, sensibility, and will" and secondly, that such conditions could be "detected, located, measured, and clearly defined."[30]

Physicians emphasizing their patients' mental rather than their physical handicaps looked instead to the work of their colleagues classifying insanity. Although most psychiatrists were only peripherally interested in idiocy, a few had tried to distinguish "amentia," the "absence of mind," from "dementia," the "loss of mind." Idiocy is not a disease, the French psychiatrist Jean Esquirol had maintained, but a "condition" in which the intellectual faculties "have never developed sufficiently for the idiot to acquire the knowledge which other individuals of his age receive when placed in the same environment." "The insane man," he explained, "is deprived of possessions which he formerly enjoyed; he is a rich man become poor; the idiot has always been in misery." Classifiers, these psychiatrists suggested, ought to be less concerned with cause than with degree of such impairment. Thus, Esquirol's writings included one brief but suggestive paragraph which proposed classifying idiots and imbeciles into five groups based upon their language skills.[31]

This idea intrigued Superintendent Wilbur. "Even the number of words used," he maintained, was "a tolerably fair test of the intelligence of a people, or an individual, under the same or similar conditions of life." Omitting only the deaf-mutes, Wilbur classified 225 cases in his own institution into eight groups, ranging from those who neither spoke nor comprehended language to those with a "fair command of language."[32]

Wilbur's classification system, however, also had its shortcomings. With no "specific distinctions to stand as metes and bounds between the different categories," he conceded, his classes "shade into each other." Moreover, he knew his idea was radical, for he was proposing that physicians adopt a "strictly mental test." Nonetheless, Wilbur wanted a system that would attempt to parallel the "growth of intelligence from infancy to manhood."[33]

Other psychiatrists, however, while agreeing with the need to classify cases by degree rather than cause of impairment, were less impressed with language systems. "The parrot can be taught to articulate," Dr. Daniel Hack Tuke concluded, "but in intelligence is far below the elephant, which cannot." Reflexes, Tuke suggested, were much better indicators of mental impairment than language. He proposed a classification system based upon quality of motor control. The lowest group consisted of those exhibiting only "the reflex movements known as the excito-motor." Those in the second group were capable of "sensori-motor" reflexes, including "those of an ideo-motor and emotional character." The highest group included "those who manifest volition – whose ideas produce some intellectual operations and consequent will." In general, Tuke's psychiatric ideas sound closest to the psychophysical assumptions adopted by nineteenth-century experimental psychologists, for they too emphasized reflexes and motor control in attempting to gauge the mind.[34]

By the early twentieth century, a third approach toward classification was gaining medical attention. This new theory was based on a broadened understanding of the meaning of "education." Of course, education for the feeble-minded had usually meant something more than intellectual pursuits. Seguin's "physiological education," for example, stressed manual and sensory training. The newer theories, however, widened this concept even further to mean the individual's ability to function socially. Thus, in his 1904 text, Dr. Martin Barr proposed an "Educational Classification" which conflated intellectual, vocational, and social skills. The lowest of his four groups included those who would need perpetual asylum care, while persons in the highest group could be "trained for a place in the world."[35]

This use of sociological criteria was most obvious in the classification system adopted by the Royal College of Physicians in London in 1908. Using starkly Spencerian language, British physicians redefined the meaning of mental deficiency in terms of fitness for surviving. Idiots, the lowest group, were those "so deeply defective . . . as to be unable to guard themselves against common physical dangers." The imbecile was "incapable of earning his own living, but is capable of guarding himself against common physical dangers." The highest grade was "capable of earning a living under favorable circumstances, but is incapable . . . of competing on equal terms with his normal fellows . . . or managing himself and his affairs with ordinary prudence." These new criteria soon won wide acceptance among physicians.[36]

Yet while the language of evolutionary sociology had allowed doctors in Britain to reach a broad (albeit still vague) consensus, American physicians remained divided. Medical meetings floundered over the absence of a common vocabulary in diverse systems stressing pathological causes, mental assessments, or social adaptability. This was the confused medical situation that Goddard encountered when he first began his "psychological work

among the feebleminded." In attempting to advance his new field scientifically, he too would have to confront the crucial question of classification.[37]

"Shifting Our Problem to Psychology"

After a year of work at Vineland, Goddard emerged with a clearer vision of
what the new psychology could contribute to psycho-asthenics. The first goal
of experimental psychology, he proposed in his first annual report, would be
to reinfuse institutions with the spirit of scientific research. By the early twentieth century, he discovered, such a spirit was sorely lacking in the world of
institutional medicine.

Like Johnstone, Goddard was contemptuous of the "custodial idea," which
dominated almost all other institutions he had visited. "In some it is positively oppressive," he reported. "One can almost read on the walls, 'The only
duty to the feebleminded is to keep them comfortable as long as they live and
be thankful when they die. They are of no use anyway.'" At Vineland, both
the missionary and the scientific spirit were still alive. "Our laboratory pays,"
Goddard insisted, "if it does nothing else than stand there as a vivid reminder
to all of us that there are problems to be solved in connection with our
work."[38]

That psychologists could solve such problems where physicians had failed,
however, was as yet unproven. Even so, within a year, Goddard tactfully began to argue the case for "shifting our problem to psychology." Psychology
could contribute to psycho-asthenics, Goddard argued, because of its close
relationship to pedagogy. Education, he asserted, was "THE PROBLEM of
the mental defective." Yet pedagogy depended upon psychology, the science
of mind. Of course, Goddard admitted, in the past this science had been "so
abstract that the would-be expert teacher got little satisfaction out of it."
Fortunately, however, times had changed for psychology.[39]

Before this "new dawn," Goddard argued, "we had almost come to the
thought that if these children were fairly comfortable as long as they lived,
all was being done." Psychologists were partly to blame for such pessimism,
he conceded, for they had "closed the door to all progress" by arguing that
the feebleminded had "no judgment, no reason, no will, etc." These theories,
however, had been based upon the now discarded belief that "judgment was
a faculty, an entity, a something within the brain that did something for us"
and "when we had an individual who had none of this faculty we had a
hopeless case." Contemporary psychologists, however, knew better. "To say
that a feeble-minded child has no judgment," Goddard now affirmed, "is
meaningless, useless or false" since the new psychology analyzed mental processes into elements "present even in fairly low-grade idiots." Thus, Goddard
was asserting that since new psychologists had reconceptualized their scien-

tific understanding of the mind, those working with the feebleminded ought to reshape their questions accordingly. The question now worth studying, Goddard explained, was not whether the elements of mind existed in the feebleminded, but whether these could be combined to produce relatively complex mental processes.[40]

Even new psychologists, however, had trouble in trying to answer such questions, for they too found it difficult just to describe the diverse cases living in institutions. Like his medical predecessors, Goddard soon turned his attention to definitions, diagnoses, and classification. "What is the situation?" he inquired. In seeking the deeper meanings of the descriptive words used in this field, he returned first to their Latin etymologies. "An idiot is by derivation a peculiar person; the imbecile is the weak one; the feeble-minded is something less than normal. All these words are very innocent," Goddard concluded, "and yet their literal meaning expresses about all we know today. 'Peculiar,' 'weak,' 'mentally feeble' children. In what way peculiar?"[41]

Evolutionary theory, he concluded, suggested a comparative response. The feebleminded person was "at the lower end of the scale, just as the genius is at the upper end," Goddard reported. This theory also suggested that the suffering of such persons resulted from their inability to adapt adequately to the surrounding environment. The psychologist's task, he argued by using a biological metaphor, would thus be to "get fruit from a plant that is an unusually slow grower." While some plants ripened annually, Goddard explained, others took longer. "Ours is the century plant," he proposed. "In our climate," he maintained, such a plant "would probably never bear fruit." Under more "favorable conditions," however, "the century plant may bear fruit in ten years, sometimes even in eight."[42]

"All are agreed," Goddard boldly, if inaccurately, asserted, that "we are dealing with a fairly healthy organism" that "lacks something which would enable it to function like the majority." The fundamental question, he argued in attacking the pessimism of his medical colleagues, "is, does it lack so much that it cannot develop at all? We have usually assumed that it does. But is it proven?"[43]

Goddard began his own studies by assuming otherwise. After a year of study, he still held the "conviction that mind is the same in them as in us, within the limits of their experience." In fact, Goddard's earliest beliefs are striking, especially in light of later controversies concerning his motives, for in starting his new research, most of his assumptions were starkly environmental – far more so than those of most of his medical contemporaries.[44]

"The differences we find," Goddard argued in 1907, "are due to difference in environment and the way in which they have reacted to it, rather than to differences in the nature of the being reacted." While such "sluggish" or "clogged" minds were clearly out of touch with their peers, new educational methods might reach them. "Our problem is something like that which Dr. Howe solved with Laura Bridgman and Miss Sullivan with Helen Keller – a problem of establishing communication," he insisted. "The feeble-minded child is a foreigner who does not speak our language and hence cannot obey our commands or follow our directions."[45]

Of course, Goddard reassured his listeners, he was not proposing that feeblemindedness could be completely cured. "Perhaps I have delayed too long emphasizing that it is a better mind and not a perfect mind that I am hoping for," he added. Yet the goal of Goddard's science was clear. The feebleminded child could not help himself. "We," he told physicians, "must do this for him."[46]

The only means of doing this was through a scientific understanding of feeble minds. Goddard knew, however, how far scientists were from such an understanding. Even the simplest assumptions were unproven. From his own work on brain dissection, for example, he knew that one could not state assuredly that the feebleminded were "deficient in brain cells." Studies of idiot brains were "far from furnishing data sufficient to substantiate such a definition," he reported, for while some showed gross anatomical defects, others were microscopically perfect. Moreover, studies of patients with brain lesions suggested that normal mentality was not "contingent upon having an intact brain." "On the other hand," Goddard added, "there is evidence that a perfect child would become an idiot if he grew up in seclusion, without training of any kind, kept in a cage and fed as is practically often done with the feeble-minded."[47]

Goddard's own model for connecting brain and behavior was still the work of Adolf Meyer. "I confidently look forward to the day," he optimistically told the association, when "on the living child we shall say after careful examination of his condition, 'He has defects of such a character in such cellgroups. Such and such exercises, such and such foods will produce the best results.' No longer guess-work," he predicted, "but treatment with scientific precision."[48]

Yet such precision depended upon the ability of scientists to describe precisely the mental condition they were treating. Like many of his psychological contemporaries, Goddard believed that laboratory research could contribute to this endeavor by merging psychiatry with psychophysics. His own comparative orientation matched most closely the approach of physicians stressing classification by degree rather than cause of impairment.

In fact, some doctors had been calling for the development of comparative assessment techniques for decades. Nearly half a century earlier, for instance, Seguin's writings had already suggested a surprising interest in new laboratory experiments designed to measure mental responses, and an acute awareness of their potential significance for his own medical field. In the future, he predicted, physicians would be able to time mental processes and thus to gauge the effectiveness of their treatments. "The improvement in these processes is capable of positive measurement," he had explained, for at first "an idiot requires several seconds to transmit an impression from without within, or a volition or order from within without," while the "normal time for these operations" was "only 1/25th of a second for the former operation, and 1/28th for the latter." Progress in "sensation, perception, volition, and even self-control" might also "become susceptible of mathematical measurement," Seguin had argued. Such assessments, however, necessarily assumed as a prerequisite a thorough knowledge of normal mental functioning. It was "paramount," Seguin had thus warned, that studies of the minds of idiots also be conducted "upon normal subjects with the strictest similarity . . . everywhere, near the abnormal, the normal; next to the shadow, the light."[49]

Dr. Wilbur had called for the development of similar measurements in 1877. "Do we not need some mile-posts along the educational path to the same end?" he had asked doctors. Wilbur wanted "some generally recognized tests of physical and mental condition" to show "the starting-point in the pupil's career, to which reference can be made from time to time to test their absolute or relative progress." Such tests would be "a form of classification," he argued, "in relation to the growth and development of the pupils." He urged physicians to note the "peculiar habits of the different shades and grades of idiocy."[50]

Yet in the thirty years separating Wilbur's work from Goddard's, physicians had made little progress in measuring idiocy's "shades and grades." Meanwhile, psychologists had continued to experiment with new laboratory techniques for measuring sensorimotor responses. What was needed, Goddard believed, was some way to bring the two fields together.

He found support for his own work in the research of Shepherd Franz, a psychologist employed in a hospital treating the insane. Like Goddard, Franz hoped to merge psychiatry with psychophysics. What was lacking in medical studies, Franz saw immediately, was "a careful analysis of the mental condition," for the very words used by doctors were vague and imprecise. "Apathetic, dull, stolid, irritable, restless, nervous, deficient memory, slow comprehension," he argued, were "general terms." For "scientific purposes and in the present state of psychology," Franz added, "the observer could and should" try to describe such conditions "more carefully." Scientists, in other words, required a more precise vocabulary in assessing the functioning of the mind – a vocabulary that psychophysics could supply.[51]

In Goddard's day, the most extensive American work on psychophysical measurement had emanated from the Columbia University laboratory of Franz's mentor, psychologist James McKeen Cattell. A student of Hall at Johns Hopkins and Wundt at Leipzig, Cattell had written his doctoral thesis

Goddard's psychological laboratory at Vineland. Among the objects shown are an ergograph, chronoscope, automatograph, target (for testing precision in motor control), and darkroom for photographic work. From Bird Baldwin, "The Psychology of Mental Deficiency," *Popular Science Monthly* 79 (1911): 83.

in 1886 on "The Time Taken up by Cerebral Operations." While in England, he had visited Francis Galton's laboratory in South Kensington, where numerous anthropometric measures of traits such as height, weight, arm span, and length of middle finger of left hand were being collected along with psychophysical measures of vision, hearing, and reaction time. In an 1890 article entitled "Mental Tests and Measurements," Cattell had proposed using a battery of ten psychophysical tests to measure the mind. By 1896, he had begun gathering anthropometric and psychometric information on all Columbia University freshmen.[52]

"It is of some scientific interest to know," Cattell reported after testing one hundred undergraduates, that such students "have heads on the average 19.3 cm. long, that 15% have defective hearing, that they have an average reaction-time of 0.174 sec., that they can remember seven numerals heard once, and so on with other records and measurements." These were "mere facts," he conceded, "but they are quantitative facts and the basis of science."[53]

Goddard hoped to make similar facts a part of the science of psychoasthenics. He immediately began gathering anthropometric data on patient height and weight from other institutions and recording the psychophysical facts acquired from laboratory equipment such as the ergograph, dynamometer, spirometer, automatograph, and chronoscope. "Child taps on an electric key at his natural rate," Goddard noted in his journal. "The taps are recorded on drum and counted. He is then told to tap as fast as he can. A f[eeble] m[inded] child has not the will power to increase his rate very much. I think perhaps this too," Goddard speculated, "may be sufficiently significant to serve in classification."[54]

Yet a program of tests and measurements designed for university undergraduates proved difficult to apply to children suffering from physical and mental handicaps. "The psychologist will appreciate the difficulty," Goddard noted, "of testing individuals who do not talk or cannot follow directions or who have little or no power of voluntary attention." He described one attempt to test a child on the automatograph, a machine designed to measure involuntary motion. "He would not try to hold it still," Goddard recorded. "I explained and illustrated, held his hand, and threatened him when he moved," he added. "No use. He could not be made to do it." Finally, Goddard put his own hand on top of the child's "for five long minutes" and held it still. "At last a gleam came into his eyes," he noted. "I removed my hand and he held his still for a minute, making a good tracing. He had at last comprehended what was wanted, found he could do it," Goddard concluded, "and immediately rose in my estimation and in his own."[55]

With ingenuity and determination, Goddard could overcome many such problems in test administration. Far more difficult, however, was the problem of assessing the meaning of his results. How, for instance, ought one to compare results gathered by special means within the institution with results found elsewhere? "Put him through all the tests," Goddard recorded about one child. "He did them well and in the absence of standards cannot attach any significance. My *impression* is he ranks rather high."[56]

Goddard's notes recording his early research experiments at Vineland show a dedicated and creative scientist facing impediments which proved increasingly frustrating. Equally evident was another problem, for undermining Goddard's attempts to study the Vineland children psychometrically was a problem which had also hampered medical work: the absence of any comparative measures assessing normal mental development in children. Without such measures, Goddard soon came to realize, the degree of his children's mental "feebleness" could not be gauged.

In his search for new strategies, Adolf Meyer had suggested another lead. "I have been told," he informed Goddard, "that Professor W. of Philadelphia has devoted more attention than any one else to the study of defectives; it might be worth your while to look up his work." "Professor W." was Cattell's student Lightner Witmer, a psychologist at the University of Pennsylvania who had also studied at Leipzig, and who was also concerned with studying children having difficulties in learning.[57]

Witmer had been working with schoolchildren ever since 1896, when a Philadelphia teacher brought a "chronic bad speller" to his laboratory. This teacher, he reported, had assumed that "psychology should be able to discover the cause of this deficiency and advise the means of removing it." Witmer solved the case: the boy needed glasses. This simple incident, he later claimed, had reoriented his entire career, for within months he had opened the nation's first "psychological clinic" to diagnose individually and treat the diverse learning disabilities of schoolchildren. In his speeches to the American Psychological Association calling for a "clinical" psychology and a "diagnostic" pedagogy, Witmer had gone much further than Hall in proposing that psychologists adopt medical means towards educational ends.[58]

Yet Witmer's work suggested no new directions for Goddard's research, for while his professional rhetoric was indeed radical, his scientific solutions were not. Prescribing eyeglasses was hardly new to students of Hall, who had been advocating eye examinations for decades. Moreover, despite Witmer's significant institutional innovation in offering schoolchildren individualized attention, he had yet to suggest any new psychological methods for diagnosing mental disabilities that Goddard might use within an institution.

More useful in linking psychology with psycho-asthenics was the work of another Cattell student, Naomi Norsworthy. In 1906, Norsworthy was an instructor in Educational Psychology at Teachers College. For her dissertation on "The Psychology of Mentally Deficient Children," she tested institutional, special class, and ordinary schoolchildren.[59]

Like Goddard, Norsworthy was keenly aware of medical debates of the day, for her study cited fourteen conflicting definitions of mental deficiency and thirteen different systems of classification. The diversity of opinions surrounding "these two fundamental matters of psycho-asthenics," she reported, were "but a sample of the confusion and disagreement that is found with reference to almost every other fact mentioned in the literature on the subject." Norsworthy also understood the fundamental need for a comparative approach in assessing the mentality of handicapped children. "Experimental evidence as to the position occupied by idiots in various mental and physical traits as compared with the position occupied by ordinary children in the same traits," she argued, was "the only means of definitely answering questions in the psychology of idiocy."[60]

Goddard found the puzzle board developed by Norsworthy to be an especially promising measure of the abilities of institutionalized children. Her other tests were "not very well adapted to our children," he conceded. Even so, he continued to use them "because we have in her book the normal standards for comparison."[61]

Yet standardizing psychometric tests solved only part of Goddard's problem. Of even more importance was finding tests whose results would prove meaningful. In fact, Goddard was now facing the same problems that had brought the entire Columbia University program in mental testing to an abrupt dead end. Even when he developed standards for tests of his own, such as needle threading or line drawing, his results were disappointing, for they correlated neither with each other nor with any other significant aspects of observed behavior.

These problems had been predicted by the most famous psychometric measurer of the day, Francis Galton, more than a decade earlier in assessing the pitfalls of Cattell's larger strategy. Whereas Cattell had conceptualized his work as gathering the quantitative facts of mind – simple facts which might form the foundation for understanding more complex mental phenomena – Galton had instead proposed that psychologists assess minds by "sinking shafts, as it were, at a few critical points." Such assessments, he argued, would be valuable only if they could be correlated with "an independent estimate of the man's powers" – in other words, if they could be validated by other independent assessments of mental capacity.[62]

In the case of undergraduates, the most obvious criteria for validating mental tests were school grades. In 1901, one of Cattell's students, Clark Wissler, undertook this task. His results, however, signaled the profound failure of Cattell's testing program, for Wissler found no correlation at all between psychophysical measurements and grades in college.[63]

Goddard's failure now paralleled Cattell's. Within the institution, he could hardly use grades to validate his work. Nonetheless, his anthropometric and psychometric results did not match the only independent criteria for assessing their validity: the staff's intuitive assessments of the children's intellectual abilities, based on years of experience in living with them. Within the institution, as within the university, anthropometrics and psychometrics produced numbers that predicted nothing.

Despite his continued dedication to experimental laboratory science, Goddard had soon exhausted the ideas of American psychologists and yet was no closer than his medical predecessors to measuring or classifying mental deficiency. "After two years my work was so poor, I had accomplished so little," he later recalled, "that I went abroad to see if I could not get some ideas."[64]

Goddard left Vineland for Europe in the spring of 1908. Seeking new directions for his own research, he spent sixty-two days traveling through England, France, Italy, Germany, Austria, Switzerland, Holland, and Belgium. During his travels, he sought advice from prominent psychologists while also visiting doctors and teachers working in nineteen different institutions and ninety-three special classes.[65]

Although he met many of the most distinguished institutional physicians of his day, his most provocative find was a short list of mental tests given to him by a Belgian doctor and special educator, Ovide Decroly. The tests had been published three years earlier by Alfred Binet, a renowned French psychologist, and his assistant, Theodore Simon, a physician. Beginning with Binet and Simon's 1905 article pointedly entitled "Upon the Necessity of Establishing a Scientific Diagnosis of Inferior States of Intelligence," Goddard found his first answer to the problems plaguing both psychologists and physicians.[66]

Goddard and Binet

Years later, in explaining the event that dramatically reoriented his career for the remainder of his life, Goddard again denied any initiative of his own and credited only luck. "My getting hold of Binet's work was the result of a series of lucky accidents," he explained. "Somehow there came into my hands a single printed sheet signed by an unknown Belgian by the name of M. C. Schuyten," he reported. "Luckily I did not throw it in the waste basket." Schuyten had been active in the European Child-Study movement, and he referred Goddard to others interested in similar problems. Through these contacts, Goddard was taken to meet Decroly, a physician who ran an experimental school in Brussels and who closely followed the literature on Child Study. When "Dr. D. came to the door," Goddard recalled, "I said I am Mr. Goddard from America. Quick as a flash, he said, 'Dr. Henry Goddard? You have written an article on the ideals of German children. My wife has translated it into French.' HE told me about the Binet tests. ALL PURE CHANCE."[67]

Goddard's meeting with Decroly may indeed have been serendipitous, but his "getting hold" of Binet's tests still suggests something more than chance discovery. Actually, Binet's 1905 questionnaire did not need discovery. After all, Alfred Binet was France's most gifted experimental psychologist. A wide-ranging thinker of prodigious intellectual energy, Binet had built his own laboratory at the Sorbonne and published his own journal, *L'Année Psychologique*. For the past fifteen years, he had been exploring various means of measuring individual differences, from comparing head sizes to gauging tactile sensations to analyzing children's abilities to define words or solve puzzles. American psychologists had long been following Binet's highly original studies of memory and suggestion, as well as his intriguing explorations of the mental differences distinguishing his own two growing daughters. Yet they had recognized nothing especially noteworthy in his 1905 scale.[68]

Goddard had not even looked up Binet when he was in France because of the discouraging reports he had received from other psychologists. "Binet's

Lab. is largely a myth," his European diary notes after a visit with Pierre Janet. "Not much being done – says Janet." Goddard had been especially disappointed in the small number of schools for the feebleminded in Seguin's homeland. "No imbecile asylums in France, no kindergartens in Germany, no Christians in Palestine! It is the same old story," he lamented.[69]

Nevertheless, upon returning home, Goddard tried out Binet's tests on the children in his own institution. Immediately, he understood their applicability. Two years of frustrating institutional experience had prepared him to see what Janet, Cattell, and even Hall, the most prescient of contemporary psychological entrepreneurs, had missed. Contained within Binet's articles, Goddard quickly realized, was an entirely new psychological approach toward diagnosing and classifying feeble minds.

The fact that Goddard became the first American to appreciate Binet's innovation is not difficult to explain. Although Goddard was in no sense Binet's intellectual peer, surprisingly these two psychologists had much in common. Most important were their relationships both to their own profession and to the professions of pedagogy and medicine. By 1908, the careers of these two very different psychologists exhibited some striking institutional parallels, for both Binet and Goddard had begun to link psychological research to the practical problems facing teachers and doctors.

In their own ways, both Binet and Goddard had grown increasingly distant from other university psychologists. Goddard's marginal professional employment, first in a normal school and then in an institution, had gradually moved him away from academic psychology. Binet too had become alienated, but for different reasons, for his own strongly independent views and frequently contentious personality had led to numerous conflicts with prominent leaders in his field, and had frustrated his attempts to secure one of the few chairs of psychology offered in France, an honor he felt he had surely earned.[70]

Partly in response to these situations, both Goddard and Binet had embraced the alternative research strategy proposed by G. Stanley Hall: they had allied themselves closely with schoolteachers. Within Child-Study organizations, both psychologists found an enthusiastic group of supporters who appreciated their leadership. Thus, while Goddard had presided over the Pennsylvania Child Study Association, Binet had become president of its French equivalent, La Société libre pour l'étude psychologique de l'enfant, called simply La Société.

Organized in 1899 by Ferdinand Buisson, a Sorbonne professor of the Science of Education, La Société had been directly modeled on Hall's American Child-Study societies. From 1900 on, Binet had become increasingly caught up in this society's activities – organizing its collaborative efforts, editing its publications, and designing its questionnaires. La Société's diverse membership included schoolteachers, directors of normal schools, university professors, priests, lawyers, doctors, sociologists, directors of special schools for the blind and deaf, and parents. As a result of these contacts, Binet had become intimately familiar with the daily problems plaguing educators. Goddard too had learned of these problems, first in his Child-Study society and then through the Feeble-Minded Club. Thus, while Goddard was meeting regularly with superintendents and special educators from Camden, Trenton, and Philadelphia, Binet was working with the director of primary teaching of the Seine and a general inspector of special education in Paris.[71]

Following Hall's model on opposite sides of the Atlantic, both men had used very similar language in defending Child Study. Both, for instance, had emphasized the ways that scientific research could change the perspective of educators, for Goddard spoke of teachers being "pulled out of ruts" while Binet hoped to break their "deadly daily routine." Both also stressed the need for collaborative research, for Goddard had told of psychologists and schoolteachers merging for "mutual help," while Binet described having "founded a kind of cooperative." Moreover, in France as in America, both schoolteachers and scientists benefited from this collaboration, for while psychologists brought pedagogy scientific legitimacy, teachers gave psychologists access to a test population. "I did not understand why a scholar like Binet could have need of us, poor teachers," one of La Société's members recalled. "Afterward," the teacher added, "I understood."[72]

It was because he was president of La Société that Binet had become involved in the question of classifying schoolchildren. Binet and Simon's 1905 articles had been written specifically to supply guidance for an educational commission appointed by the Paris Minister of Public Instruction in 1904 to set standards for "special class" admissions. And while acknowledging that this was actually "a work of administration, not a work of science," Binet nonetheless took it seriously. To "be a member of a special class can never be a mark of distinction," he insisted; such important assessments ought to be based upon scientific standards.[73]

Of equal significance were Goddard's and Binet's relationships with physicians. Unlike most of their university peers, both of these psychologists were very familiar with medical theories of psycho-asthenics. Goddard had learned about this field after starting his research at Vineland. Binet too knew this literature, for he emanated from a distinguished family of physicians and had himself begun and then abandoned the study of medicine. More important, his collaborator, Theodore Simon, was a physician who had interned at the Perray-Vaucluse colony, an institution that housed more than two hundred mentally handicapped boys.[74]

Although Binet's express purpose in designing his 1905 intelligence scale was pedagogical, his writing actually suggests that he had a medical audience in mind as well. While he now ignored Cattell's psychophysical measurements, he closely scrutinized the classification systems devised by a variety of prominent physicians, among them Seguin, Ireland, and Esquirol. In fact, embedded within Binet's articles was a blistering attack on medical expertise. One had to guard, Binet warned readers, "against intuition, subjectivism, gross empiricism, decorated by the name of medical tact, and behind which ignorance, carelessness, and presumption hide themselves."[75]

Binet easily mocked the writings of his medical contemporaries such as D. M. Bourneville, a highly respected French physician, who had distinguished a "fugitive" from a "fleeting" attention in separating idiots from imbeciles. "The vagueness of their formulas reveals the vagueness of their ideas," Binet noted contemptuously. "They cling to characteristics which are by 'more or less,' and they permit themselves to be guided by a subjective impression which they do not seem to think necessary to analyze. . . . " Such attitudes, he argued, impeded scientific progress.[76]

Binet, like psychologist Shepherd Franz, was thus attacking the very vocabulary that doctors used to describe their patients. His criticisms, however, implied far more than a mere debate over precision in language. Those who believed that "at bottom this is only a question of terminology," Binet insisted, were deceiving themselves. "It is very much more serious," he maintained. The real problem, he bluntly suggested, was that physicians did not know what they were talking about.[77]

Where physicians went awry, Binet argued, was in their failure to distinguish sharply the body from the mind, and thus physical from mental impairment. If a child was diagnosed an idiot, Binet emphasized, it was not because he "does not walk, nor talk, has no control over secretions, is microcephalic, has the ears badly formed or the palate keeled." By stressing such physical criteria, doctors had missed the point. "The child is judged to be an idiot because he is affected in his intellectual development," Binet insisted. This was "so strikingly true," he argued, "that if we suppose a case presented to us where speech, locomotion, prehension were all nil, but which gave evidence of an intact intelligence, no one would consider that patient an idiot." Mental deficiency, Binet concluded, must therefore be classified by purely psychological criteria – criteria that assessed only comparative intellectual development.[78]

Ironically, by stressing intellectual impairments over physical disabilities, Binet's new conceptualization of mental deficiency actually harkened back to the much older lay meaning of this term. While Seguin had tried to use the science of his day to link minds with muscles, Binet had once again divorced the two. Instead, his research now linked age with understanding. Thus, once again, as English common law had proposed centuries earlier, an idiot was one who, despite being of lawful age, could not answer an "easie question."

Like his psychological contemporaries, Binet also saw the need for more precise measurements in dealing with the mind. Differences in degree, he argued, were of no value unless they could be measured, even if only crudely. Yet in pursuing his own research on individual differences, he had developed a concept of psychological measurement that differed markedly from Cattell's.[79]

He was not seeking measurements "in the physical sense of the word," Binet had written in 1898 in explaining his own conceptualization, "but only a method of classification of individuals." Psychologists could not "*measure* one of their intellectual aptitudes in the sense that we measure a length or a volume," he argued. Thus, if one individual could remember seven digits, and others six or eight, this was not the same as measuring "wooden beams, to say that one is six meters long, one seven, the other eight." In the latter case, "one really measures," but in the former, one cannot know "whether the difference between a recall of six digits and a recall of seven digits is or is not equal to the difference between the recall of seven digits and the recall of eight." Psychologists "do not measure," Binet therefore concluded, "we classify!" It was just such a form of classification, however, that institutional physicians had long been seeking.[80]

By 1905, Binet had begun to study closely both the methods used by classroom teachers and those used by institutional physicians. The best way to assess the intellectual level of institutionalized children, he now argued, was by comparing them to normal children. That "which especially strikes us," Binet remarked, "is the resemblance between young normals, and subnormals very much older." Such resemblances were "so numerous, and so striking that truly, one could not tell by reading the reactions of a child whose age is not given, whether he were normal or subnormal." To Binet, this observation was extremely suggestive, for it meant that both normal and feeble-minded children were following the same developmental sequence – an assumption that confirmed the broader evolutionary theories of the day. It also suggested a new means of classifying institutional cases.[81]

Binet's new approach to classification involved arranging a series of simple tasks and questions according to their degree of difficulty, as experienced by normal children of ages three, five, seven, nine, and eleven. The simplest involved following a lit match with one's eyes, the most difficult answering abstract questions. Other test items contained in this scale asked children to identify objects, to make rhymes, to compose sentences, or to compare lines of different lengths.[82]

Binet's scoring procedure was quantitative in a different sense as well. Unlike Cattell, Binet now counted neither seconds nor centimeters; instead he recorded the number of tasks or questions completed satisfactorily – a strategy that owed far less to physics than to pedagogy. Binet was thus inventing a new type of psychological examination, for his mental tests blended the techniques used by scientists with those used by schoolteachers.

Using this strategy, Binet had devised a new means of determining degrees of feeblemindedness. Institutional patients labeled idiots, he argued, did not seem able to go beyond the sixth of his thirty tests, while those labeled imbeciles rarely got beyond the fifteenth. These results, Binet proposed, suggested a crude new system for classifying institutional cases.[83]

By the time Goddard returned to America and published an account of this 1905 scale in his institution's modest journal, the *Training School Bulletin,* Binet had made another major advance. In his 1908 article, "The Development of Intelligence in the Child," Binet supplied a battery of tests, this time arranged in groups and standardized for normal children of different ages. Binet now noted, for example, that in answering the question, "Are you a little boy or a little girl?" three-year-olds sometimes made mistakes, but "a normal four year old child will always answer correctly when asked its sex," thus indicating that "between the third and fourth years a marked change takes place in the mental state of the child." This simple test thus became one of his measures of normal four-year-old intelligence. Binet established similar norms by asking children of different ages to compare weights, to recognize colors, to remember a series of digits, to count coins, to describe pictures, or to make comparisons between objects, among other tasks. "When one has been struck by a playmate who did not mean to do it what must one do?" asked one test question. Another presented children with a statement such as "I have three brothers, Paul, Ernest and myself," and then asked, "What is silly in this sentence?" While these tests were eclectic, Binet believed that most of them were measuring judgment skills. By comparing an individual child's responses against norms established empirically for children between the ages of three and thirteen, he argued, one could measure a child's relative "mental level."[84]

Goddard at first expressed skepticism at such a bold claim. "It has always been easy for the writer to be tolerant of the critics of the Binet tests," he would later recall in 1933, "because when he first read the 1908 article, he laid it aside with the mental comment that there was nothing in it because it was 'impossible to measure intelligence in any such way.' During the past twenty-five years we have become more familiar with that refrain," Goddard added, "than anything in print except the Star Spangled Banner." Yet he soon changed his mind. "The writer quickly realized," Goddard reported, "that he was in no position to pass judgment upon this instrument." After all, he had been seeking a means of assessing degrees of mental deficiency, and "here was a psychologist of world-wide reputation who offered this plan and claimed that it had at least some value. It was not good sense to throw it aside," he concluded, "because one reading of it was not convincing."[85]

Goddard did much more than give the article a second reading: he translated it and tried it on all the Vineland children. His confidence in Binet's methods grew as he increasingly noted the correspondences between the "mental levels" ascribed to the children by Binet's tests and the staff's intuitive assessments of their abilities. "It met our needs," Goddard stated simply. "A classification of our children based on the Scale agreed with the Institution experience."[86]

In the years that followed, Goddard would become America's most avid convert to Binet's "psychological method" of classifying the feebleminded. From his position at Vineland, he would also be able to explain his new practices to institutional physicians. The effects would be profound, for Binet's new concepts would ultimately lead to a major transformation in the medical understanding of mental deficiency.

The "Happy Blending of the Pathological and Psychological"

After working with the new tests for one year, Goddard was ready to present Binet's ideas to American doctors. At the 1909 meeting of the American Association for the Study of the Feeble-Minded, the professional organization for institutional superintendents, Goddard broached the subject by discussing the broader problem of classifying cases. Using far more tactful language than Binet, Goddard too now challenged medical systems of classification.

First, Goddard attacked systems based on pathology. "We have our medical classification such as microcephalic, hydrocephalic, Mongolian, etc.," he remarked, but "what is the value of this classification? How closely can we classify?" he asked physicians. "How small must a head be before it is microcephalic?" Such a classification may offer "a convenient way to shelve these cases, perhaps, but does it help us in our dealing with them?" Goddard inquired. "Is a microcephalic defective limited definitely in his powers?"[87]

Goddard next questioned sociological systems. Barr's "educational classification," he reported, was certainly valuable, and Goddard had no intention of discrediting it. Still, this system was "of no use until we have had the child in the institution long enough to find out how trainable he is." Physicians, he maintained, needed a faster means of diagnosis.[88]

Abandoning psychiatric models stressing sensorimotor skills, however, proved much more problematic for Goddard. Admitting that psychophysical tests had proven no more successful than other approaches would certainly have been difficult, for such tests provided the rationale for his own laboratory, with all of its expensive equipment. Goddard still held out hope that such equipment might yet prove useful in indicating mental capacity. "I have a feeling," he told the association in 1909, "that motor control – how the child handles its muscles – may be ultimately a stronger basis of classification than the mental process, – the more purely psychological." After all, he explained, psychologists no longer considered mind an entity in itself; it was the sum of processes, including movements which "we certainly can and do measure." Even so, Goddard had to concede the superiority of Binet's new approach and thus informed physicians of Binet's "tentative set of mental tests which might serve as a basis for classification."[89]

In the discussion that followed, physicians again expressed their own dissatisfaction with existing systems. "In connection with the so-called pathological classification," Dr. Charles Bernstein remarked, "I think we have no right to consider it of any worth at present. Here is a case of microcephalis or hydrocephalis – what does it mean?" he asked. Such terms indicated no "definite pathological condition," this doctor conceded; "they refer to size. Mongolian does not mean anything definite pathologically, it does not suggest any underlying condition. We have no intelligent pathological classification at present," Bernstein concluded.[90]

The discussion ended with a call for consensus. "It seems to me we ought to get together," one doctor remarked, "the institutions, anyhow – and have a basis of classification." Toward this goal, the association designated a Committee on Classification to study all available methods. Dr. Fernald was appointed its chairman, and Goddard was made a member.[91]

Like many such committees, this one got off to a sluggish start. "I beg to call attention," Fernald wrote members less than a month before their report was due for the next annual meeting, "that we are on a committee on classification." His own recommendation to committee members suggested just how effective Goddard's arguments had been, for Fernald now proposed avoiding pathology and instead adopting a system "based entirely upon the degree of intelligence present." Classes should be "so descriptive that they are obvious and intelligible to the well-educated general practitioner," Fernald stated. This was important, he explained, for the "non-institution man" had long failed to realize that "pathological types may present any degree of mental defect."[92]

Most members responded vaguely; Goddard, however, now had a specific agenda in mind. "I have felt just exactly as you express it in regard to classification for sometime," he tactfully reported to his physician colleague, "but I feared that I was a heretic and that no one would agree with me." Since trying Binet's tests, he wrote Fernald, he had been "constantly amazed" at the agreement between "the mental age of these children as shown by these tests and what we know of them from experience." At least temporarily, Goddard suggested, the association ought to consider using Binet testing, a set of purely psychological criteria, to classify feeblemindedness. "The difficulty now," he wrote, "is that we are hardly any two of us agreed. The old classifications . . . are so illogical, based as they almost all are on more than one basis of classification, and consequently leading to confusion throughout."[93]

This committee took no action. Since Goddard was the only committee member who attended the 1910 meeting of the American Association for the Study of the Feeble-Minded, he was asked to submit a report. He decided to present his own classification system based on Binet testing "and let the Association decide."[94]

By 1910, however, Goddard was no "heretic," for there were by then several other Hall students employed in institutions. These included psychologists Edmund Huey at the Lincoln institution and Frederick Kuhlmann at Faribault, both of whom had also found Binet's tests useful. Goddard himself presented the strongest evidence. His paper on "Four Hundred Feeble-Minded Children Classified by the Binet Method" showed how institutions such as his own could adopt a precise language supplied by a psychologist to classify all their cases.[95]

Goddard's paper actually contained more than just the results of his tests; it also described his own preliminary attempt to validate these results by comparing them with the one criterion in which physicians had faith: institutional experience. Thus, after dividing his children into lists according to their Binet scores, Goddard then called upon "the experience of all those in the institution who knew these children and had known them for some years." At a meeting, he asked the heads of each of the institution's departments if "the children given in any one list seemed to them to be all of about the same mental capacity; whether any in the ten-year-old list, for example, seemed to them to be much higher or much lower than others in that list." Objections were raised, Goddard reported, but these were "always answered by others in the room, that the objection was not a valid one, that the child in question was of about the same grade as the general run of the group." The outcome of such a procedure was that "no child was entirely thrown out of the group by even a majority of those present, to say nothing of a universal condemnation of the result." Goddard next tried the same method with the school's teachers. "Precisely the same thing happened . . . ," he reported.[96]

Using this method, Goddard had devised his own crude means of confirming Binet scores. Like most of the statistical procedures of his day – procedures which were themselves being invented alongside the new measuring techniques – this one would hardly withstand later methodological scrutiny. Even Goddard conceded that such results were "not wholly unlooked for by us"; moreover, he knew that appealing to "the memory of a group of people, however well they may know the children," was "not quite conclusive."[97]

Nonetheless, Goddard still believed that he now had in hand a method of assessing mental abilities that matched institutional experience far better than any tried previously. One could not use Binet's tests within the institution, he reported, "without becoming convinced that whatever defects or faults they may have, and no one can claim that they are perfect," they still came "amazingly near what we feel to be the truth in regard to the men-

Three Vineland children. From top to bottom, their chronological ages are twenty-three, sixteen, and twelve, but all were diagnosed as having a mental age of eight on the Binet scale. The middle boy is "Keith," who had "no signs of defect in his face." From H. H. Goddard, *Feeble-Mindedness: Its Causes and Consequences* (New York: Macmillan, 1914), facing p. 124.

tal status of any child tested." Whether he could prove so or not, Goddard now saw Binet's work as grounding institutional diagnoses upon a basis of science.[98]

If Goddard's primitive attempt to correlate intelligence tests with institutional experience reflects a flawed statistical strategy, it also suggests a shrewd professional strategy. After all, Binet's tests did more than measure: they also redefined feeblemindedness in new psychological terms. By using these terms, both Binet and Goddard were actually challenging more than the meaning of mental deficiency. Also being challenged was the power of the medical profession to define mental illness.

Most striking in these challenges were the different responses they evoked – responses that may have resulted from the different personalities of these two psychologists as well as from the diverse ways in which they conceptualized their own relationship to the medical profession. Whereas Binet was frequently abrasive, Goddard was conciliatory. Thus, Binet forcefully presented his tests as a challenge to the faulty expertise of physicians, whose imprecision had masked their own inability to grasp fully the nature of the condition they were studying. Goddard, however, presented these same tests as scientifically confirming medical intuition, thus suggesting that physicians had been correct all along, but had simply been unable to express their views scientifically.

And whereas Goddard and Binet were both educational insiders as a result of their Child-Study connections, only Goddard was an insider within the institutional world. His appeals to his medical colleagues were carefully phrased in terms of what "we" needed to know – or what "we feel to be the truth." The consequences of such a strategy were apparent, for while Binet was initially rebuffed or ignored by most physicians within his own country, Goddard quickly gained a sympathetic American medical audience. Perhaps Binet's tests "do not tell us anything we did not know before," Dr. Fernald thus concluded, "but they tell us in an hour what otherwise it takes us six months or a year to find out; and they tell us more accurately and consistently."[99]

One day after Goddard presented his paper, the American Association for the Study of the Feeble-Minded unanimously adopted his "New Classification (Tentative) of the Feeble-Minded" along with his recommendation that mental deficiency be diagnosed by Binet–Simon tests. "It was agreed," the *Journal of Psycho-Asthenics* reported, "that the Binet mental tests afforded the most reliable method at present in use for determining the mental status of feeble-minded children." An idiot was now defined as one testing below a mental age of three on the Binet scale, and an imbecile as one testing between mental ages three and seven.[100]

The association even accepted Goddard's proposal for renaming the highest group, since the term most commonly used, "feebleminded," had also become the generic term for all three classifications. In seeking a term that would match the older lay understanding and yet have no negative connotations, Goddard proposed a word of his own invention that again showed the imprint of his classical education. "The term 'fool' in the old English sense, of a person lacking in intelligence or common sense," he explained in defining the word for the *Century Dictionary* a few years later, "was exactly what we wanted but is unfortunately too harsh a term to be used." Instead Goddard coined a new term from the Greek root found in English words such as "sophomore" and "oxymoron." His new word, "moron," was defined specifically as "a person of arrested mental development, with an intelligence comparable to that of the normal child between 8 and 12 years inclusive." For the rest of his life, Goddard would regard this opportunity to add a word to the English language as one of his greatest accomplishments.[101]

Far more important than this new word, however, was the new idea of comparing chronological with "mental" age. Such an idea appealed immediately to American institutional physicians, for it objectified their intuitive understanding of the differences between normal and subnormal child development. "Who is there that does not have a mental picture, always in view, of the activities and capacities of normal children at different ages?" an editorial by Dr. Rogers noted in explaining the new system. "What more natural and rational than to compare the mind, backward in development, with a normal one?"[102]

Of course, in emphasizing such comparisons, this association certainly had no intention of "belittling the pathological basis of the mental defect itself," Rogers added. Instead, Rogers saw the association's actions as illustrating "the natural and logical blending of medical and psychological influences." Pathological terms, such as hydrocephalic or Mongolian, would still be used as descriptive adjectives (e.g., Mongolian imbecile). The result, he proposed, would offer "a happy blending of the pathological and psychological descriptions." "As to the matter of emphasizing a psychological basis for classification rather than a pathological one," their journal noted, these doctors saw "no serious objection" if it allowed them to determine "quickly even an approximate estimate of the child's mental ability by some system . . . that presents to all . . . a common mental standard."[103]

To American institutional physicians, Binet's achievement, especially as explained to them by Goddard, was nothing short of remarkable. "If it had been easy to advise a classification of general application," Rogers conceded in recognizing Binet's accomplishment, "it would have been done long ago." Many students "familiar with medical and pathological studies" with "plenty of material on which to work" had repeatedly tried and failed to do exactly that. For the first time, American medical superintendents of institutions for the feebleminded shared a precise diagnostic vocabulary – a vocabulary whose key concepts had been redefined for them by psychologists. Mental deficiency now meant deficient in "intelligence"; it could best be diagnosed and classified by the use of intelligence tests.[104]

At the time, Binet's success in redefining mental deficiency, and Goddard's in recognizing and promoting his accomplishment, attracted little notice. Neither the wider medical establishment nor the larger psychological community as yet paid much attention to this surprising new development within the marginal world of institutions for the feebleminded. It would be years before either profession would fully realize the immense import of this decision by doctors to use intelligence tests to measure minds – a decision that would ultimately generate intense conflict once again over the very meaning of mental deficiency.

To Goddard, however, its immediate significance was absolutely clear. He had demonstrated the value not only of his own laboratory but also of scientific psychology. Both, he believed, had finally proven their social worth. In the aftermath of such success, Goddard's confidence both in his own methods and in his profession soared.

Yet even in announcing their decision to adopt Goddard's proposed classification system in 1910, Dr. Rogers also sounded several notes of caution, which in themselves suggested some of the battles that would be fought in the future. In protecting the field against "unprincipled charlatans," he warned, one had to guard against "the idea that anybody without special training can diagnose and classify mental defect." Moreover, the entire question of psychological classification rested upon psychologists' abilities to supply what they had promised – "a concise summary of the intellectual expressions of the mind of the child in groups corresponding to its different ages." Binet had laid the foundation for such work, Rogers believed. There remained, he added optimistically, only the "securing of well-worked out normal data from American school children." Such a task would soon consume the attention of American psychologists.[105]

Within four years of starting his "psychological work among the feebleminded," Henry Herbert Goddard's efforts had had a major impact upon American medical practice. Using Binet's ideas, he had convinced institutional doctors to redefine mental deficiency in terms of intelligence. With his first stunning professional victory for applied psychology firmly in hand, Goddard once again turned his attention to the schools.

4

Psychological Work in the Schools:
The Statistical Meaning of "Subnormality"

Burning Questions

"How intensely must one be diseased – physically, mentally, or morally – to be educationally quarantined?" Philadelphia schoolteacher Francis Burke Brandt had asked the National Education Association in 1901. "How extensively must one be defective . . . to be thrust from the school?" he wondered. "How completely must one be degenerate . . . to be excluded properly from the training primarily designed for normal children? Under the operation of compulsory school laws," this teacher concluded, "these are becoming burning questions." Seeking answers, by 1906 Brandt had become a regular member of Vineland's "Feeble-Minded Club."[1]

Brandt was not the only teacher to ask such questions in the early decades of the twentieth century, or to turn for answers to the Training School at Vineland. Although Henry Herbert Goddard had become increasingly involved in medical debates, he had never abandoned either his Child-Study connections or his sense of himself as an educator. Over the next decade, both he and Edward Johnstone maintained close ties to school personnel at all levels; they also became active participants in a wide variety of educational conferences, including the meetings of the National Education Association.

Even more importantly, both men continued to host the semi-annual gatherings of the "Feeble-Minded Club" on their campus. What pedagogy needed most of all, Goddard strongly believed, was to be regrounded on a foundation of science. Yet in order for scientists to make a real contribution to this field, he insisted, they would have to understand and address the concerns most troubling to ordinary classroom teachers. By providing an opportunity for schoolmen and scientists to share their ideas in a setting promoting mutual respect and cooperation, Goddard and Johnstone hoped to effect a gradual transformation in both scientific understanding and educational practices.

Largely as a consequence of this collaboration, the Vineland Training School came to play a role within American education starkly out of proportion to its small size and humble origins. Between 1906 and 1916, this little school's accomplishments would indeed be remarkable. In a single decade, Goddard and Johnstone were able to alter dramatically the relationship between the institution and the public school. At the same time, their attitudes and ideas would inspire a new generation of educators eager to use psychology to aid pedagogy. As a consequence, Goddard's influence on American education would be felt for decades to come.

Even more long-lasting would be his influence in disseminating psychological tests of intelligence into American public schools. In a few short years, Goddard would be able to introduce this new science into school systems throughout the United States. Once again, his actions would lead to a significant rethinking of the very meaning of mental "subnormality," this time within the schools. They would also help to establish a new social role for applied psychology in the professional middle ground between medicine and education.

As a result of these activities, Vineland itself would be transformed from an unknown training school into a world-renowned educational institution. Such a transformation hardly seemed possible in the opening decade of the twentieth century. Yet from his new base in the marginal world of institutional pedagogy, Henry Herbert Goddard would soon leave a lasting mark on American school practices.

Promoting Special Education

In many ways, the Vineland school's transformation paralleled the equally spectacular growth of the movement that American teachers in 1902 had christened "special education." During the following decade, a growing number of public schools began to establish separate classrooms not merely for "delinquent" but also for "defective" students – students apparently suffering from physical or mental handicaps that impeded their school progress. Among those arguing most vociferously that the state owed such students an education, and that they therefore should not be "thrust from the school," were special educators.

During these years, teachers active in the NEA's Special Education Department lobbied extensively to force state governments to educate physically and mentally disabled children in public classrooms. Among the most active of these lobbyists were Johnstone and Goddard. Through their frequent speeches, strong ties to normal schools, and close working relationships with schoolteachers, both men quickly gained national reputations as leading advocates of the rights of the handicapped to a state-funded education, and powerful promoters of the need for special education within public schools.

Special education, however, meant special equipment, special teacher training, and special classes ideally limited to fifteen, according to Goddard, at a time when classes of fifty were common – in short, special allocations of resources that school systems were reluctant to make at a time of burgeoning enrollments. In response, lobbyists phrased their arguments in language most likely to appeal to cost-conscious bureaucrats: school efficiency.

"No school system of 500 children can afford *not* to have its special class," Goddard insisted. Segregating the slowest students in classrooms offering a special curriculum would not only help them, he argued, but would also increase the productivity of the teacher and the remaining students. Even if a school had to hire a special teacher for only five pupils, Goddard proposed, the gains both to these pupils and to the rest of the class would "repay any seeming excess of expense."[2]

Lobbyists for special classes stressed not only school efficiency, but also the inefficiency of wasted human potential. American schools, one advocate declared, were letting children "drag along year after year learning practically nothing, acquiring habits of idleness, losing whatever self-respect they might have originally possessed and becoming possessed by a feeling of bitterness and resentment toward the school in which they did not fit." "Special" children instead needed special attention. "Do you not see then the boon this special class is to this hopelessly handicapped child?" a Rochester medical inspector argued. Such a child would no longer be "the conspicuously dull member of his class who is the object of jeers and taunts and those many cruelties that children know how to inflict upon him." Having his work "adapted to his abilities" would provide the "greatest stimulus to further effort, namely the successful accomplishment of a labor undertaken. As one of our little boys said one day in class," he recounted, "'Gee, I didn't know I could do anything.'"[3]

These strong appeals to economy and to emotion gradually expanded educational obligations. "It is a generally accepted principle," an NEA report stated in 1908, the year Edward Johnstone became president of its Special Education Department, that the state was "in duty bound to provide an education for all children capable of intellectual improvement." By 1911, the U.S. Commissioner of Education had recognized this movement as one of the most significant of the previous decade, for special classes had been started in nearly one hundred American school systems.[4]

Yet while the practice of starting special classes solved some immediate problems, it created many others. What, teachers wondered, should be taught in such classes? Should special classrooms be essentially remedial, offering a slower version of the normal curriculum and striving to return students to ordinary classrooms as quickly as possible? Or should they follow the model of institutions for the handicapped and employ a strikingly different curriculum? Most crucial of all, who belonged in these classes?

Such questions led to no easy answers. Teachers concerned with establishing some sort of consensus on these issues were soon frustrated, for while
schools often employed a similar language in starting "backward," "special,"
or "ungraded" classes, these classes differed markedly both in their curricula
and in the criteria used for student classification. In "no two of four cities –
Boston, New York, Baltimore, and Washington," one teacher reported, did
"the term 'ungraded class' mean the same thing."[5]

School physicians, moreover, were proving of little help in remedying this
confusion, for in diagnosing children, they too tended to use a range of terms
with no consistent connotations. "Feeble-minded children, atypical children,
subnormal children, mentally deficient children, mentally defective children,
arrested mental development, developmental inhibition, and backward children," physician Mary Pogue told the NEA in 1905, were "terms used by
the laity and profession alike." Unfortunately, she concluded, these terms
conveyed "no definite meaning."[6]

This semantic confusion had serious educational consequences. With no
common selection criteria, one teacher lamented, special classes were becoming "a sort of dumping-ground for all those children with whom the regular
grade teacher does not know what to do." This was especially evident in the
diverse composition of special classes. One Boston class of fifteen, for example, was described by a physician as follows: "two had had rickets, six
convulsions, one epilepsy, three were seriously deaf, four had difficulty with
the ordinary movements of walking and skipping, ten spoke with defective
articulation, two had deformed palates and only three had good teeth."[7]

Johnstone blamed many of these problems on medical colleges, which still
offered doctors "no instruction in the diagnosis and prognosis of mental deficiency." In 1909, Goddard sent out a questionnaire to discover just what
medical and normal schools were doing "to educate Physicians and teachers
to understand the defective child." With "a few partial exceptions," he reported, "the answer in both cases is 'NOTHING.'"[8]

The very fact that medical and normal schools were paying these problems
so little attention, however, again left the field wide-open for psychologists.
Since applied psychologists tended to see their discipline as bounded by psychiatry on one side and pedagogy on the other, the mental problems of
schoolchildren offered practitioners an especially suitable subject matter. In
addressing such problems, the psychologist might prove "even more important than the physician and of even more importance than the teacher," Johnstone predicted, for he was capable of advising "in both directions." The psychologist, Johnstone surmised, "really represents the point at which the two
lines, teacher and physician, should meet."[9]

The American psychologist who had done the most to institutionalize a
new role for his profession at precisely this meeting point was Goddard's

contemporary Lightner Witmer. In 1907, the University of Pennsylvania expanded Witmer's innovative "clinic" for schoolchildren, and Witmer began publishing reports of his cases in his new journal, *Psychological Clinic.* Witmer's description of a typical clinic day suggests the range of problems confounding educators. Thus, one afternoon he saw "a child six years old, who comes because he cannot talk" and diagnosed the case on sight as "a mongolian imbecile." His next case concerned a fourteen-year-old girl "who also does not talk," and while she too was thought to be feebleminded, Witmer instead found her "almost totally deaf." A third case concerned a boy with "well set up parents" who complained that their son had "no ambition." Psychologists, Witmer argued, ought to diagnose and treat each of these cases on an individual basis.[10]

Johnstone, however, favored another strategy. Teachers could learn to recognize and to educate mentally handicapped children themselves, he suggested, in the same way that he and Goddard had learned: by experience. His "summer school" offered special educators the chance to live with and learn from such children by spending a summer in an institution. Moreover, with Goddard's laboratory operating on the premises, teachers could also learn the latest psychological developments.[11]

Goddard seconded this. Where was "the model school for the teachers of mental defectives?" he asked. "Nowhere except in the institutions for the feeble-minded." Through their summer school, Goddard and Johnstone encouraged teachers to seek answers to the "burning questions" posed by special education through a combination of institutional experience and psychological science.[12]

Vineland Spirit

Vineland opened its "summer school" for schoolteachers in 1904. In its first class were five teachers. The school's goal was to allow these teachers to learn from the institution. "We brought them right into the life of The Training School," Johnstone explained.[13]

This strategy, however, led Johnstone to discover a serious problem impeding this work: the schoolteachers were frightened of his children. "I doubt if any one of them really unpacked her trunk until after the first week . . . ," he observed. Their experiences on this campus would markedly change their attitudes, for by the summer's end, these five teachers had been transformed into the "loyal and enthusiastic alumnae" of Vineland's "class of '04."[14]

By the time Goddard joined Johnstone and Alice Morrison Nash, this institution's main teacher, on the summer school staff in 1907, enrollments had tripled. He was soon immersed in this new work, for at Vineland, teacher training took precedence even over research. By 1909, this school had drawn teachers from Buffalo, Toledo, New York, Worcester, Jersey City, Cincinnati, and Washington, D.C. Enrollments in 1910 reached thirty, and a twelve-week "Winter Training Class" was added. By 1911, with sixty teachers accepted and applications exceeding one hundred, three new cottages at Vineland had to be constructed along "Summer School Lane."[15]

What attracted so many teachers to Vineland was the same thing that had first caught Goddard's attention – its successful methods of educating even seriously handicapped children. Like other institutions, this school's curriculum was based on the methods of "physiological education" first elaborated by Edouard Seguin. Such methods fascinated public schoolteachers who had rarely taught much besides the three R's.[16]

"To spend years teaching a child the very beginning of reading, writing, numbers, etc., and daily becoming more convinced that the little he ever will learn, he will never be able to use either for his own help or happiness, or to benefit others," Johnstone explained, seemed "not only foolish but positively wicked." Much better results came from untraditional methods. "The children's gardens are more valuable in training than most school rooms," he argued, as was "learning to saw and hammer, to mix mortar and to use the common tools with the engineer, plumber, and electrician." Farm tasks and housework Johnstone saw as equally educational. "The secret," he asserted, "is to find the thing the child *can* do, then working and studying become a joy."[17]

Goddard supported Johnstone's strategy. "Our greatest *duty*," he explained, "is perhaps to preach the *gospel of works* for the feeble-minded." Too many persons believed that "salvation lies in the ability to read books, to write letters and to count millions – salvation not only for the soul but for the mind," he lamented. Instead, the feebleminded should be taught things to "make life pleasanter for them." Toward this end, "the training of games, of athletics, of doing things" was "absolutely essential to mental efficiency later." Such a curriculum, he argued, made sense in terms of evolution as well as the gospel. "Why should we not expect," he asked, that feebleminded children lacked "the power to utilize elaborate systems, systems which the human race lived without for ages?" The feebleminded, in other words, resembled man at an earlier evolutionary stage, and ought to be educated accordingly.[18]

A similar curriculum, Goddard believed, should be adopted by special classes. Addressing the NEA in 1910, he offered advice starkly opposed to the philosophy that many decades later came to be called "mainstreaming," for instead of trying to make the special child's curriculum as "normal" as possible, Goddard advocated accenting the differences. "As long as the motto of the special class is, 'as nearly like the regular class as possible,'" he argued, "everybody will be dissatisfied and discouraged." What "spoils the whole work," Goddard believed, was holding teachers to a standard "which could not be accomplished."[19]

Such remarks illustrate just how much institutional life had gradually changed Goddard's own thinking. While Witmer's work with schoolchildren led him to try to close the gap between normal and subnormal performance, Goddard's work with more seriously impaired cases had left him far less sanguine about the possibilities for psychological intervention. He had begun his work in 1906 with the strong belief that mind was essentially "the same in them as in us"; by 1910, however, he had come to regard his charges as "emphatically different from normal children." The feebleminded child could not be "appealed to in the usual way," he now argued. Such a child had only a few years for growth and development. If "wisely handled," Goddard reported, he might learn enough to make his life "more happy and comparatively useful." To maximize this possibility, he concluded, the curriculum offered in special classes ought to stress manual training.[20]

Of course, parents, Goddard admitted, might object to such a curriculum. "They fear that it will be called the 'fool-class,' and will bring them or their associates into disrepute," he conceded, "or will, as they say, put a stigma on the child for life." If run badly, he predicted, it would mean "all of this and more"; if handled well, however, parents would be "delighted at the results." Crucial to its success was its curriculum. "Make the special class like the regular class in everything but rate of progress and you emphasize the dullness of the pupils," Goddard argued. Make it "different in outward appearance as well as inward plan – a class where the children are happy because they succeed – and you emphasize the pleasant side." This new emphasis, he maintained, would alter the attitudes of all involved.[21]

The key to such education was, of course, the special teacher. To attract the best people, Goddard advised schools to select those "who love children and have all the heart-qualities" – and pay them 25 percent more than normal teachers, thus inducing "good people to prepare themselves for this work." According to the numerous teachers who visited this school, no one demonstrated these "heart-qualities" more fully than Edward Johnstone, Alice Nash, and Henry Herbert Goddard, for in working together, these three educators had created a unique learning environment at Vineland.[22]

As part of their summer program, Johnstone, Nash, and Goddard worked tirelessly to influence attitudes as well as methods. Teachers who came for the summer found themselves most deeply touched by this institution's idealism. Especially striking were its welcoming environment, its remarkably warm relationship between children and staff, and its unique blend of scientific and Christian enthusiasm – all of which came to be known to contemporaries as the "Vineland Spirit."

This spirit was embodied both in the school's motto, "Happiness first, and

all else will follow," as well as in the many original methods Johnstone invented to reassure his children that they were wanted and "belonged" somewhere. "We have a little secret society," Johnstone explained to outsiders, "in
which the password is, 'We belong,' and the signal is a smile. Wherever we
go about the school, if anyone looks cross or sad, someone is sure to look at
you and smile in your face, saying, 'Do you belong?' and you simply have to
smile back because you are just as human as they are." Such behavior led
one visitor to christen Vineland the "Village of Happiness."[23]

Life within this "village" taught powerful lessons of its own. "Day by day,"
one summer school graduate reported, teachers had been led to "think in
the open, unbiased spirit that characterizes the truly scientific work of Dr.
Goddard; to observe with the honest, just, sympathetic attitude of Mrs.
Nash, and to walk in the paths of encouragement and happiness as demonstrated by Professor Johnstone." Her Vineland diploma meant that she was
"prepared, yea in honor bound, to go forth doing honest work, arousing
deep interest and radiating 'Vineland Happiness.'"[24]

Other graduates described an equally intense sense of "spirit." One class
even wrote a school song. "For years we have always wanted / To find a splendid school like this . . . ," they sang. "But no college did we find / That quite
took up all our mind / Till in the school at Vineland we took rest." Their
song's chorus reiterated Vineland's themes of happiness, smiling, and belonging:

> It's a school where they have no frowning,
> And I think I can plainly say,
> That we are all quickly learning
> To be happier every day. "Oh yes!"
> It's a smile here at every turning
> At the same time we have to delve
> But we haven't cared – for *we belong* –
> To the Class of 1912.[25]

Teachers left Vineland with a deep sense that they too belonged to a community with a mission. Their letters back to the school, which Johnstone
circulated through the *Training School Bulletin,* described their successes and
frustrations in spreading this spirit throughout the country.

"I thought perhaps you might like to know how and what we are doing
down here among the children who require especial training," one graduate
wrote upon returning to Richmond, Virginia. "There have arisen many puzzling things in our work, but I have tried to meet each with 'The Vineland
Spirit,' which has proved such an untold help to me." She was enthusiastic
about her results, for her pupils "would hardly be recognized as the same sad
uninterested little ones who came to us four months ago. They are happy and

Vineland Summer School Class, 1913. Number 26 is future psychologist Florence Goodenough. Courtesy of the Archives of the History of American Psychology, University of Akron.

eager about their work and attend well. I cannot help feeling encouraged and hopeful, in spite of the great odds against which I know we still have to work," she concluded. "I also know that when the Vineland spirit is once planted, all must sooner or later give way to it, so I am endeavoring to plant that 'spirit of live and help live' which fills the very air you breathe in your school."[26]

Teachers also told of their dealings with frequently desperate parents. "Without an exception I have come away with the parents begging me to take *their* child," a Toledo teacher reported. "One even said . . . I don't care if you don't teach her a thing, if you will just let my Mary be with you and make her happy." "The parents have responded beautifully," wrote another graduate. "There have been days when my heart ached sorely for the poor parents." A Massachusetts teacher told of using Vineland's methods to train children and finding them "greatly helped by so doing," she wrote, for "several parents have spoken of the noticeable usefulness and self-reliance at home." Parents were "delighted with every good result," she reported, "and very anxious to help if they are told what to do."[27]

Most profoundly affected by their Vineland experiences were the teachers themselves. "I am grateful for a taste of practical Christianity," one graduate wrote. "I came to study an Institution and I find myself studying myself." Others reported equally intense feelings. It was hard to "describe in words the impressions which have so deeply affected me as to change my entire attitude toward the world in general and children in particular," wrote another teacher. "There is a sort of beautiful atmosphere here which can only be due to the beautiful personalities with whom we come in contact." A "self-sacrificing life," she concluded, "is to me a religion which is hard to find in churches and has a deeper effect upon me than any sermon."[28]

So many alumni offered emotional personal testimonies to the changes wrought by this school's "true Christian spirit" that Johnstone himself began to wonder. "Granting that there is a just noticeably greater amount of 'true happiness,' 'Christian spirit' here than elsewhere," he inquired, "where did it come from?" "We did not bring it with us," he noted of his work with Goddard and Nash. "We were all average people when we came. The places we left have not sunk noticeably in the moral scale since we left."[29]

Johnstone found his own answer in the words of another graduate. "The thing that has impressed me most is the influence resulting from work with and for these children – the reflex influence on the teachers and attendants," she noted. "It seems to develop the very best that is in a man or woman." "That is it," Johnstone declared. "It is the children. It is they who have developed in us whatever of beauty is to be found there." His words would surely have pleased Reverend Garrison, the school's founder. "Fathers, mothers," Johnstone exclaimed, "you have lent us your children to bring our lives to perfection."[30]

The Vineland experience meant more than religious mission, however; it also meant faith in scientific research. Such faith was evidently harder to sustain once teachers returned to their schools. "In research work we do nothing," and problems were "not always approached in a spirit favorable to the development of ideas," a frustrated graduate reported. "'It can't be done,' is an excuse we too often give for a thing's not being done. If we undertake to keep tab on a thing and find it in bad shape the solution is very simple – 'don't keep tab on it.' Perhaps I shouldn't make this so strong," this teacher concluded, "but the Vineland folks must share with me the blame if my standard is too high."[31]

Nevertheless, summer school graduates did learn to diagnose cases resembling those in institutions. "I have found two mongolians, one cretin and a case of hydrocephalus," a Michigan teacher wrote. "Every time I find a typical case I feel so 'Vinelandy.'" A Pennsylvania teacher recognized a "Microcephalus type." "Christopher is slowly improving," she wrote. "I often have him in a class with the older ones, but only give him the work he is capable of doing." "Two of my boys are almost deaf, and two are institutional cases," a Toledo schoolteacher wrote. "One is certainly 'Brant No. 2' [a Vineland boy]. I am very much interested in them, and like the work."[32]

By 1913, Johnstone, Goddard, and Nash had personally trained more than three hundred special educators. Their program proved so popular that at least eight colleges and universities established similar summer training programs for special class teachers. This school's indirect impact was even more significant, for nearly an entire generation of special educators would be affected by Vineland's messages.[33]

Many Vineland graduates later became influential leaders in special education programs throughout the country. In city after city, Vineland-trained teachers gradually assumed powerful positions, either in public school systems or in private institutions. Among this school's alumni, for instance, were Meta Anderson (class of '08), who became Director of Special Education in Newark; Alice Metzner ('10), Supervisor of Special Education in Detroit; George Snow Gibbs ('10), Director of the Psychological Clinic for the Public Schools of Salt Lake City; Anastasia Vaughan ('13), Supervisor of Clinical Examinations for Special Education in Philadelphia; Robert B. Irwin ('14), Director of Research and Education for the American Foundation for the Blind; Mrs. James (Devereaux) Fentress ('10), founder of the Devereaux Schools; Mrs. John R. (Woods) Hare ('14), founder of the Woods Schools; and Florence Goodenough ('13), psychologist at the Institute of Child Welfare, University of Minnesota. "Their missionary work," Johnstone wrote proudly of his graduates, "has had a profound influence on public education in all its branches."[34]

Such teachers disseminated Vineland's influence for decades to come. They also subtly transformed the institution's relationship to the society surrounding it. At last "'the stone which the builders refused has become the headstone of the corner,'" Johnstone concluded. "That much despised institution, the idiot asylum," he explained, had become "the laboratory for the public schools." Perhaps a Seattle special teacher best expressed this new relationship between school and institution when she wrote that as summer approached, "we instinctively turn our faces toward the East and Vineland, whence cometh our help."[35]

The Statistics of School Failure

Had Goddard confined his concerns to training schoolteachers and shaping the special curriculum, his influence in the public schools might have remained marginal. After all, despite the remarkable growth of the special education movement, policies toward the handicapped had always been a relatively peripheral school issue. Even Goddard saw the special pedagogy that he was advocating as affecting at most perhaps 1 or 2 percent of American schoolchildren.[36]

Goddard's educational influence soon increased dramatically, however, in relation to a second issue gaining attention during the same decade. The new issue focused specifically on children older than they ought to be for their grade in school. The "overage" problem, as it came to be called, affected perhaps as many as one-third of America's urban schoolchildren. In responding to it, educators began to reconceptualize the very meaning of mental "subnormality" in new ways – ways that placed a particularly strong emphasis upon school statistics.[37]

The overage problem was first identified in 1904, when New York City school superintendent William Maxwell reported that 39 percent of his city's elementary school children were at least two years behind their proper grade. Included within Maxwell's report was a small yet highly significant statistical innovation. Instead of reporting the number of children failing school, Maxwell had calculated the *percentage* of schoolchildren who were overage. He had thus made comparisons with other school systems possible.[38]

Maxwell's simple innovation caught the immediate attention of other superintendents, who soon reported comparable findings – 47 percent in Camden, for instance, according to superintendent James Bryan (a member of the Feeble-Minded Club), and 37 percent in Philadelphia, according to superintendent Oliver Cornman (who later joined). By 1909, educational researcher Leonard Ayres had published his highly influential national study, *Laggards in Our Schools,* which found the percentage of school "laggards" to vary widely, from a low of 7 percent in Medford, Massachusetts, to a high of 75 percent in Memphis, Tennessee. By 1910, the overage problem had become a matter of serious educational concern.[39]

The emergence of "overage" as a problem in itself suggests widespread changes in the perception of childhood. After all, classrooms containing children of mixed ages were hardly a new development. Such mixtures had been a hallmark of nineteenth-century one-room schoolhouses, and while admittedly expedient, they had never been alarming. Yet in calling attention to the high percentages of overage children, these superintendents now implied that something was awry. Their writings embodied a new understanding of the relationship between child development and school learning: what mattered was not only what children learned, they suggested, but when they learned, for education ought to be commensurate with children's physiological and mental growth.[40]

These reports also had serious implications for advocates of compulsory education. While such statistics mattered little to those who saw the new laws as simply keeping children off the streets, they alarmed others who had hoped to guarantee all citizens a rudimentary eighth-grade education. Too many children, superintendents now claimed, were spending their required

The Feeble-Minded Club, which included psychologists, teachers, and doctors, meeting at Vineland, 1916. Among those pictured are; front row: psychologist Bird Baldwin (third from l.); H. H. Goddard (far right); second row: psychologist Carl Brigham
(fourth from l.); Goddard's assistant Edgar Doll (fifth from l.); third row: Philadephia
medical inspector Walter Cornell (far left); Child-Study promoter Earl Barnes (third
from l.); Philadelphia superintendent Oliver Cornman (fifth from l.); back row: E. R.
Johnstone (far right). Courtesy of Archives of the History of American Psychology,
University of Akron.

school years repeating, rather than progressing through, the lower grades. At
best, these findings meant that the schools were seriously "inefficient"; at
worst, that compulsory education was failing.[41]

Yet what exactly did such numbers mean? Were Kansas City's schools,
with nearly 50 percent overage, more than twice as inefficient as Boston's,
with only 22 percent overage? Had New York really improved its efficiency
by nearly 25 percent when it reduced the number of overage children from
39 to 30 percent in one year? And what about the fluctuations within school
systems? Some Philadelphia schools, for instance, promoted over 90 percent
of their pupils, and others under 60 percent.[42]

While superintendents pondered the meaning of these statistics, they did
concur in one conclusion: American schools were, in both senses of the word,
failing too many children. In the process, they were wasting both public funds

and student minds. Worst of all, according to Ayres, they were teaching perhaps as many as six million children "the habit of failure." But did the problem lie with the school or with the child? "Overage" statistics in themselves suggested no answers.[43]

These studies did spur additional investigations, however, which soon suggested many reasons for such problems. The main reason that so many children were "too old" for their grade was simply that they started school too late. This was especially true of immigrants. A 1908 New York study of 980 fifth-graders, for instance, illustrated this clearly. While such students should ideally have been eleven-year-olds in their fifth school year, these pupils ranged in age from nine to eighteen, and had been in school from two to ten years. In this sample, 456 fifth-graders (46%) were overage; of these, however, 165 (36%) had actually received less than five years of schooling.[44]

Language problems, chronic illnesses, and irregular attendance explained the slow progress of many others. Taken together, such studies provided educators with a profile of the typical school "laggard": the child who failed was most likely to be male, foreign-born, from a non-English-speaking family, already older than his classmates, marked with a poor attendance record, and suffering from more than the "average" number of what contemporaries classified as physical and mental defects.[45]

In assessing these defects, most educators had yet to make stark distinctions between hereditary and environmental factors. Thus, among the causes of slow school progress they listed malnutrition, defective vision or hearing, enlarged adenoids or glands, defective teeth, "mental" defects and "other" defects. Nevertheless, teachers did try to distinguish children whose mental development had been permanently "arrested" from those only "backward" or temporarily "retarded," in the newly emerging educational vocabulary of the day.[46]

Similar distinctions were endorsed by the two psychologists most actively involved with diagnosing school disabilities – Lightner Witmer and Henry Herbert Goddard. Retardation, Witmer argued, referred only to a child's relatively slow progress and was thus "a mental status, not a brain disease or defect as is idiocy." Goddard too saw retarded children as those suffering from "sickness, physical impairment, or unfavorable environment. When the cause is removed the child progresses at a normal rate." Feebleminded children, on the other hand, were those who became "increasingly below the normal child of corresponding age, finally becoming completely arrested." These were children for whom ordinary schoolwork was "not adapted" and "does not bring out any latent powers that they may have" – children who instead needed the special curriculum of the special class. Thus, to both educators and psychologists, the terms "retarded" and "feebleminded" had acquired distinctly different meanings.[47]

In establishing close working relationships with school superintendents, psychologists Witmer and Goddard both came to play major advisory roles within American public education in the first two decades of the century. By 1910, Goddard had become the nation's foremost psychological expert on educating feebleminded children, and Witmer its authority on the far larger group of retarded children.

By 1911, a survey of school policies relating to such children had been prepared for the U.S. Bureau of Education by Witmer, Ayres, and school superintendent James Van Sickle. Entitled *Provision for Exceptional Children in Public Schools,* this report examined the myriad ways that American educators were dealing with children who did not progress at the normal school rate. Among the children it considered, for instance, were those who progressed much faster than their peers. Most of this study, however, focused on slow students. Included were analyses not only of special classes but also of the "school evils known as repetition, retardation, and elimination" – failing a grade, falling behind, and leaving school – evils which these educators saw as affecting "from 10 to 50% of the entire school membership."[48]

Especially striking is the way that these researchers tried to use statistics to gain a better understanding of school successes and failures. Incorporated into their report was a detailed explanation of four different methods that schools might adopt to gauge the extent of the overage problem. The simplest was an "Age and Grade Table," a chart that schools could prepare to show the class standing of children of different ages. Alternatively, schools could chart the grade level achieved by all thirteen-year-olds, thus assessing what compulsory schooling had accomplished. A third method counted the percentage of children repeating each grade. The most sophisticated method explained here, however, was the "Age and Time-in-School Table" – a chart that allowed teachers to gauge the progress of schoolchildren while attempting to control for the most important variable identified in explaining "overage": number of years in school.[49]

Taken together, these methods show educators striving for something beyond simply the collection of data. Equally important was the development of better analytic techniques to make school statistics both comparative and meaningful. Such techniques, they believed, would eventually transform pedagogy into a science. And central to this science would be new methods of measurement.

Reliable measurement, these teachers insisted, held the key to making pedagogy a legitimate field for scientific study. It was most important, Ayres declared, that "the old criteria of 'good' and 'poor' and 'striking' and 'appealing,' make way for quantitative standards of measure and comparison." Philadelphia superintendent Oliver Cornman argued a similar point when he compared educators' attempts to measure retardation with scientists' success in measuring latitude. Measurement was crucial to science, Cornman asserted. The "science of heat was made possible by the thermometer," he explained; "astronomy was mere astrology before the application of the pendulum" and "chemistry was but alchemy prior to the perfection of the analytic balance." Similarly, measurements were an essential prerequisite for advances in education. Without them, Cornman warned, "pedagogy will remain, – well, pedagogy will remain pedagogy." [50]

It was just such a measuring device that Henry Herbert Goddard believed he had found by 1910. The Binet–Simon intelligence tests, he told the NEA that year, could be used to distinguish the retarded, who required remedial help, from the feebleminded, who required an entirely new curriculum. Such tests could thus help to supply a scientific standard for special class placement – the very purpose for which Binet had originally invented them. [51]

Even more broadly, however, the overage problem also suggested another direction for psychological testing – one which again emphasized the common interests of school and institution. Within institutions, medical superintendents had used Binet's tests to compare age with "mental age." Meanwhile, school superintendents were searching for a means of comparing age with grade in school. Goddard now saw a way of again bringing the two together: Binet testing, he believed, could be used to compare "mental age" with grade in school.

Binet Testing in the New Jersey Schools

In the fall of 1910, Goddard decided to experiment further with Binet's scale. What he wanted most of all was the opportunity to try out his new tests on a large population of ordinary schoolchildren. Through his contacts in the Feeble-Minded Club, he had come to know the superintendent of a local school district with several thousand pupils. Their friendship notwithstanding, Goddard realized that such an unusual request might prove a problem, for any schoolman would begin to worry about "all the things that might happen." [52]

"I also knew that there was a big difference between asking a favor of a man, and offering him a favor," he later related in explaining the subtle stratagem which he adopted to gain access to public schools. So he avoided asking for permission, and instead casually informed the superintendent about the valuable results he had gotten from testing children within his own institution. Perhaps, he then added offhandedly, the superintendent would also like to have his students tested. If so, Goddard told him, he could "probably get permission from the Superintendent of the Training School to give the test to every child in the Public Schools. He immediately expressed his thanks and gratitude," Goddard reported, "and we went to work the next Monday morning." Through his intuitive understanding of how such bureaucracies worked, and with his consistently cooperative and nonconfrontational style, Goddard had achieved another major coup for his profession. In 1910, he introduced Binet-Simon tests of intelligence into American public school-rooms for the very first time.[53]

The following year, he published his results in an article entitled "Two Thousand Normal Children Measured by the Binet Measuring Scale of Intelligence." While presenting his findings, Goddard also took pains to describe his method in detail. It was this article, he later declared, which soon "aroused the schoolmen of America."[54]

The Binet tests had been administered by five research assistants, Goddard explained, "all expert in the use of these tests, having been trained in the laboratory." He reported only negligible differences in results attained by different examiners. Moreover, children being tested were *"never scolded"* and *"always encouraged,"* he reported. Above all, Goddard was proud of his sample, which was nearly ten times the size of that used by Binet in 1908. "There is every reason to believe, and statisticians confirm this," he confidently (if incorrectly) pronounced, "that any group of two thousand children may be taken as a fair sample of conditions to be found in any number of children in any country."[55]

Even more than his methods, Goddard was pleased with his results – for the first six grades, that is. While only 36 percent of the 1,547 elementary school children he tested scored a mental age equal to their chronological age, he reported, 78 percent tested within a one-year range. Another 4 percent tested more than a year above their chronological age; these were probably the children whom psychologist Guy Whipple had called "gifted," Goddard asserted. The 15 percent testing two or three years behind their chronological age were probably *"merely backward,"* he argued, and "not permanently arrested"; these children deserved remedial help to "come up to the normal." The lowest 3 percent testing more than three years behind, however, were probably feebleminded and thus suffering from mental defects that "can never be overcome." Ideally, Goddard believed, these children belonged in institutions; with spaces limited, however, special classes were the next best alternative.[56]

In some ways, Goddard's results matched other contemporary studies. For instance, one 1910 study had found 4 percent of schoolchildren advanced in grades beyond their age – a figure matching Goddard's.[57] In finding only 15 percent "backward," however, Goddard's results were markedly low, for schools frequently cited retardation rates double this amount.[58] On the other hand, Goddard's figure of 3 percent feebleminded was high, for medical surveys generally estimated the feebleminded to constitute less than one-half of 1 percent of the population.[59]

Surprisingly, Goddard did not try to validate intelligence testing by citing other studies. Instead, he offered an argument of a different sort:
Binet–Simon tests had validated themselves, he insisted, for most striking
about these statistical results was their distribution into something approxi-
mating a Gaussian or normal curve – the kind of bell curve that Francis
Galton had obtained in plotting the distribution of anthropometric variables
like height. To Goddard, this offered all the proof he needed, for such a curve,
he believed, had a meaning of its own.

"To a person familiar with statistical methods," Goddard explained, this
bell curve alone constituted "practically a mathematical demonstration of
the accuracy of the tests." It was "almost beyond the bounds of possibility"
to get such a distribution, he insisted, unless Binet's tests were "amazingly
accurate." To Goddard, such a curve in itself meant that "intelligence" was a
measurable variable normally distributed among the school-age population.
If this were so, then Binet–Simon testing could become much more than
simply a method for identifying the feebleminded.[60]

It was in dealing with "normal" children that Goddard pushed his new
assumptions furthest. "A question of considerable interest," he wrote, "is to
what extent have our ordinary methods of classification in the schools more
or less consciously and accurately recognized the actual mental standing of
the child." How accurate, in other words, was school grading? He answered
this by introducing the new variable measured by his tests, "intelligence,"
into the overage debate. In a series of charts, Goddard now compared age
with grade in school, while this time controlling for "mental age."[61]

This new variable did appear to bear some relationship to school progress,
for while age matched grade for only 30 percent of these children, according
to Goddard, "mental age" matched grade for 43 percent. Nevertheless, for
most children, Goddard concluded, mental testing and school grading "do
not agree at all."[62]

Such a conclusion might have been disappointing had Goddard intended
to use school grades to validate testing. Goddard, however, was far from
discouraged by his results. Had Binet mental ages simply matched school
grading, he now argued, then these new psychological measurements would
have proven superfluous, for they would have demonstrated only that "we
somehow by our ordinary methods size up children about right and do not
need any such thing as the Binet Scale in order to set us right." Such, however,
was evidently not the case.[63]

For instance, Goddard now assumed that eight-year-olds scoring a mental
age of eight "would naturally all be in the third year of school"; instead he
found forty-three "behind what they should be" and another nine "ahead of
what their mental endowments would probably permit." Of fourteen ten-
year-olds still in the second grade, only half were "where they ought to be –

the rest are unjustly treated." Such statements made Goddard's underlying assumptions clear: in each case, he suggested, the tests were right and the schools wrong.[64]

These conclusions allowed Goddard to indict school practices. How much "injustice," he asked, "is being done these children by the ordinary school routine?" Such determinations were critical, Goddard argued, for school injustices had social consequences. Children unfairly left behind frequently became incorrigible, he maintained, while those pushed beyond their capabilities would break down later.[65]

Yet while Goddard emphasized these discrepancies between "actual mental endowments," as measured by Binet tests, and school advancement, his article simultaneously inferred that his tests had confirmed educators' experiences in another way. "Nothing could be clearer than the way in which these figures demonstrate what we all know from experience must be true," he reported, "that is, that we drag the dull child up, trying to keep him up to his grade and hold the bright child back to keep him to the same grade, thus doing gross injustice to both." Thus, just as he had previously argued to physicians that Binet testing confirmed their intuitive experiences, he now made a similar argument to teachers. Grading children incorrectly was "*precisely what we all know does happen in the school,*" Goddard asserted. Testing may not have matched the grading procedures of the schools, Goddard was suggesting, but it still matched what schoolteachers intuitively knew to be true.[66]

If, however, the results gathered from Binet testing were "amazingly accurate," as Goddard vehemently maintained they were based on their normal distribution, then educators now had in hand a new scientific measuring device for the schools. Binet-Simon testing offered educators "a wonderfully valuable method of measuring our efficiency and our accuracy in the grading of children," Goddard now proposed.[67]

Unfortunately, Goddard could not yet offer this improvement to teachers beyond the sixth grade, for his tests for older children failed completely. Nearly all students above the sixth grade passed the twelve-year tests, he reported, while the thirteen-year tests proved too difficult. This failure in itself offered "a hint" of what higher tests might look like in the future, he proposed. Instead of seeking tests "of universal application," Goddard speculated, psychologists might test a "special gift" that was "possessed by some people but not by others although one would not dare say one was of higher intelligence than the other" – an idea that embodied a very different conception of the meaning of intelligence than the hierarchical one he was now employing. For the present, however, he conceded that tests for higher grades did not work, and he therefore "based nothing upon their value." After the sixth grade, Goddard concluded, "we lose the curve."[68]

Goddard's final task was to revise his scale in accordance with the results acquired from his new sample. By then, however, Binet too had published some further revisions. By combining Binet's new suggestions with his own New Jersey findings, and by removing or relocating questions failed by too many "normal" children, Goddard developed his own 1911 "Binet–Simon Revision."[69]

He announced his research results to the NEA that summer. At the same time, Goddard could also make a second major announcement. The New Jersey legislature, he proudly told teachers, had just passed the nation's first law mandating special education classes in public schools.[70]

Goddard had helped to draft Chapter 234 of New Jersey Public Law of 1911, which required local school boards to start classes for deaf and blind children whenever there were ten or more such students, and to begin special classes, limited to fifteen, whenever a district had "10 or more pupils who are three years or more below the normal." As drafted, it also stated that the Commissioner of Education "shall, with the advice and consent of the State Board of Examination, prescribe the Binet or such other methods as to him may seem best for ascertaining what children are three years or more below the normal." Such a law, Goddard declared triumphantly, gave the Binet–Simon scale of intelligence "official recognition in the state of New Jersey."[71]

Goddard's legal triumph was short-lived, however, for he soon had to add a small disclaimer to his paper. At "the last moment," he conceded, "the word 'Binet'" had been "stricken out of the law." While evidently disappointed, Goddard was still satisfied with the new law's wording. In defining children requiring special class instruction as those "three years or more below the normal," Goddard argued, this law offered de facto recognition of the new concept of "mental age." Binet testing, Goddard rightly maintained, had been "recognized in reality, tho not named."[72]

By 1911, the special education movement had thus entered a new phase. In New Jersey, state law now required school districts to determine which children were in need of special education. Such determinations, Goddard argued vehemently, could best be made by using psychological tests of intelligence. Psychologists had thus found another important social application for their expertise. In introducing intelligence testing into American public schools, Goddard had apparently scored another stunning victory for applied psychology.

Critics of Testing

Viewed retrospectively, Goddard's New Jersey school study is certainly open to serious criticism. Later generations would find his bold assertions about the representativeness of his rural New Jersey sample to be naive; his failure to consider any other variables (e.g., ethnicity, economic background, language proficiency, or even number of years in school) disturbing; and his confident conclusions about the meaning of his normal curves simply wrong. Such a study would hardly withstand modern methodological scrutiny.

Goddard's crude mathematical methods and faulty assumptions partly reflect his ignorance of newer statistical techniques largely instituted in the decade since he left graduate school – problems common to the first generation of American social scientists. Less easy to understand (but also not uncommon in writings of this era) are the arithmetic errors which permeate this 1911 study.[73] For instance, Goddard's chart accounts for only 1,536 children, not 1,547, as he claimed. The number of children testing at age was 576, not 554 (37 instead of 36 percent). Similar mistakes can be found in nearly every number he reported. While Goddard based so much of his argument on his normal curve, eleven of the twelve totals used to construct this curve were added incorrectly. These errors were all too small to alter either the shape of Goddard's curve or his conclusions; even so, they still cast serious doubts upon the quality of such research, especially in a study which placed so much emphasis on numbers.[74]

While Goddard's sloppy arithmetic evidently escaped the scrutiny of his contemporaries, his new procedures did not. Experimental psychologists employed in university laboratories had yet to pay much attention to the new methods being promoted by Binet and Goddard. Those working in applied areas, however, understood their ramifications immediately. Within one year, detailed critical responses to Goddard's article had appeared in the three American journals most dedicated to developing a new working relationship between psychology and education: Hall's *Pedagogical Seminary,* Witmer's *Psychological Clinic,* and the latest addition to the field, the *Journal of Educational Psychology,* begun in 1910 by academics seeking a "common meeting ground for the psychologist and the educator."[75]

Especially in light of the sharp polarization evident in later testing controversies, these first American reviews are striking, for most of them were mixed. Despite their strong interest in new quantitative techniques, these reviewers did not simply accept Goddard's conclusions or his methods uncritically; neither, however, did they summarily reject them. Instead, both Binet's and Goddard's writings provoked a complex, sophisticated, and multidimensional exploration into the very meaning of the concept now being called "intelligence." In the process, reviewers for the first time raised many questions about intelligence testing – questions that would continue to be asked for the rest of the century.

Typical of such responses was Leonard Ayres' review in *Psychological Clinic.* As the nation's foremost authority on school "laggards" and a strong proponent of quantification, Ayres might have been expected to embrace Goddard's findings enthusiastically. Instead, he expressed serious reservations. In attempting to assess intelligence, Ayres argued pointedly, Binet–Simon tests were measuring the wrong types of abilities.[76]

Ironically, Ayres summarized his critique by using one of Binet's questions written for ten-year-olds: "Why should you judge a person by what he does rather than by what he says?" That deeds reflect more than words, Ayres argued, was axiomatic to the "literature of every age" – so axiomatic, in fact, that "Binet and Simon have rightly assumed that it forms a part of the knowledge of every normal ten-year-old child." Yet two-thirds of the Binet scale tested only language ability, thus assuming that "native ability to do can be tested by testing the ability to use words about doing." At issue was "the fundamental fact that the motivating stimuli which shape one's actions in coping with a real problem in life are invariably multiple and complex," he explained, "whereas those which determine his answer to a hypothetical question are simple, few, and different in quality."[77]

Ayres illustrated this by giving "a number of intelligent adults of demonstrated practical ability" another Binet question written for ten-year-olds: "What ought one to do before taking part in an important affair?" "The writer's experience in putting this question to business men is not encouraging," he reported sarcastically. "A few answers have been received ranging from 'Take a bath,' and 'Put on your best clothes,' to 'Take some money from the bank' and 'Transfer your property to your wife,'" while most had replied "with energetic expressions of short and ugly words and emphatic protestations that the question is unanswerable." Ayres' facetious example actually illustrated a serious point: that answers to many Binet questions reflected "daily environmental experiences" as much as "intelligence" – experiences which "differ radically among different children," while others depended "directly on the excellence of the child's schooling."[78]

Ayres also noted a more specific problem with Goddard's results that in later years would prove extremely significant: Binet's tests for young children were evidently too easy, and those for older children too hard. By combining the results from children of all ages, these errors had canceled each other out, thus producing Goddard's "normal curve." Using the same procedure with his own data on school grading, Ayres too could produce a normal curve. If, as Goddard claimed, such a curve alone proved the accuracy of Binet testing, then Ayres' curve offered equal proof that school grading was "neither too hard nor too easy, but almost exactly right." The truth, Ayres added, was that such a curve actually hid wide variations between grades and cities, just as Goddard's hid variations from age to age.[79]

Along with his criticisms, however, Ayres admitted that Binet's tests had much to commend them. While other tests had remained largely restricted to "workers in psychological laboratories," Binet's had been widely endorsed by "practical teachers and workers with children." The reason for this was clear.

Everyone knew "what is meant when one says that a given child shows intelligence equal to that of a ten-year-old normal child," he remarked. On previous scales, "no one knew what the steps . . . meant in terms of anything else." This advantage, Ayres decided, "outweighs the shortcomings of the tests themselves." The "writer does not wish to appear as an antagonist of the Binet–Simon Measuring Scale of Intelligence, for he is not," Ayres concluded. Nevertheless, he warned that the "paramount problem" with Binet's tests was whether "they really measure native ability."[80]

Serious questions concerning the abilities being measured also informed psychologist Clara Schmitt's review in *Pedagogical Seminary*. Having worked with psychiatrist William Healy in his "Juvenile Psychopathic Institute," the country's first "psychiatric clinic" for delinquents, Schmitt too sought psychological solutions for social problems. In Schmitt's judgment, however, Binet's tests offered "an inadequate measure of mental ability," for they failed to account for multiple types of mental skills.[81]

Schmitt illustrated this by describing a boy brought to Witmer's clinic because of reading problems. This six-year-old, who liked to help his father assemble electrical equipment, had scored a mental age of seven on the Binet scale. Schmitt then summarized all the evidence about this boy's mental state: to the school, he was a "dullard"; to his father, "all right"; to his playmates, a "molly coddle"; and to Binet testers, "somewhat precocious." "No two of these judgments were the result of the same set of data," Schmitt insisted. Should his school and social disabilities continue, she added, "he certainly would not escape being considered a defective." If Binet were correct, then "we would have to believe that the majority of seven year old boys possessed his mechanical efficiency and his academic inefficiency, which is not true," she argued. "Children of six can learn to read," Schmitt declared, and "if children of seven can assemble the parts of a gas engine," as this child could, then "no one knows it." What Schmitt wanted were separate tests of social, mechanical, and, above all, school skills. Thus, while Ayres had argued that Binet's tests were too scholastic, Schmitt found them not scholastic enough, for they placed too little emphasis on specific skills that schoolchildren needed – skills like reading and writing.[82]

Schmitt also objected to Goddard's conclusions about the meaning of his curve. Since these tests were actually measuring "many qualities of children of different ages," she insisted, the resulting curve was "merely a happy or an unhappy accident." As for Goddard's charges of widespread school injustices, she replied, teachers might well retort that since they had known the child far longer, their judgments should be considered "more accurate than that arrived at by a ten or twenty minute examination over very little of the matter with which the school concerns itself." Testing, in other words, ought to be judged against the experiences of the school, and not vice versa.[83]

Not all reviewers were so skeptical, however. The most enthusiastic response came from psychologist Lewis Terman. Like Goddard, Terman was
a Hall student; after graduating from Clark, he had worked first as principal
of a San Bernardino high school and then as an instructor at the Los Angeles
Normal School. By 1911, he had become a professor of education at Stanford
University. Unlike Goddard, however, Terman had long been fascinated by
the problem of measuring minds, for in his dissertation he had tried to develop tests to distinguish "bright" from "stupid" boys.[84]

Binet's concept of age scaling, Terman believed, constituted a critical scientific breakthrough. All previous mental testing had been "fruitless," he now
conceded; even from his own work he "could hardly count value received."
With older methods, diagnosing the child's mind "hardly transcends guess-
work." Binet scaling, however, was "rapidly making possible a clinical child
psychology."[85]

Yet despite his enthusiasm, even Terman criticized Goddard's curve. By
"lumping all the ages together," Terman complained, Goddard had con-
cealed "the very facts we wish to know." What Terman wished to know, however, was something very different from what either Ayres or Schmitt wanted,
for while they were most concerned with the content of Binet's tests, Terman
hoped to learn "how nearly accurate the scale is at every point." Like Goddard, Terman believed that a normal distribution was proof in itself that
Binet's measures were valid. A more skillful statistician than either Binet or
Goddard, he now began working on his own revision by expanding the test
population, improving the scoring procedure, and producing even better normal curves for children at each age level.[86]

Terman also accepted without question the idea that "innate intelligence"
was a quality distinct from school success. It was now "possible for the psychologist to submit, after a forty-minute diagnostication," he maintained, "a
more reliable and more enlightening estimate of the child's intelligence than
most teachers can offer after a year of daily contact in the schoolroom."
This was not because educators exercised bad judgment, this former teacher
argued, but because "all human estimations are relative to some standard"
and teachers shared no objective standards, while psychologists now did.
Like Goddard, Terman believed that schools ought to measure their accomplishments against psychologically established norms.[87]

These charges and countercharges concerning the proper relationship between psychology and the schools brought even more attention to Binet testing. Within a few short years, the literature on intelligence testing had begun
to burgeon, both in the United States and in other countries. Among psychologists and educators interested in pedagogical questions, Binet's ideas were
quickly gaining a growing audience.

"Perhaps no device pertaining to education has ever risen to such sudden prominence in public interest throughout the world as the Binet–Simon measuring scale of intelligence," J. Carleton Bell announced in a 1912 *Journal of Educational Psychology* review of the writings of Binet, Goddard, Ayres, Terman, and others. While Bell agreed with Goddard that many students were probably placed in the wrong grade, he was not willing to draw such a conclusion simply from Binet testing. A child's school grade, Bell reasoned, ought to be based upon school subjects. It was "quite beyond the mark," he argued, "to carry these tests directly over to school work and make them a basis or a criterion of school grading." After all, Binet's tests were "supposed to be tests of *native intelligence*," Bell insisted, "and to neglect as far as possible all that has been learned in school."[88]

Yet this in itself raised a more fundamental problem. In reviewing this new literature, Bell had been "impressed by the fact that not a single investigator has raised the question 'What is native intelligence? What does the term signify?'" The testing debate, he declared, was proceeding backward, for testers were claiming success in measuring a variable that they had not yet defined. Was it "not time," he wondered, "to begin at the other end" by first analyzing "the complex, popular term intelligence" and then devising tests to measure it? Such a task, Bell concluded, would be worthy of "the best efforts of any experimentalist."[89]

Perhaps the most astute of the early critics to question the relationship between "intelligence" and school success was Binet himself. As Bell noted, Binet's writings had failed to define intelligence explicitly. "We have not attempted to treat, in all its scope, this problem of fearful complexity, the definition of intelligence," Binet had admitted in 1908. "We do not measure the intelligence considered separately from a number of concrete circumstances – the intelligence which is needed for understanding, for being attentive, for judging. It is something far more complex that we measure."[90]

Still, Binet too worried about distinguishing "intelligence" from "scholastic aptitude," for in his tests, he wrote repeatedly, he sought to measure only "the natural intelligence of the child, and not his degree of culture, his amount of instruction." In Binet's conceptualization, intelligence and schooling were far from synonymous. After all, many children had been deprived of instruction. They may have lived far from school, he speculated, or simply been kept at home to "rinse bottles, serve the customers of a shop, care for a sick relative or herd the sheep." Even without schooling, Binet insisted, such children might still be highly intelligent.[91]

Nor did intelligence alone guarantee school success, Binet argued, for schoolchildren also needed qualities dependent upon "attention, will, and character," including "a certain docility, a regularity of habits, and especially continuity of effort." In fact, even more than Goddard's, Binet's writings expressed a subtle hostility toward school injustices and pedagogical rigidity.

(Binet would educate his own two daughters at home.) His tests, Binet declared, would at last allow psychologists to "free a beautiful native intelligence from the trammels of the school."[92]

This distinction between "innate" and "scholastic" abilities had ramifications not only for schoolchildren; it was equally significant for psychologists. Especially important were its professional implications. Both Goddard and Binet were committed to developing a new profession of applied psychology positioned precariously between medicine and pedagogy, and yet subservient to neither. Binet's insistence upon the psychologist's newfound expertise as a diagnostician of mental subnormality – a role previously held by doctors and teachers – was proving especially useful in establishing such a profession. Through his new form of measuring the mind, Binet was also asserting a new social identity for professional psychologists.

Thus, as early as 1905, Binet could already delineate three different methods, all useful and yet distinct, for diagnosing "the intellectual level among subnormals." These were the *"medical method,"* which revealed *"possible* signs of defect"; the *"pedagogical method,"* which revealed *"probable* signs of defect"; and the *"psychological method,"* which was "almost *certain* to reveal the signs of defect." In effect, Binet was redefining the child's mind as an entity related to but yet distinguishable from both his body and his schooling; simultaneously, he was also distinguishing the expertise of the psychologist from that of the physician and the schoolteacher.[93]

Yet while Binet had found little difficulty in distinguishing psychological from medical assessments, separating psychology from pedagogy proved far more problematic. He continued to believe that "intelligence" was a distinct and ultimately measurable variable. As early as 1908, however, Binet had begun to suspect that his own measurements were really conflating four variables: "the intelligence pure and simple"; "extra-scholastic [environmental] acquisitions"; "scholastic acquisitions"; and "acquisitions relative to language," which depended "partly on the school and partly on the family circumstances." Despite his ongoing efforts to revise his tests to separate "intelligence" from school learning, Binet thus conceded that the two in fact overlapped.[94]

Goddard, however, conceded nothing of the kind. Perhaps because he still based his own understanding largely on findings from children living within a single institution, he paid no attention at all to other variables such as family circumstances or differential schooling. While acknowledging that Binet's scale was still imperfect, Goddard expressed no serious doubts that these tests were measuring innate abilities only. Neither the cautious reservations of other testers nor the sharp comments of his critics did anything to dampen his enthusiasm for this new psychological invention. By 1911, Goddard expressed far fewer qualms about using Binet testing than Binet.[95]

Nevertheless, the cumulative effects of such criticisms were apparent in a new school battle that soon embroiled Goddard. It began when this psychologist was invited to participate in the New York City School Survey of 1911. The ensuing controversy would ultimately call into question nearly all of the ideas that Goddard had been promoting for the past five years, including his beliefs about the relationship between schools and institutions, the special class curriculum, the training of teachers, and, most crucially, the role that intelligence testing ought to play in diagnosing schoolchildren. In the process, the larger issue of psychological expertise in the schools would itself be called into question.

Binet Testing in the New York Schools

"I was rejoiced to know you had gained an entree into the New York City School System and hope you will be able to help out there," psychiatrist William Healy wrote Goddard upon learning of his appointment to New York's school survey. Such an entree was indeed another coup, both for applied psychology and for Goddard personally, for it meant that within three short years, Goddard had taken Binet's tests from an institutional school of under 400, to a public school district of under 2,000, to the nation's largest school system – a system with more than three-quarters of a million students. Goddard had thus gained the opportunity to demonstrate the relevance of psychological testing to public schooling on a scale far surpassing anything yet attempted by his contemporaries.[96]

The New York City School Inquiry of 1911–1912 was the first of the massive school surveys that would soon become a hallmark of progressive education. It had emerged after a decade of conflicts between this city's Board of Education, which had continued to expand its social agenda under the strong leadership of Superintendent William Maxwell, and its Board of Estimate and Apportionment, which hoped to contain the ever-increasing school budget. Impressed with Ayres' investigation of ways to decrease "laggards" and increase efficiency, in 1910 the Board of Estimate decided to sponsor its own comprehensive examination of all aspects of school functioning – a project that provoked the antagonism of Maxwell. To head its "School Inquiry Committee," the Board of Estimate chose prominent city politician (and future mayor) John Purroy Mitchel. To oversee the "educational aspects" of this study, the committee hired Paul Hanus, Professor of Education at Harvard University.[97]

Surveying the New York schools was, according to Hanus, a "colossal task," for even more formidable than this city's massive educational bureaucracy was its political intrigue. Fearful that his work would be undermined by self-serving politicians, Hanus strove for a scientific study emphasizing "statistical, comparative, and experimental" methods. By 1911, he had hired eleven educational experts to help him. Among them was Henry Herbert Goddard, who was asked to survey New York's program of "ungraded" or special education.[98]

At the time, special education in New York City had entered its second decade. Under Maxwell's guidance, this city had established various innovative programs to combat the "overage" problem, including "C" classes for children needing English language instruction, "D" classes for slow students approaching the end of their compulsory schooling, and "E" classes offering rapid advancement for late starters hoping to catch up to their peers. Also opened was a public School for the Deaf, special classes for crippled children and children with serious vision problems, and "open air" classes for those considered susceptible to tuberculosis.[99]

In 1900, New York City had also opened its first "ungraded" class for children apparently suffering from mental impairments. Its teacher, Elizabeth Farrell, hoped to use her experiences from a one-room schoolhouse to give such children individualized instruction. Within six years, this city had established 14 such classrooms, and Farrell was promoted to "Special Inspector for Ungraded Classes." In 1907, the Board of Education hired its first physician with psychiatric training, Dr. Isabelle Thompson Smart, as "Medical Examiner for Ungraded Classes." By 1912, New York City was educating about 2,500 children in 131 ungraded classrooms.[100]

As a member of Hanus' survey team, Goddard paid visits to nearly all of these classes. By 1912, he had produced a highly critical report. The New York program, he proposed, ought to be radically enlarged; it needed a new curriculum, more equipment, more supervisors, better teacher training, and higher salaries. If the Board of Estimate had hoped for cost-cutting recommendations, they were surely disappointed, for in every area, Goddard recommended more spending.[101]

At the heart of Goddard's survey was the premise now guiding his own pedagogical research: the institution ought to be the laboratory for the special classes. Yet while institutions had largely adopted a manual curriculum, he reported, this city's ungraded class teachers still wasted about half their time teaching the three R's. Further impairing instruction was the shortage of skilled teachers, for many Goddard met were "painfully aware of their own lack of training and their own inability to do for the children what they feel might be done." Such teachers ought to be trained by institutions; otherwise, they were "left as the physician would be who has gone through his medical course but has had no laboratory or hospital experience." Those who did not know institutional cases found their work handicapped by "the impression that the children are really normal, or will yet prove normal."[102]

The special class ought to follow the institution, Goddard insisted, since

An ungraded class in P.S. 110, Manhattan, displays its handiwork, 1911. From H. H. Goddard, "Ungraded Classes," in Paul Hanus, ed., *Report on Educational Aspects of the Public School System of the City of New York. . .* (New York: City of New York, 1911–1912), facing p. 366.

both were teaching many of the same types of children. In New York's ungraded classrooms he had found "imbeciles of mongolian type, microcephalic idiots, hydrocephalic cases, cretins," and "a large number of middle and high grade imbeciles." Also present were many less impaired children – cases that Goddard had christened "morons."[103]

To his dismay, Goddard reported, he found the New York program plagued by misdiagnoses. The wrong children were frequently placed in special classes, he wrote, because teachers could not judge the "mentality of the child." In ungraded classes Goddard saw "children who are really almost normal, but have been mistaken by the teacher because she has been unable to understand them." Teachers often misread "only temporary or individual idiosyncrasies" as signs of mental impairment. Using Binet's tests, he found children whose "mentality ranges from that of a three-year-old to the mentality of a normal child" in the same class.[104]

Far more problematic, Goddard insisted, were the thousands of feebleminded schoolchildren whom teachers had failed to recognize. Such children were likely to prove a costly burden, both to the school system and to society, he warned, unless properly diagnosed. In order to identify these cases, he proposed, the schools needed psychological expertise.[105]

Goddard's most controversial conclusion concerned the number of feeble-minded children in the New York schools. In the time allotted, he conceded, he had been unable to test students "extensively or systematically"; in fact he had tested only 268, most of them in ungraded, D, or E classes. Nonetheless, since he believed that his previous study had produced a sample whose results could be applied anywhere, he simply generalized from New Jersey to New York. "The most extensive study ever made of the children of an entire public-school system of two thousand," he wrote of his own work, "has shown that 2 per cent. of such children are so mentally defective as to preclude any possibility of their ever being made normal and able to take care of themselves as adults." Goddard thus concluded that New Yorkers should expect to find 2 percent of their schoolchildren mentally defective as well. If he were correct, then New York City ought to have been providing special education not for 2,500, but for over 15,000 children.[106]

Goddard's conclusions, like those of others working on the Hanus study, provoked publicity, protest, and controversy. Among the skeptics was Superintendent Maxwell, who resented the entire survey as a costly exercise in political interference in the schools. "Dr. Goddard," he noted sarcastically, "after testing about 268 children . . . reaches the conclusion that 15,000 . . . are mentally defective." Such arguments "suggest the propriety of requiring all future school investigators, before engaging in this difficult and delicate work, to pass a thorough examination on Inductive Logic."[107]

Even more outraged was Elizabeth Farrell, New York's "Inspector of Ungraded Classes." Proud of both her city's and her own accomplishments in promoting special education, Farrell defended her programs. In a sophisticated and lengthy reply, she attacked Goddard's survey for its faulty research methods, erroneous assumptions, and unwarranted conclusions. At the same time, Farrell directly challenged Goddard's beliefs about the relationship between the institution and the special class.[108]

What worked in institutions, Farrell argued, was not necessarily best for public schools. Academic work had "a value beyond its use to the child," which was "not recognized by those whose attention is focused upon children in institutions," she insisted. While Goddard had stressed the similarities between ungraded and institutionalized children, Farrell saw her own role as "emphasizing the points of resemblance and minimizing the differences between the regular grade child and the ungraded class child," thus fostering "the self-respect of the unfortunate one." In opposing Goddard, she was articulating an early version of an argument which would reemerge many decades later in the movement promoting "mainstreaming."[109]

Equally suspect to Farrell were Goddard's complaints about the qualifications of New York's special educators, many of whom had been given high ratings by school officials. "The question here," she asserted in contrasting Goddard's expertise to that of educational administrators, "is the value of the judgment of men and women whose business is the supervision and rating of teachers, as compared with the opinion of a research student in psychology."[110]

The most serious challenge pitting educational against psychological expertise, however, came in response to Goddard's most damaging charge: that the schools contained too many cases of "mistaken diagnoses." Since Goddard had based his diagnoses on Binet-Simon testing, and since he conceded that his 2 percent estimate "stands or falls with the validity of the scale," Farrell took direct aim at intelligence testing.[111]

In challenging testing, Farrell gathered comments from both European and American reviewers who had expressed serious reservations about Binet's work. Among those she quoted was Lightner Witmer, who now openly opposed what he saw as Goddard's overreliance on Binet's tests. "No Binet-Simon tests, nor any other tests, will inform us as to what children we shall consider feeble-minded," Witmer stated bluntly, for feeblemindedness ought to be defined not psychologically but socially. Witmer even recommended dispensing with the very term "mentally defective" and substituting the term "socially defective."[112]

Farrell found her strongest ally in New York's Medical Examiner for Ungraded Classes, Dr. Isabelle Thompson Smart. If Goddard's claims concerning mistaken diagnoses were correct, then Smart too bore responsibility, for she and Farrell had personally diagnosed all ungraded class children. Having studied psychiatry in Europe, however, Smart regarded her own diagnostic skills as far superior to Goddard's.[113]

Smart had long been concerned that psychologists were trying to usurp the role of school physicians. Those claiming that "the psychologist is all that is necessary," she had written in 1908, were surely taking a "very narrow stand." "The psychologist is needed, unquestionably," she had conceded, "but ordinarily what does he know of the physical life and physical needs of the child, or – for that matter – of the mental life either, where pathologic conditions exist and need to be eliminated?" To Smart, medical expertise had to remain paramount, for "no reputable physician will treat a case which has not come up for personal examination and diagnosis in the first place." Responding to the school survey, she contrasted her own diagnoses to Goddard's; thus, in the case of one boy whom Goddard had found feebleminded, Smart blamed his poor school performance on a speech impediment.[114]

Farrell also gathered comments from medical superintendents. While many of these, at Goddard's urging, had conceded that testing could indeed be used to confirm institutional experience, they were still reluctant to see it supplant medical diagnoses. "Theoretically the Binet test is of equal value in classifying cases of mental defect," Superintendent Walter Fernald admitted – a position that in fact echoed Goddard's – "but I have had twenty-five years of experience in the diagnosis of thousands of these people," he added, and therefore it was "not necessary for me or for my assistants to use a Binet test for classification." "I do not believe," Fernald concluded, "that any merely psychological measurements will take the place of practical medical training and experience." Dr. Martin Barr voiced similar sentiments. Binet's tests were not "infallible by any means," Barr argued; instead, as a physician he was "always careful to get the family history and note the stigmata of degeneration."[115]

Farrell also expressed a more general complaint, for she had evidently been expecting something far different from Goddard. Ironically, even her criticisms reflected the high esteem with which most special educators regarded this psychologist. "The service given Rousseau to general education, by Pestalozzi to the education of poor children, by Horace Mann to public education in the United States, is similar to that expected from Dr. Goddard for the education of mentally defective children . . . ," she wrote. In her eyes, he had failed.[116]

Instead of emphasizing the dynamic force driving this movement forward, Farrell argued, Goddard had produced "a work without a past and with no future." His survey offered only short-term prescriptions, not a "broad vision of the function of the school in this its latest problem." Most significant in addressing such problems, Farrell maintained, was "our attitude towards them"; yet readers would search this report "in vain for a philosophy upon which to found their practice." What Elizabeth Farrell had obviously hoped for from Henry Herbert Goddard was a broadly supportive statement praising the efforts of teachers such as herself to expand the horizons of the public school; instead, she got a report criticizing her department for specific administrative shortcomings. "To be unable to see the forest for the trees is sad," she concluded. "To have missed the vision is sadder still."[117]

Farrell's sharp reply must have surprised Goddard. For the first time, this psychologist's consummate social skills had failed him, for in this instance he had seriously misgauged his professional audience. Perhaps because of overconfidence, Goddard had become critical rather than cooperative. By stressing "mistaken diagnoses," he had directly pitted his own skills in diagnosing feeblemindedness against those of the teacher and doctor charged by the city with this function. In the process, he had provoked a fiercely argued counterattack which questioned both his own expertise and his faith in testing.

By 1913, both Goddard's critical report and Farrell's angry reply had reached New York's Board of Education. In order to consider all the suggestions in the massive Hanus survey, this board appointed separate committees to review the conclusions of each surveyor and recommend reforms. In 1914, the committee studying Goddard's survey issued its report. Its conclusions suggested a compromise which incorporated some of Goddard's suggestions, and some of Farrell's. "We agree with Dr. Goddard," members declared, "that 'the enormous growth of these classes is in itself sufficient indication of the size of the problem and the reason for many shortcomings.'" Also adopted were Goddard's recommendations for granting ungraded class teachers salary bonuses and leaves for additional training. Other recommendations made by Goddard, however, were rejected. This committee could find no "exact determination of the best kind of curriculum." As for Goddard's estimate that 2 percent of city schoolchildren were feebleminded, committee members supported Farrell's contention that such a figure was completely unsubstantiated.[118]

Even more pointedly, while admitting the need for "frequent medical and psychological tests," this report made no mention of Binet–Simon testing. Instead of recommending intelligence testing as its main criterion for ungraded class placement, as Goddard had forcefully proposed, it explicitly endorsed the "educational" classification system devised earlier by Dr. Barr – a system classifying children according to their social and vocational aptitudes. Moreover, while Goddard had recommended that the schools appoint "five more examiners (psychologists and physicians)," this report instead proposed hiring "three physicians who shall be skilled in nervous diseases." Evidently, these educators still trusted medical far more than psychological expertise. In May 1914, these and other recommendations were formally adopted by the Board of Education, thus officially ending debate over Goddard's survey.[119]

After a string of successes, Farrell had presented Goddard with his first serious opposition. She had effectively countered his conceptualization of the relationship between the institution and the public school and raised major challenges to his claims to expertise. She had also stopped the school board from officially endorsing intelligence testing. Ironically, Goddard's survey led New York City to hire more physicians, not psychologists. Despite his best efforts, by 1914 this Board of Education remained unconvinced of the value of Binet–Simon intelligence testing.[120]

Teacher Testers

If Goddard's goal had been to get intelligence testing officially adopted by the New York school board, then he had clearly failed. On the other hand, if the testing movement had suffered a major setback in New York City, one would hardly know it. To the contrary, the very fact that intelligence testing had been tried in this city further advanced its legitimacy. In addition, Goddard's report exacerbated public fears that the feebleminded were far more prevalent than previously believed – a fact that in itself suggested the need for more testing. And while Goddard had clearly stated that he had not *found* 15,000 feebleminded children, but had instead *estimated* this number, the distinction between an estimate and a finding was quickly lost in the publicity that ensued. "Finds 15,000 Pupils are Feeble Minded," a *New York Times* headline announced a day after Goddard's report was made public.[121]

Moreover, despite the warnings of their critics, psychological tests were rapidly gaining widespread acceptance. By 1914, intelligence tests had been introduced into hundreds of schools – not because school boards voted to adopt them, but because teachers simply started using them. Since 1910, Goddard's summer school graduates had been spreading more than just "Vineland Spirit"; they had also been disseminating the Binet–Simon scale throughout the United States.

"This splendid North Dakota air has done as much for me as I hoped," one Vineland graduate wrote in 1910. "I am just ready now to turn my attention to applying the Binet tests; so far I have only tested three." Other alumni were soon doing similar work. "The Binet tests are being used by fifteen special teachers," reported a Washington, D.C., graduate by 1911.[122]

Teachers told of using these tests to discover their own cases of "mistaken diagnoses." "Would you believe that I find idiots in a public school?" a Vineland graduate wrote. "Well, I have and more than one. A twelve-year-old girl tested two and three points, and was enrolled in the first grade for the fifth time." Others reported that the wrong children were in special classes. A Newark teacher found "four children who are not feeble-minded at all but were so badly treated in the schools that they became 'incorrigible,' and refused to learn anything. It is very slow work and very discouraging at times," she added, for "though your hopes are sky-high one minute, they are in the depths the next."[123]

Through contact with such teachers, administrators also learned about these new psychological measures. One New Jersey official endorsed the work of a teacher when test results proved "wonderfully consistent with the conclusions already formed by the principals, though they had no scientific reasons for such deductions." "Both the Superintendent of Schools and the Master of our building are most hearty in cooperation and much interested in what was gained from the summer at Vineland," wrote a Massachusetts teacher.[124]

Despite the forceful objections of Elizabeth Farrell, Leonard Ayres, Lightner Witmer, and others, Goddard's influence on New York teachers proved especially powerful. By 1914, he had personally designed the curriculum in four of seven institutions providing training for this city's special educators. Besides the Vineland summer school, the most important of these was New York University's School of Pedagogy, where Goddard had been a visiting lecturer since 1907. In 1912, NYU began a "summer course on the Education of Defectives," with Goddard as director. Working as associate director was a younger psychologist named Arnold Gesell.[125]

Like Goddard, Gesell was a former schoolteacher who had earned a doctorate under Hall; after graduation, he had taught alongside his former Clark classmate, Lewis Terman, at the Los Angeles Normal School. Visiting Vineland in 1909, Gesell too had been deeply touched by the "informality and sincerity of the work," both in its "scientific and humanitarian aspects." Moving to Yale's education department in 1911, Gesell opened what later became the Clinic of Child Development. Between 1912 and 1915, he spent his summers working with Goddard to train New York City schoolteachers.[126]

By 1912, the impact of these programs was already evident, for thirty-six New York teachers had graduated from Vineland's summer school, while sixty-three had attended classes at NYU. By 1913, NYU students could also enroll in a "Laboratory Course on Tests of Intelligence." "Those who complete this course satisfactorily," this school's Summer School catalogue announced, "should be able to give the Binet tests correctly."[127]

Yet this very success in spreading testing soon raised a new problem, for not everyone was as pleased as Goddard to see teachers becoming so able. Among those most critical of the growing phenomenon of teacher testers were psychologists following the "clinical" career paths tenuously carved out by Witmer and Goddard – careers either in clinics associated with the public schools or in institutions. For these new professionals, teachers using Binet's tests posed a serious challenge to their own expertise.

Typical of such critics was Clara Harrison Town, director of the Laboratory of Clinical Psychology at the Lincoln State School. "A peculiar situation has developed," she noted in 1912. Binet had offered a new method to "his confreres" which had "the virtue of simplicity – all the apparatus required beyond a few pictures, some bits of wood, and a little money, being a mind well trained in psychological theory and methods." Educators had seen its value. "Accustomed to the complicated apparatus of a psychological laboratory," she reported, "the laity were pleased to find it unnecessary and overlooked entirely the fact that the psychologist himself was not unnecessary." The "laity," in other words, were embracing psychological testing, but not psychologists.[128]

Psychologist J. E. Wallace Wallin was even angrier about this situation. A gifted scientist with an abrasive temperament, Wallin could barely hold a job in the fragile new field of clinical psychology. Goddard had helped him gain employment as the first psychologist in an institution for epileptics in Skillman, New Jersey, in 1910, but he soon resigned after repeatedly clashing with its medical superintendent. In 1912, Wallin became director of the Psychological Clinic at the University of Pittsburgh, and in 1914, director of the Psycho-Educational Clinic in St. Louis.[129]

Like Witmer, Wallin envisioned a broad clinical role for psychologists modeled on medical practice; consequently, he deplored the spread of testing to teachers. "The opinion seems to prevail," he lamented, "that *any person of intelligence is qualified to make mental diagnoses,* but that none but qualified physicians can make general physical or neurological diagnoses." Wallin was "quite certain that many diagnoses of teachers or nurses based purely upon the Binet tests will be very misleading, often humorously absurd, and at times pernicious." By 1913, he had become the most outspoken critic of "amateur" testers.[130]

"I was not fighting a man of straw," he later recalled. Responses to a 1913 questionnaire sent by Wallin to school boards confirmed both the widespread dissemination of testing and its use by those besides psychologists. Of 103 school systems responding, 72 percent offering any psychological testing used only Binet tests or puzzle boards. "Of 115 examiners or testers," Wallin wrote, "52 were special class teachers, 11 were principals or supervisors of special classes, four were superintendents, five were alienists, 22 were medical inspectors or physicians, eight were psychologists, and 13 were clinical psychologists." Excluding only those with psychiatric or psychological training, Wallin thus contemptuously classified 74 percent of these examiners as "mere 'Binet testers.'"[131]

Psychologist Guy Montrose Whipple noted similar problems in a 1912 editorial on "The Amateur and the Binet–Simon Tests." "It is fair to assume that the Binet scale will be used," he concluded, "but who will conduct the examinations?" Most schools had not yet seen "the imperative need of establishing psychological clinics or bureaus of child study, or even of employing a consulting psychologist." Consequently, tests fell to "medical inspectors, which is bad enough," or to "classroom teachers, which perhaps, is still worse." If improperly used, he warned, testing would become "a farce which can but bring discredit upon psychology and retard the movement for its application to educational practice." Yet Whipple knew that the supply of psychologists could not possibly meet the schools' demands. At a minimum, he proposed, examiners ought to receive "special drill in the handling of what is really an exceedingly delicate psychological instrument."[132]

Responding to such criticisms, Lewis Terman rose to the defense of teacher testers. There were dangers, he admitted, in taking too seriously the mental diagnoses made by "school nurses, teachers, and doctors"; still, he warned his contemporaries against an "unduly contemptuous attitude toward the efforts which are put forth by the average worker." More pointedly, Terman chided the sudden "expertise" of psychologists in a method barely a few years old. Clinical psychology was "still in an embryonic stage of development," he reminded psychologists, and "no one, however sagacious or well trained in psychological technique, can rightfully lay claim to any great degree of psycho-clinical expertness." The gap between "amateur" and "expert," Terman was suggesting, was far smaller than his professional colleagues were implying.[133]

This new controversy put Goddard in an especially awkward position. Of course he defended teacher testers, a good portion of whom he had trained personally. Yet even he straddled the issue in seeking a professional strategy that would allow teachers to use testing while encouraging schools to hire psychologists. Thus, while insisting that only well-trained persons should decide "borderline cases," Goddard still maintained that even novices could get valuable rough estimates.[134]

More significantly, while Witmer and Wallin pressed for the establishment of psychological clinics offering a wide range of diagnostic services, Goddard saw Binet testing as "of such remarkable accuracy that it supersedes everything else." The growing acceptance of testing as a diagnostic tool, both in institutions and in public schools, overshadowed the warnings of critics. "So rapidly has this conviction spread and so widely has it extended," he wrote in 1913, that "the criticisms that from time to time appear only arouse a smile and a feeling akin to that which the physician would have for one who might launch a tirade against the value of the clinical thermometer." Like the thermometer, Goddard believed, Binet testing could provide information useful to both professionals and amateurs alike.[135]

Meanwhile, Goddard continued disseminating intelligence tests to eager teachers in much the same manner that he had distributed eye charts through the Child-Study movement a decade earlier – he simply mailed them to anyone who requested them. By 1914, tests were in use in Altoona, Baltimore, Cambridge, Cleveland, Denver, Houston, Little Rock, Los Angeles, Minneapolis, New Orleans, Raleigh, Saginaw, Spokane, and dozens of other cities. By the time Lewis Terman published his 1916 Stanford–Binet Revision, which largely replaced Goddard's 1911 Revision, the little Training School at Vineland had disseminated 22,000 copies of the Binet–Simon scale.[136]

The consequences of this dissemination, moreover, could already be felt. In their conferences and journals, psychologists and educators were just beginning to explore Binet's new concept of "intelligence." They asked difficult questions about its meaning, its susceptibility to measurement, and its relationship to schoolwork. They wondered about the existence of multiple types of intelligence. They also questioned the very possibility of ever distinguishing "innate" from school or environmental learning. All of these issues would continue to haunt the testing movement in the decades that followed.

Neither educators nor psychologists had much time to ponder their answers, however, for intelligence testing as a social phenomenon had already started to effect a transformation – in school policies, in psychological practices, and in the relationship between the two. Also being transformed was the school meaning of mental "subnormality." In the new statistical language of the day, the subnormal schoolchild increasingly came to be defined not simply as one who failed in school, but as one whose test score placed him at the bottom of a bell curve – a curve devised for schoolteachers by psychologists. The criteria for special class admission had indeed begun to change.

Many of these changes could be directly traced to the small institutional school at Vineland. By 1914, Henry Herbert Goddard had used Binet's ideas to help draft the first special education legislation in the nation. He had challenged school grading policies in New Jersey and diagnostic practices in New York. He had inspired hundreds of teachers from all parts of the country, while also training them to administer intelligence tests. And he had designed the curricula for several universities that would train hundreds more. By promoting testing in public schools, Goddard had begun to institutionalize a new role for psychologists as diagnosticians of the normal and the subnormal – a role whose repercussions would be felt for the remainder of the century.

5

Causes and Consequences:
The Kallikak Family as Eugenic Parable

A New Biology and a New Sociology

The "problem of the feeble-minded," Henry Herbert Goddard had written in 1907, "is a psychological and educational problem." Seven years later, these problems had doubled. There were "four lines along which investigation must proceed – four problems to be solved," Goddard wrote in 1914, for the problems of the feebleminded were not only psychological and pedagogical, but biological and social as well. And each of these problems, he insisted, required research.[1]

Most striking in this expanding research agenda was Goddard's growing confidence in his own ability to ask and answer broad questions of importance to his field. In 1907, Goddard's incursions into what was then considered medical territory had been cautious, with their claims for psychological expertise grounded narrowly on pedagogical experience. Yet in the following years his pronouncements gradually grew bolder. Spurring him on were his dramatic successes in promoting special education and intelligence testing as answers to pedagogical and psychological questions. Within seven years, Goddard believed that he had discovered equally important answers to biological and social questions – answers that would once again leave their mark on American society.

By the time Goddard published an expanded version of his New York school survey as *School Training of Defective Children* in 1914, he had already written two new books which would make him world famous. Both explored the problems of the feebleminded from a biological and social standpoint. Both brought Goddard and Vineland immediate international acclaim.[2]

The most important of these, at least to Goddard, was *Feeble-Mindedness: Its Causes and Consequences,* a massive research project which documented the patient records and family histories of each of the more than three hundred cases at Vineland. Published in 1914, this work would constitute the most ambitious accomplishment of Goddard's career. Far more influential, however, was a much shorter 1912 study which chronicled a single family's history. Entitled *The Kallikak Family: A Study in the Heredity of Feeble-*

143

Mindedness, it followed the lives of a young Vineland woman, pseudony-mously named Deborah Kallikak, and her relatives. Forever after, Henry Herbert Goddard's name and reputation would be most closely connected to the fate of this single text – even more so, in fact, than to the fate of intelli-gence testing.[3]

To many psychologists, physicians, biologists, and sociologists who read it in 1912, *The Kallikak Family* seemed to suggest a major scientific break-through. This book's publication quickly catapulted Goddard into the most respected ranks, both nationally and internationally, of scientists studying mental deficiency. Of equal importance was its impact on the public, for God-dard also wrote this book with the general reader in mind. Over the next three decades, *The Kallikak Family* would be reprinted eleven more times; its story would also be told and retold in scientific textbooks, court cases, politi-cal speeches, public exhibits, and popular magazines.

As the century progressed, however, Goddard's *Kallikak* monograph would also inspire a growing body of criticism. To generations of historians who would read it decades later, this study's shortcomings would be all too apparent, while its earlier acceptance as a work of science would prove both puzzling and troubling. Such acceptance could only be explained, they would argue, by emphasizing the political and social prejudices of the period.[4]

At the time they were written, however, Goddard's theories on the causes and consequences of feeblemindedness generated a very different reception from the scientific community. Both *The Kallikak Family* and *Feeble-Mindedness* suggested major advances in the field of psycho-asthenics, his contemporaries believed, for both seemed promising works of research which used new methods to find answers to important questions of the day.

Central to both books was the most important question of all, a question that had perplexed physicians, reformers, and the public for centuries: what caused feeblemindedness? Since many of the Vineland children had been ill since birth, this question suggested an even broader one: what caused birth defects? How, Goddard asked, could a scientist explain the debilitating men-tal conditions afflicting feebleminded children? What was their cause?

According to Goddard's medical predecessors, these questions could usu-ally be answered with two words, which encompassed just about all they knew: "poor inheritance." Yet what exactly did this phrase mean? Just what did these children inherit – from their parents, their family, or their society? And how exactly did this process work? Beginning in 1909, Goddard turned his scientific attention toward answering these crucial questions. By 1912, he believed he had the answers in hand.

In his immensely successful monograph, Goddard captured both scientific and popular ideas about the larger meaning of heredity in a moment of tran-sition. Specifically, he was able to graft new biological theories about how heredity worked onto an older set of medical beliefs that still permeated psycho-asthenics – beliefs that emphasized the interconnections between medical and social "pathology." In the process, his research linked the new biology to an older sociology. Despite its deceptively simple framework, Goddard's *Kallikak* study actually contained a complex blend of new and old which his contemporaries found compelling. Such a mixture can only be understood, however, by examining the ways that this generation addressed an issue crucial to the study of mental deficiency: the causes of defective births.

Sins of the Fathers

In many ways, ideas about the causes of birth defects at the time that Goddard began his Vineland research still reflected the profound influence of an older set of Christian beliefs. The most widely held of these explained childhood infirmities as punishments for parental sins. After all, the Bible, many Americans believed, was explicit on this point. The "iniquity of the fathers," it warned, would be visited "upon the children unto the third and fourth generation." [5]

Goddard's nineteenth-century predecessors studying the feebleminded had rejected interpretations of this biblical passage that suggested divine intervention in human affairs. Even so, their understanding of the "lawfulness" of natural phenomena still tended to reaffirm, rather than challenge, religious assumptions. Most physicians and reformers agreed with ministers, for instance, when they argued that many physical ailments could be avoided by adherence to Christian codes of conduct. Through such arguments, they tended to reconcile enlightened medical explanations with older religious injunctions in explaining children's afflictions.

This reconciliation was evident, for example, in the speeches of influential nineteenth-century reformers such as Samuel Gridley Howe. Idiocy was not a special Providential dispensation, Howe had explained; neither, however, was it an accident. Instead, it was "merely the result of a violation of natural laws," which, "if strictly observed for two or three generations, would totally remove from any family, however strongly predisposed to insanity or idiocy, all possibility of its recurrence." Such medical "chastisements" had been "sent by a loving Father to bring back his children to obedience to his beneficent laws." [6]

Like other natural laws, these had been shrouded in darkness for centuries, but clues to their discovery were "written upon every man's body." Man now knew that intermarriage of relatives, intemperance, attempts at abortion, and above all "self-abuse" could lead to tragic consequences. "Can there be so sad a sight on earth," Howe wondered, "as that of a parent looking upon a son deformed, or halt, or blind, or deaf" and knowing "that he himself is the author of the infirmity"? In obeying or defying God's laws, man chose to make his world a heavenly home or a living hell. "Talk about the dread of a material hell in the far-off future!" he exclaimed. "The fear of that can be nothing to the fear of plunging one's own child in the hell of passion *here.*"[7]

Similar warnings punctuate the writings of prominent nineteenth-century physicians such as Edouard Seguin. Addressing doctors, Seguin too hoped that research on the "intimate, even secret, even criminal, causes of idiocy" might make young couples "aware of the dangers incurred in their posterity by any breach of the laws of moral health and society." Such an awareness would prove useful, he argued, "particularly in spreading the dread of heredi- tary punishment set forth in the Bible."[8]

The connection between immoral causes and medical consequences affected more than young couples, however, for man was also married to society. "God has joined men together, and they cannot put themselves asun- der," Howe warned. Private immorality, in other words, had public conse- quences. It was the rapidly changing nature of society that was proving in- creasingly alarming both to reformers and physicians. By the 1840s, a wide range of European and American writers had begun to associate what they saw as the deteriorating conditions of urban, industrial life with a decline in private morality and a consequent rise in defective births. Theories connect- ing these social, moral, and medical "pathologies" would dominate discus- sions of feeblemindedness for the next hundred years.[9]

These discussions took place most frequently within organizations where physicians and reformers met to consider problems of public welfare. The most important was the Conference of Boards of Public Charities, a forum organized in 1874 to bring together persons active in state programs. Re- named the National Conference of Charities and Corrections (NCCC) in 1882 (and the National Conference of Social Work in 1917), these meetings drew personnel from public health departments, almshouses, and prisons, as well as medical superintendents from institutions for the insane and the feebleminded.[10]

In the last quarter of the nineteenth century, the NCCC grew in signifi- cance as a meetingplace for the exchange of ideas about managing institu- tional populations. Participants pressed as well for increased government involvement in identifying those requiring institutional care. The group's im- pact is apparent, for instance, in the special 1880 census report on "Defective, Dependent, and Delinquent Classes of the Population of the United States" prepared by prison expert Frederick Wines.[11]

According to physicians, "defectives, dependents, and delinquents" were not really distinct classes; instead, these categories overlapped. Feeblemind- edness, doctors argued, was simply one of many possible manifestations of a more general and frequently inherited physiological weakness – a weakness that left individuals susceptible to social failure. This weakness might appear in other forms as well, such as insanity, alcoholism, pauperism, criminality, or sexual promiscuity.

This theory came largely from the religious ideas of a French physician, Benedict Morel. According to Morel, both physical ailments and antisocial behavior were evidence of man's gradual "degeneration" from the perfect state of health that had been his "before the fall." In his highly influential 1857 tract, *Traité des dégénérescences physiques, intellectuelles et morales de l'espèce humaine,* Morel had insisted that a deteriorating environment could impair a man's ability to provide his children with a good inheritance, and could thus lead the species to degenerate – physically, intellectually, and morally – if social reforms were not implemented. By the end of the century, Morel's work had inspired a wide range of "degeneration" theories connecting feeblemindedness to other forms of medical and moral deviance.[12]

These diagnostic interconnections were obvious, for instance, in the writings of medical superintendents. "How many of your insane are really feebleminded or imbecile persons," Dr. Isaac Kerlin had asked the NCCC in 1884. How many "incorrigible boys" were really "moral idiots"? "And pauperism breeding other paupers," he wondered, "what is it but imbecility let free to do its mischief?" In "every possible manner," Kerlin warned in words that again suggested the secret sins of the fathers, such persons had "retaliated on their progenitors for their origin and on the community for their misapprehension."[13]

In the future, Kerlin noted, such misapprehension would end. The "correlation of idiocy, insanity, pauperism, and crime will be understood as it is not now," he predicted. "There will be fewer almshouses, but more workhouses," he prophesied. "Jails, criminal courts, and grog-shops will correspondingly decrease," to be replaced by larger institutions, or, as Kerlin called them, "'villages of the simple,' made up of the warped, twisted, and incorrigible, happily contributing to their own and the support of those more lowly." In the midst of an increasingly complex urban, industrial society, these institutions for the "simple" would become "'cities of refuge,' in truth; havens in which all shall live contentedly, because no longer misunderstood nor taxed with exactions beyond their mental or moral capacity."[14]

Such segregation would allow the society of the future to control more than behavior, Kerlin suggested; it would also stop such persons from passing their weakened physical and moral conditions on to any more progeny. They "'shall neither marry nor be given in marriage' in those havens dedicated to incompetency," Kerlin argued elliptically in using a biblical phrase that analogized such asexual "havens" to heaven itself.[15]

As Kerlin's utopian vision makes clear, regulating the reproduction of the feebleminded had become a central institutional concern. According to doctors, some sort of sexual control was desperately needed, for they saw degenerate births as increasing. This problem concerned social reformers as well, for according to their research, "defectives, dependents, and delinquents" were troublesome, socially expensive, and above all prolific.

The most frequently cited proof of this in Goddard's day was an 1877 study produced by Richard Dugdale. While studying prison reform in Ulster County, New York, Dugdale had discovered six convicts related to each other. Tracing this family's lineage back to their colonial ancestors, he uncovered hundreds of other prisoners, paupers, and "moral degenerates," including one he christened "Margaret, the Mother of Criminals." Most startling, Dugdale calculated that this single family, whom he called "The Jukes," had already cost the state of New York $1,308,000 in social services and lost revenues. The conflation of moral, social, and medical problems is apparent in Dugdale's title, *The Jukes: A Study in Crime, Pauperism, Disease and Heredity.*[16]

The involvement of heredity in these issues of crime, pauperism, and disease also caught the attention of this era's new "social scientists." Like the "new psychologists," specialists in the emerging disciplines of criminology, anthropology, and sociology grounded many of their new theories on Darwinian analogies. The transmission of traits from one generation to the next, they argued, was as critical for the survival of a society as for a species.

While physicians saw a weakened physiological inheritance as linking defectives to dependents and delinquents, social scientists drew similar conclusions from different premises. Social change, they argued, paralleled biological change, for in both processes, life evolved from the simple to the complex. Social aberrations of various sorts might thus be symptoms of hereditary "atavism" – the behavior of persons whose simple inheritance had left them unable to adapt to their complex environment. In their evolutionary analogies, the feebleminded, as well as the insane, paupers, and criminals, resembled both primitives and children.[17]

By the end of the century, both physicians and social scientists were losing faith in their abilities to counter defective inheritance by improving the environment. Society, they proposed, should instead begin to protect itself scientifically by learning to recognize the outward signs of medical and moral degeneration. The most popular of these theories came from Italian criminal anthropologist Cesare Lombroso, who claimed that criminal behavior could be correlated with distinct physiological stigmata – canine teeth and birdlike noses in the case of murderers, for instance.[18]

To a psychologist like G. Stanley Hall, all of these theories were suggestive. The new psychology, he argued, citing both Morel and Lombroso, ought to include a subfield which Hall labeled "studies of decadents." This would include the study of "idiots, paupers, tramps, blind, deaf, and other defective classes, and even monstrosities." Man's infirmities, both medical and moral, could now be explained within an evolutionary paradigm. "All human degeneracies, whether individual or inherited," Hall concluded in 1894, "are being substituted, in the world's great algebra of morals, for the almost unknown symbol, sin." At the same time, "the word 'health' is again approaching its old and larger Biblical sense of holiness."[19]

In the long run, the most influential of the social theorists studying heredity proved to be Francis Galton, a cousin of Charles Darwin. In unequivocal language, Galton pressed the Darwinian revolution to its logical conclusion: if man truly belonged within the animal kingdom, then selective breeding could improve this species as much as any other. Scientists, Galton argued, ought to eschew moral pronouncements and instead breed a better man. Coining a new word from a Greek root meaning "good in birth," in 1883 he christened his new science "eugenics."[20]

Whereas Hall's writings continually reconciled Darwinism with Christianity, Galton's contained more than a whiff of religious skepticism. "I have no patience," he admitted, "with the hypothesis occasionally expressed, and often implied, especially in tales written to teach children to be good, that babies are born pretty much alike, and that the sole agencies in creating differences between boy and boy, and man and man, are steady application and moral effort." Man differed from man from the moment of birth, Galton insisted. Eugenics, he argued, would literally do away with the need for religious admonitions by scientifically breeding a physically, intellectually, and morally superior species.[21]

By the century's end, the development of simple new surgical techniques brought Galton's eugenic vision more firmly within the realm of medical possibility. Cures for feeblemindedness, or for numerous other conditions believed to be hereditary, appeared as distant as ever; "prevention," however, now seemed a possibility, for doctors had apparently developed safe new methods for sterilizing persons thought to be carrying a degenerate inheritance.[22]

The social potential of these new procedures caught the attention of institutional physicians frustrated by their own failure to check what they saw as the rising tide of medical and moral deterioration. Goddard's contemporary, Dr. Martin Barr, was especially clear on this point, which he explained in a botanical metaphor: "We must tap the main root if we would destroy the evil," he proposed, by "the free – I speak literally not metaphysically – the free use of the pruning knife."[23]

In the decades that followed, the consequences of allowing doctors to "tap the main root" provoked fierce debate. By the time Goddard wrote his monograph on the Kallikaks, sterilization had won grudging acceptance as a legal option in eight states, including New Jersey. Meanwhile, questions concerning the role of heredity as a cause of medical and social failure continued to dominate institutional discourse.[24]

Ironically, the terms of this discourse have been obscured by their very ubiquity. The writings of nineteenth-century physicians and reformers abound with references to "heredity" and "environment" – the same terms which came to shape scientific debate during the twentieth century. Nonetheless, when doctors used these words in 1900, they meant something subtly different than what they came to mean only a few decades later. At the time Goddard began researching the causes and consequences of feeblemindedness, the concept of "heredity" was itself on the verge of a major redefinition.

The Medical Meaning of Heredity

In discussing the causes of feeblemindedness, Goddard's medical colleagues shared a common understanding about the processes controlling inheritance. When these men spoke of "heredity," they did not necessarily mean to exclude "environment." Nor did their frequent references to "nature" ignore defects in "nurture." To the contrary, for most physicians, nature and nurture remained interconnected – as interconnected as medicine and morality.

Fusing them together were doctors' assumptions about just how heredity worked. Despite biologist August Weissmann's controversial proofs in the 1880s that acquired traits could not be inherited, most physicians still believed the contrary. Above all, they found it difficult to accept the idea that environmental factors could not significantly affect pregnant women and their offspring. Instead of embracing "Weissmannism," doctors in the early twentieth century still clung to an older set of ideas propounded by French biologist Jean Baptiste Lamarck, who had claimed that an organism's environment could alter its inheritance.[25]

These Lamarckian legacies are obvious in medical literature of Goddard's day. Dr. Barr, for instance, repeatedly blamed mental deficiency on poor inheritance; his understanding of heredity, however, included many elements that would later be considered environmental. "Poverty, hard work, not infrequent intemperance, and many anxieties," Barr explained, might reduce an expectant mother "to a state of quasi-imbecility." If she also suffered from "that exhausted vitality from a child-life in the factories, of which so much was heard in England," or from the "excitement of competition" common in America, such influences might create a condition "so abnormal as to constitute direct transmission by her to offspring of weakness – mental, moral or physical." These weaknesses led to "a lowering of moral tone," Barr explained, "as indicated by indulgence in petty vices, irresponsibility or consequent inability to attain success in life." Such constitutions were "almost sure to develop idiocy or imbecility in offspring."[26]

When these doctors spoke of "hereditary" or "congenital" factors, they meant factors affecting parents not only prior to conception, but also during conception and during pregnancy. Poor heredity, for instance, was the price paid by an Elwyn child "conceived in a debauch." Barr knew of another boy "born a veritable Esau, with a thick growth of reddish hair on his back and chest" after his pregnant mother had been "chased by a cow." In fact, in the same article recommending the surgical use of the "pruning knife" to stop hereditary degeneration, Barr also warned expectant mothers to avoid French naturalist novels, for they were filled with "frightful moral monstrosities" that might take their toll on the next generation. Such prenatal influences made the mother's hereditary responsibility much more critical than the father's.[27]

Moreover, when physicians spoke of hereditary illnesses, they did not mean the inheritance of a specific disease. Mental illnesses, doctors argued, did not resemble infectious diseases; instead, they were "chronic" or "constitutional" conditions for which one inherited a predisposition or "prepotency" that might or might not be realized. Doctors often diagnosed patients with "neuropathic" constitutions – those predisposed toward nervous or mental disorders. "It must be recognized," Barr explained, "that it is not necessarily a *specific* neurosis that is transmitted"; instead, one inherited a "disordered arrangement of nerve tissue" that might present itself in many different forms "according to the degree of prepotency in the mingling of the parental elements." Thus, "an insane parent may bear an epileptic child, or an epileptic parent a child who is a profound idiot."[28]

This broad conception of hereditary causation pervaded all forms of medical writing in Goddard's day, from case studies to medical statistics. A Berlin study of 1893, for instance, had listed the most important cause of mental defects as "neuroses." J. Langdon Down, reporting on 2,000 cases, had found 45 percent caused by "neurosis in parents." Barr, studying 1,044 idiots, attributed 38 percent to "hereditary insanity or imbecility," and 57 percent to "other neuroses."[29]

Both Goddard and Johnstone used similar language in explaining the Vineland cases. "One of the most prolific sources of mental dullness or deficiency," Johnstone wrote in 1907, was "parentage in the great Neuropathic family." Within this family, Johnstone included "the idiot, the imbecile and the feeble-minded; the insane, epileptic, and paralytic; some of the consumptives, blind and deaf; some kleptomaniacs and recidivists or habitual criminals, sexual perverts and prostitutes, tramps and paupers, chronic inebriates and petty criminals." While children produced by such persons "do not necessarily have the defects of their parents," he reported, they often developed "some other of the forms mentioned above," thus producing "a great circle in which we find cause degeneracy, result degeneracy."[30]

Johnstone too incorporated environmental influences into the very meaning of weak inheritance. "The old war between heredity and environment is still on," he wrote, "and the questions as to whether this condition is caused by nature or poor nurture have not been satisfactorily settled." For Johnstone, the most important causes "belong to both classes." Thus, not only frights, accidents, or excessive worry but even the "general characteristics of the family as to ability to get along, thrift, etc.," might disturb the pregnant woman and "leave their mark upon the child." Indeed, Johnstone believed that "any characteristics in the family may be transmitted, the bad with the good and the former often intensified in the transit." He blamed many cases of mental dullness on "improper food and poor environment, particularly the constant living and sleeping in unsanitary surroundings, where even the air is unfit for adults to breathe." Even school stress could lead to "the weakening of the stock." Environmental disturbances thus became an intrinsic part of what a child inherited.[31]

Such an open-ended conceptualization of the meaning of heredity offered Goddard's contemporaries a wide latitude in accounting for otherwise inexplicable conditions. Heredity, their writings suggested, was a "force" of nature – something perhaps analogous to gravity. It could be weak or strong, potent or deficient. Doctors as yet had no mechanism for explaining exactly how or when such a force might strike; even so, they insisted that it operated according to natural law. This law, however, was vague, idiosyncratic, and essentially untestable. Thus, while Barr concluded that "heredity is law – a law verified by accumulated evidence gathered in every department of science that treats of organic life," all he could offer were "examples of the force of this law."[32]

It was this force that Goddard too hoped to study experimentally. Although he shared his medical colleagues' sense that the feebleminded probably contributed more than their share of society's problems, and that, if left untrained, they might become a social "menace," Goddard was neither a "hereditarian" nor a "eugenicist" in any sense at the time he began working at Vineland. Neither his dissertation nor his Child-Study writings showed any interest in the doctrines of Morel, Galton, Lombroso, or other hereditarian thinkers of his day.

To the contrary, what had attracted Goddard to the Vineland school had been its pedagogy. His earliest addresses express a striking optimism about environmentally induced treatments and a criticism of the pessimism of his medical colleagues. The feebleminded, Goddard had suggested in 1907, might simply be suffering from poorly functioning neurones. It was "yet to be proved," he had insisted, "that the improvement of such an impaired neurone is impossible."[33]

Nevertheless, after spending several years with institutionalized children, Goddard's hopes for their improvement became far more circumscribed, while his interest in the causes of their condition grew. By 1909 he believed he had found a strategy for studying this problem scientifically: his research would be based upon a new set of scientific "laws" – laws elaborated not by doctors, but by practitioners in the newly emerging field of experimental biology. In following these biologists, Goddard's work would incorporate a fundamental redefinition of the meaning of heredity.[34]

The Biological Meaning of Heredity

Goddard's new method for studying heredity emerged as the result of a very short letter mailed to Superintendent Johnstone in March 1909. "Dear Sir," it stated in its entirety,

> Have you at your Institution heredity data concerning feeble mindedness? If you have published any such data in your reports, I shall be glad to receive such.

The letter was signed by experimental biologist Charles Benedict Davenport. In answering it, Goddard began a correspondence that would alter the rest of his career as profoundly as had his discovery of Binet's tests.[35]

What Goddard could tell Davenport in 1909 was not very promising. The data on heredity, he explained, were "meagre and so unreliable that we have not considered it worth while to make any tabulated study of these records." In attempting to accumulate evidence on cause, Goddard faced a serious problem, for information supplied by parents was so unreliable that it was hardly worth collecting.[36]

This unreliability resulted largely from parents' genuine ignorance about the causes of their children's ailments. It was compounded, however, by a reluctance to divulge any personal information that might suggest what they and their doctors often suspected was the real cause – secret sin. In Wines' census study, for instance, parents had attributed their children's mental handicaps not only to diseases but also to snakebite, lightning, overeating, loss of property, homesickness, and frozen feet, among other causes. Such results had led one physician to distinguish "cause" from "*ascribed* cause" in handling patient records.[37]

Getting complete family histories, Johnstone lamented, was "often almost impossible." Goddard quickly learned that the data he could collect were doubly questionable in light of the motives of those supplying them, for parents were more than willing to report whatever they thought might aid their child's chance of being admitted to the institution, or to suppress whatever might hinder it.[38]

Even so, Goddard told Davenport, he had conceived of a new means of circumventing some of these problems. First of all, information gathering had to be separated from admissions. Instead, he had begun sending an "after-admission form" to families and physicians. In addition, he also wrote to "parties who know the family in question and are not too intimate to tell the truth about them." Thus, Goddard too apparently suspected that the real causes of mental deficiency were being suppressed, and that medical infirmity was linked to family secrets.[39]

These letters to acquaintances were "yielding fruit slowly." He could already cite one promising case of information supplied by a "correspondent." It concerned a Vineland child born in an almshouse to an unwed mother who had since remarried several times. Goddard had "not yet gotten back to the grandparents," he told Davenport. Within three years, he would trace back six generations in the family history of this child, who would eventually become known to the world by her pseudonym, Deborah Kallikak.[40]

In 1909, however, a shortage of funds was hampering his research. A year earlier, Goddard had applied to the Carnegie Institution for $25,000 to fund his laboratory. Among other uses, he had hoped to pay the salaries of assistants who would "collect data on heredity." Full patient histories could only be obtained, he had argued, "by the visit of an expert to the home and neighborhood." His fund-raising had thus far been unsuccessful, but as soon as possible, Goddard told Davenport, he hoped to send researchers "on the road" to "quietly and tactfully draw out all the information that it is possible to get." Yet this lay in the future. Apologizing for his "long statement, for which you did not ask," Goddard added that he intended to study this problem soon "on rather extensive grounds."[41]

"I can hardly express my enthusiasm," was Davenport's immediate reply. He was delighted to find someone interested in doing "extensive work into the pedigree of feeble minded children" and excited about its scientific promise. A month later, in April 1909, Davenport visited Vineland and met Goddard for the first time.[42]

Surprisingly, these two men had much in common. Both had been born in 1866 into pious New England families whose roots could be traced to colonial times. Even more significant than their analogous "pedigrees," however, were their parallel professional histories. The academic credentials of these two scientists were certainly not comparable, for while Goddard had found employment at West Chester State Normal School, Davenport had been an outstanding researcher at Harvard and the University of Chicago. Yet despite the gap in their scientific reputations, by 1909 Davenport and Goddard were in oddly comparable positions, for each had taken the unusual step of abandoning teaching to forge a full-time career in experimental research. Davenport had left the University of Chicago to direct the Station for Experimental Evolution at Cold Spring Harbor, New York, a laboratory funded by the Carnegie Institution, two years before Goddard left West Chester for his much humbler laboratory at Vineland. These two nonacademic researchers quickly developed a supportive working collaboration.[43]

What Goddard wanted most from Davenport was scientific information. Above all, he needed a "clear account of Mendelian theory." Since 1900, scientific journals had been filled with references to the rediscovery of the experiments of Austrian monk Gregor Mendel, first published in 1866. Davenport had been the first American biologist to recognize their significance. He recommended that Goddard read *Mendelism,* by British biologist R. C. Punnett. Through such works, Goddard came to understand heredity as biologists had begun to conceptualize it. In his new writings, heredity soon came to mean something very different from what it meant to his medical contemporaries.[44]

This new conceptualization owed its origins largely to Francis Galton, who had been experimenting for decades with ways of measuring inheritance. Toward this end, Galton advocated a definition of heredity that pointedly contradicted most of the major premises guiding medical thought. First and foremost, he unequivocally rejected the inheritance of acquired characteristics. When Galton spoke of heredity, or "nature," he meant only tendencies parents had themselves received at birth; all else he called "nurture." Thus, Galton advocated a much sharper division between heredity and environment than did most physicians.[45]

In deeper ways as well, Galton radically altered the meaning of heredity. To Galton, heredity was not a vague "force" that might or might not strike; it was a relationship between one generation and the next. Most intriguing to Galton was the possibility of measuring exactly what had been transmitted from parent to child. In his own research, he concentrated on studying the inheritance of discrete, measurable characteristics, such as height.[46]

Galton's science also placed far less significance on the role of the mother, for in devising his own innovative mathematical formulas to compare traits in parents and children, he treated the influences of both sexes equally. This was implicit in the asexual words he used in speaking of family relationships – words like "sibling" to refer to either brothers or sisters, or "mid-parent" to describe a statistical composite averaging the traits of fathers and mothers.[47]

Galton's ideas had proven crucial in helping to establish a working agenda for experimental biologists. Despite scientific ignorance concerning the agents of hereditary transmission – for as British biologist William Bateson conceded in 1900, "no one has the remotest idea how to set to work on that part of the problem" – Galton's formulas for measuring relationships between generations had at least shown biologists how to study the "outward facts of transmission." Biologists, Bateson concluded, were "beginning to see how we ought to go to work."[48]

For Bateson and his followers like Punnett, however, Mendel's experiments soon offered an even better model of how to work. Mendel, they believed, had actually discovered the "laws of hereditary transmission." To biologists such as Davenport, these laws seemed almost as inspirational as Darwin's work had been to an earlier generation.[49]

Mendel's laws, as Davenport explained them, could be reduced to three basic principles. The first stated that qualities appearing in an organism could be analyzed into distinct "unit characters" inherited independently. The second proposed that there was "a molecule or associated groups of molecules called a *determiner*" transmitted in the germ plasm, and that these were "the only things that are truly inherited." The third principle explained that there were two determiners for each unit character. These determiners for each trait were then segregated from each other in the formation of "gametes" (sperm or eggs). Through the process of sexual union, new offspring received one determiner from each parent for each unit character.[50]

Each of a parent's two determiners, Davenport stated, had a random chance of being transmitted to each offspring. Yet the resulting characteristics were not due to blending; instead, the "dominant" determiners would consistently prevail over the "recessive" ones. Davenport noted that Mendel had even predicted the frequencies with which certain traits would be seen. For instance, in cases where both parents had themselves inherited two recessive determiners, the recessive trait would appear in all offspring; if each parent had one dominant and one recessive determiner, the recessive trait would be visible in about one of four offspring. Such findings were startling, for they suggested that the laws governing heredity were both understandable and predictable.

Goddard learned about these new biological developments, as Davenport suggested, through Punnett's writings. Like Galton, Punnett too strictly separated nature from nurture, for his book sharply distinguished biological from social inheritance. Man "stores knowledge as a bee stores honey or a squirrel stores nuts," he conceded, but this was not heredity. Progress, Punnett wrote, meant that the "hoard has been improved, and is of more service to man in his attempts to control the surroundings." Yet such hoarding had "nothing more to do with heredity in the biological sense than has the handing on from parent to offspring of a picture, or a title, or a pair of boots." Heredity meant only "the transmission of something intrinsic" from "gamete" to "zygote," or, in Goddard's words, from the germ plasm to the individual. The "participation of the gamete" thus became for Punnett the defining "criterion of what is and what is not heredity."[51]

Goddard found Punnett's book "as fascinating as a novel and as instructive as true science." Most exciting of all, Punnett believed that these laws could be applied to man. Davenport had already begun using Mendelian concepts to explain the inheritance of human traits like hair color. In the process, he had also come to embrace Galton's eugenic dream of improving the human race by selectively breeding out harmful traits and breeding in desirable ones. Davenport's eugenic studies, however, had been seriously limited by a dearth of material documenting human "pedigrees."[52]

After reading Punnett and Davenport, Goddard too became convinced that Mendelian biology would ultimately supply new answers to long-standing questions about the causes of inherited conditions, among them mental deficiency. The next step for scientists, he believed, was to gather the facts in support of Mendel's laws. Over the next few years, he tried to organize his own heredity studies the way a biologist would.

But how would a Mendelian proceed? Clearly the first step was to record the appearance of specific traits over several generations. Davenport assumed that feeblemindedness was a recessive trait that would either appear or not appear in the ratios prescribed by Mendel. It was especially desirable, he wrote Goddard in 1909, to "collect facts supporting the conclusion that the offspring of two imbecile parents are all imbecile."[53]

Yet recording "imbecility" as a biological "trait" again raised the critical question of classification. By 1909, Goddard had already begun classifying institutional cases by Binet testing – a method that stressed degrees of intelligence according to relative "mental ages" along a continuum. Mendel, on the other hand, had preferred to work with variables that "permit of a sharp and certain separation" – variables that could thus be clearly dichotomized as either "dominant" or "recessive." Fusing Binet with Mendel would pose problems.[54]

These difficulties in classification were not peculiar to the traits Goddard was studying; they played an equally central role in a bitter controversy that had divided British scientists studying heredity into two opposing camps. On one side were the followers of Galton and statistician Karl Pearson, who had remained skeptical of Mendel's laws and who preferred to study traits whose distribution could be plotted along a bell curve – traits like height, for instance. The central role of statistics in their studies led this group to be christened "biometricians." Opposing them were the "Mendelians," led by Bateson and including Punnett, who, following Mendel, preferred to divide traits into discrete categories. Thus, while Galton measured the actual height of peas, Mendel, in Goddard's words, had classified "a quality like tallness, as contrasted with dwarfness."[55]

Even better than studying traits like height, Mendelians believed, were traits like color, for Mendel's peas were classified as either yellow or green. Yet even this classification still posed problems, the biometricians argued, for peas often shaded gradually from yellow to green, and many were "greenish yellow."[56]

Bateson's response to such charges is significant for what it suggests about the work that Goddard was about to undertake. This problem of classification, Bateson argued, had arisen not because of Mendelian theory but because neither Galton nor Pearson had been "trained in the profession of the naturalist." Naturalists, the Mendelians contended, would properly see traits that might be missed by those trained only in statistics. Classification required "experience," for statistics alone, in Bateson's opinion, could never "substitute for the common sieve of a trained judgment." In fact, "nothing but minute analysis of the facts by an observer thoroughly conversant with the particular plant or animal, its habits and properties, checked by the test of crucial experiment," Bateson insisted, "can disentangle the truth."[57]

Experience was equally relevant to the kind of classification that Goddard now had in mind for studying feeblemindedness. Goddard's psychometric definition of intelligence, based on Binet testing, should have left him as skeptical as the biometricians about the relevance of applying Mendelian ratios to the "trait" he was examining; after all, his own graphs presented "intelligence" as a variable distributed along a bell curve – a curve that very much resembled Galton's curves for height. Nevertheless, in following the Mendelians, Goddard now began to conceptualize intelligence less as a score along a continuum and more as the presence or absence of a single trait. In order to use Mendelism to explain heredity, Goddard had to ignore Binet's carefully measured distinctions between idiots, imbeciles, and morons, and instead simply dichotomize intelligence as either "normal" or "feeble."

Goddard's adoption of this dichotomous model is apparent, for in discussing causes, he left his bell curves behind. Instead, on his new "heredity charts," developed with Davenport's assistance, intelligence literally was transformed into a color – a white symbol for normal intelligence, or a black symbol for feeblemindedness. Mendelism, as Goddard understood it, permitted no grays.[58]

Classifying mental traits thus became a critical part of Goddard's new biology. If, as Bateson suggested, a skilled naturalist could reliably classify plants or animals based on experience, why couldn't the expert studying human mentality do the same? The new research into heredity would require the services of experienced human "naturalists." Such persons would have to classify the human animal in his natural surroundings – that is, man within society. Between them, Goddard and Davenport carefully defined this new profession, which they called the eugenic "field worker."

Field Work

By the end of 1909, Goddard had gathered enough funds from Vineland's contributors to hire his first heredity field worker. Davenport, having solicited far more funding from his own wealthy benefactors, opened a "Eugenics Record Office" in 1910 to coordinate the collection of human "pedigree" information. By 1911, training field workers had become a central part of Davenport's program for introducing eugenic science into American society.[59]

According to his plan, field workers would spend a summer at Cold Spring Harbor studying Mendelian science. They would then spend several weeks at institutions like Vineland gaining crucial experience. This combined training, both Davenport and Goddard believed, would produce human "naturalists" capable of classifying traits like mental deficiency when they encountered them "in the field."[60]

In many ways, the field workers produced by this training program greatly resembled their contemporary cousin, the social worker, whose professional role was also being defined in the same decade. Both positions involved information gathering through "friendly visiting"; both also depended upon "sympathetic and confidential relations" between worker and client. Perhaps because of these qualifications, as well their low pay, both jobs largely fell to idealistic middle-class women.[61]

Despite their similarities, these two groups of professionalizing women were working toward different objectives, Davenport emphasized. Social workers were gathering information to alleviate environmental hardships. Field workers, by contrast, were struggling to "unravel the laws of inheritance"; their task, he asserted, was to "work out the gametic nature of each individual studied." To do this, they needed to produce human "pedigrees" by tracing "collaterals, descendants and consorts of all individuals the make-up of whose germ plasm it is desired to understand." Although the field worker might try to leave her subjects "a little happier" than she found them, she was warned to be "neither a missionary nor a reformer – her sole business is to do a work of science, which in this particular case, is the appreciation of mental states."[62]

Of the women whom Goddard hired for such work, surely the most unusual was Elizabeth Kite, who came to Vineland in 1909 at the age of forty-five with a unique set of qualifications. Kite had been born into a prosperous Philadelphia family. Like Goddard, she had been raised by Quaker preachers and educated in boarding schools. She had begun traveling widely as a child accompanying her aunt on Quaker missions. As an adult, she had continued her travels, for Kite had been a Philadelphia school principal, a science teacher in San Diego, and a teacher of French and German in Nantucket; in between, she pursued university studies in Germany, France, Switzerland, and England. In 1905 Kite earned a teaching certificate, the Diplome d'Instruction, Primaire-Superieur, from the Sorbonne. She spent the next two years studying English history at the University of London. In 1906, Kite converted to Catholicism and became deeply involved in church activities. Returning to Philadelphia, she planned to open a "fresh air" retreat for city children in a home owned by family friend Samuel Fels. A Jewish philanthropist who had made his fortune manufacturing soap, Fels was also a regular contributor to the Vineland school and had agreed to help pay the salary of a field worker. It is likely that he recommended Kite for the new position.[63]

Kite certainly seemed well qualified to Goddard, for her education and worldly independence made her a most promising field worker. Her proficiency in French was a strong asset, for she could read Binet; by 1916, she had translated two volumes of Binet's articles, which the Vineland Laboratory published. Her historical training was equally valuable for tracing family pedigrees. Goddard soon came to appreciate another asset as well: Kite's dramatic writing style. This was evident in colorful letters she had published years earlier in Quaker magazines describing her travels through North Africa, where she had seen "indescribably picturesque beggars crouched in every corner," and experienced "mystery and calm so profound that those used to our madly rushing civilization can hardly hope in a few days or weeks even to begin to comprehend it."[64]

Such experiences, however, had hardly prepared Kite for the work she was about to undertake. Starting field work in 1910, she found herself in an environment more foreign to her than any she had yet visited – the poverty-stricken backwoods and industrial slums of New Jersey. It was this world that Kite would try to comprehend in a few weeks of "friendly visits."

Kite began by visiting the relatives of Vineland children. She found country people easy to interview, for she could drop by for a friendly exchange "without in any way betraying the real object of our visit." Since the Vineland school was highly regarded and parents glad to hear news of their children, most entertained her hospitably. "Spontaneous human sympathy brings out the very best that is in one," Kite noted, "so that even the defective has experienced the truth of the saying that it is more blessed to give than to receive." In asking questions about long-dead relatives, Kite sometimes suggested that she was a historian gathering information about the Revolutionary War era. By such subterfuges, she began recording her impressions of the mentality of the people she met, and of their ancestors.[65]

Of course, having read Binet, Kite knew well that this was not in any way the type of mental diagnoses that he had intended. Above all, Binet had insisted that objective standards be substituted for the subjective assessments of doctors and teachers. Yet intelligence testing was an extremely impractical means of gathering the information needed to confirm Mendelian ratios. Kite did test several children in local schools; however, she certainly had no pretext for testing their parents. It was "impossible to apply Binet's precise scientific scale in every case," Kite conceded. "Subjective appreciation of mental states," she admitted, "in spite of the fact that Binet rigorously opposes, and in season and out of season reiterates his antagonism to this method – does enter into the diagnosis of most cases." Kite was thus conceding a major point – that Goddard's new research methods were violating both the letter and the spirit of Binet's writings.[66]

She justified these violations with a simple explanation. "[W]e of the Training School, having had an insight into the problem guided by Binet's definite lines of demarcation," she argued, "feel that our diagnoses rest upon a comparatively solid basis of fact." Binet's ideas, Kite insisted, could be boiled to a minimum when mixed with experience. The "fundamental organ" of intelligence, she summarized by quoting Binet, was "the judgment, otherwise called, good sense, the practical sense, initiative, the faculty of adaptation. Judge well, understand well, here lies the essential force of the intelligence." If individuals showed good judgment, she asserted, they could never be misdiagnosed as feebleminded. "Armed with this central thought," Kite concluded, "the Vineland field worker goes forth."[67]

Still, Kite was going forth with what was "primarily a social test," not Binet's psychological test.[68] Furthermore, her task was complicated by the new requirements of Mendelian research. In order to decide if characteristics encountered "in the field" appeared in accordance with expected Mendelian ratios for recessive traits, Kite needed to determine whether the poor social judgments she observed were due to environmental or mental limitations. Was the field situation simply the result of "unfortunate environment," or was the individual she was interviewing mentally incapable of overcoming obstacles? The field worker, in other words, had to distinguish weak "nature" from destructive "nurture"; she had to determine which was cause and which consequence. In this research plan, Kite thus had to decide on the spot what Goddard was supposedly trying to prove – that mental conditions were the result of heredity. To Goddard, however, the real proof would lie with numbers – in this case with Mendelian ratios.

Separating causes from consequences was hard enough in determining the proper "color" (white for normal, black for feebleminded) to portray relatives on Goddard's new heredity charts. Mendelian research, however, also required knowledge of the "color" of past generations. How could field workers determine the mentality of relatives long dead? This apparently insurmountable problem had led contemporaries to question the entire enterprise. "Again, it has been asked, 'How can you decide that a subject who died fifty years ago was feeble-minded?'" Kite noted. "Surely you cannot give the Binet test to him!"[69]

The solution to this problem came, as it frequently did within the new science of eugenics, from the writings of Galton. Although he had never studied hereditary feeblemindedness, Galton had faced an analogous situation in assessing what he called "hereditary genius." The only way to gauge the intellectual abilities of those long dead, he had concluded, was by gauging their reputations. "Is reputation a fair test of natural ability?" Galton had asked. "It is the only one I can employ – am I justified in using it?" His answer was clearly yes, for he could see no reason for not "accepting high reputation as a very fair test of high ability." If reputation tested high ability, what about low ability? Was low reputation an equally fair test? To Goddard and Kite, the answer was an equally clear yes.[70]

Like Galton, Goddard knew that in basing his conclusions on reputation, he was following historical, and not psychological, procedures. "In determining the mental condition of people in the earlier generations (that is, as to whether they were feeble-minded or not)," he explained, "one proceeds in the same way as one does to determine the character of a Washington or a Lincoln or any other man of the past." Of course, unlike Washington or Lincoln, most of the people he was studying had left few written documents; many had even failed to record their marriages. In such cases, field workers had to depend upon the memories of the living to reconstruct the lives and judgments of the dead. By asking questions about the ways that past relatives had earned their living, raised their children, and conducted themselves socially, field workers could determine "with a high degree of accuracy," Goddard argued, "whether the individual was normal or otherwise."[71]

Mendelian science, however, encompassed more than merely gathering pedigree information; scientists also needed to test the laws of heredity through experimentation. Mendel had experimented with peas. Davenport bred canaries and chickens at Cold Spring Harbor. Goddard too hoped to conduct experiments that would elucidate the laws of heredity; yet he certainly could not "experiment" with the Vineland children in any such way.

Instead, he adapted another of Galton's strategies. To prove the power of nature and the "impotence of nurture," Galton had recommended studying unusual arrangements in which either nature or nurture had apparently been held constant. For instance, in seeking a situation in which environment remained constant while heredity varied, he had compared "sons of eminent men with the adopted sons of Popes." If social help mattered most, Galton hypothesized, then these adopted sons would "attain eminence as frequently, or nearly so as the sons of other eminent men." Galton's cavalier attitude toward gathering evidence to answer his question would also come to typify much eugenic research. "I do not profess to have worked up the kinships of the Italians with any especial care," he conceded, "but I have seen amply enough of them, to justify me in saying that the individuals whose advancement has been due to nepotism, are curiously undistinguished." Far more influential over the next century was Galton's suggestion for a natural experiment that would control for heredity while varying environment: studying identical twins who had been reared apart.[72]

Galton's search for unusual human situations obviously influenced Goddard, for he too began to look for "nature's experiments." While his contemporaries frequently claimed that the Jukes study had proven the hereditary basis of crime, pauperism, and promiscuity, Goddard disagreed. This study had shown a "startling array of criminals, paupers, and diseased persons, more or less related to each other," he observed. Yet the "most that one can say," Goddard reasoned, was that if such a family were left alone, it was likely "not to improve, but rather to propagate its own kind and fill the world with degenerates of one form or another." Although he strongly suspected that this degeneracy was hereditary, he conceded that Dugdale had not proven "the hereditary character of any of the crime, pauperism, or prostitution that was found."[73]

Other family studies touted by eugenicists, such as Albert Winship's 1900 study comparing the descendants of Puritan divine Jonathan Edwards with those of his Jukes contemporary, "Old Max," had similar limitations. Such a study did not prove the power of heredity, Goddard argued, for Winship could not show that "had the children of Edwards and the children of 'Old Max' changed place, the results would not have been such as to show that it was a question of environment and not of heredity."[74]

Goddard instead sought a study with a natural control group, like those proposed by Galton. He saw possibilities in the cases of persons producing children with several different spouses. One such case involved the mother of the Vineland child first brought to Goddard's attention by a neighbor's correspondence. By the end of 1909, Goddard had begun tracing this family's history, as well as those of other Vineland children, in more detail.[75]

The resulting studies suggest the degree to which eugenic scientists like Davenport altered the direction of Goddard's research. Instead of medical Lamarckism, Goddard now enthusiastically adopted the new biological Mendelism. His contacts with biologists had supplied him with a different working definition of what heredity meant, how it operated, and what scientific proof of its power would look like. All he needed now were facts gathered "in the field" to prove the relevance of Mendel's laws in determining the causes of feeblemindedness. Field workers like Kite, he firmly believed, would help make such research possible by blending biological understanding with psychological insight gained from institutional experience.

Publicizing Eugenics

Goddard was certainly not the only researcher fascinated by the Mendelian breakthrough in understanding heredity and intrigued by the potential of an interventionist biology. Davenport had been equally successful in wooing many of the most prominent medical, biological, and social scientists of his day to the eugenics cause. By the end of the decade, he had organized his diverse group of followers into an integrated institutional network.

He supervised this network through the auspices of the American Breed-

ers' Association, an organization founded in 1903 to bring university biologists and agricultural breeders together. Davenport's pathbreaking research
on breeding chickens made him a natural leader within such an organization;
when his interests shifted toward the study of human heredity, he was able to
move the Breeders' Association in the same direction. Perhaps nothing better
signified the intent of evolutionary biologists to place man squarely within
the animal world than the establishment of the ABA's "Committee on Eugenics," with Davenport as its secretary, in 1906. For the first time, human ailments, like those afflicting plants and animals, would be studied by an organization presided over by the Assistant U.S. Secretary of Agriculture.[76]

By 1909, Davenport had begun recruiting his own members to the ABA
and organizing them into a series of specialized research-oriented subcom-
mittees, each functioning under his leadership. By the year's end, Goddard
had become secretary of the ABA's new subcommittee studying feeblemindedness. Its members included Wistar Institute biologist H. H. Donaldson as
well as several powerful medical superintendents: Walter Fernald of Waverley;
J. C. Carson of Syracuse; J. M. Murdock of Polk, Pa.; and A. C. Rogers of
Faribault, the subcommittee's chairman. "How fascinating the idea of producing evolution of animal and vegetable life as desired!" Rogers exclaimed.
"To be able at will, not only to reforest the globe and restore the fauna of the
nearly extinct species, but to produce the individual characteristics desired."[77]

"Rogers is a fine fellow," Goddard wrote Davenport; yet he himself was
less sanguine about what this subcommittee could accomplish. "The men in
charge of institutions either never had any scientific interest or have had it
killed out by administrative labors," he reported. Most would "not be able to
do anything," he predicted, but they would "strongly encourage the fellow
who is willing to do the work."[78]

Goddard, however, would certainly not be working alone, for in following Davenport, he found himself in the company of the most prestigious
research-oriented scientists of his day. Among the enthusiastic new members
of the Committee on Eugenics, for instance, was Alexander Graham Bell,
whom Goddard would have known as the past president of the NEA's Special Education Division. Bell, whose wife was deaf, had spent several decades
studying the offspring of marriages between deaf and non-deaf persons; since
his results resembled those predicted by Mendelians, he avidly joined Davenport's subcommittee on the inheritance of deaf-mutism, and served as well
on the Board of Directors of the Eugenics Record Office. Goddard's former
mentor and role model, Dr. Adolf Meyer, who had since become head of
the Phipps Psychiatric Institute of Johns Hopkins University, chaired the
subcommittee studying hereditary factors in insanity. Experimental psychologists Robert Yerkes of Harvard and Edward Thorndike of Columbia Teachers College joined M.I.T. biologist Frederick Adams Woods on the subcom-

mittee studying the inheritance of mental traits. Psychiatrist William Healy joined University of Chicago sociologist Charles Henderson on the subcommittee investigating the hereditary basis of crime.[79]

Davenport hoped to spur his followers' efforts forward by encouraging them to publish their results quickly. In order to "hold the field," he wrote only months after his first visit to Vineland in 1909, he urged Goddard to present "a preliminary report concerning our knowledge of inheritance of feeble-mindedness" at the next Breeders' Association meeting. That December, heredity charts in hand, Goddard gave his first paper focusing on the causes of mental deficiency.[80]

Goddard's paper illustrates theories of heredity in a moment of transition, for his explanations suggest not only the newer ideas of Galton and Mendel, but also the lingering legacies of older theories. Still in evidence is language describing a general inherited weakness. The offspring of an alcoholic man, for instance, offered "what may be called a natural experiment of extreme suggestiveness," Goddard announced. With his first wife, who suffered from tuberculosis, this man had produced five normal children and six who had died young. With a second wife, whom Goddard labeled alcoholic and feebleminded, at least three of seven children were feebleminded. "It seems to be fairly clear in this case," Goddard argued in blending the medicine and the biology of his day, "that the father's alcoholism may have caused the physical weakness that led to so many early deaths in the first family, but the mother's defect has been directly transmitted in the second. . . . " Another of "nature's experiments" concerned an alcoholic man with defective fingers whose marriage to a "normal woman of good family" had produced normal children. Children from a second marriage to a "feebleminded and alcoholic woman," however, had defective minds as well as fingers. "Apparently the first wife was prepotent," Goddard hypothesized, while the second "was not prepotent in that she allowed him to transmit his physical defect, although she transmitted her mental condition." In describing other cases, however, Goddard adopted a newer language which treated feeblemindedness as a single recessive trait. One mother of a Vineland child, for instance, had herself been "one of three imbecile children born of two imbecile parents. The results here," Goddard explained, "could, of course, be nothing but defectives."[81]

Not all the charts proved feeblemindedness to be hereditary. Mongolism, Goddard argued, was "clearly not hereditary," since the charts of such children showed normal families (except for one relative reportedly "insane with religious mania"). Instead, Goddard concluded that this condition must be "an arrest of development resulting from some cause acting *in utero,* perhaps about the second month."[82]

Goddard's progress in tracing such families delighted Davenport. It evidently impressed Assistant Secretary of Agriculture William Hays, the secretary of the Breeders' Association, as well, for according to Davenport, Hays had found the paper "excellent" and wanted to publish it in the *American Breeders' Magazine.*[83]

This good news, however, raised new problems. While such publications might advance Davenport's interests, Vineland's Board of Directors regarded them with more suspicion. They had no qualms about Goddard's work in publicizing special education; publicizing data on "human matings," however, was another matter. The public, one board member told Goddard, was hardly ready for such sexual explicitness.[84]

Goddard had reservations of his own. If he published his findings on only a portion of Vineland families, those remaining to be interviewed might become wary of disclosing information, thus jeopardizing his research. Johnstone too was worried, for he was not willing to compromise his institution's good standing by violating patient confidentiality.[85]

Goddard tactfully broached these issues with Davenport. What would happen, he inquired, if his heredity charts were seen by "newspaper or magazine writers who would be very glad of an opportunity to exploit the matter?" Publicity might backfire. "We all know that there is a morbid interest on the part of the general public in all matters pertaining to sex and the sexual relations," he noted, and this research dealt with sexual relationships in a "very plain and definite way." Such stories might "'queer' the whole country, so to speak, in regard to this whole matter of Eugenics," Goddard worried. "So strong is this feeling of danger in the minds of some people," he continued, "that they think the publication of such a magazine is untimely, fearing that the public is not ready for such a discussion and presentation of the facts of human breeding."[86]

Davenport immediately assuaged Goddard's fears. While he personally shared Goddard's reticence on the subject of sex, he felt no qualms about promoting knowledge concerning the sexual transmission of biological weaknesses. *American Breeders' Magazine,* he replied, was hardly a public forum, for its readers were "almost entirely experts." Goddard's article would be in no more danger of being exploited by those seeking "a sensational story to sell" than would the *Vineland Annual Report.* Moreover, Goddard had so disguised his sources that they "might just as well be on the planet Mars."[87]

Davenport's response also addressed a deeper issue raised by publicity, for at stake, he believed, was the social role of the scientist. Like the physicians who had preceded him, Davenport saw no separation between biology and sociology. His own position was unequivocal: scientists ought to advocate policies to advance public welfare. "There is no use in collecting facts," he told Goddard, "unless they can be put before at least the scientific public" and thereby "form the basis of eventual action." Davenport pressed Goddard to publicize his findings so that "even the people may be informed as a means for creating a proper sentiment which would lead to proper legislation." It was "our duty as men of science," he wrote Goddard, to "steer our course perfectly directly." He urged Goddard to "be bold and pursue the even tenor of our way, despite the dangers which surround us."[88]

These arguments evidently reassured both Goddard and his institutional employers, for *American Breeders' Magazine* published his paper, with his charts, in 1910. More pointedly, Goddard even reprinted one of Davenport's addresses on "Fit and Unfit Matings" in the *Training School Bulletin*. And although it was written for physicians in language that laymen might have found difficult, this article still offered Mendelian advice for the first time to those of Vineland's supporters who might find themselves "drifting towards marriage."[89]

Despite Goddard's forebodings, Davenport's movement was beset less by exploitative opponents than by enthusiastic supporters. Goddard's warnings reveal the degree to which he had misjudged the temper of his own times, for the eugenics movement's willingness to deal with the "facts of human breeding" in a "very plain and definite way" came to be one of its most powerful appeals. For precisely this reason, eugenics quickly attracted the attention not only of biological and social scientists but also of feminists and sexual radicals such as Margaret Sanger and Havelock Ellis.[90]

In the years that followed, this movement also received support from individuals desperately seeking scientific answers to questions of birth defects. To parents who had already given birth to a handicapped child or a child who had died young, this new science seemed to offer their best hope for gaining crucial information. For the remainder of his career, Goddard would continue to supply such parents with what could only be called an early version of genetic counseling.

Davenport's stress on superior and inferior bloodlines drew to his side other supporters as well. Included were a wide spectrum of American nativists, racists, and imperialists who had long sought a sociology dividing the "fit" from the "unfit," and who hoped to use eugenic science to promote new legislation supporting their views. Since Davenport saw no need to separate the new biology from its social applications, he readily welcomed such supporters into his eugenics organization.[91]

Among the most influential of these would be Harry Laughlin, Davenport's Cold Spring Harbor associate who would help direct the Eugenics Record Office. In 1911, Laughlin became secretary of the ABA's "Committee to Study and Report on the Best Practical Means of Cutting off the Defective Germ-Plasm in the American Population." Although not a member, Goddard served as one of this committee's many expert advisors. And while the committee would ultimately recommend institutional segregation or sexual sterilization as its most effective strategies, Laughlin himself would also become a fierce campaigner for other solutions that emphasized racial or ethnic differences. Among those he advocated would be a new policy to restrict immigration by country of origin.[92]

Goddard had yet to show any interest in such a policy. Instead, his speeches grafted the new biology onto an older sociology espoused by physicians and reformers for more than half a century. Like his predecessors, he too stressed the connections between medical and moral deviance, and the rising public expenditures that were their consequence.

This continuity is evident in Goddard's 1910 speech presenting six heredity charts to the New Jersey State Conference of Charities and Correction. Although his research methods might be new, he conceded, his conclusions were not, for this audience had long known that "feeble-mindedness is at the root of probably two-thirds of the problems that you as a charity organization have before you," and the root cause of such problems was "defective ancestry."[93]

Once again, Goddard's charts mixed medical theories about the inheritance of a neuropathic condition manifesting itself in many forms with newer theories positing a single recessive trait following Mendelian laws. Thus, in one case, Goddard described an alcoholic woman and her tuberculous sister as showing "that those two girls inherited from their parents a weakness that has manifested itself in one way in one sister and the other way in the other." On another chart, however, he assumed that a Vineland child's nine siblings were all "undoubtedly feeble-minded," for the father had been an imbecile and the mother an idiot, and "two feeble-minded parents never have anything but feeble-minded children."[94]

Still intriguing Goddard was the chart of the child whom he had mentioned in his very first letter to Davenport. Born illegitimately in an almshouse, this girl was of the "high grade type which the ordinary person does not recognize"; because their condition usually went undetected, such persons often did "the most damage in the world." What now struck Goddard was society's shortsighted morality. He scorned the idea that "the worst thing in the world that can happen in a community is to have an illegitimate child born." "My friends," he told welfare workers, "it *is* rather disgraceful; but there are worse things than that. This is a worse thing: those people hunted up the father, and he was a drunken, epileptic, feeble-minded man. But they must be married . . . and they married that man and that woman and they had two children."[95]

Of course, the results proved disastrous, he claimed, for this woman and her next two husbands had left the state with at least eight feebleminded children to care for. Such problems had a long history, he lamented, for the "grandfather of the whole group was feeble-minded, and committed incest with his daughter." Worst of all were the prospects for the future, for except for the Vineland girl and her grandfather, who had been jailed, the rest of this family remained "at large."[96]

The only good news was that Miss Kite was already on this family's trail. Her reports "from the field" supplied Goddard with a steady stream of anecdotal evidence concerning the mentality of this girl's relatives, both alive and dead. Kite's stories chronicling this family's medical and moral failures shocked Goddard deeply; both for their own and for society's sake, he increasingly came to believe, such persons ought to be institutionalized. In response to these social tragedies, he now argued, something on a larger scale had to be attempted.

Over the next two years, Goddard's sense of the need for social activism increased, while his fears about publicity correspondingly evaporated. By 1912, he was ready to publish his own appeal to the public. In his first book, Goddard brought together all the influences he had been assimilating over the previous four years: Binet's measurements; Mendel's laws; Galton's calls for an experiment with natural controls; and Kite's reports from the field. The resulting monograph, entitled *The Kallikak Family: A Study in the Heredity of Feeble-Mindedness,* would prove more popular than anything else generated by the American eugenics movement.

The Kallikaks

Goddard prefaced his book by explaining the purpose of his Vineland laboratory. Since 1906, he told readers, he had been trying to determine the "different grades and types" of feebleminded children and to keep records on what they were capable of doing. He had also made "a definite start" toward determining the cause of this condition by sending field workers to "learn by careful and wise questioning the facts that could be obtained."[97]

"It was a great surprise to us to discover so much mental defect in the families of so many of these children," he reported. While readers might wonder how he had gotten "such definite data in regard to people who lived so long ago," Goddard reassured them that his judgments were sound.[98]

Addressing "the scientific reader," he took a different tone. Although he believed his information "accurate to a high degree," he admitted that his book made "rather dogmatic statements" and drew "conclusions that do not seem scientifically warranted from the data." Such statements, however, were for "the lay reader." "We would ask that the scientist reserve judgment," he pleaded, for his next book would offer fuller proof of these statements.[99]

The text that followed was obviously aimed at "lay readers," for it opened in storybook form. "One bright October day, fourteen years ago," Goddard recounted, "there came to the Training School at Vineland, a little eight-year-old girl." He told how the child had been born in an almshouse to an unwed mother, how her father had failed to support her, and how, because she "did not get along well at school and might possibly be feeble-minded," she had been sent to the Training School.[100]

Goddard then summarized the child's institutional records. "Deborah" had arrived in 1897, nearly a decade before Goddard's laboratory had opened. On her admission form, institutional personnel had recorded a wide range of traits relevant to their diagnosis:

> Average size and weight. No peculiarity in form or size of head. Staring expression. Jerking movement in walking. No bodily deformity. . . . Understands commands. Not very obedient. Knows a few letters. Cannot read nor count. . . . Power of memory poor. . . . Can use a needle. Can carry wood and fill a kettle. Can throw a ball, but cannot catch. . . . Excitable but not nervous. . . . Not affectionate and quite noisy. . . . Obstinate and destructive. Does not mind slapping and scolding. Grandmother somewhat deficient. Grandfather periodical drunkard and mentally deficient. Been to school. No results.[101]

In the following years, Deborah's teachers traced her progress. Her temperament, "impudent and growing worse" by 1899, improved sporadically. By 1912 this twenty-three-year-old excelled in needlework, wood carving, basketry, gardening, and music, and was helpful in handling children. Academically, however, she remained a poor reader, counted on her fingers, and had difficulty mastering abstractions. "She can put the right number of plates at the head of the table, if she knows the people who are to sit there," Goddard explained, "but at a table with precisely the same number of strangers, she fails in making the correct count."[102]

Goddard had tested Deborah on the Binet–Simon scale in 1910, when she was twenty-one years old, and found her to have "the mentality of a nine-year-old child with two points over." She failed tests that involved repeating digits, counting stamps or coins, making rhymes, rearranging words into sentences, and defining abstract terms. "The reader will see that Deborah's teachers have worked with her faithfully and carefully," he argued, "hoping for progress, even seeing it where at a later date it became evident that no real advance had been made."[103]

The most serious problem, however, was not Deborah's past but her future. "To-day if this young woman were to leave the Institution," Goddard asserted, "she would at once become a prey to the designs of evil men or evil women and would lead a life that would be vicious, immoral, and criminal, though because of her mentality she herself would not be responsible." The problem was thus not only Deborah's mentality, but also her sexuality. Instead of chastising Deborah, he posed "the ever insistent question," this time with "good hope of answering it." This question, Goddard wrote, "is, 'How do we account for this kind of individual?'"[104]

The answer was simple: "'Heredity,' – bad stock." The human family, Goddard explained, contained stocks that "breed as true as anything in plant or animal life." And whereas in the past, such a statement would have been "a guess, an hypothesis," he now had "what seems to us conclusive evidence of its truth."[105]

This proof came from field reports gathered by "women highly trained, of broad human experience, and interested in social problems." In Deborah's case, Kite had interviewed the concerned neighbor who told her about Deborah's mother; she had then traced other relatives across the state. "The surprise and horror of it all," Goddard recounted, "was that no matter where we traced them . . . an appalling amount of defectiveness was everywhere found."[106]

This story also contained a mystery, for Kite occasionally found herself "in the midst of a good family of the same name." Both families, apparently unrelated, traced their ancestors back to eighteenth-century forebears named "Martin Kallikak." The link between these two families was finally discovered, Goddard announced dramatically, when a relative "revealed in a burst of confidence the situation." This relative told Kite that Martin Jr. "had a *half-brother* Frederick, – and that Martin never had an own brother 'because,' as she now naively expressed it, 'you see, his mother had him before she was married.'"[107]

With this crucial bit of information, the mystery was solved, for Kite was able to reconstruct the tale of Martin Jr.'s illegitimate parentage. Martin Sr. had been only fifteen when his own father had died, she explained, "leaving him without parental care or oversight." After joining a Revolutionary War militia, the boy visited a tavern and met "a feeble-minded girl by whom he became the father of a feeble-minded son." Named Martin Kallikak, Jr., by his mother, this illegitimate child "handed down to posterity the father's name and the mother's mental capacity."[108]

Kite's task had then been to assess the mentality of all 480 descendants linking Martin Kallikak, Jr., to his institutionalized great-great-granddaughter, Deborah. These assessments, Goddard assured readers, had not been made in haste. "Oftentimes a second, a third, a fifth, or a sixth visit has been necessary," he explained, in order to induce these families "gradually to relate things which they otherwise had not recalled or did not care to tell."[109]

Of these 480 relatives, Goddard labeled 291, or more than 60 percent, as "undetermined" – a term used when "we could not decide." Although these were frequently "people we can scarcely recognize as normal," for they were not "good members of society," Goddard still lacked the proof necessary to

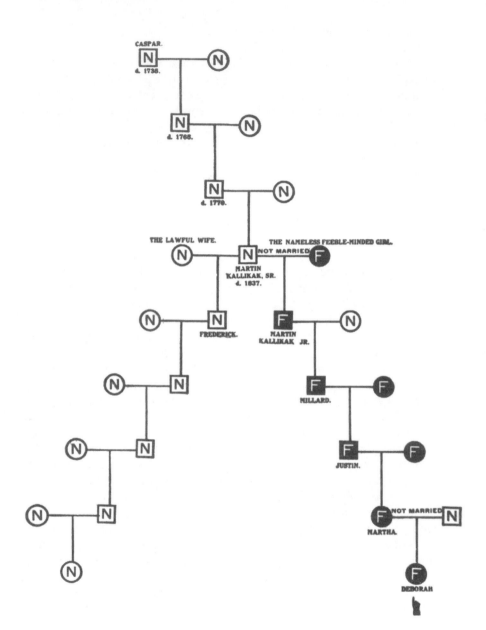

A heredity chart showing the line of descent of the Kallikak family from their first known colonial anscestor to their institutionalized descendant, Deborah. Note how Martin Kallikak, Sr., divides this family into a "good branch" and a "bad branch." From H. H. Goddard, *The Kallikak Family* (New York: Macmillan, 1912), p. 36. Squares indicate males and circles females; N indicates "normal" mentality and F feeblemindedness.

classify them as feebleminded. Nevertheless, of those he felt competent to judge, he certified only 46 as "normal" while claiming "conclusive proof" that 143 were feebleminded.[110]

Most significant, in Goddard's estimation, were the social costs attached to such statistics, for the Kallikak relatives had left a legacy of moral and medical failure in excess of their numbers. "Among these four hundred and eighty descendants," Goddard chronicled,

thirty-six have been illegitimate.
There have been thirty-three sexually immoral persons, mostly prostitutes.
There have been three epileptics.
Eighty-two died in infancy.
Three were criminal.
Eight kept houses of ill fame.

Even worse, these descendants had married into other families, thus affecting a total of 1,146 persons. Once again, Goddard classified a majority of these cases (687) as "undetermined." Of those remaining, he labeled 197 normal and 262 feebleminded.[111]

These statistics were most striking when compared to those calculated for the legitimate side of the same family, started a few years later when Martin "straightened up and married a respectable girl" from a good Quaker family. Kite's task here had been considerably simplified, for one of these descendants supplied her with an up-to-date family genealogy. This relative even spent a day preparing the "normal chart" that appeared in Goddard's book. Kite evidently saw no conflict of interest in relying on such a source. "No praise can be too high for such disinterested self-forgetfulness in the face of an urgent public need," she wrote of her new assistant. Based on these data, Kite classified all 496 relatives resulting from this union as "normal people" (even including two alcoholics and one "sexually loose" man).[112]

The contrast between the two sides of this family was stark, for while the illegitimate side had produced "paupers, criminals, prostitutes, drunkards, and examples of all forms of social pest with which modern society is burdened," the other side produced "no illegitimate children; no immoral women . . . no epilepsy, no criminals, no keepers of houses of prostitution." Instead, these children had shown "a marked tendency toward professional careers." They had "married into the best families in their state, the descendants of colonial governors, signers of the Declaration of Independence, soldiers and even the founders of a great university." This branch of the family had produced "nothing but good representative citizenship. There are doctors, lawyers, judges, educators, traders, landholders, in short, respectable citizens, men and women prominent in every phase of social life."[113]

Kite's discovery of the relationship between these two branches was espe-

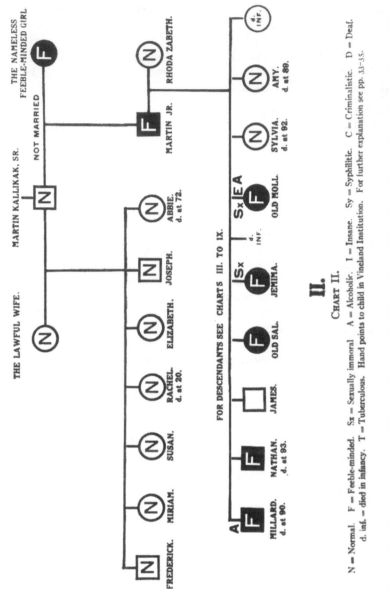

N = Normal. F = Feeble-minded. Sx = Sexually immoral A = Alcoholic. I = Insane. Sy = Syphilitic. C = Criminalistic. D = Deaf.
d. inf. = died in infancy. Hand points to child in Vineland Institution. For further explanation see pp. 33–35.

II.

CHART II.

A heredity chart showing the children of Martin Kallikak, Sr., by his wife and by the "nameless feeble-minded girl." From H. H. Goddard, *The Kallikak Family* (New York: Macmillan, 1912), p. 37.

cially serendipitous, Goddard believed, for to him it suggested a natural control group. "We have, as it were," he argued, "a natural experiment with a normal branch with which to compare our defective side." Environment, he believed, had been held constant in this "experiment." "Both lines live out their lives in practically the same region and in the same environment," he argued, "except in so far as they themselves, because of their different characters, changed that environment." Based on this premise, the good side of the family became "our norm, our standard, our demonstration of what the Kallikak blood is when kept pure, or mingled with blood as good as its own," and could be compared with "the blood of the same ancestor contaminated by that of the nameless feeble-minded girl." Goddard devised the family's pseudonym to reflect this unusual natural situation, for he invented the name "Kallikak" by combining the Greek root *kallos* (meaning beauty) with *kakos* (bad).[114]

The Kallikak study, Goddard concluded, thus satisfied the need for a "natural experiment" distinguishing nature from nurture, for while previous family studies had been inconclusive, his own he declared "thoroughly logical." In other studies, it was impossible to know what good environment might have produced. "Fortunately for the cause of science," Goddard asserted, "the Kallikak family, in the persons of Martin Kallikak Jr. and his descendants, are not open to this argument," for they had been diagnosed as feeble-minded, and "no amount of education or good environment can change a feeble-minded individual into a normal one, any more than it can change a red-haired stock into a black-haired stock." Feeblemindedness, in other words, according to Goddard's increasingly inflexible hereditary construct, was a trait passed on much like hair color. "The biologist," he insisted, "could hardly plan and carry out a more rigid experiment or one from which the conclusions would follow more inevitably."[115]

Goddard's next task was to use his facts to confirm Mendel's laws. He illustrated the presence or absence of the "trait" of feeblemindedness on two dozen heredity charts, filled with their ubiquitous black and white symbols – circles for females and squares for males, colored black for feebleminded and white for normal mentality. Taken together, they charted Deborah's feeble-minded inheritance back six generations.[116]

Charts alone, however, would never appeal to the "lay reader," for facts and figures, Goddard argued, do not suggest "flesh and blood reality." As a supplement, he added "pictures from life" as "graphically written up by our field worker." Kite's skill in classifying those she met, and those she heard about, lay at the crux of Goddard's argument. He evidenced his own faith in her "excellent judgment" by quoting long sections from her reports verbatim. By including such passages, he wrote ingenuously, readers would see "something of her method," and could thus judge for themselves "the reliability of the data." To Goddard's later chagrin, these passages have allowed future generations to do just that.[117]

Kite's reports were written with her usual dramatic flair. "On one of the coldest days in winter," she began, "the field worker visited the street in a city slum. . . . "

> She had previously tested several of the children of these families in the public school and found them, in amiability of character and general mentality, strikingly like our own Deborah, lacking, however, her vitality. There was no fire in their eyes, but a languid dreamy look, which was partly due no doubt to unwholesome city environment. A boy of eleven . . . was standing by the fire with a swollen face. He had been kept home on this account. In a rocking-chair, a little girl of twelve was holding a pale-faced, emaciated baby. In the corner two boys were openly exposing themselves. . . . The entrance of the field worker caused no commotion of any kind. . . . Another younger girl was at school, the family having been at last able to provide her with shoes. The girl of twelve should have been at school, according to the law, but when one saw her face, one realized it made no difference. She was pretty . . . but there was no mind there.

"Stagnation," Kite argued, "was the word written in large characters over everything. Benumbed by this display of human degeneracy," she concluded, "the field worker went out into the icy street."[118]

Further visits to other branches of this family disclosed episode after episode of "human degeneracy." One relative had been a horse thief; another had run off with gypsies. Above all, the family had a sordid history of sexual misconduct. A father nicknamed "Old Horror" had lived with his daughters in a dilapidated dwelling where "great scandals" had taken place. A woman had "an illegitimate son by a man who was high in the Nation's offices." Deborah's half-sister had been placed in thirteen foster families, and in one had set a barn on fire. "When found by our field worker," Goddard wrote, she "had already followed the instinct implanted in her by her mother, and was on the point of giving birth to an illegitimate child."[119]

Such episodes had made this family memorable wherever they went. "'Simple,' one neighbor remarked about Martin Jr.," Kite reported, "'not quite right here,' tapping his head, 'but inoffensive and kind.'" From these assessments, Kite drew "the positive assurance that environment without strict personal supervision made little difference when it was a question of the feeble-minded."[120]

Kite also illustrated the "respectability and usefulness" evident on the good side of this family. Such stories, she admitted, could never match "the bizarre experiences of the abnormal." In her duties as a naturalist, however,

Deborah's relatives identified as "Great-grandchildren of 'Old Sal.'" From H. H. Goddard, *The Kallikak Family* (New York: Macmillan, 1912), facing p. 88.

she included them, beginning with the graveyard of Martin Sr., "lying picturesquely on the crest of a hill." In Kite's telling, even the family furniture spoke of the benevolent influence of past generations, for their ancestral farm still contained "the same fireplace, the same high-backed chairs, the clock, desk, and china cupboard." Descendants returned each summer literally to enjoy the fruits of the family tree, for they could still consume "the luscious grapes and other fruit planted by their ancestor."[121]

Kite's "pictures from life," elaborated through anecdote, dialogue, and metaphor, offered the public a dramatic account of the difference between good and bad inheritance. She illustrated this further with photographs taken "in the field." In front of their dilapidated wooden shacks stood the "bad" Kallikaks. In their ragged clothing and shabby surroundings, they indeed made a pitiful sight – especially when contrasted with their apparently more fortunate cousin, Deborah. Photographed in a lovely dress, a big bow in her hair, a book in her hands, and a cat lying contentedly across her lap, Deborah Kallikak's institutional portrait graced the book's frontispiece.[122]

The *Kallikak Family,* however, was intended to be more than a popular tract. Goddard also saw it as a serious work of science. This simple family

Deborah's relatives identified as "Children of Guss Saunders, with their Grand-mother." From H. H. Goddard, *The Kallikak Family* (New York: Macmillan, 1912), facing p. 88.

study, he believed, would ultimately prove "of remarkable value to the sociologist and the student of heredity." After compiling his charts and telling his stories, Goddard carefully drew out all of its meaning for his readers.[123]

Science as Sermon

In *The Kallikak Family,* Goddard crystallized his new hereditarian theories, developed through his contact with Davenport and other biologists. Instead of vague references to the "force of heredity," which characterized the work of his medical predecessors, he now spoke with increasing specificity. Feeble-mindedness, this book suggested, was probably a "unit character" determined by the germ plasm, passed from generation to generation in accordance with Mendelian principles, and immune from environmental influences. The social implications of this new theory were specific as well. With his simple dichotomies distinguishing normal from feeble, fit from unfit, competent from incompetent, moral from degenerate, Goddard's charts suggested a stark new sociology, illustrated in black and white.

Equally suggestive, especially in light of later interpretations of this popular monograph, was what Goddard left out. Despite its dire warnings about low Binet scores and bad bloodlines, Goddard's appeal to the laity was free

Deborah Kallikak in her institutional home, 1912. From H. H. Goddard, *The Kallikak Family* (New York: Macmillan, 1912), frontispiece.

from the nativist and racist sentiments that pervaded the writings of many other eugenicists. This book mentioned no threats from racial miscegenation; it said not a word about white supremacy or Anglo-Saxon superiority; it expressed no fears of foreign immigration. To the contrary, Goddard's most threatening of social troublemakers, the Kallikaks, were themselves white, Anglo-Saxon Protestants who had been living in America since the time of the Revolution.[124]

Instead of the new social fears alarming many of his eugenics contemporaries, Goddard was still worried about the old ones: promiscuity, adultery, incest, crime, alcoholism, and idleness – the perennial concerns of nineteenth-century Christian moralists. Much of this book's popular appeal can be explained less by its promotion of eugenics than by its fusion of new Mendelian science onto an older and still powerful moralism. (While Goddard made one disparaging allusion to "euthenists" who believed "that environment is the sole factor," he never even used the word "eugenics" in this book.)[125] And whereas Galton had envisioned eugenics as an alternative to traditional religion with its morality tales, Goddard and Johnstone still spoke in the accents of secularized Protestant evangelism, while Kite was a passionate believer and an active member of the International Catholic Truth Society. Her own attitudes are evident in the title of an article on the Kallikaks she published in the *Survey:* "Unto the Third Generation."[126]

Buried just below Goddard's new scientific language linking Binet's tests with Mendel's laws, one can still hear strong echoes of the older synthesis linking children's ailments with their parents' secret sins. *The Kallikak Family* grafted the scientific concept of cause and effect onto a biblical understanding of causes and their consequences. The resulting blend would certainly have surprised both Binet and Galton, had either lived long enough to see it published (both men died in 1911), for Goddard had incorporated their new science into an old-fashioned Christian parable. Nowhere was his own family's missionary heritage more evident than in Goddard's explanation of the meaning of the Kallikak story.

"We have here," Goddard wrote, "a family of good English blood of the middle class" which for four generations had maintained "a reputation for honor and respectability." Yet when a "scion of this family, in an unguarded moment," had stepped from "the paths of rectitude," he had started a line of mental defectives. Martin Sr.'s career offered "a powerful sermon against sowing wild oats." He had done "what unfortunately many a young man like him has done before and since, and which still more unfortunately, society has too often winked at, as being merely a side step in accordance with a natural instinct, bearing no serious results." Martin may have believed his act "atoned for, as he never suffered from it any serious consequences." Even his contemporaries had failed to appreciate the "evil that had been done." Its real consequence, however, was the production of a "race of defective degenerates who would probably commit his sin a thousand times over."[127]

It had taken six generations to understand "the havoc that was wrought by that one thoughtless act," Goddard argued in his own homiletic voice. "Now that the facts are known," he intoned, "let the lesson be learned; let the sermons be preached; let it be impressed upon our young men of good family that they dare not step aside for even a moment. Let all possible use be made of these facts," he proclaimed, "and something will be accomplished." In gathering the facts to prove Mendelian laws, Goddard had thus produced a scientific sermon against promiscuity.[128]

The new biology, however, marked Goddard's sermon as different from those of his predecessors. Even had Martin Sr. stayed "in the paths of virtue," Goddard acknowledged, there remained "the nameless feeble-minded girl." For her the question of virtue was irrelevant, for the feebleminded, he believed, were largely incapable of self-control. In order to ensure its safety, Goddard now argued, society would have to exercise control in other ways. But how far could scientists go in controlling such socially irresponsible persons?[129]

In sharply dichotomizing the responsible from the irresponsible, Goddard's book helped pave the way for public consideration of new solutions to such questions – solutions that institutional physicians had been advocating for decades. His *Kallikak* study widened the social chasm separating a hypothetical "them" – "social pests" of all sorts – from an unspecified "us" – respectable citizens such as Goddard and his readers. This was obvious in the questions he phrased for his audience: "Why do *we* not *do* something about it? . . . What *can* we do?"[130]

The answer to the last question, at least according to Goddard's eugenics contemporaries, was quite a lot. In handling the most severe cases of idiocy – the type that Goddard now called the "loathsome unfortunate" – some had even "proposed the lethal chamber." Goddard personally did not endorse such "solutions." Humanity, he believed, was "steadily tending away from the possibility of that method" and he saw "no probability that it will ever be practiced."[131]

Society's real problems, Goddard instead argued, came not from idiots – not even "loathsome" ones, for such persons rarely reproduced – but from the moron type. Deborah Kallikak was typical of the "high-grade feebleminded person, the moron, the delinquent, the kind of girl or woman that fills our reformatories." While teachers clung to hopes that such girls might still "come out all right," Goddard saw this as a delusion, for girls like Deborah usually got into trouble, "sexually and otherwise."[132]

In the past, Deborah and her relatives had rightfully "won the pity rather than the blame of their neighbors." Society, however, had yet "to suspect the real cause of their delinquencies, which careful psychological tests have now determined to be feeble-mindedness." Instead, it had blamed "viciousness, environment, or ignorance." With his own blend of Binet testing and institutional expertise, Goddard believed he had largely solved this diagnostic problem. Experts, he asserted, could now recognize even the moron.[133]

This recognition made another solution possible: institutional segregation. Had Deborah's promiscuous ancestress been institutionalized, he argued, all that followed could have been prevented. But how practical a solution was institutionalization? According to every medical estimate of the day, as well as Goddard's own studies, only a small portion of the feebleminded – perhaps as few as one-tenth of those purportedly needing care – were in institutions. Such a solution would surely be very expensive.[134]

Goddard's book did contain an appeal for more funds. Institutionalization, he argued, was "not by any means as hopeless a plan as it may seem to those who look only at the immediate increase in the tax rate." Such institutions would largely replace almshouses and prisons, Goddard predicted, thus saving money in the long run. Good institutions could even train the feebleminded to "go out into the world and support themselves," were it not for one problem: "the terrible danger of procreation, – resulting in our having not one person merely, but several to be cared for at the expense of the State." The new biology, like the medical theories it supplanted, again saw sexuality as a central concern of social policy.[135]

A solution to this predicament did exist, Goddard told his readers, for some of his contemporaries had proposed "to take away from these people the power of procreation." Surgical sterilization, he explained, could eliminate the chances of childbearing. Goddard evidently still had reservations about such methods. "Objection is urged that we do not know the consequences of this action upon the physical, mental, and moral nature of the individual," he admitted. And while good results had been claimed, "it must be confessed that we are as yet ignorant of actual facts."[136]

The most serious difficulty, however, would be deciding who to sterilize, for scientists were "still ignorant of the exact laws of inheritance. Just how mental characteristics are transmitted from parent to child is not yet definitely known," he conceded. Deciding "beforehand that such and such a person who has mental defect would certainly transmit the same defect to his offspring" thus became "a serious matter."[137]

Still, he reported, recent biological breakthroughs were moving society closer to such solutions. Goddard introduced this new science to his readers by explaining Mendel's "law of inheritance" and his discovery of "unit characters." "We do not know that feeble-mindedness is a 'unit character,'" he admitted. "Indeed, there are many reasons for thinking that it cannot be." Yet "for the sake of simplifying our illustration," Goddard hypothesized that it was and examined his data accordingly.[138]

The results, he conceded, were inconclusive, for he did not have enough data to prove that feeblemindedness was behaving like a single recessive trait. In his next book, he hoped to accumulate the evidence to prove his case more convincingly. "Enough is here given," he proposed, "to show the possibility that the Mendelian law applies to human heredity." If it did, then this indicated the necessity of "understanding the exact mental condition of the ancestors of any person upon whom we may propose to practice sterilization."[139]

But was sterilization really the answer? Sterilizing Martin Jr. would have saved society from "five feeble-minded individuals and their horrible progeny," Goddard wrote, "but we would also have deprived society of two normal individuals" who "became the first of a series of generations of normal people." While Goddard was willing to endorse sterilization as "a makeshift, as a help to solve this problem because the conditions have become so intolerable," he clearly preferred institutionalization. In the "present state of our knowledge," his book concluded, institutional segregation seemed "the ideal and perfectly satisfactory method."[140]

By ending his book with an open discussion of sterilization and segregation, and by incorporating a popular explanation of Mendelian genetics into his text, Goddard clearly advanced the interests of new eugenic scientists such as Davenport. At the same time, his sermon illustrating how one sinful act could wreak social havoc upon future generations also reaffirmed older biblical warnings. Fusing the two together, Goddard had created a powerful new message of his own. By presenting his findings within a narrative structure with which he himself was intimately familiar, he had found a way to explain his science to the public: like his Christian forebears, Henry Herbert Goddard was teaching the people through parables.

Moral Lessons

Goddard's scientific parable about the good and bad Kallikaks quickly caught the attention of his contemporaries. Scientists, physicians, reformers, and popular audiences responded immediately to his story. Each of these groups found something to admire in his research; each also found something to support their own views in his findings.

To Davenport and his followers, this story offered a means of explaining Mendelian genetics to nonscientific audiences. Goddard's book would long remain the most popular publication associated with the American eugenics movement. It quickly took its place alongside Dugdale's *Jukes* study as proof of the connection between poor heredity and social problems. In the decades to come, the very phrase "Jukes and Kallikaks" would enter into the American vernacular to connote troublemaking families of dubious lineage.

Medical superintendents were equally pleased with its conclusions, for along with the new, Goddard had affirmed the old: the diagnostic expertise of institutional personnel, the need for more institutions, and the connections between medical and moral deviance. Goddard's research, Dr. Rogers concluded in the *Journal of Psycho-Asthenics,* was nothing short of "epoch-making," for he declared this experiment "one of the most enlightening and instructive contributions to heredity that has ever been made." Other doctors were also impressed with Goddard's efforts to ground their intuitive knowledge upon a foundation of scientific law. "This carefully worked out analysis and painstaking study of a concrete example of the workings of the laws of heredity as related to feeble-mindedness," stated a book review in the *Medical Record,* "is highly praiseworthy. . . . " This journal recommended the book to "both the medical and the lay reader," for it was "as interesting as a romance" while also conveying "a moral and a great lesson."[141]

Reviews by psychologists were a little more mixed. The *American Journal of Psychology* praised the book. "Dr. Goddard has been fortunate enough, as the archaeologists say, to make a 'find,'" its reviewer noted, "and he has also had the training which enables him to utilize his discovery to the utmost." Yet psychologist James McKeen Cattell, reviewing it for *Popular Science Monthly,* evinced at least some skepticism. Goddard's results were interesting as "a contribution to our knowledge of the workings of heredity and as a proof of the need of practical measures for eliminating feeblemindedness and lessening vice and criminality in the community," he conceded; yet this was "scarcely the natural experiment in true heredity which Dr. Goddard claims it to be." Even a normal son left illegitimate and neglected might have married with the "degenerate and feeble-minded," Cattell argued, while a feebleminded child left well cared for might have had no descendants.[142]

Popular reviewers, on the other hand, clearly appreciated this book's blend of science and sin. *The Kallikak Family,* reported *The Dial,* was "a scientific study in human heredity, a convincing sociological essay, a contribution to the psychological bases of social structure, a tragedy of incompetence, and a sermon with a shocking example as a text." *The Independent* agreed. The *Kallikaks* was "the most convincing of the sociological studies brought out by the eugenics movement," its reviewer asserted, for it "would hardly be possible to devise in the laboratory experimental conditions better adapted to produce a clear and decisive influence of heredity"; at the same time, one could hardly find "a more impressive lesson of the far-reaching and never-ending injury done to society by a single sin."[143]

Goddard's most important audience was, of course, the "lay reader." Here his book found its strongest response, for his tale contained many of the elements of a melodrama – colorful characters, tragic figures, a mystery, and a moral lesson. In February 1913, popular poet Edwin Markham prepared an excerpted version of this story for a "Book of the Month" selection in *Hearst's Magazine.* "And a strange story it is," Markham exclaimed, "with every element of pity and terror, as we see ignorance and degeneracy stalking among men, leaving destruction and death in their wake." Goddard had recounted the "ghastly record of the descendants of Martin Kallikak, the reckless soldier, and the nameless, witless girl that followed the camp." And while "poor Deborah's relations were smitten" with feeblemindedness, Martin "doubtless forgot the girl and her child; but society for generations has been paying for the evil he set in motion."[144]

An even more unusual indicator of this book's broad appeal, and of its melodramatic potential, was an offer Goddard received from a prominent Broadway agent interested in securing the rights to his story for Joseph Medill Patterson. Mr. Patterson had "written plays with ideas back of them," she told Goddard. The wealthy descendant of a famous publishing family and a former socialist, Patterson was coeditor of the *Chicago Tribune;* years later, he would gain fame as the editor of the New York *Daily News* and the innovator of American tabloid journalism. Although Patterson's Kallikak play was apparently never written, Goddard's response to his offer is still suggestive: he would be interested, Goddard replied, only if "assured that the play would carry the moral lessons which the book is intended to convey."[145]

Goddard did not specify which "moral lessons" he wanted to see dramatized. His short monograph, however, was filled with new lessons for his lay readership – lessons about Binet's tests and Mendel's laws; about the connections between bad blood and bad behavior; about the social irresponsibility of the biologically unfit, and the need for the socially responsible to do something about it; about the new solutions then being considered, such as sterilization or segregation; about the happy lives that such persons could lead within institutions. Perhaps the book's most powerful moral lesson, however, was still its oldest. Its message was best expressed not by any anecdotes concerning family members, but by Kite's account of what had happened to the Kallikak family Bible.

Purchased in 1704, this Bible had been handed down from generation to generation within the good side of the Kallikak family. It was currently in the possession of a descendant who had become a minister. After more than two centuries, this family heirloom was "still in an excellent state of preservation." On the flyleaf, a father had bequeathed the book to his son by inscribing a verse; it bid the son, "So oft as in it he doth looke," to remember that his father's life had "aye been guided by ye precepts in this booke." In words that conveyed their own ages-old message of causes and their consequences, and of the deeper meaning of a goodly inheritance, this father had then enjoined his descendants to "walk in the same safe way."[146]

6

The Biology and Sociology of "Prevention": "Defectives, Dependents, and Delinquents"

Progressive Science

"It gives me very great pleasure to have a few minutes in which to tell you a little about our work – I am going to call this, 'A New Method of Studying Causes of Dependency and Crime,'" Henry Herbert Goddard proudly told the New Jersey State Conference of Charities and Correction in 1910. For this "new method," he added humbly, the conference was indebted to his institutional employer, Edward Johnstone, for he had insisted that "if the State had to take care of these children, these people, these unfortunates," it ought to get "some return in the way of knowledge." The State evidently agreed, for a year later, in the first act of its kind, the New Jersey legislature allocated $2,000 "for research" into the causes of dependency and delinquency. This research would be conducted by "field workers" sent out from the Vineland Training School for Feeble-Minded Girls and Boys.[1]

New Jersey's small legislative grant was significant – not for the funds it supplied, but for the political legitimacy it granted to Goddard's new psychological science. By 1910, Goddard had begun to take his science far beyond his own institution. Psychologists, he proposed, ought to study social problems such as dependency and delinquency. They should focus not on poverty and crime but on the pauper and the criminal. Toward this end, Goddard could introduce two new research methods into contemporary debates: first, intelligence testers could measure the minds of dependents and delinquents; secondly, field workers trained in Mendelian science could trace their family histories.

These methods were indeed new. The theories behind them, however, were not. Goddard's social research was built upon a premise long accepted by the American reform community: that there was a causal connection between mental deficiency, economic dependency, and social delinquency. To Goddard, this connection could be expressed simply: "People get into the jail; people get into the almshouse; people get into insane asylums," he believed, because "they or their parents were feeble-minded."[2]

The idea that members of the "defective, dependent, and delinquent classes" shared much in common had been a long-standing belief expressed by both charity workers and their social science counterparts for more than half a century. To many physicians, all three conditions were simply different manifestations of the medical phenomenon still called "degeneration" – a process which caused the inheritance of a vague "neuropathic weakness" that often led to social failure. A "degenerate" inheritance, doctors argued, could manifest itself not only in feeblemindedness, pauperism, or crime, but also in epilepsy, insanity, alcoholism, or sexual immorality. This older medical theory blended surprisingly well with the newer theory being avidly proposed by biologists promoting eugenics – a theory which linked both physical ailments and antisocial behavior to the direct transmission of inferior "germ plasm" inherited according to Mendelian ratios.[3]

Both theories found broad support from the social research of the day. In study after study, survey after survey, researchers had concluded that defectives, dependents, and delinquents were indeed members of a common class – a class of persons more prone to become paupers, more likely to land in prison, and more vulnerable to illnesses, both physical and mental. To contemporaries, these studies presented an alarming portrait of the social dangers posed by this population. At the same time, they also offered grim testimony to the suffering of the lowest stratum of American society.

This widespread suffering in itself challenged older and more optimistic ideas about the concurrent progress of science, technology, and society. To many social reformers who considered themselves "progressives," the striking presence of sickness, crime, and poverty amidst American industrial plenty suggested a need for action. By the early twentieth century, one fact seemed incontestable: social progress had as yet left millions behind.

In the past, problems such as crime and poverty had largely been dealt with through punitive measures, religious admonitions, and acts of charity. Such methods, progressives now claimed, had proven both inadequate and ineffectual. New approaches would be necessary to solve the massive social problems associated with urban, industrial development. New methods would be needed.

For many reformers, there was one outstanding example of successful intervention in solving urban problems: the public health campaigns that had mandated preventive measures to control epidemics. Deadly diseases such as cholera could not as yet be cured, they conceded; they could, however, be prevented, if the state were willing to assert its power in new ways. Similar strategies ought to be tried with social problems, reformers contended, for poverty and crime were also reaching epidemic proportions. And while cures might be elusive, "prevention" was a possibility. Society, in other words, needed to learn to prevent the ills it could not cure.[4]

Goddard too saw preventive medicine as a model for his work. "If a new disease were to appear and sweep away 2 per cent of our children," he argued, "we would immediately set to work to prevent that disease." Feeblemindedness, he now argued, was "a condition affecting 2 per cent of the children, not, indeed destroying their lives, but rendering them incapable of living with the rest of us on equal terms." Scientists knew "no way of curing the condition, and we cannot destroy its victims. We might prevent," he added.[5]

Prevention, however, required an understanding of causation. By 1914, Goddard believed he had found the solution to this problem through Mendelian science. In his 1912 monograph, *The Kallikak Family,* he had introduced this new science to the public. Within two years, he had also completed a far more complex scientific undertaking: determining a cause for the mental condition of each person in his institution. He compiled his findings in a massive 600-page tome entitled *Feeble-Mindedness: Its Causes and Consequences.* Throughout his career, this work would remain Goddard's most comprehensive contribution to his science, as well as his most elaborate statement about the need to prevent both the biological and the sociological consequences of poor heredity.[6]

"A True Relation of Cause and Effect"

Goddard's 1914 volume, *Feeble-Mindedness,* was indeed ambitious, for within its pages he presented a brief portrait of each of the 327 persons then residing in the Vineland school. Included were summaries of each child's medical history, school records, behavioral patterns, and vocational aptitude. Illustrating this text were photographs of many children as well as samples of their penmanship, spelling, arithmetic, letter writing, and artwork. Goddard also reported each person's mental age, in many cases supplying specific examples of Binet questions passed or failed.

At the heart of this study was a broad investigation into the question of causation. Goddard began by examining "assigned" causes – the reasons given by parents or physicians at the time of a child's admission. These ranged widely – from childhood illnesses to unusual injuries to bizarre experiences. "Mother shocked by sight of woman with hare-lip," stated one parent. The "child swallowed a button," stated another. "Frightened by dull nippers of barber at first hair cut," another parent reported.[7]

Goddard treated all parental reports with skepticism. Only by tracing the family histories of all of his cases, he now believed, could the true causes of feeblemindedness be understood. To conduct such research, he hired two additional field workers, Jane Griffiths and Maude Moore, to join Elizabeth Kite. Using the same methods employed in the Kallikak study, these three women spent several years supplying Goddard with hundreds of family charts constructed from information gathered "in the field."[8]

The primary purpose of these charts was to document the presence or absence (marked by black or white symbols, respectively) of feeblemindedness over several generations. Also reflected were broader concerns about poor inheritance, for recorded as well were various other conditions – physical, mental, and social. In addition to N (for normal mentality) and F (for feebleminded), Goddard's field workers marked their charts with an alphabet of their own representing a range of human maladies:

A	Alcoholic – meaning decidedly intemperate, a drunkard				
B	Blind	I	Insane	Sy	Syphilitic
C	Criminalistic	M	Migrainous	Sx	Sexually immoral
D	Deaf	Neu	Neurotic	T	Tuberculous
E	Epileptic	Par	Paralytic	W	Wanderer, tramp
G	Goitre				

Also noted were miscarriages, infant deaths, cases of incest, illegitimate births, and consanguineous marriages.[9]

The evidence for these diagnoses was gathered from various sources: field worker judgments, medical records, memories of relatives, and local gossip. Also recorded were miscellaneous characterizations or activities – anything, in fact, that might prove noteworthy. "Ugly and heartless" read a caption describing one relative; "brute," stated another; "inherited money," reported a third.[10]

Closely supervising this process was Goddard, for information gathered "in the field" was regularly mailed back to Vineland, where it was carefully copied onto charts. Amidst all this evidence, Goddard kept his main objective firmly in mind: to gather the data necessary to prove that feeblemindedness was a recessive trait inherited according to Mendelian laws.

He also made sure that his field workers knew what they were looking for. Examining one chart submitted by Miss Griffiths, for instance, Goddard noted that two parents had been judged defective, and two children normal. In such cases, he reminded her, "all of the children should have been defective." Of course, Goddard was not directly asking his researcher to change her diagnosis. "I do not mean that you should have marked him so because of course we do not unless we have the objective evidence of it," he added quickly. In this case, the children were young; their feeblemindedness might appear later, he wrote reassuringly.[11]

With such working hypotheses held so firmly in mind, it is hardly surprising that many cases did indeed conform to Mendelian expectations. Of 327 families investigated, Goddard classified half (164) as clearly illustrating "Hereditary Feeble-Mindedness," and another 34 as "Probably Hereditary." Within these categories, he placed families with charts "dotted with the black symbols." Even if a Vineland child had only one feebleminded relative, however, this was often sufficient for such a classification, especially if the overall "tone of the family" was low.[12]

Once a diagnosis of "poor heredity" had been established, all other information became secondary. A child may have had spinal meningitis, convulsions, or paralysis. Its mother may have been alcoholic or syphilitic. Its childhood injuries may have been traumatic. Such factors were consequences, Goddard now argued, for the "real cause" had been discovered: "hereditary taint." "One glance at the chart," Goddard wrote of one such case, "shows that we need not call in acute disease or alcoholism or anything else except the defective germ plasm to account for this condition."[13]

For another 37 cases, the best explanation still seemed to be "Neuropathic Ancestry." These families had histories not of feeblemindedness but of other nervous ailments, including insanity, paralysis, epilepsy, or migraine. Such conditions, Goddard hypothesized, might have caused feeblemindedness by affecting the "metabolism of the offspring" instead of the "germ cells of the parents."[14]

Only when a good family chart had eliminated both "poor heredity" and "neuropathic ancestry" as causes would Goddard consider other possibilities. Of those that remained, he classified 57 as "Accidents" – a broad category for "preventable" conditions that had injured the child before, during, or after birth. These included prenatal illnesses, injuries, or traumas to the mother, difficult deliveries, and miscellaneous diseases or accidents occurring in childhood. In one case, a mother had taken drugs to induce abortion; in another, Goddard suspected lead poisoning, for the father was a pottery worker. He blamed 16 cases on spinal meningitis, and another 10 on scarlet fever, typhoid fever, convulsions, infantile paralysis, beriberi, whooping cough, and measles. Also within this category were 11 cases of "Mongolism," a condition which Goddard saw as "congenital" but not "hereditary," for it was found even in the best of families.[15]

For 8 cases, even Goddard had to list "No Cause Discovered." These were respectable, intelligent, and healthy families who could think of nothing that would explain their child's condition. Finally, 27 cases were "Unclassifiable," for field workers had been unable to gather enough data to make a diagnosis.[16]

After presenting all his findings, Goddard analyzed them in a series of tables, charts, and graphs. His most important challenge was answering two questions: was feeblemindedness a "unit character," and was it inherited "in accordance with the Mendelian Law"? To find answers, he combined all the data on "matings" within the 164 charts marked "hereditary." His method here, he summarized, had been "to assume the law, then see if the conditions as we find them can be accounted for on this assumption." The results were clear: the answer to both questions was "yes," for his data were "strongly confirmatory of the theory." In this sense, he argued, adopting the Mendelian hypothesis and then confirming it "'in the field" had led to a scientific breakthrough, for discovering "a true relation of cause and effect" was "science itself."[17]

Yet while most of Goddard's book dealt with biological causes, far more influential were his discussions of sociological consequences. The "discovery of the moron," he reported to both scientific and popular audiences in his numerous publications and speeches, had serious implications for the study of social questions. Between 1910 and 1915, Goddard was able to blend together the "new psychology" of intelligence testing with the "new biology" of eugenic field work. The result was a sweeping new sociology which profoundly influenced contemporary understanding of the causes of poverty and crime.

"Why Is a Pauper?"

Of the social questions requiring answers in Goddard's day, none was more pressing than determining the causes of poverty. Only by understanding cause and effect, progressive reformers reasoned, could society alleviate the suffering of the poor and, more importantly, prevent it in the future. This emphasis on prevention could be found in progressive writings advocating solutions which spanned the political spectrum – from social work to socialism, "euthenics" to eugenics.[18]

Like many of his contemporaries, Goddard believed that American society could learn to prevent poverty. Charity, he suggested, was an emotional, not a scientific, response. "A beggar stops me on the street and asks for alms," Goddard recounted. "My whole heart goes out toward him, and I am inclined to hand him a coin. But my science comes to my aid," he added, "and I realize that by giving him promiscuously in this way I am not helping him, I am really doing him an injury. I am tending to continue the condition of life in which he lives, to promote beggary and improvidence. I, therefore, refuse to give him the coin and go my way."

In Goddard's telling, the point of this story was neither to promote self-reliance nor to defend laissez-faire capitalism. In doing nothing, "I am doing something that is equally wrong," Goddard believed. "I simply become the cold scientist instead of the humane fellow-being." The "real duty" of a scientist was "to go behind the conditions as I meet them, and discover what underlies; discover why that man was begging, and remove the cause." Scientists, in other words, should study pauperism. The first step, of course, was for researchers to gather the facts; thereafter, the laws of cause and effect would become visible.[19]

Goddard's own ideas about causes and effects, however, differed markedly from those of many other reformers attempting to explain poverty in his day. Among the most systematic of contemporary fact gatherers, for instance, was Robert Hunter, a Hull House social worker who had embraced socialism a year after publishing his influential 1904 study, *Poverty*. Americans, Hunter had charged, were guilty of "the grossest moral insensitiveness" for their laxness in identifying poverty's victims as well its causes. Much of this he blamed on religious doctrine. "Did not the Lord say, 'The poor always ye have with you'?" he observed. While Hunter believed that the "sins of men should bring their own punishment, and the poverty which punishes the vicious and the sinful is good and necessary," he still saw the "mass of the poor" as "bred of miserable and unjust social conditions, which punish the good and the pure, the faithful and industrious, the slothful and vicious, all alike." Such misery, Hunter had argued, had been caused by "social and economic forces," and the "wrongful action" of these forces was "a preventable thing."[20]

Unlike Hunter, Goddard saw no "social and economic forces" as causing the poverty of his day; yet he too tried to distance himself from older religious explanations blaming individual sin or a failure of will. Politicians, Goddard believed, needed new answers to an age-old question. He rephrased this question in a new "scientific" way: "Why is a Pauper?"[21]

Implicit in Goddard's phrasing was the traditional distinction between poverty, an unfortunate but still respectable economic condition, and pauperism, its more disreputable and chronic cousin. The pauper, Goddard explained, was usually seen in two ways. To some, he was "a person who will not work sufficiently to earn his living, – he is lazy and prefers to live at the expense of someone else." (Even Hunter described pauperism as "analogous to parasitism in biological science.") To others, he was a person "overtaken by misfortune."[22]

Goddard then spelled out his own ideas. "We will not here contend for what might be considered an extreme view," he maintained, "that people overtaken by misfortune are seldom allowed to become paupers, that humanity is so kind, so philanthropic that it is always willing to help the person who is desirous of helping himself and that the misfortunes of life are overcome by this mutual helpfulness." Nor did he believe that any person who seemed unwilling to work was "by that very fact defective either physically or mentally." Still, he conceded, there was "much truth in each of these statements."[23]

Few persons, Goddard believed, had become paupers solely from misfortune. Yet "laziness" was an equally unscientific explanation, for there were many reasons to believe that a lazy person might have something "fundamentally wrong with his mind or his body." It was "not natural for a child to be lazy," Goddard insisted, for children by nature were "active and industrious." "The lazy boy is a diseased boy or a defective boy," he argued. If such a child "seems to be lazy, there is a cause for it and this cause must be sought out and removed."[24]

Such ideas seemed especially promising in 1914, for medical researchers had recently found one cause that explained apparent "laziness" in just such a way. Since 1909, Dr. Charles W. Stiles had been conducting a massive campaign, funded by the Rockefeller Foundation, proving that hookworm infestation was widespread in American southern states. This condition, previously undetected, had rendered its victims anemic and thus unable to work. To contemporaries, Stiles' discovery suggested the long-sought explanation for the "lazy South." "Now, for the first time," progressive journalist Walter Hines Page had announced in 1912, "the main cause of their long backwardness is explained and it is a removable cause." To Page, the explanatory power of this "germ of laziness" could hardly be exaggerated. "The hookworm has probably played a larger part in our Southern history," Page had concluded, "than slavery or wars or any political dogma or economic creed."[25]

Goddard too was impressed by this discovery, for it showed how the helpless victims of hookworm disease had been misdiagnosed as "good-for-nothing, shiftless, lazy people, people who might earn their living if they would." In an analogous way, he hoped to prove feeblemindedness another unrecognized cause of apparent laziness and concomitant poverty.[26]

Of course, doctors diagnosing hookworm had brought a message of hope to sufferers across the country, for this disease could be cured. By contrast, those diagnosing feeblemindedness spoke far more pessimistically. For persons suffering from a feeble inheritance, both biologists and sociologists predicted, the future often led only to pauperism.

The evidence supporting such a conclusion was overwhelming, contemporaries argued, for numerous surveys had shown that persons suffering from severe physical or mental ailments were indeed likely to become paupers. Typical of this research, for instance, was the "field work" conducted by Dr. Anne Moore for the Public Education Association of New York City, a group advocating progressive reform. Moore had investigated the home life of children leaving the city's special classes. Entitled *The Feeble-Minded in New York,* her study was published in 1911 by the State Charities Aid Association.[27]

As was common in such studies, Moore conflated medical and moral "pathologies"; thus, she expressed equal alarm at finding mental illness, alcoholism, drug addiction, sexual immorality, child abuse, "self-abuse," and "vile language." The "horrors" attending feeblemindedness, she concluded, had "in no way been exaggerated," for mental deficiency led to "poverty, degeneracy, crime and disease." Many of the city's feebleminded lived in families poor enough to merit the attention of welfare agencies. One hardly needed more proof, eugenicists argued, that the defective and the dependent were often one and the same.[28]

Such a connection seemed plausible enough. After all, for persons struggling to maintain a minimal livelihood, Moore concluded, the birth of a feebleminded child might in itself plunge a family into "pauperism." Mothers "governed by especial maternal tenderness" were particularly likely to deplete a family's resources and overtax their strength, she reported. And while most parents had fiercely opposed placing these children in institutions, Moore tried to convince them otherwise. Institutionalization, she explained, would free a poor mother from "the burden of an unwise decision to care for the weakling herself."[29]

Institutions would also protect feebleminded children from the actions of less devoted parents, Moore argued, for her interviews revealed many cases of cruelty and neglect. These included children who had left the special classes only to find themselves ignored by their families, abused at home, or abandoned to the streets.

Among these, for instance, was "Philip," a "defective of the very lowest type," who had been locked out of his house. "Donato" had been "treated most brutally" and had himself begun to exhibit brutality. "Mamie" had been raised by an alcoholic father who "beats his wife, uses vile language and is not a fit person to have the charge of children." This case had already led to tragedy, for this girl had "set on fire a younger sister who was a paralytic," burning her to death.[30]

Many of these problems, Moore claimed, were the fault of feebleminded parents. Among the worst of the mothers she met was "Frieda." Despite the best efforts of numerous welfare agencies, including the Charities Organization Society, the Society for the Prevention of Cruelty to Children, the Association for Improving the Condition of the Poor, the Salvation Army, several churches, and various hospitals, Frieda's first seven children had all died from "neglect and starvation." Prospects for an eighth child looked no better, for at the time of its birth Frieda and her husband were once again homeless "vagabonds."[31]

To Goddard, such surveys proved one point: feebleminded persons frequently became paupers. But even so, was the converse equally true? Were many paupers feebleminded? Moreover, did one condition cause the other? Nowhere was the need to distinguish causes from consequences, or heredity from environment, more critical to state policy than in assessing these connections between mental deficiency and economic dependency.

The relationship between the two conditions, however, was complex, for inability to support oneself economically formed a key part of the social definition of mental deficiency most frequently cited by doctors – the definition adopted by Britain's Royal College of Physicians in 1908. Even the highest grade of the feebleminded – the type that Goddard christened the "moron" – was defined as a person "capable of earning his living under favorable circumstances," but "incapable from mental defect" of "competing on equal terms with his normal fellows" or of "managing himself and his affairs with ordinary prudence." In other words, the feebleminded were *by definition* those most likely to lapse into the pauper class. Undoubtedly, many Americans were finding it difficult, if not impossible, to earn a steady living. But was this incapacity due to "mental defect"?[32]

Goddard conceded that the circumstances surrounding earning a living were indeed becoming more difficult. Even so, he expressed no essential dissatisfaction with the economic environment of his day. Instead, his paeans to opportunity and upward mobility offered a hereditarian version of the popular nineteenth-century rags-to-respectability stories of Horatio Alger. "The world is full of people who have started out with as little capital in the way of education as can be imagined," he declared, "and yet the something within them has pushed them forward. Their inborn intelligence has enabled them to master the work of a trade and they have steadily forged to the front." Conversely, in his writings he frequently expressed a subtle contempt for anyone who long remained a "farm hand," or a "shop-girl," or a "common laborer on the railroad."[33]

Perhaps this attitude came from his own traumatic experience with nineteenth-century genteel poverty. After all, Henry Herbert Goddard knew firsthand how misfortune (such as the death of one's father) could leave a family destitute. Yet he still interpreted such experiences within the small-town context he himself had known as a child – the kind of community in which a poor but capable boy could still be educated enough to rise in the world, largely through his own efforts and aided only by the generosity of concerned neighbors and church groups. Whatever his reasons, Goddard turned a deaf ear to charges taken far more seriously by more radical reformers of his day – charges which claimed, for instance, that dangerous industrial workplaces, or exploitive labor practices, or corporate greed were responsible for a rise in American poverty, which in turn was causing more sickness of all sorts, including more birth defects.[34]

Even more pointedly, Goddard ignored evidence suggesting that at least some parents were leaving children in institutions because they could no longer support them. Perhaps the saddest Vineland case which indicates such an economic motivation concerned "Gussie," who came to this institution from an almshouse. Gussie's troubled childhood can be gauged from his own brief autobiography:

> I was born in 1894 in a boat house I lived with my mother, my Father ran away when I was 3 1/2 years old. then my mother and I left together. I was 3 1/2 years old when we left————. My birth place which is a half of a mile from————then my mother and I traveled we went lots of different places when I was near seven years old we went to the poor house
>
> My mother worked in a laundry there. I was there nearly a year. then

I lived in Metuchen for several weeks. then I came here when I was 8 years old. and my mother went to New York City now she is working and getting along good. I am going to school in the afternoon and working at the Taylors trade in the morning I started to learn the trade at near fourteen.[35]

While Vineland teachers had found Gussie a pleasant and relatively capable child, at about age fourteen they reported a change for the worse. Gussie "did not take as much interest in his work – had silly spells and laughed at nothing; became slow and fussy, not as full of life as formerly," his records stated. By sixteen, his writing had become more rambling and painfully slow; yet he continued to work in the tailor shop and developed "a great notion of going out and earning his own living." His changed behavior puzzled the staff. To Goddard, these changes suggested an incipient insanity. "On the other hand," he conceded, they might also indicate "a strong desire to get out in the world and work for himself." Even so, if "uncontrolled by any good judgment," Goddard declared, this too was "the working of a feeble-minded individual of rather high grade."[36]

Unfortunately, Goddard reported, he could learn very little about Gussie's family history. All that was known about this boy's father was that he had been "criminalistic." Field workers did track down his mother, however, and promptly pronounced her "feebleminded and sexually immoral." A letter from mother to son, with its misspellings and missing words, makes her own predicament clear.

My dear son Gussie
    You are better make up your to remain where you are for another year. And try and be contented. As I would have no busness of out now it is terrible warm in New York the heath is just killing Gussie the buiding where I am working is going to be pulled down so I will be out of work in a few week. So I cant have no home so be a good boy now and try and be contented for another year an other year will not be long slipping so the next time I go out there I will take you right out I remain your fond mother

                                                                        Nora G.

To Goddard, such a letter was itself suggestive. Gussie's mother, he concluded, "seems to have about the same mentality as her son."[37]

Perhaps it was the comparison with his own mother's history that made Goddard respond so coldly to Gussie's mother's plight. After all, Sarah Goddard, like Nora G., had been forced to raise a young son without a husband or an income. Both women had chosen to leave their sons in boarding schools, albeit of very different types. (Gussie, however, would not long remain in the Vineland school; shortly after receiving his mother's letter, he took his fate into his own hands and ran away.) Yet despite their roughly comparable situations, Sarah G. and Nora G. had led very different lives. After all, Mrs. Goddard, poverty notwithstanding, had always written articulate letters to her son. Even more significant to Goddard, he could hardly have imagined either his mother or his two older sisters manifesting "sexual immorality," no matter how poor the family had become. Such behavior, he believed, demonstrated a lack of self-control that in itself suggested the workings of a feeble mind. In fact, Goddard knew of no other way to explain the failure of women such as Nora to internalize society's standards of self-discipline – standards that shaped his intuitive understanding of "normal" moral, and consequently mental, development.

A similar logic is evident in Goddard's diagnoses of many other cases as well. Martin Kallikak, Jr.'s alcoholism, for instance, proved the same point, for this case too showed a striking absence of self-control. "'Old Martin could never stop as long as he had a drop,'" Goddard reported in recounting a story told to his field worker. "'Many's the time he's rolled off of Billy Parson's porch. Billy always had a barrel of cider handy. He'd just chuckle to see old Martin drink and drink until finally he'd lose his balance and over he'd go!'" Such a tale, Goddard believed, spoke for itself. "Is there any doubt," he ingenuously asked his contemporaries, "that Martin was feeble-minded?"[38]

Hardest of all for Goddard to accept as "normal" was the social resignation that marked the behavior of poor country people his field workers encountered, especially in chronically depressed and isolated rural areas like the New Jersey Pine Barrens, where many "Kallikaks" lived. In many ways, the portrait which Goddard presented of Deborah Kallikak's mother is itself a model of middle-class frustration with the fatalism of the poor.[39]

Deborah's mother had shown "no malice in her life nor voluntary reaction against social order," Goddard wrote, "but simply a blind following of impulse which never rose to objective consciousness." Her life had "utterly lacked coordination," for there was "no reasoning from cause to effect, no learning of any lesson." She had "never known shame," he concluded.[40]

Most troubling was the absence of any aspirations toward upward mobility. "Her philosophy of life is the philosophy of the animal," Goddard surmised, for there was "no complaining, no irritation at the inequalities of fate. Sickness, pain, childbirth, death – she accepts them all with the same equanimity as she accepts the opportunity of putting a new dress and a gay ribbon on herself and children and going to a Sunday School picnic." There was no "comprehension of the possibilities which life offers or of directing circumstances to a definite, higher end." While this woman showed "a certain fondness for her children," Goddard believed her "incapable of real solicitude for them," for she was "utterly helpless to protect her older daughters, now on the verge of womanhood, from the dangers that beset them, or to inculcate in them any ideas which would lead to self-control or to the directing of their lives in an orderly manner."[41]

Such persons, Goddard argued, made up "the type of family which the social worker meets continually and which makes most of our social problems." To him, these cases offered individual confirmation of the "general statement that every feeble-minded person is bound to be the victim of his environment because he has not intelligence and judgment and will-power enough to control that environment."[42]

Goddard was aware of the consequences following from such a conclusion. He now claimed, for instance, that no amount of work in the slums would succeed "until we take care of those who make the slums what they are." Even if all the slums were replaced by "model tenements," he argued, "we would still have slums in a week's time, because we have these mentally defective people who can never be taught to live otherwise than as they have been living." Not until this class found their lives "guided by intelligent people," Goddard insisted, would society "remove these sores from our social life."[43]

Such arguments could be used both to defend the status quo and to curtail welfare spending. Goddard, however, saw himself as doing neither. He was not arguing that institutionalization replace slum clearance; unless "the two lines of work go on together," he concluded, "either one is bound to be futile in itself." Moreover, unlike many of his colleagues in the eugenics movement who resented spending tax dollars on welfare programs (among them Charles Davenport), Goddard consistently pressed both the public and the state legislature to spend more – on institutions, on diagnostic services, and on special education for blind, deaf, and feebleminded children. In his own mind, Goddard believed he was preaching not laissez-faire capitalism but a new version of social responsibility.[44]

Nor did he believe that he was blaming the poor for their own condition. By rejecting older explanations which portrayed paupers as lazy or vicious, and instead adopting a new hereditarian determinism, he thought he was doing just the opposite. The real problem, he insisted repeatedly, was not that such persons would not behave, but that they could not. "The emphasis here is on the word 'incapable,'" he wrote. Defective dependents were incapable of adapting to the complex economic environment of the day.[45]

It was within this context that Goddard interpreted Alfred Binet's injunctions about "intelligence" being a relative, and not an absolute, measurement. "As Binet points out," he noted, "normal intelligence is a relative matter," for "that which is sufficient for a French peasant out in the country is not sufficient for a Frenchman in Paris." Many socially irresponsible persons were "of a type that in the past and under simpler environments have seemed responsible and able to function normally but for whom the present environment has become too complex." Thus, to Goddard, the relativity of intelligence meant that some were capable, and others incapable, of modern urban living. "They are where they are thru no fault of their own," Goddard contended, "but because their burdens – those of making a living – were too heavy for them. Society," he insisted, "should have protected them."[46]

But in order to protect such persons in the manner that Goddard had in mind, social scientists first had to know how to "draw the line between responsibility and irresponsibility." By 1914, Goddard had devised a solution to this problem by fusing psychological with social definitions of feeblemindedness. "We start out with an hypothesis somewhat new," he explained, that "may be stated as follows: there are all grades of responsibility, from zero to the highest; or, there are all grades of intelligence from practically none up to that of the genius or the most gifted. Responsibility varies according to the intelligence."[47]

If Goddard's hypothesis were correct, then Binet testing could be used to assess not only intelligence but an even more crucial variable as well: social responsibility. Testing, he now proposed, ought to be taken out of the schools and into the community. "We must measure the intelligence," Goddard insisted. "Knowing the grade of intelligence we may know the degree of responsibility. Knowing the degree of responsibility we know how to treat."[48]

The kind of treatment that Goddard foresaw would require society's most intelligent citizens to take responsibility for their less intelligent brethren. Thus, while some progressive reformers were arguing for a more egalitarian social order, Goddard envisioned one more stratified. Society would consist of "persons of sufficient intelligence to function in the simplest environment, those of higher intelligence who can function in a more complex environment and so on to the most complex." In fact, he added, "this is what we attempt to do now," but it was done in a "bungling way." Intelligence testing would make society's task of grading responsibility "less bungling and more practical."[49]

For the lowest grades, the simplest environment was, of course, the institution. Here persons of limited capacity would find their lives significantly improved, Goddard believed, for they would no longer have to engage in an economic struggle for survival. Instead of being a burden, they might even contribute to society by working (under supervision) at simple tasks – as farmhands, handymen, or housekeepers. Of equal importance, society's future needs would be met as well, for through strict policies of sexual segregation, the feebleminded would finally be prevented from passing their inferior inheritance on to any more progeny.

Once again, Goddard's argument for institutionalization represented a mixture of new and old – new methods confirming old theories. The diagnostic role he prescribed for intelligence testing was certainly new. In essence, however, his understanding of mental testing represented a restatement of an older evolutionary paradigm which still informed much social science. Goddard's science drew an analogy between "defectives" and more "simple" peoples of the past. Both groups, moreover, could be compared to children.

To Goddard, this science had clear social implications. Like children, those deemed mentally, and therefore socially, inferior deserved loving care and protection. Like children, they also ought to abrogate political, economic, and sexual independence, for these functions should be the province only of socially responsible adults. According to Goddard's paternalistic sociology, paupers diagnosed as "defective dependents" ought to be granted the social status of permanent children.

Defective Delinquents

Goddard's new theory linking social irresponsibility with low intelligence had implications not only for studying poverty, but also for explaining crime. In several ways, his paternalism fit well within the larger progressive campaign to reform the American justice system. In the two decades following 1899, nearly every American state established a separate system of "juvenile courts" to deal with young offenders. Inherent in these reforms were new ideas about the very nature of child development.

Most important was a growing awareness that young minds worked differently than those of adults. Consequently, juvenile offenders should to be treated differently. According to progressives, they deserved more parenting, less punishment, and fewer civil liberties. This could be accomplished through an expanded interpretation of the doctrine of *parens patriae* – the state's ability to act as parent if the well-being of a minor was at risk.[50]

Delinquency, like dependency, these reformers argued optimistically, might be prevented if society were willing to study its causes and apply their findings in adjudicating cases. Such arguments called for a closer relationship between social scientists and the justice system. Among the pioneers promoting this new relationship was psychiatrist William Healy, who in 1909 opened a "Juvenile Psychopathic Laboratory" to study cases appearing before Chicago's Juvenile Court. Goddard too hoped to introduce psychological science into the justice system. In dealing with delinquency, he suggested, the expert on mental deficiency ought to become a useful ally to the court.[51]

Such alliances made sense to many contemporaries, for here too Goddard was building upon an older medical tradition. While defectives and dependents had long been linked by definitions stressing economic adaptation, the even older concept of "moral imbecility" linked defectives with delinquents. "Moral imbeciles," according to physicians, were persons in whom the "moral sense" had never developed (as opposed to the "morally insane," in whom this sense had stopped functioning). Doctors often used this concept to explain the behavior of emotionally disturbed children who often wound up in institutions.[52]

Dr. Isaac Kerlin of the Elwyn institution, for instance, had written of children given to "violent explosions of temper." Such cases, he reasoned, might be a younger version of the criminally insane. His successor, Dr. Martin Barr, had focused even more attention upon those whom "neither coaxing nor discipline had the slightest effect." Barr reported children who were "self-willed, obstinate, given to unclean habits, delighting in sulking, in annoying other children, and in torturing animals." These cases were hard to miss, for they continually caused problems for the staff. "I wish I could say something good of M.," an institutional teacher wrote of one child, "but I cannot, for he is wicked in every way and gives constant trouble." "Capable of discriminating between right and wrong," Barr observed of another, "but generally prefers wrong."[53]

Among the Vineland children, Goddard found twenty-five manifesting "tendencies that would make them criminals if they were responsible." Most striking to him was not their small number, but their similar test results, for at least twenty tested to a mental age of nine or ten – a result which could "hardly be accidental." To Goddard, this indicated a relationship between "criminal tendencies" and "arrest of development at about nine." At this age, he hypothesized, while the normal child began to acquire the mental capacities necessary for making moral judgments, the subnormal child fell behind.[54]

Since such children often resembled juvenile delinquents, he now wondered, was it possible that many persons in reform schools were actually mental defectives? If so, they were victims of misdiagnosis. Such cases called for the expertise of the psychological tester.

To test his hypothesis, Goddard first sent a questionnaire to reform school superintendents. Among other questions, he asked them to report the number of cases they considered mentally defective, the methods used to treat them, and their success rate. His inquiry drew thirty-four replies, many confirming his suspicion that reformatories housed children who did not belong there.[55]

New York's Elmira State Reformatory, for example, housed a "residue of such low type that very little can be accomplished with them," its superintendent reported. The head of California's Preston School of Industry believed a tenth of his four hundred students "ought not to be sent here," but since the home for the feebleminded was "full to overflowing," there was no other place for them. As for treatment, he could offer kindness, medical attention, and a disciplined schedule of work, sleep, and recreation – methods which had brought "considerable success" in producing "a fairly well equipped boy." They were "certainly better off here than they would be running the streets," he concluded.[56]

Yet estimates varied widely. One superintendent believed almost half his charges feebleminded; another knew of no such cases. On average, these schools reported nearly 15 percent of their charges as mentally defective children who were not getting, in the words of one superintendent, "that kindly care which is impossible for them to receive in a reformatory institution, whose discipline is intended to be of a very different nature."[57]

These boys, Goddard argued vehemently, did not belong in reform institutions, for their childish minds had been "entirely misunderstood both by their parents and their teachers and the criminal authorities." The real danger, he explained, was that misdiagnoses were creating a criminal class. "This is the material out of which we make our adult criminals," Goddard insisted, "since there is no other course open to them." Instead of punishing such children, the courts ought to be "sorting them according to their mental capacity, and consequently, responsibility, and treating them in accordance with their mental condition."[58]

Once again, Goddard argued that the behavior of such persons itself proved intellectual subnormality. Many crimes were "foolish and silly," he noted. "Judge and jury are frequently amazed at the *folly* of the defendant – the lack of common sense that he displayed in his act. It has not occurred to us that the folly, the crudity, the dullness," he argued, "was an indication of an intellectual trait that rendered the victim to a large extent irresponsible."[59]

By 1914, he could confirm this hypothesis, for Binet testers were reporting high rates of mental deficiency within reformatories. At New Jersey's Rahway Reformatory, for instance, testers had found 40 percent to be mental defectives. This figure had reached 70 percent at the Ohio Boys and Girls Schools, and 79 percent at three Virginia reformatories.[60]

To Goddard, these numbers suggested that crime was connected to an inherited condition – a condition not likely to improve much despite the best of loving intentions. In many ways, such views were sharply at odds with the broader emphasis upon environmental remedies that largely dominated the new movement to reform the courts. He found a receptive audience, however, among those working with persons proving resistant to reformation. Prison wardens seconded his scientific conclusions, for example, as did persons frustrated by their inability to convince prostitutes to change their ways. Equally interested were those dealing with other mental "abnormalities." Perhaps, they argued, Binet testing could shed new light on cases sent, for lack of a better alternative, to institutions for the criminally insane.

As early as 1911, Goddard had been invited to test twelve criminals committed to the New Jersey State Hospital for the Insane. These cases, which ranged in age from eleven to fifty, offer good examples of the types of persons most perplexing to the courts – persons who often wound up in institutions of one sort or another. They included a child molester; a woman who had killed an infant; a woman threatening to kill her family; a kleptomaniac; a pyromaniac; and a prostitute. In each case, authorities had found the behavior in question both inexplicable and dangerous – too dangerous to let loose upon society.[61]

The case of the prostitute illustrates once again the frustration that Goddard and his contemporaries faced in explaining the behavior of persons who failed to internalize accepted social morality. If there had been "a good moral character showing in her eyes," Goddard observed, this girl "might be considered pretty." And although her family history was filled with "insanity, alcoholism, illegitimacy and libertinism," it was not her background that had led the courts to question her mental state, but her unrepentance. Married before the age of fifteen, she openly favored another lifestyle. "Married life was too slow for me," she stated frankly. "I grew sick of it. He used me fairly well, was better to me than any man has ever been since, but I hated the sameness and wanted attention from others," she explained. "It was all my fault," she conceded. "Patient talked freely of her past life," Goddard recorded in diagnosing this case. "Shows no sign of shame or embarrassment; said she saw no harm in living as she did." Goddard's professional judgment was clear: "Moral sense lacking, probably never developed."[62]

Perhaps in no area were moral prescriptions stronger, and reformers more frustrated, than in dealing with "wayward girls." When a probation officer allowed Goddard to examine a group of such girls between the ages of fourteen and twenty, he readily obliged. These test subjects were "healthy, physically well-developed young women, who in many instances are quite attractive," he reported. They were "potentially capable of much that is good" but they "do not conform to the conventions of society." And since they had "committed perhaps the worst offense that a young girl can," Goddard felt "compelled to say, almost with a groan, 'Cannot something be done to save such girls?'" The answer was now at hand, Goddard suggested, largely through Binet testing. "Is it possible that we can determine by these tests," he asked, "whether these wayward girls are responsible," or did they behave as they did "because they have not mind enough to do differently?"[63]

While Goddard had expected to find many girls feebleminded, even he was surprised by the results. Only four of fifty-six girls passed the Binet tests for age thirteen. Goddard classified the rest as "clearly mental defectives" who, "had they been taken early enough," could have lived happy and useful lives within institutions. "As it is," he concluded, they "must always be a disappointment, incapable of bearing the responsibilities that have been put upon them." Even worse, Goddard reported, many had already become mothers of children like themselves.[64]

Goddard's study supplied more than Binet scores, for he also described family backgrounds. To Goddard, these chronicles again proved that social problems ran along family lines. His own reports, however, contained more than enough evidence for any environmentalist.

Most evident was a pattern of seriously disturbed home environments. "Neglected and abused at home" was a common complaint. These girls frequently had alcoholic fathers and prostitute mothers. "Mother, very dangerous, immoral woman," read one account. "Works in mill. Goes away at 6:30 A.M.: locks the children out of the house." Many girls were runaways. Their siblings often wound up in institutions as well – be they reformatories, prisons, or insane asylums.[65]

Typical of such cases was the nervous fifteen-year-old whose mother was an "alcoholic and an opium fiend" and father "a confirmed drunkard." Goddard described her as "stubborn, excitable, sensitive, cries easily." After living on the streets for several months, she had been committed to a reformatory. "Expelled from school one year before commitment," her records stated, "as her influence was considered bad for the other children. Mother sent her into the streets to beg for money from men to be used to buy drink. The child was unchaste with them and took the money gotten in this way to her mother."[66]

Even reformers well aware that these girls suffered from bad environments, and willing to provide them with new homes, were finding such cases intensely difficult to "save." One girl had been tried in eleven different probationary homes in one year; she was finally sent to an industrial school. Such failures were not hard to explain, for according to institutional accounts, these girls were temperamental, argumentative, unruly, and angry. "When given some sewing to do," a report stated of one girl, "she tore it into rags. When told to iron some of the officer's clothes, she deliberately burned them."[67]

The reasons for their anger were also evident. "Mother immoral – living with man not her husband – keeping house of ill-fame, using her children for gain," read one history. This girl's alcoholic father had "attempted rape upon his own children." One sister, previously in prison, was "now living a vicious life," Goddard noted; another had been declared insane. "Committed at 15 as beyond control, immoral and a runaway," the girl's records stated. "Had child; father unknown. Wholly incompetent to care for it. Some days wants to give baby away and next day would not part with it for the world. Very nervous and moody."[68]

Most alarming of all to reformers, these girls were promiscuous. One was "constantly attracting the attention of men." "Boy crazy," said the record of another. "Nina," a "well-developed girl, of unusual beauty" at age fifteen when Goddard met her in the Newark Juvenile Court, had a typical family history: "Father an alcoholic, degenerate; mother a prostitute; elder sister a prostitute; sister and brother had gonorrhea." Nina had become impossible to control. "This girl absolutely incorrigible, steals, associates with the commonest type of men, even yelling to them from the House of Detention, absolutely immoral," Goddard observed. "Cannot associate her acts with punishment," he concluded.[69]

Goddard's theories now suggested a reason for the failure of even well-intentioned reform efforts. After all, even a good home could not cure a feeble inheritance. More broadly, Goddard's findings about the mental status of "wayward girls" were relevant to another issue capturing increasing public and political attention during the Progressive era: prostitution.

"Why Is a Prostitute?"

"Perhaps there is no problem looming larger at the present time," Goddard observed in 1914, "than prostitution with its attendant horror the so-called white slave traffic." In the years between 1909 and 1914, a variety of political and social reformers had argued for new strategies to combat this "Social Evil." Among the most publicized were investigations into the trade in "white slaves" – innocent girls reportedly trapped or lured into becoming prostitutes.[70]

This movement quickly caught the attention of persons working with the feebleminded. The belief that many prostitutes were not only of low character, but also of low intelligence, had been widely accepted for decades. Feeble-mindedness was often cited by reformers as one of prostitution's main causes. Even so, Goddard argued, "nowhere has it been given the prominence that is due it." Once again, Goddard saw science as offering a solution to an age-old problem. "One of the most hopeful indications of a better future," he observed, "is that society is beginning to ask *why?*" This could be asked of "nearly all the individuals or types of individuals who are making our great social problems." Thus, Goddard asked the question: "Why is a prostitute?"[71]

The "unthinking," he replied unselfconsciously, "almost invariably have a single answer which suffices to account for the entire condition." Yet since "one person's guess is as good as another's," he offered his own estimates of prostitution's causes.

> "Pure wickedness" – doubtful if any; possibly 1 per cent. "Sex perversion" – probably 2 per cent. is a liberal allowance. "Inordinate sex appetite uncontrolled" – not over 2 per cent. "Innocent girls betrayed" – 10 per cent. is probably high. "Poverty and the industrial situation" – it may be 15 per cent. "Involuntary – white slaves" – doubtless less than 10 per cent.

These reasons explained about 40 percent of prostitution. His explanation for the remainder was simple and unsurprising. "In a word," Goddard concluded, "they are feeble-minded, and not responsible for their mode of life."[72]

This conclusion, Goddard explained, was based on an understanding of both the economic and the moral weaknesses of the feebleminded. "Fortunately," he argued, "most people can earn a living by faithful industry, even though it be of a very menial sort." Persons of little intelligence, however, were driven to do this "dishonestly or immorally." And while the feebleminded boy might fall into a life of crime, the girl "drifts more naturally into a life of prostitution." Such a girl had "normal or nearly normal instincts, with no power of control." Her mental state made her "dull to the moral perceptions," for she could not appreciate "the wrongfulness of such a life." And while some feebleminded women had sought this life, others had simply proven "easy victims of the cadet, the white slaver or the madame," for there were "always those willing to take advantage of her condition."[73]

Yet these arguments needed proof. "Now, what are the facts?" Goddard asked. "Is any considerable percentage of prostitutes feeble-minded?" While Goddard had "as yet no accurate statistical studies," he did have estimates which again suggested strong links between mental and moral deficiency.[74]

First of all, Goddard argued, many women in institutions had illegitimate children. "Though not perhaps professional prostitutes," he admitted, they had still "shown their tendency to go in that direction." Even more worrisome, many Vineland girls were quite attractive; if left alone, their very innocence would make them especially vulnerable to sexual exploitation. "Nellie," for instance, was "a handsome young woman of 19 with the appearance of a girl of 15 and the mentality of a child of 9." She offered a "striking illustration of the type of woman who, out in the world, becomes quickly victimized."[75]

Equally suggestive in linking low intelligence to sexual immorality, Goddard argued, were data gathered from women's reformatories. While women might be sent to such institutions for a variety of crimes, most had been committed solely for sexual offenses. According to contemporaries, these women too were scoring poorly on the Binet scale.[76]

Perhaps the best evidence of this came from research conducted at the State Reformatory for Women in Bedford Hills, New York. Headed by progressive reformer Katharine Bement Davis, this facility became famous as "perhaps the most scientific institution of its kind in the world." In trying to determine which of her charges were fit subjects for rehabilitation, Davis too decided to experiment with mental testing. "The desirable thing is to know beforehand – to diagnose before instead of after treatment," she argued. "If this is reasonable in medicine, why not in penology?"[77]

Toward this end, in 1911 Davis hired Dr. Jean Weidensall to begin testing the mentality of the women at Bedford Hills. A psychologist who had previously worked at Healy's Psychopathic Laboratory, Weidensall spent several weeks at the Vineland school, working with Goddard and "studying his methods." By 1912, she could report her data on one hundred Bedford Hills inmates, seventy-one of whom had been committed solely for sexual offenses. The results were startling, Superintendent Davis reported, for Weidensall had found "not one who is normal mentally." While these women ranged in age from sixteen to twenty-nine, their average mental age was ten. To Davis, this helped explain why her own reform efforts had failed in so many instances. "It is quite obvious that the methods of an educational institution designed to train young women for competition in the world," she concluded, "are quite other than those of an institution designed for the kind, custodial care of women who will never have need for self-direction."[78]

Other Binet studies conducted in women's reformatories confirmed these findings. Of 500 girls at a Geneva, Illinois, reformatory, for instance, more than 74 percent had been committed solely for "immorality." Tests of 104 who had led an "immoral life" showed that 97 percent were below their proper mental age. "This does not by any means indicate that 97 percent of prostitutes are feeble-minded," Goddard cautioned in summarizing this study, for it was "only natural to expect that the feeble-minded ones would be the ones to be caught." Instead, he estimated that half of all prostitutes might be feeble-minded.[79]

High rates of feeblemindedness were also found at the Massachusetts Reformatory and the New Jersey Home for Girls. To Goddard, these findings were convincing. Since "every institution of this character that has been investigated has shown results of this character," he considered it "all but a foregone conclusion that all the other institutions will show the same thing." After all, Goddard wrote, "one does not need to see all the crows that are flying in any season to conclude that crows are black. One may have seen less than a hundred out of the thousands that are alive," he argued, "but if all that one has seen are black, one has a logical right to conclude that all crows are black. . . . " To Goddard the analogy was clear: all investigations to date had found that "a large percentage, in all probability more than half of these girls, are feeble-minded"; new studies were likely to show the same.[80]

Goddard could add personal evidence as well. Although such women appeared normal "to the uninitiated," he claimed, the expert could recognize their mental defect at a glance, in spite of "their physique, in spite of their pretty faces, in spite of their often winsome ways and sometimes even a kind of shrewdness." Goddard had made such on-the-spot diagnoses himself, for along with two physicians, he had conducted "a brief observation of three disorderly houses in Philadelphia," and had "no hesitation whatever" in certifying twenty of thirty women present as "proper persons to be in institutions for the feeble-minded."[81]

The impact of such theories was soon evident in the writings of the political bodies most concerned with curbing prostitution in the Progressive era: vice commissions. If Goddard were correct, then very little could be done to "save" such women besides institutionalization. As the Chicago Vice Commission concluded, "you can rescue the intelligent girl, but the subnormal girl you cannot rescue."[82]

Goddard's influence on these bodies was made even more explicit when George Thacher, a member of Portland's Vice Commission, began referring to four related women living in his community as "Mrs. Kallikaks." "There is no indication of feeble-mindedness on the part of Mrs. Kallikak No. 1," Thacher reported, "but I know of performances of hers *that indicate abnormality.*" "Mrs. Kallikak No. 2" had "a husband who is a divekeeper" and "a daughter who is a prostitute. *She also has a feeble-minded child,*" he emphasized. "Mrs. Kallikak No. 3" had "a *mentally defective child,*" while "Mrs. Kallikak No. 4" was "the matron of a disorderly house." While Thacher had been unable to test this family, he still incorporated Goddard's psychology into his report, for he described such women as "victims of arrested mental development" who had "stopped growing mentally at eight, nine or ten years of age." Most girls who drifted into a "life of shame" were "below normal in intelligence," Thacher argued. This could be proven by simply talking to them, for while their faces might not indicate their mental state, their conversation surely did. "I have listened to their talk many times," Thacher reported, "and very rarely has there been any sparkle of wit or cleverness." During his investigations, he had met "only eight women with whom it was interesting to talk." The rest, Thacher concluded, were "simply poor *children.*"[83]

For Thacher, the need for action was clear. He hoped to establish a "morals court . . . removed from politics" to "cut out the abnormal degenerates from their positions as instructors in the art of love." The most immediate need, however, was proper diagnosis, for society ought to "determine the mental capacity of the offending individual and the consequent responsibility" before determining treatment. It was "folly," Thacher concluded, to treat such persons "as adults having full control of their powers." Goddard concurred. Mr. Thacher, he responded, had demonstrated "an intelligence and judgment that is unusual, and of course his experience is great." His conclusions were therefore "of no little significance and force."[84]

Such solutions were still only second best, however, for to Goddard, "prevention is better than cure." The best course for society would be to diagnose defective individuals "before they have learned what an immoral life is." Such diagnoses would allow many prostitutes to become "inoffensive, pleasant and happy children throughout their lives."[85]

After summarizing his evidence, Goddard now refined his own estimates. "To repeat our causes of prostitution," he explained,

> we still say wickedness – very little; sex perversion – very little; inordinate sex appetite – very little; what seems like this is really lack of con-

trol, due to feeble-mindedness. Innocent girls betrayed – very large, 50 or 60 per cent. perhaps, *but these girls are betrayed because they are feeble-minded* and do not know how to protect themselves. Poverty and industrial conditions – very large; but these people are unable to earn a living because *they are mentally defective,* so they fall into this easy kind of life. Involuntary – the white slaves – very large; but again they are easily the victims of the white slaver *because they are feeble-minded.*

Feebleminded women "yield too easily," Goddard summarized, for "they were but children in mentality and did not understand how to protect themselves."[86]

In this instance, Goddard explicitly used his psychological explanation to challenge economic explanations of prostitution. "Much has been made of the condition of the under-paid shop-girl who is compelled to supplement her earnings by immoral acts," he reported. In many cases, such a girl, "while not perhaps a moron," was "not very many degrees from it, – a girl who has not been able to learn enough to enable her to earn a larger wage." These girls were "merely ignorant, not mentally defective," he cautioned, but "where a fair degree of intelligence is present, the girl does not remain ignorant."[87]

Intelligence testing, Goddard believed, would soon settle questions concerning the mentality of prostitutes. "Some day," he wrote in 1914, "a vice commission or a progressive court will arrest a typical group of prostitutes and test their mentality by approved methods." Only then would the problem "begin to approach solution." Goddard predicted what they would find: a large number feebleminded, many others of low intelligence, a few "sexually abnormal," and those remaining "probably victims of circumstances."[88]

By the time his manuscript was ready for publication, such a study had been conducted, for Binet tests formed a part of the report of the Massachusetts "Commission for the Investigation of the White Slave Traffic, So Called." In fact, the working assumption linking feeblemindedness to prostitution was apparent in the very composition of this commission, for heading it was Dr. Walter Fernald of the Waverley institution. Its findings included an examination of 300 prostitutes conducted by four women: two physicians, one field worker, and one psychologist "of wide experience with the Binet and other laboratory tests."[89]

According to this commission, 154 (51 percent) of these prostitutes suffered from mental defects "so pronounced and evident as to warrant the legal commitment of each one as a feeble-minded person or as a defective delinquent." To Fernald, many seemed indistinguishable from persons in his own institution, which housed "an equal number of women and girl inmates, medically and legally certified as feeble-minded, who are of equal or superior mental capacity." Even those labeled normal were "of distinctly inferior intelligence," Fernald reported, for "only a few ever read a newspaper or book, or had any real knowledge of current events, or could converse intelligently upon any but the most trivial subjects." To Goddard, these facts proved his point. *"Not more than 6 of the entire number seemed to have really good minds,"* Goddard observed in quoting this report, adding the emphasis himself.[90]

Once again, mental testing and family histories seemed to confirm one another. These women "as a class came from shiftless, immoral and degenerate families," Fernald summarized. They were "industrially inefficient, as shown by the low wages received," as well as "very deficient in judgment and good sense." And once again, it was the women's unrepentance which best confirmed this diagnosis. "The general moral insensibility, the boldness, egotism and vanity, the love of notoriety, the lack of shame or remorse, the absence of even a pretence of affection or sympathy for their own children or for their parents, the desire for immediate pleasure without regard for consequences, the lack of forethought or anxiety about the future, – all cardinal symptoms of feeble-mindedness, – " Fernald explained, "were strikingly evident in every one of the 154 women."[91]

For problems such as prostitution, Goddard concluded confidently, a solution was finally at hand. "Attack the problem from the standpoint of feeble-mindedness," Goddard suggested, and "it can be reduced at least 50 per cent. in a very short time."[92]

Witness for the Defense

Goddard's arguments concerning the social irresponsibility of persons deemed intellectually inferior had serious implications not only for preventing crime, but also for broader questions of criminal justice. If the feeble-minded were not responsible, Goddard argued, then they ought not be held liable for their actions. In 1914, Goddard took to the witness stand on two occasions to try to explain this point to juries. His efforts put him in the most unpopular positions of his career, for in both cases, Goddard testified against imposing the death penalty on murderers whose crimes had inflamed the wrath of the public.

The very fact that Goddard had been called to testify in such cases was itself innovative, for in 1914 the role of psychologists within the American legal system had yet to be established. The most influential of Goddard's contemporaries promoting such a role was Hugo Munsterberg, Harvard's outspoken and controversial émigré academic, who had tried to win for American psychologists the legal status that his colleagues in Germany and France had already begun to enjoy through their studies of hypnosis, suggestion, and the reliability of witness testimony. In 1896, one of Munsterberg's former co-workers had apparently become the psychology profession's first "expert witness" by testifying in a Munich murder case that trial publicity had led witnesses to "retroactive memory-falsification."[93]

Munsterberg's intercessions into the American legal system had brought him maximum publicity, but minimal judicial success. His opinion had first been sought in the case of Richard Ivens, an apparently feebleminded Chicago man who confessed to a murder and then retracted his confession. Suspecting mental abnormalities, the defense mailed Munsterberg a copy of the confession, which he promptly dismissed as a "rather clear case of dissociation and auto suggestion." His involvement backfired, however, for local newspapers wrote of "Harvard's Contempt of Court" and saw psychology as "another way of possibly cheating justice." Meanwhile, Ivens was executed before record crowds, who cheered approvingly outside the jail. In his second case, Munsterberg sided with the prosecution, with no better results. This time he interviewed labor activist Harry Orchard, who had confessed to killing a former Idaho governor, thus gaining immunity for himself while implicating labor leader "Big Bill" Haywood. After administering nearly a hundred psychophysical tests, Munsterberg declared Orchard's testimony "true in every word." "As far as the objective facts are concerned," he reported in his typically arrogant manner, "my few hours of experimenting were more convincing than anything which in all those weeks of the trial became demonstrated." Apparently the jury disagreed, for Haywood was acquitted.[94]

Neither these courtroom failures, nor the criticisms by lawyers that his writings constituted "yellow psychology," nor the press's mocking references to his psychological apparatus as a "lying machine" dampened Munsterberg's confidence that the courts would soon come to appreciate psychological expertise. "To deny that the experimental psychologist has indeed possibilities of determining the 'truth-telling powers,'" Munsterberg insisted, "is just as absurd as to deny that the chemical expert can find out whether there is arsenic in a stomach. . . . "[95]

Goddard seconded Munsterberg's belief that applied psychology belonged in the courtroom. Yet he interpreted his own role as an expert witness in a different way. Goddard's model was not the chemist but the psychiatrist. And whereas Munsterberg's brash behavior incurred the hostility of lawyers and judges, Goddard blended more easily into courtrooms, for his actions closely resembled those of institutional physicians whose testimony had been accepted for decades.

In a broader sense, both Goddard and Munsterberg concurred in arguments concerning the causes of crime. In his popular 1908 book, *On the Witness Stand,* Munsterberg had explicitly rejected Cesare Lombroso's criminal anthropology. "The psychologist," he warned, "is to disburden society of its responsibility for the growth of crime, inasmuch as he is called to testify that the criminal is born as such." In such cases, Munsterberg told his colleagues, psychologists should "refuse to furnish evidence." Still, Munsterberg did suggest that inherited tendencies might be involved indirectly. "In short," he explained, "there are minds which are born slow or stupid or brutal or excitable or lazy or quaint or reckless or dull – and in every one of such minds a certain chance for crime is given." Such conditions were not in themselves determining, for "to be born with a mind which by its special stupidity or carelessness or vehemence gives to crime an easier foothold than the average mind," according to Munsterberg, "certainly does not mean to be a born criminal." Yet he was still suggesting that certain inherited traits might lead to a higher susceptibility toward crime.[96]

Goddard's writings made an analogous distinction between strict determinism and increased susceptibility. In "the light of present-day knowledge of the sciences of criminology and biology," he argued, "there is every reason to conclude that criminals are made and not born." Yet the "best material out of which to make criminals" was, of course, feeblemindedness.[97]

To Goddard, this indirect connection still suggested that such persons were not fully responsible, for they did not possess intelligence enough to understand their own actions. "We are at once possessed by a feeling which is akin to that which we have for the two-year old child," he wrote, "who has pulled the cloth and pushed the lamp off onto the floor, thus burning up the house. . . . " Such behavior ought to inspire an "attitude of pity and sympathy, not of revenge." By using Binet tests in court, Goddard believed, psychologists would be able to diagnose the mental age of the accused, and thus to assess the degree of responsibility.[98]

The first case in which Goddard testified concerned a homicide committed near the village of Poland, New York. Goddard's arguments did little to endear him to the local community, for both prosecutor and public, he reported, saw this crime as "a carefully planned, premeditated, cold-blooded murder of the most atrocious character, committed with a fiendishness seldom seen among human beings." Ironically, the case in question was also especially symbolic for Goddard: it involved a troubled pupil who had murdered his teacher.[99]

The facts of the case were uncontested. On the morning of March 28, 1914, the bloody body of Lida Beecher, a twenty-year-old schoolteacher, had been found. The same day, sixteen-year-old Jean Gianini was apprehended. Although all evidence was circumstantial, the boy had confessed, telling without remorse how he had stabbed his teacher to death because she had repeatedly humiliated him in class.[100]

Jean's father, a prominent member of the community, believed his son insane. To defend him, he hired John F. McIntyre, a well-known New York City attorney, who tried a novel defense: although this boy had sometimes earned good grades, Gianini was still an imbecile, for he did not understand the quality of what he had done, and was therefore not responsible. Among

Jean Gianini, c. 1914. The picture in the upper right corner was taken in jail. From H. H. Goddard, *The Criminal Imbecile* (New York: Macmillan, 1915), frontispiece.

those he called to testify were three nationally prominent experts. Dr. Carlos MacDonald, president of the American Psychiatric Association, saw in Gianini the physical "stigmata of degeneracy," including an oddly shaped head and palate. Dr. Charles Bernstein, superintendent of the Rome, New York, institution, confirmed this diagnosis. Most crucial of all, however, was Goddard's testimony, for he could diagnose mental deficiency through intelligence testing.[101]

"Is Jean Gianini an imbecile?" Goddard asked. The answer, of course, depended upon the definition of this term (used in law as a generic synonym for feeblemindedness). Gianini was a *"high-grade* imbecile," he concluded, the kind that rarely showed their defect in their faces and could thus be recognized only by experts. In this regard, Goddard explained to the jury, his own work with Binet's tests resembled the work of a physician whose instruments allowed him to discover diseases that eluded laymen.[102]

And while the public might regard this diagnosis as "absurd," Goddard argued that it was absurd to believe anything else. In ignoring such diagnoses, "we insist upon believing the unbelievable," he maintained. "We view a crime like the one under discussion and say frankly, 'It is unbelievable that any reasoning, intelligent person could commit such an atrocious act,' and yet we believe that this boy did," and that "such a grade of villainy exists and that it can suddenly appear in a boy who never before manifested anything approaching it." To Goddard, this made no sense. "The fact is," he concluded, "that our instinctive revulsion against such a thought is the correct view."[103]

Unless one believed in the unscientific concept of "villainy," Goddard argued, the crime itself was "the strongest kind of evidence that he is not a normal boy." Gianini's feeble attempts to cover his crime, the language of his confession, and his childish actions in the past all supported this diagnosis, Goddard proposed. So did Gianini's behavior toward his defense counsel. When his lawyer brought in several experts to help the boy, he had proven far more interested in eating his dinner. "As one of the experts testified," Goddard remarked, "'As between soup and safety, Jean prefers soup.'"[104]

Goddard knew of other psychological theories that also explained this boy's behavior. There was, for instance, the theory of "adolescence," made popular by his former mentor, G. Stanley Hall. "Jean was sixteen years old, an age when sexual passion is strong," Goddard recounted. "The new physiological function of sex is established, great psychic changes have occurred," he explained. In a boy "morally well-endowed," this meant "the evolution of ideals, ambitions, moral and religious ideas, attention to dress and appearance, interest in the opposite sex," and a new physical potency which found outlet through "polite and friendly association with his girl friends, in chivalric attentions and devotions, with more or less definite plans for future marriage and parenthood." In persons with "little or no moral principle," however, "we see the impulse leaping over the social conventions and attaining complete sexual gratification illegally." Such impulses might also be expressed violently, ranging from "rough horseplay with girls, such as pushing, pulling, . . . up to physical injury, torture, and even murder." The scientific literature on this subject was growing. "Volumes could be written – indeed volumes have been written," Goddard exclaimed, " – showing the tremendous force of this sex impulse at this age, and the multifarious ways in which it expresses itself. . . . "[105]

Goddard's best evidence, however, and the reason he had been called, concerned Binet tests, which showed Jean to have a mental age of ten. "It was somewhat difficult to estimate his mentality with the usual exactness," Goddard had to concede, "since others had already used the tests," thus making it "impossible to say how much Jean had learned from his previous examinations." Among these were psychiatrists brought in to rebut Goddard's testimony. Thus, in the Gianini trial, answers to Binet's questions were interpreted by both the prosecution and the defense.[106]

In this battle over Gianini's mental state, his lawyer insisted, it was Goddard's judgment that ought to prevail. "Now Goddard," he told the jury, ". . . swears with the utmost degree of positiveness, without halt or hesitation or equivocation . . . that he found this boy in the ten year old class." Such testimony represented the latest in scientific expertise. "Goddard says that this test is utilized everywhere in the schools; they have laboratories in the primary school; they use it here; they use it everywhere; that he has been using it for years; and that it requires expert knowledge to use it," McIntyre added. "And gentlemen," he reiterated, "Goddard says he is in the ten year class."[107]

Ironically, while the experts argued over Binet tests, the jury seemed equally influenced by family history. Jean's deceased mother had been diagnosed as alcoholic and insane, and a deceased brother probably an idiot – all evidence of "neuropathic ancestry." Once again, both the psychological and the biological evidence pointed toward mental deficiency.

Whatever this jury's reasoning, its decision brought Goddard – and applied psychology – its first American legal victory, for the defendant's life was spared "on the ground of criminal imbecility." Instead of being executed, Gianini was committed for life to New York's Matteawan Asylum for the Criminally Insane.[108]

As expected, this action was received bitterly by the local community. "Not infrequently have verdicts in murder trials been unacceptable to the populace," Goddard conceded, and many considered this one "unaccountable." Goddard, however, found it remarkable. "Probably no verdict in modern times has marked so great a step forward in society's treatment of the wrongdoer," he boasted. "For the first time in history," he declared, "psychological tests of intelligence have been admitted into court and the mentality of the accused established on the basis of these facts." The significance of this, he exclaimed, "cannot be overestimated."[109]

The Gianini case, according to Goddard, had set a legal precedent by recognizing that "*weakness* of mind" was "of the same importance as *disease* of mind." It meant that feeblemindedness, like insanity, needed to be considered "in all discussions of responsibility." As a corrolary, it also suggested that psychologists be accorded the same legal status as psychiatrists in assessing the criminal mind.[110]

Goddard's legal precedent was short-lived, however, for in his second case he suffered a setback. This time he testified in a criminal trial in Media, Pennsylvania. Once again, Goddard was called upon by the defense to gauge the mind of a murderer who had committed an apparently senseless crime.

Accused of the November 7, 1913, murder of Lewis Pinkerton, a farm manager, were George March, a dairyman, and nineteen-year-old Roland Pennington, a farm laborer. The state's case against both men rested on Pennington's confession, which had led the defense to question his mentality.[111]

Pennington had related March's jealousy over his common-law wife's rela-

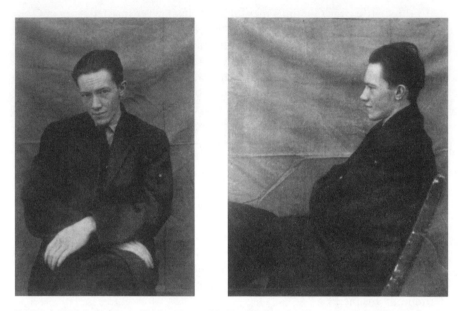

Roland Pennington, c. 1914. From H. H. Goddard, *The Criminal Imbecile* (New York: Macmillan, 1915), facing p. 42.

tionship with his employer, Pinkerton. Apparently, March had convinced Pennington to help him commit murder by telling him that Pinkerton had a thousand-dollar bill in his pocket. The boy's description of the crime that ensued was chillingly mundane.

> . . . George and I were separating the milk down at the milk house before breakfast, and George said, "Well, Lew will have that bunch of money on him to-day. Let's get it." I said, "What do you mean?" He says, "Why, do away with him." I says, "What? Kill him?" He says, "Yes." I says, "No. I won't kill him." He says, "Well, you start it and I'll finish it. . . ." I didn't say anything.
>
> That afternoon . . . George . . . handed me the blackjack and said, "Here's the blackjack; you can do it with that." I put it in my pocket. We then went to the barn. From then up to about five o'clock, while we were working about the barn, George kept saying to me, "Don't lose your nerve. The first chance you get after the workmen are gone, get him." Several times, he said, "Don't miss your chance – Don't forget."

Pennington then told how he had struck his surprised employer in the head, and how March had then helped him bludgeon the man to death and bury the body. Afterward, Pennington seemed to forget about the money which was supposed to be his motive.[112]

The defense attorney was baffled, for he found his client "unfeeling and without proper appreciation of the enormity of the crime." Since circumstantial evidence suggested the workings of a childish mind, and since Pennington had a cousin in nearby Elwyn institution, Dr. Barr sent his medical assistant, Dr. P. M. Kerr, to join the defense.[113]

Kerr's first interview with his client was revealing. "Did you like Lew?" he asked Pennington. "Yes, Lew was a good fellow. Me and him never had any trouble," was the response. "Then, why in the name of all that is natural, did you kill him?" Kerr inquired. "I don't know, I guess because George told me to," the boy answered without emotion.[114]

Since medical tests revealed no serious abnormalities, Kerr improvised his own version of an intelligence test. "Roland, you studied United States History; what were the causes of the Civil War?" he asked. "To free the slaves," Pennington replied. "Who was the commander in chief of the Revolutionary War?" the doctor asked. "I don't know that one," the boy answered. "Was it Washington?" Kerr prompted. "Oh, yes," his client answered.[115]

Perhaps because such answers proved nothing, the defense instead turned toward Goddard's tests. In defending Pennington, Goddard took another unpopular stand. "Here again is a crime so abhorrent in its details that it is unbelievable," he conceded, for while March's behavior could be explained by jealously, Pennington had "no adequate motive." His behavior was "absolutely incomprehensible on any theory that assumes that he is a normal boy of nineteen years," Goddard insisted. As expected, Binet tests confirmed his suspicions, for Pennington tested to a mental age of eleven.[116]

In this case, Goddard saw his conclusions as congruent with medical evidence. Even so, Dr. Kerr's writings suggest a subtle rivalry between physician and psychologist. Intelligence tests might prove of interest to psychologists, Kerr argued, but they did not measure "real intelligence." More importantly, their usefulness to doctors was "negligible," for to "the initiated" like himself, he contended, such tests were "superfluous."[117]

Yet intelligence tests did have one major advantage over medical methods, even Kerr conceded: they were "intelligible to a layman." Such persons "may not appreciate just what low, middle, and high grade imbecility means," he explained, but they did know what to expect from an eleven-year-old child.[118]

This advantage proved especially crucial with juries. "During the trial, therefore," Kerr reported, "we deemed it wise to declare to the jury: 'Roland is a high-grade imbecile or moron and his mentality is that of a normal boy of eleven.'" Yet because he was citing them, Kerr found himself in the ironic position of having to defend Binet's tests on the witness stand. "One would think that I am the culprit," he complained in responding to the prosecutor's "bully ragging." "The attorney sneers at the Binet tests (his questions clearly show his entire ignorance of their very existence). . . . "[119]

Once again, while both sides argued over test results, an equally crucial piece of evidence proved to be family history. So important was it to challenge Kerr's claims of "neuropathic family inheritance" that the prosecutor called four physicians to dispute the mentality not only of Pennington, but of his institutionalized cousin as well. These "medical mercenaries," Kerr noted contemptuously, were merely general practitioners and thus "pseudo-experts." They had given the jury the impression that "a normal boy could be incarcerated against his will." Such an absurd proposition, Kerr insisted, proved only "the unscrupulousness of some physicians, the assininty [*sic*] of some attorneys, and the patience of a certain Judge."[120]

Against such charges, Goddard rallied to Kerr's defense. "It was unfortunate, indeed," he argued, that such men had been "allowed to pass as experts." The knowledge of institutional personnel, Goddard insisted, far surpassed that of any other physician. In this case, institutional physician and institutional psychologist banded together in a common defense of experiential expertise.[121]

Their defense of Roland Pennington, however, failed to convince the jury. Goddard attributed the guilty verdict to the jury's fears that declaring Roland an imbecile would have invalidated his testimony against George March as well. Thus, instead of risking an acquittal for both men, they sentenced both to death.

The verdict outraged Kerr. "Experience of centuries has taught us to synonymize irresponsibility and imbecility; we know that the average imbecile is a potential criminal – protect him!" he exclaimed. "He needs a harbor, not a dungeon." Roland had received "the same brand of justice that has been meted out since the days of Medievalism." Kerr's contempt for general practitioners was exceeded only by his contempt for the jury. "Behold then, the expert," Kerr surmised, in a statement far more bitter than any made by Goddard, "holding forth in all sincerity and faith, and in manner deferential – expatiating matters psychological to plow-boys and negro coachmen!"[122]

Goddard too was deeply disturbed by the verdict. It was "abhorrent," he argued, "to think of a child (in mind) going to the electric chair" for a deed committed "under the influence of a superior intelligence." Pennington's weakened mind, he argued, had made him especially susceptible to suggestion. In executing him, society was in effect murdering a child. During the following year, Goddard worked closely with the local Quaker community to try to overturn the verdict and spare Pennington's life. He wrote a strong letter to the governor of Pennsylvania asking him to commute the sentence. Pennington's lawyer even had hopes of taking his appeal all the way to the

Supreme Court. These efforts did not succeed, however, and in 1916 Pennington was executed. His arguments had failed, Goddard wrote, to "save the commonwealth the shame of officially putting to death a person who had only a child's intelligence."[123]

Goddard may have lost the Pennington case, but his theories were beginning to win converts within the American criminal justice system. His influence could be seen, moreover, at many levels. Goddard's ideas caught the attention, for instance, of powerful politicians such as James Cox, Ohio's progressive governor. By 1914, Goddard had helped Cox create the Ohio Bureau of Juvenile Research, an agency charged with investigating cases and recommending sentencing.[124]

More broadly, Goddard's arguments were beginning to raise doubts in the minds of lawyers and judges. This was especially striking in the changed career of Harry Olson, Chief Justice of Chicago's Municipal Court. As an aggressive prosecutor, Olson had won the death penalty in many cases, including that of Richard Ivens, whose confession Munsterberg had questioned. "I noticed that most criminals appeared to be both physically and mentally deteriorated," he later recalled, "but believed they were wicked and depraved" and "if they wished they could have been moral and honest." This was the "traditional view of the legal profession." Over the years, however, Olson began to question "the wisdom of some of our laws" as well as "our method of crime suppression." The Ivens execution haunted him. "This case so impressed me with the inadequacy of our methods," he confessed, that "I concluded, when soon afterwards I became head of the Municipal Court," to "create a psychopathic laboratory." Olson also became an ardent eugenicist. In 1914, he opened his laboratory; on its staff was Samuel Kohs, a psychologist who had worked with Goddard at Vineland.[125]

Goddard also tried to use his talents as a popular writer to bring his message to a wider public – the types of persons who might serve on juries. In 1915, he published *The Criminal Imbecile: An Analysis of Three Remarkable Murder Trials.* Included were popular accounts of the lives and trials of Jean Gianini and Roland Pennington. By this time, Goddard could add evidence from a third trial, held in Portland, Oregon, in 1914, in which a psychologist had used Binet testing to diagnose the mind of a murderer. By 1915, one verdict was clear: psychological tests of intelligence had entered into the American criminal justice system.[126]

"Marvelously Fruitful" Hypotheses

Goddard's novel fusion of psychological and social diagnoses suggested new answers to society's problems. By preventing the spread of hereditary feeble-mindedness through early diagnosis and institutional segregation, he maintained, pauperism, crime, and prostitution could be reduced by as much as 50 percent within one or two generations. These preventive measures would also have an impact on other problems perplexing reformers. Thus, by 1915, Goddard had elaborated a sociology linking intelligence testing to just about every important social question of his day.

The temperance question, for instance, could now be reconsidered. Physicians had long speculated on the relationship between alcohol and birth defects. In Goddard's data, alcoholic families had more than twice as many miscarriages and infant deaths as others. At least eighteen Vineland children had alcoholic mothers (and sixty-two alcoholic fathers); nearly all these families were in the "hereditary" group. "A decided relation between Alcoholism and Hereditary Feeble-mindedness!" Goddard concluded. "Is it a causal relation? If so, which is cause and which is effect?" Goddard's own answer was clear: there was no need to cite alcohol as a cause, for "the feeble-mindedness of the ancestors is the all-sufficient cause." To Goddard, chronic drinking was simply another consequence of diminished intellectual capacity and its corollary, diminished social responsibility. "Indeed one may say without fear of dispute," he announced, "that more people are alcoholic because they are feeble-minded than vice versa."[127]

"We may say that every feeble-minded person is a potential drunkard," Goddard now proposed, for such persons had "no control over their appetites," and given temptation, "they easily yield." These individuals could be "preached to and profess conversion, only to be found in the gutter tomorrow." Their children were "told about the effects of alcohol, and then go out on the street and into the saloon." This explained why educational efforts had failed to reduce the "drink-bill of the nation." Thus, Goddard could now offer a scientific explanation for the failure of the temperance campaign – a campaign that had been strongly endorsed by his own family as well as his Maine community.[128]

High truancy rates might be explained in an analogous way. "Many a school child becomes a truant because he *cannot* succeed in school," Goddard noted. Even the annoying group of "undesirable citizens" that Goddard called the "ne'er-do-wells" – those who "while not paupers . . . often have to receive assistance from others; while not criminal, prostitute or drunkard, are still shiftless, incompetent, unsatisfactory and undesirable members of the community" – could now be considered in a new light. In the past, such persons had been called "wilful, wayward, or ignorant." Goddard's new explanation disparaged free will and emphasized inherited incapacity. "Is it not possible that they have not learned better manners and habits," he asked, "because they *could* not?"[129]

In speech after speech and article after article, Goddard stressed mental incapacity as cause, social irresponsibility as consequence. His arguments in

each instance followed the same logic. First he described the social characteristics of persons already institutionalized – their inability to support themselves, for example, or their potential for violent or delinquent behavior, or
their lack of sexual self-restraint. His hypotheses then posited the converse:
that noninstitutionalized persons exhibiting similar antisocial traits might in
fact be mentally defective. Finally, he used Binet tests to prove that such
persons were frequently feebleminded. In each instance, test results showed
the same thing: many had a mental age of twelve or less. Goddard had thus
apparently discovered the underlying factor connecting defectives, dependents, and delinquents.

For those who thought his solution too simple, he had a ready response.
Why, he asked rhetorically, had this connection "not been discovered
sooner?" The answer was due to a common misunderstanding. In the "popular mind," Goddard explained, feeblemindedness was "synonymous with idiocy or imbecility" – grades of defect often manifested "in the faces of the
individuals." Yet most social problems emanated from the highest grade, the
morons, whose defect could be diagnosed only by experts. It was this discovery that had allowed Goddard to "formulate working hypotheses which have
proved marvelously fruitful." If Goddard were correct, then society urgently
needed the diagnostic skills of the psychological tester to help prevent crime
and poverty.[130]

Goddard's arguments had clearly proven persuasive to many of his contemporaries. In a few short years, he had taken psychological tests of intelligence far beyond institutions for the feebleminded, and even beyond the
classroom. Between 1910 and 1915, intelligence testing had been introduced
into a wide range of American institutions, among them reformatories, insane asylums, prisons, and courtrooms. Binet's questions had been asked of
paupers and prostitutes, delinquent boys and wayward girls. Their answers
had influenced decisions made by physicians, administrators, vice commis-
sioners, judges, and juries. Even more broadly, Goddard had succeeded in
focusing the nation's attention on a set of potentially menacing characters:
the feebleminded pauper, the defective delinquent, the subnormal prostitute,
the criminal imbecile. In the literature of the day, these characters would
appear again and again in discussions of the causes of crime and poverty.[131]

During the next five years, Goddard and his fellow testers would expand
their influence even further, thus reaching far beyond the population of "defectives, dependents, and delinquents." Such actions, however, were not without their critics. By 1915, Henry Herbert Goddard would find himself
embroiled in an intense new debate over the very meaning of the term "intelligence" – psychologically, biologically, and socially. It was this debate that
would begin to call into question nearly all of his previous findings.

7

Psychological Work and the State:
Reformers, Professionals, and the Public

"A Wise Mother State"

"Among the social tasks that confront state governments today," noted a 1914 editorial in the *Survey,* the social work journal which perhaps best captured the spirit of progressive reform, "none is more pressing than the care of the feeble-minded." This was true not because the feebleminded were so dangerous in themselves, this journal argued, but because their involvement complicated just about every other social problem. Even so, caution was necessary, for many a scheme designed to usher in a "golden age of civic righteousness" had instead led only to "new evils." After all, it "frequently happened in the history of reform movements," the *Survey* reminded its readers, "that a remedy, like the nostrum of the quack doctor, has actually caused an exacerbation of the very ill it was designed to abate." Anyone doubting this, it added, had only to look at the dismal historical record of legislative bills entitled *"an act to amend an act, entitled an act to amend an act, entitled, etc."*

Fortunately, in caring for the needy, these editors predicted confidently, this was not likely to happen, for in dealing with this social problem "we are on firmer ground." Reformers now knew that the blind, deaf, insane, and epileptic were "unable to protect themselves" and therefore "dangerous and disturbing elements in orderly society, which a wise mother state will care for and control for their own best good and for the benefit of normal citizens. We make the plea for the feeble-minded," they continued, "that they also are understood" for "as to them, the day of experiment and question has passed."

This new understanding could be expressed simply. "Feeble-mindedness," the *Survey* concluded, "acts alternately as cause and as consequence of alcoholism, immorality, pauperism, criminality, insanity, epilepsy." It was the "visible sign of degeneracy, that master evil of civilization, which, if unchecked – no less than war, and with a persistence which war itself fails to match – threatens the supremacy of the leading races of mankind."[1]

To the reform-minded editors of the *Survey,* by 1914 American society had learned enough about the feebleminded. It was now ready for action. After years of studies and surveys, scientific research and social experiment, re-

222

formers knew what had to be done. All that was needed was to inform the public to create the proper climate for political reform.

Intertwined in this editorial were the two themes that would dominate public discussions of this issue throughout the Progressive era – humane concern and social alarm. Although later generations would see these themes as inherently contradictory, to reformers of this era they were complementary. The feebleminded deserved kindly care, they insisted; if they did not receive it, they would wreak untold havoc on society. Americans could deal with this issue now, they warned, or pay a much higher price later in degenerate offspring.

Evident as well was the mixture of old and new – old fears about the dangers of degeneracy mixed with new calls for expanding state action. Degeneration, these reformers argued, meant regress, the very opposite of progress. It could best be opposed by empowering the state to provide more care for and control over the health, education, and welfare of its citizens – the kinds of care that might be provided by "a wise mother."

Progressive America might be considered a "mother state" in a second sense as well, for this era saw a new generation of women reformers assume positions of political influence. Their very presence blurred the meaning of older divisions between private and public, home and state, and brought a new legitimacy to issues concerning not only women but also children. Perhaps nothing symbolized their new role better than the 1912 appointment of Julia Lathrop to head a new federal agency, the Children's Bureau.[2]

In an age preoccupied with conserving natural resources, many women focused on the most precious resource of all: the "conservation of the child." Some worked under a label which still carried a missionary connotation, "child saving"; others operated within the traditional framework of female reform societies and women's clubs; still others sought influence through their newer roles as professional "social scientists." All these efforts had broad ramifications for the population now classified as society's permanent children, the feebleminded.[3]

Much of the massive attention being focused on the feebleminded in the decade before World War I could be traced directly to Henry Herbert Goddard and the Vineland Training School. As feeblemindedness gained in social and political importance, so too did the professionals who now claimed to be best able to diagnose it – intelligence testers. In the years between 1910 and 1917, Goddard and his fellow testers proved increasingly successful in convincing city, state, and federal agencies to use their expertise. In the process, they disseminated intelligence testing ever more widely to the American public.

Such a rapid dissemination took even the psychology profession by surprise. As diverse segments of the reform community increasingly embraced testing, the community of professional psychologists struggled to reach a scientific consensus about just what it was that they were promoting, and how it might be regulated. Equally interested in regulating testing would be the medical profession, for doctors too were soon drawn into the controversies generated by this new psychological science.

During the prewar decade, psychologists, physicians, reformers, and the public would all become participants in a series of debates concerning the new science of the mind and the new role of the state. Implicit in shaping these debates was a widely held assumption – that social reform ought to be grounded in knowledge gained from social science. Equally crucial was a set of evolving relationships – between reformers and scientists, between scientists and professionals, and between professionals and the public. Each of these relationships would be deeply affected by ideas developed by Henry Herbert Goddard in working with the feebleminded. Each would also be restructured within the broader political context of progressive efforts to create a "wise mother state" – a state with more power over and more responsibility for the health, education, and social well-being of its citizens.

Health Reformers

In the decade preceding World War I, the problem of feeblemindedness was debated within a political context strikingly different from that found in postwar decades. Still largely absent was the sharp polarization pitting reformers who believed in improving "heredity" against those wishing to reshape the "environment." Of course, arguments emphasizing the importance of heredity were still being widely promoted, especially by leaders of the eugenics movement. Yet even eugenics enthusiasts such as Goddard did not feel a need to declare their allegiance to reforming only one of the two. Instead, many reformers of this era felt quite comfortable in calling for improvements in both.

Their attitudes were captured in the language of the day, for more frequently invoked by reformers than either "nature" or "nurture," "eugenics" or "euthenics," was the more ambivalent and complex term "hygiene." To Goddard and his contemporaries, this term incorporated both medical and educational, biological and sociological components. It also meant that political reformers had absorbed and were now extending one of the most powerful lessons learned from the public health campaigns: that what a society could not cure it might still prevent. In placing a strong emphasis upon prevention, both eugenicists and social workers, physicians and educators, scientists and politicians frequently found common ground.

During the prewar decade, several American reform movements were organized explicitly around this concept. The National Committee for Mental

Hygiene, for example, was created by physicians, philanthropists, and re-formers committed to promoting a preventive approach toward mental ill-ness. A similar coalition founded the American Social Hygiene Association with the aim of preventing both prostitution and venereal disease. Even the public health movement occasionally found itself rechristened "public hy-giene." Child-Study advocates experienced a similar rechristening. "The new interests and endeavors in the conservation of children," psychologist Arnold Gesell reported, "we group under a new and pregnant phrase, Child Hy-giene," a fact legitimized when the National Education Association changed the name of its "Department of Child Study" to "Department of Child Hy-giene" in 1911.[4]

The child hygiene movement proved especially significant for the emerging profession of "clinical" psychologists, many of whom found themselves working at the interstices of public school and public health reform. Among its leaders was Lewis Terman, who became a strong advocate of free, state-sponsored medical and dental clinics for schoolchildren. Terman's concern with reforming nurture as well as nature is apparent in the very titles of his three earliest books: *The Teacher's Health: A Study in the Hygiene of an Occu-pation* (1913); *The Hygiene of the School Child* (1914); and *Health Work in the Schools* (1914). In addition to testing, Terman promoted open-air classrooms for children with tuberculosis. His motives here are not hard to discern, for he too suffered from terrifying bouts of this disease. "There is an old saying that if you scratch a health reformer," Lewis Terman wrote of himself, "you will find an invalid."[5]

Goddard too saw himself as a health reformer working to bridge the gap between psychology and medicine. In 1909, he was one of many psychologists invited to participate in the famous conference organized by G. Stanley Hall to commemorate the twentieth anniversary of Clark University. On this occa-sion, Hall once again made history by sponsoring Dr. Sigmund Freud's only visit to America. Yet while Freud's lectures on psychotherapy suggested the direction that clinical psychology would take in the future, Goddard's brief address at the same event spoke to a far more mundane and humble present. "Research in School Hygiene," his title stated, "in the Light of Experiences in an Institution for the Feeble Minded."[6]

What psychology could best contribute to child hygiene, Goddard be-lieved, was research. Scientific studies, he argued, offered the only means of supplying definitive answers to questions of social welfare. His efforts to study the feebleminded were strongly seconded by his employer, Edward Johnstone, who supported investigations into anything that might "materi-ally help any one of these children." Answers were needed, Johnstone argued, to tell society "what to do to ameliorate the condition as we find it, what is necessary in order to prevent it, and how to present the facts so that even the

most ignorant shall heed and follow." These questions concerning remedia-
tion, prevention, and public education could only be answered by scientific
research.[7]

Toward these ends, in 1910, while its field workers were busily engaged in
conducting eugenic research, the Vineland school also opened a "Division of
Medical Research" dedicated to "the betterment of the physical and mental
condition of our children by discovery of curable or partially curable defects."
This new department conducted blood examinations, urinalyses, and tests
for tuberculosis and syphilis.[8]

By 1914, Goddard's research department had been expanded to include
three divisions: psychology, psychopathology, and the brand-new field then
calling itself "biological-chemistry." By 1915, laboratory work at this institution filled thirty rooms. Even university scientists were impressed by its psychological, medical, biological, sociological, and pedagogical studies. "It
may be said, indeed," Gesell commented, "that an embryo university has
sprung up within this institution for the feeble-minded." Despite the pessimism implicit in many of Goddard's hereditarian pronouncements, his small
laboratory instigated more research into possible cures for mental disabilities
than any other American institution of this decade.[9]

Ever the optimist, Johnstone envisioned a future when man could prevent
birth defects. This would be possible, he believed, if reformers would commit
themselves to supporting scientific research. "In my dream," Johnstone wrote
in 1914, "I see the determination of the cause of that unfortunate type called
the Mongol. . . . " "Perhaps," he added in foretelling of brighter days to
come, "we may even examine the body fluids of the expectant mother, and by
administering whatever may be lacking, prevent the birth of such a child."[10]

Yet such remedies remained dreams. In the meantime, both Johnstone and
Goddard believed, the state would need to find other ways of preventing
feebleminded persons from passing on their condition, with all that it implied
for society, to their children. The most effective methods for stopping this
propagation, they argued, were institutional segregation or sexual sterilization. To Goddard and Johnstone, these actions could be justified as public
health measures necessary to ensure the physical and social well-being of
the next generation. Both methods were potentially promising in helping to
eliminate feeblemindedness, they argued; both, however, would also involve
difficult questions of state power and public acceptance.[11]

The question of sterilization was especially problematic. In an era when
even disseminating information about birth control was illegal, Goddard
joined his medical colleagues in supporting the legalization of involuntary
sexual sterilization. Goddard still had qualms, however, about advocating the
widespread use of this procedure to eliminate feeblemindedness. As a practical measure, Goddard argued, sterilization policies were "fraught with

difficulties." Like many of their contemporaries, Vineland personnel worried that by removing the fear of pregnancy, sterilization might actually encourage immorality and prostitution (and spread venereal disease). Even more important was public resistance to the use of such surgery, as well as lingering questions concerning its safety, consequences, and, above all, its constitutionality. The sterilization law passed in New Jersey in 1911, for instance, contained several provisions which Goddard personally opposed, for it allowed for sterilizing not only feebleminded and epileptic persons, but also rapists and convicts – provisions which he considered more punitive than medical. By 1913, its legality had been effectively challenged in state courts. In sum, despite Goddard's frequent discussions of this topic, no sterilizations were performed in New Jersey institutions during the years when he worked in this state.[12]

A far better way to stop the propagation of feeblemindedness, Goddard believed, was through institutional segregation. Yet this solution too was problematic. After all, only a very small fraction of the feebleminded population currently resided in institutions. Convincing parents to commit their children to such facilities, Goddard knew, would be a formidable task. Even more difficult would be funding such a plan, for many more state institutions would have to be built.[13]

Convincing the public to fund more state institutions was also of particular importance to Vineland personnel. In order to prevent their institution from becoming a custodial home for adults rather than a training school for children, the Vineland staff knew that they would have to place older residents in other facilities. For older girls (including Deborah Kallikak), a solution existed, for they could be sent to the State Institution for Feebleminded Women, which was also in the town of Vineland. Finding permanent homes for older boys, however, was more difficult.[14]

To solve this problem, the Vineland school joined other facilities in experimenting with newer and more cost-effective forms of permanent custodial care. Among the most popular of these were "farm colonies." In 1912, the Training School purchased an additional five hundred acres of uncleared oak scrub located south of its main campus. Renaming it the "Menantico Colony," Johnstone planned to use this land to provide his older boys with challenging work and a new home. Their goal would be to transform this wasteland into a thriving agricultural community; at the same time, they could also demonstrate the latest methods of scientific farming and animal husbandry. Working closely with agricultural researchers from Rutgers University, this colony began experimenting with new methods of peach growing, egg laying, and poultry breeding with great success; in later years, new research on poultry vaccines would be started in the Vineland school's laboratories. By 1914, Johnstone had convinced the State of New Jersey to establish a similar farm colony for feebleminded boys in Burlington County. It too would soon become a model of progressive cooperation between institutions and scientific agriculture.[15]

Such colonies, Johnstone dreamed, might one day become self-supporting. Goddard too saw them as "the ideal solution" to the problem of the feebleminded. In the meantime, however, even these institutions still required massive financial support from the state. To gain this support, both Goddard and Johnstone turned their attention toward lobbying both the legislature and the public in new ways. Establishing a comprehensive state policy to deal with the feebleminded, these reformers believed, would require a more extensive political effort and a much broader educational campaign. During the pre-World War I decade, the Vineland school would try to supply both.[16]

Extending the Work

As early as 1910, Johnstone was ready to begin taking his case to the people. That January, he allocated $500 of his school's philanthropic funds to create a new "Committee on Provision for the Feeble-Minded." The "time had come," wrote Joseph Byers, former New Jersey Commissioner of Public Institutions and later the head of this committee, "for the instruction of the public and its legislatures."[17]

This committee's first goal was to increase New Jersey's institutional expenditures. Remarkably, they accomplished this through a sophisticated and intensely well-organized lobbying campaign – placing articles in local newspapers, writing to every state legislator, urging families with children on waiting lists to contact their representatives, and asking businessmen who knew these families to write letters as well.[18]

Letters from parents seeking a child's admittance to the Training School provided especially compelling testimony about the need for the state to create more such institutions. "We are sorry that our boy is losing so much time," one New Jersey father wrote. "He is the only pupil excluded from the public school, and there are no private schools here in Newton through which he might learn something." "It is now about two years since first I made application for my son's admission," wrote another. "He is now 15 years old, and instead of getting better, as I had hoped, he is getting worse. . . . I feel like a condemned man who is carrying his case to the last court of appeals. If you cannot help me, then, indeed, there is no help for me except that of God almighty." Such needs could not be met by private charity alone, the Committee on Provision insisted; instead, it was the state's role to care for and educate all of its needy children.[19]

The result of these efforts, Byers reported, was "large and immediate": whereas New Jersey had spent about $350,000 on institutions for feebleminded and epileptic persons over the preceding twenty-two years, the Vineland campaign generated an additional $211,000 in only twenty-two weeks. Of these state funds, the Vineland school itself received, in Byers' words, "no share except the gratification of the missionary who sees his work, and not his own head, crowned with success."[20]

This success quickly caught the attention of other institutions across the nation. Facing tight state budgets and often recalcitrant legislators, they saw new hope in Vineland's example of how to effect political change. There were "so many cries of 'Come over and help us,'" Byers recounted, "that the missionary spirit of the Village was stirred to answer them." In the following years, Johnstone enlarged the Committee on Provision, alternately labeling it the Training School's "Extension Department," for its mission was to "extend" the work being done at Vineland from the institution to the community, and from New Jersey to other states. To manage the new department, he hired his brother-in-law, Alexander Johnson, an experienced social worker and former institutional superintendent who had previously served as secretary of the National Conference of Charities and Correction.[21]

By 1914, the new department had taken an active role both in educating the public and in lobbying state legislatures throughout the country. Its efforts proved especially important in the South, where education, health, and welfare institutions of all sorts were often severely underfunded – or nonexistent. Vineland's staff worked diligently to support the opening of the first public institutions for the feebleminded in North Carolina and Virginia; they received requests to address the state legislature in South Carolina, and to help similar movements in Georgia, Arkansas, Missouri, Tennessee, and Texas. Recounting the sad plight of the feebleminded before the Southern Sociological Conference, Johnson urged state governments to do more. "They are children always," he explained. "The mother State must take them into her good motherly arms and care for and control them as the best thing for her and by far the best thing for them."[22]

Goddard too made every effort to disseminate his message as widely as possible. Both *The Kallikak Family* and *The Criminal Imbecile* had been written explicitly for the "lay reader." Both told graphic tales with clear moral messages which could be easily summarized in popular lectures or pamphlets. In other ways as well, Goddard rarely lost an opportunity to educate the public in order to create a proper sentiment for social reform.[23]

After watching a Broadway play called *Erstwhile Susan* in 1915, for instance, Goddard wrote a personal note to its actress. "One sentence struck me in particular," he reported, "where you said that it was the right of every woman to become the mother of one child at least. . . . " He had been working for many years to reduce the propagation of feeblemindedness, he reported. "Feeble-minded girls are victimized by men . . . ," Goddard ex-

One panel from a five-panel pamphlet illustrating the story of the Kallikak family. Although the organization producing it is not identified, the entire pamphlet was captioned "There are Many Girls in Montana Needing the State's Protection—1919." From File M614, Goddard Papers. Courtesy of the Archives of the History of American Psychology, University of Akron.

plained. "In the interest of humanity," he pleaded, ". . . I am minded to ask you if you would not be willing to add to the sentence that I have quoted above, these words, 'Unless she is feeble-minded.' I do not believe it would mar your play," he added; "indeed I think it would make a hit, and I am sure it would help us in our great struggle to make the public understand the menace of the feeble-minded girl." In this instance, his artistic intervention proved successful. "I shall introduce it in the play," the actress replied.[24]

Even more important in educating the public were large traveling exhibits prepared by a variety of organizations. Designed as a means of mass education, they were frequently displayed in the lobbies of large public buildings, or as part of county fairs, or in conjunction with professional meetings taking place nearby. For instance, while the Fifteenth International Congress on Hygiene and Demography was being held in Washington, D.C., in 1912, thousands of daily visitors could see its exhibits, at least three of which touched on the problem of feeblemindedness: eugenics, sex hygiene, and mental hygiene (which included a display about the Vineland children). There was "almost an epidemic of exhibits now operating throughout the country," Goddard's assistant, Edgar Doll, reported. The Washington conference in particular had proven "most gratifying in the progressiveness manifested," he added, "and in the desire for unification in all the branches of science represented."[25]

The effects of these mutually reinforcing publicity campaigns emanating both from Vineland and from organizations promoting eugenics, mental hygiene, social hygiene, public health, and child welfare were soon felt. "The whole country seemed to be becoming feebleminded-conscious," Byers recalled. "Clubs and associations of all sorts, especially women's clubs, churches, schools and universities were asking for information, help and guidance. Their common plea," he remarked, "was, 'Tell us what we can do.'"[26]

Responding to these increasing demands on their staff's time and energies, in 1915 Vineland's Board of Directors approved a new "campaign of publicity and education" by merging the Extension Department into an enlarged national "Committee on Provision for the Feeble-Minded." Although the new organization soon moved its headquarters to Philadelphia, Vineland's influence was apparent, for this committee hired Joseph Byers as its executive secretary, Alexander Johnson as its "field secretary," and Elizabeth Kite as a field worker. Among the traveling exhibits they helped prepare was one sponsored by Pennsylvania's Public Charities Association; in Philadelphia alone, it was viewed by 100,000 persons. Equally significant were lectures by traveling staff members. By the time this committee was dissolved in 1919, Alexander Johnson estimated, he alone had given 1,100 lectures in 350 cities and towns in 33 states, reaching an audience of about 250,000.[27]

In all their speeches, Byers insisted, committee members had "preached the gospel of the feeble-minded" which consisted of "but one doctrine – that of a better understanding of the feeble-minded and larger and better provision for their care, with eternal emphasis on the need for scientific research in the field of prevention."[28] This "gospel," however, actually contained not one but two doctrines. The first stressed the need to protect the feebleminded, God's most helpless children, from the dangers posed by an indifferent society. The second emphasized the need to protect society from the dangers posed by the feebleminded.

It was this second doctrine that increasingly captured the most public attention. "The menace of the feeble-minded" was not simply "a figure of speech," Goddard reminded his audiences; it was a concrete and pressing social issue which needed to be addressed. "The feeble-minded person is a menace to society at the present moment," he stated frequently. Most dangerous of all was the moron, who was "responsible to a large degree for many, if not all, of our social problems." By controlling the feebleminded, Goddard argued, the state would simultaneously reduce crime, poverty, prostitution, alcoholism, illegitimacy, and numerous other social ills. Such arguments proved especially effective in convincing state legislatures to increase their institutional allocations.[29]

These popular campaigns mixing humanitarian concern with social fear had an impact not only on state legislatures but on city governments as well. New York City's attention was drawn to the problem, for instance, by local issues such as Goddard's school survey. Equally influential were national events, such as the assassination attempt on former president Theodore Roosevelt in 1912. Although the gunman, John Schrank, was later found not guilty by reason of insanity, to many New Yorkers this case too seemed to prove just how menacing persons loosely labeled "mentally defective" could become.[30]

Schrank, three city psychiatrists argued in a letter to the *New York Times,* belonged to "the class of mental defectives who are potentially criminals, liable at any moment to commit deeds of violence upon the slightest provocation." Many such persons were still "at large," these doctors declared. "And why?" they asked. "Because there is as yet established no system whereby competent experts can see such individuals and judge whether the community should be safeguarded against them." What the city ought to have was a preventive program which would keep "scientific records" of such persons, thus bringing about "their complete classification." What it needed was an agency which would combine medical treatment, scientific research, and social protection.[31]

In the following months, the *Times* joined the campaign to establish such an agency. Shortly thereafter, with help from a private endowment provided by the Women's Municipal League, a clinic at New York Post-Graduate Hospital was expanded to become this city's official "Clearing House for Mental Defectives." Directing it was Dr. Max Schlapp, psychiatrist for the Society for the Prevention of Cruelty to Children. Hired to help him was psychologist Leta Hollingworth. In 1914, the Department of Public Charities briefly took over its funding, thus making Hollingworth New York City's first civil service psychologist. Within five years, this clinic had administered both medical care and mental examinations to more than ten thousand persons referred to it by the city departments of Health, Police, and the Children's Court, as well as by educators, social workers, and charity organizations.[32]

Thus, in New York City, concerns about feeblemindedness led to an expanded public role for diagnostic psychology. This role owed its origins both to those who hoped to broaden the city's medical and educational services and to those who feared political chaos. "The names of Schrank, Gallagher, Czolgosz, Guiteau, and J. Wilkes Booth come to mind," the *New York Times* observed in summarizing such fears, "in considering the function of the Clearing House for Mental Defectives. . . . "[33]

If ever a "clearing house for mental defectives" existed, however, it was the Vineland institution itself. Anyone interested in learning about feeblemindedness could begin with this school's small journal, the *Training School Bulletin.* Included within its pages were touching descriptions of the Vineland children, personal accounts from special class teachers, alarming reports from eugenic field workers, shocking statistics concerning rates of crime and poverty, notices of new legislation, summaries of the latest advances in medicine and biology, and psychological studies conducted with Binet tests.

In less than a decade, Johnstone and Goddard had transformed their obscure little institution in rural New Jersey into a center of international influence – a model school famous for its advocacy of special education, scientific research, and social reform. In the midst of a busy schedule of entertaining children with birthday parties and school plays, musicals and picnics, auto rides and stereopticon shows, this school entertained visitors by the thousands. Among its guests were psychologists, physicians, biologists, and educators, ministers and missionaries, governors, senators, and state delegations, as well as visitors from foreign countries including Canada, Germany, Denmark, Spain, Brazil, and Japan.[34]

The Vineland school had evolved into a unique educational institution in its own right, not only for its students but also for the larger society. It hosted tours for the National Education Association, the American Psychological Association, and the National Conference of Charities and Correction. Professors from nearby colleges, including Princeton, Bryn Mawr, the University of Pennsylvania, and New York University, regularly brought their classes on field trips. Local farmers visited this school to observe the latest methods of scientific agriculture. The local medical society held meetings there, as did the State Science Teachers Association. Special educators took courses there and spoke of it as their "alma mater." (The graduation speaker for the 1911 summer school class was New Jersey's progressive governor, Woodrow Wilson.) And, of course, Goddard and Johnstone continued to host the semi-annual meetings of the "Feeble-Minded Club." During the 1916–1917 school year alone, Johnstone estimated, his school had entertained and educated 2,500 visitors.[35]

Also visiting the school were parents desperately seeking information about their children's disabilities. For these families, institutions often provided the only information available, for local physicians frequently knew little about diagnosing or treating such children. "Many children are brought only to be tested," Johnstone reported in 1912. Telling parents "the truth about a child's condition" was "a hard thing to do," he admitted. "Only one who has experienced it can appreciate the pathos of the situation." In addition to these visits, the Vineland staff was deluged by hundreds of letters from families requesting advice.[36]

Other requests for information found their way to Vineland indirectly. Many mothers of mentally handicapped children wrote letters to the Children's Bureau in seeking help. "I am trying to locate a reliable school . . . for subnormal children," a Nebraska mother wrote Julia Lathrop. "I know such schools exist, but I do not know where." "We have a little boy 14 years old who is abnormal mentally," wrote an Arkansas mother. "Can pay a moderate price and make most any sacrifice for him." "Is there no place where parents of small means can educate a defective child?" asked a mother from Georgia, a state which as yet had no institutions. In replying to such requests, Lathrop often suggested that parents write to Johnstone. In response to others, she recommended literature, including works by Goddard.[37]

Especially interested in the social problems of feebleminded children were women's organizations. "We are organizing a class for the Study of Feeble-Mind[ed]ness – its prevention, care and research phases," stated one request sent to Lathrop by a Charleston woman's club. "We want to be educated!" Lathrop too would try to educate women through her own lectures. Addressing the Washington Alliance of Jewish Women in 1913, for instance, she recounted in detail Goddard's shocking discovery of the Kallikak family, Dr. William Healy's work with delinquents, and Madam Montessori's new ideas about special education. All were important, Lathrop concluded, in suggesting practical methods for safeguarding the lives of America's children.[38]

Also reaching the Children's Bureau were more specific inquiries. "Please let me have some information as to where I can procure sets for making Binet's tests. . . , " wrote a Kentucky doctor. "Is it necessary for a physician to take a course to use the Binet or Simon Tests, or could I myself . . . ?" asked a Virginia man who ran a home for boys. "Can you tell me where I can find out about the 'Binnet' test for defective children?" wrote an American missionary working in India. These requests too were forwarded to Goddard, who, according to Lathrop, had probably "developed this work as fully as anyone in the United States."[39]

As such letters suggest, extending the work for the feebleminded was also extending intelligence testing. Amidst a mounting barrage of publicity concerning the growth of special classes, the need for mental diagnoses in courtrooms, and the social costs of degenerate families, Binet's tests were being disseminated to an ever wider public.

Among the many exhibits traveling about the country, for instance, were several live demonstrations of Binet testing. At the Washington Hygiene and Demography conference, Cornell University psychologists demonstrated Binet testing before thousands of daily visitors. In Washington state, a Child-Study exhibit did the same, while the Vineland school contributed a similar live exhibit to a New York Mental Hygiene conference. Testers also tried other means of gaining publicity. "Considerable effort was made to have accurate reports published in the newspapers," Doll declared, "and this was attended with some success."[40]

Doll was correct, for newspapers too were soon disseminating Binet testing to the public. In April 1915, the *Chicago Tribune* devoted an entire page to publishing the complete Binet–Simon scale, including its pictures, puzzles, and correct answers. The text encouraged parents to give the Binet tests to their own children annually. "Here's a pageful of information which will enable *Tribune* readers to keep track, year by year, of the mental development of their children," it began. "Cut out the page – save it," the newspaper advised. In its headline, this story proclaimed the current status of intelligence testing. "Here's Chance to Test Your Mental Development According to Famous Binet System," it reported, adding in a subtitle, "How Human Intelligence is Determined by Science."[41]

Yet if this newspaper's editors believed that scientists had agreed upon a system to determine intelligence, they were mistaken. As intelligence testing continued to gain publicity, it also drew increasing scrutiny from within the community of psychological scientists. In the years between 1910 and 1917, the activities of Goddard and his fellow testers drew American psychologists into an intense and multifaceted debate over just what sort of science their profession was now disseminating, and what their own social role ought to be. Their questions gradually took shape as psychologists struggled to understand a scientific product whose popularity was rapidly outstripping their ability to control it.

Intelligence Testing and the Psychological Community

To Goddard's psychological peers, the growing popularity of intelligence testing suggested opportunities as well as dangers. Public acceptance, they realized, meant power; it could ultimately open many occupational doors. Yet so much public knowledge was also problematic, for it called into question the crucial distinction between amateur and expert – a distinction which formed a central component of both scientific and professional identity.[42]

To university psychologists, the idea of mental testing in itself was hardly a new phenomenon. To the contrary, the Binet scale was introduced into an academic context in which mental tests of all sorts were already flourishing. Since 1895, the APA had maintained a Committee on Physical and Mental

Chicago Daily Tribune, April 29, 1915, p. 5, showing the entire Binet-Simon scale, including its pictures, puzzles, and correct answers.

Measurements to promote standardization and collaboration in gathering laboratory data from sensory, motor, and mental tests.[43]

Psychologists interested in applied work were also deeply involved with testing. For instance, those promoting a more scientific pedagogy (most prominent among them Edward Thorndike) had been standardizing tests to gauge student achievement in reading, arithmetic, handwriting, and other subjects. Meanwhile, industrial psychologists Walter Dill Scott and Walter Bingham were developing tests of vocational aptitude. By the second decade of the twentieth century, the pages of experimental, educational, and applied psychology journals were glutted with research reporting test results of one sort or another.

Yet even within such a context, the Binet scale stood out, for this test had acquired a popular identity and a social utility unlike any other. Within a few short years, intelligence testing became the subject of far more articles than any other type of testing. When psychologist Guy Whipple compiled his first *Manual of Mental and Physical Tests* in 1910, he included Binet's scale; preparing a second edition in 1915, he omitted it, for this type of test, he explained, required a volume all its own. By 1913, a bibliography on Binet testing cited over 250 international publications; by 1917, more than 450 new citations had been added.[44]

Although such widespread discussion and dissemination was encouraging, it was also disturbing, for what was actually being described in this massive new literature was a practice with few clear guidelines. During this decade, intelligence could be tested by psychologists, doctors, teachers – or anyone else. They could use Binet's 1908 scale or his 1911 revision; Goddard's 1911 revision; or the translations or revisions supplied by numerous others. With each revision, questions were usually changed or moved from one age level to another. Different testers used different criteria for scoring their results; they drew different conclusions from their findings. The result, professional psychologists agreed, was confusion.

To address these problems, intelligence testers sought opportunities to come together. Among these was a conference organized by Lewis Terman as part of the Fourth International Congress on School Hygiene, held in Buffalo in 1913. Such a framework again emphasized testing's role within the broader progressive social agenda, for alongside sessions discussing "Crippled Children," the "Conservation of Vision," "Tuberculosis Among School Children," and "Child Labor" was the nation's first "Conference on Binet-Simon Scale." Participants in such conferences had yet to recognize each other as polarized adversaries pitting heredity against environment; instead, they still largely related to one another as experts emphasizing different parts of a common reform process.[45]

For some psychological testers, however, such conferences only reemphasized the need for a sharper distinction – not between hereditarians and environmentalists, but between "amateurs" and "professionals." Intelligence tests, they argued, should be administered by no one but professional psychologists. Moreover, their results should not be interpreted alone; instead, they should be considered within a context supplied by other psychological assessments.

Among the most consistent and outspoken advocates of this position were psychologists J. E. Wallace Wallin and Guy Whipple. It was "an entirely questionable, unscientific and dangerous procedure to permit amateurs to brand children as feeble-minded solely upon the basis of intelligence tests," Wallin insisted. If skill in psychological diagnosis could really be mastered after "a six weeks' summer course in a training school," he added in an obvious allusion to Vineland's training program for schoolteachers, then "it would be safe to set it down as humbug." Whipple was disturbed to learn that medical inspectors were using testing. Allowing anyone "without extensive psychological training to attempt to diagnose the precise mental status of a school child," he argued, "is about as absurd as for a mere psychologist to attempt to diagnose incipient tuberculosis."[46]

Other psychologists, however, were less worried about who was doing the testing than about what all of these new findings meant. Some were troubled by the very concept of measuring "intelligence," a variable far more complex than any they had previously studied in the laboratory. Perhaps psychologists were making too great a leap, they worried, in adopting this new form of mental testing.

The "Binet people, including yourself, are not sufficiently self-critical," University of Iowa psychologist Edwin Starbuck warned Goddard privately. He recommended returning to simpler and more well-established psychological procedures, such as testing sensory discrimination and motor skill – tests for which there was "some definite measuring stick." Yet Goddard's reply again emphasized usefulness. These older tests might be "exceedingly interesting and valuable if we could use them," he conceded; however, "feeble minded children cannot do these psychophysical tests," he reminded Starbuck. By contrast, Binet's tests were already proving their usefulness daily in diagnosing the feebleminded, he added.[47]

To Goddard, these criticisms from his psychological contemporaries represented a new version of an old story. There were two ways of reaching a conclusion, he summarized: the "philosophical" and the "practical." The first, Goddard explained, was theoretical; the second, empirical. "History is full of instances where these two do not agree," he added. For example, until the recent work of Wilbur Wright, many theorists had argued that man would never fly. To Goddard, the lesson was clear: the "practical man who has a problem to solve, does not hesitate to try a proposed method simply because the scientist assures him that it will not work."[48]

The same distinction between theory and practice was evident in the testing controversies, Goddard believed. "There are still psychologists who assert that intelligence cannot be measured," he reported, "and others who have said that even if it can be measured, the Binet scale cannot and does not measure it. *The fact is that intelligence is being measured,*" Goddard added with emphasis, "and *is being measured by the Binet scale,* and that *practical results of far-reaching importance and significance are continually being obtained from the use of this scale.*" Like flying, Goddard argued, intelligence testing had already proven its usefulness to the American public.[49]

By 1914, Goddard could claim support for his own "practical" position from at least one group of psychological theorists. There was "a school of English writers who incline somewhat strongly toward the view of what has been called 'A general intelligence,'" he wrote. Goddard was referring to the followers of British psychologist Charles Spearman. Like his colleague, Francis Galton, whose earlier studies of "hereditary genius" had assumed the existence of an inherited and measurable quality of mind, Spearman too had turned his attention toward mental measurement. In 1904, he had proposed that tests gauging specific intellectual skills also suggested the existence of an underlying, all-pervasive, and innate entity that he christened "general intelligence," or "*g.*" Yet while academicians had been analyzing Spearman's mathematical proofs for the existence of "*g*" over the past decade, Goddard seemed remarkably unaware of and uninterested in such research. Only when Spearman's follower Cyril Burt included these ideas in his studies of children did Goddard take note of them. "At least it is comforting," Goddard now concluded, "to find that the existence of a general intelligence has already been arrived at by an entirely different method of approach." For Goddard, data gained from practical application were far more valuable than any abstract theoretical arguments.[50]

It was psychologists working in applied settings, however, who produced some of the most astute early criticisms of intelligence testing. Their arguments grew out of their personal experiences in schoolrooms, courtrooms, hospitals, and clinics. Such work led some applied psychologists to question the relationship between the new psychological variable called "intelligence" and other variables, including schooling, home environment, social class, and language skill. It inspired others to seek better methods of test construction and data interpretation. Instead of a simple battle between advocates and opponents, testers were soon engaged in a complex exploration of the Binet scale's strengths and weaknesses as a practical diagnostic tool.

Unfortunately, the data and arguments produced by these practitioners have been almost entirely overlooked by later generations, for they were soon overshadowed by the more overtly and intensely politicized conflicts of the following decades. Yet notwithstanding their later invisibility, these debates are important and intriguing, for in the less polarized climate of the prewar decade, the earliest American intelligence testers had begun to investigate some of the very issues that would haunt their profession for the rest of the century.

Among the most insightful of these early practitioners to raise serious questions about the meaning of test results was Grace Fernald, who had worked with Dr. William Healy in his Chicago Psychopathic Laboratory. Like Goddard, Fernald agreed that Binet's tests could be used "with reasonable accuracy" in the first eight school grades. Even so, she wondered about her results. There was no "final standard" against which to gauge their validity, Fernald complained. Comparing test results with school results did not solve this issue, she argued. "One group of people seem to be using the school as a check on the Binet test," she concluded, while "another group is using the tests as a check on the school." To Fernald, the problem was clear. "Don't we go in a circle?" she asked pointedly.[51]

Even more troublesome were studies that took testing beyond schools. In working with delinquents in both Chicago and Los Angeles, Fernald had become skeptical. There was no evidence, she insisted, to prove that tests designed for "the more regular, orderly, school-attending children" worked equally well with "those who live very different sorts of lives." Street life, Fernald argued, produced "a type, sharp, active, original, but not given to verbal abstractions. Words are something to be used when necessity demands," she explained, "but definitions, comparisons, etc. are unnatural." Such a child "acts rather than talks."

Answers Fernald had received to Binet's questions confirmed this preference for deeds over words. When she asked one boy to define a fly, for instance, he simply caught one for her instead. This delinquent, an unschooled Polish immigrant graded feebleminded by another examiner, had been "the moving spirit in a burglary that required the utmost ingenuity." Such children were scoring poorly on the Binet scale, Fernald argued, because they lacked "the leavening influence of over-much schooling."[52]

Goddard, however, repeatedly rejected such arguments. Following Binet's example, his 1911 revision eliminated any question that explicitly required school-taught skills, such as reading. For Goddard, this solved the problem of school influence. "These are not tests of school training; they are tests of mental development," he insisted. Anyone living "in any sort of average environment" should be able to do them, "even though he has never been to school, even for a day." Two years later, his views remained unshaken. "Many people think that the school child has an undue advantage," Goddard observed, but this was not true, for the tests were based not on school learning but on a developmental sequence.[53]

Yet Fernald was unconvinced. As Goddard's very own studies showed, she retorted, many delinquent children were growing up in environments rife with sickness, crime, alcoholism, and parental abuse – environments which could hardly be called "average." Such children might well prove feebleminded, Fernald conceded, but there was as yet "not sufficient evidence to so classify them."[54]

Grace Fernald was not the only American practitioner to see the need to take environmental differences into account in interpreting test results. "The Binet norms," psychologist Edmund Huey of the Lincoln Institution had written in 1910, had been devised for "the working classes of Paris." They may need revision, he proposed, for "other social classes and other regions." William Healy also wondered about using Binet testing on "a cosmopolitan population" such as the one he faced in Chicago. "The system was built up for a homogeneous people," he argued, "and largely turns in many ways upon the ability to use and understand language." As a result, students "glib in the use of language" were being graded too high, and others too low. "Those who think that this scale measures general ability apart from schooling and other advantages," Healy wrote, "should read Binet himself on this subject," for Binet had correlated "easy circumstances" with "higher intellectual development."[55]

Healy was correct; Binet had begun to consider precisely such questions shortly before his death in 1911. His curiosity had been piqued by researchers trying his tests in Europe and America. Republished in Binet's journal, *L'Année Psychologique,* for instance, was a translation of Goddard's New Jersey research, which reported its results enthusiastically. Other studies, however, reported results which led Binet to reconsider his conclusions.[56]

Among these was a study by Belgian educator Ovide Decroly, the physician who had first introduced Goddard to Binet's work. Decroly's Brussels students had scored more than a year above their French counterparts. These results, both Binet and Decroly hypothesized, were probably due to one main difference: Binet's subjects had been "poor without being indigent," while Decroly's lived in "easy circumstances." Binet confirmed this by examining tests results from two Parisian schools with students from different social backgrounds. The results again favored the wealthier students, this time by three-quarters of a year.[57]

"Is this a matter of heredity? Is it a matter of education?" Binet had asked pointedly. For him, it was "difficult to establish a difference between the two factors which are here operating in conjunction." Other investigators had already found that poor children "are shorter, weigh less, have smaller heads and slighter muscular force, than a child of the upper class; they less often reach the high school; they are more often behind in their studies." Some of these "inferiorities," Binet speculated, "probably are acquired" and others "probably congenital."[58]

Examining his results more closely, Binet found the gap between rich and poor most striking in questions emphasizing language. This "verbal superiority," he concluded, must have come from better schooling and family life. Rich children were "in a superior environment from the point of view of language," he argued; "they hear a more correct language and one that is more expressive."[59]

Goddard, however, paid no attention to this aspect of Binet's writing; in his own work, he had yet to recognize the deeper problem that had ultimately perplexed even Binet – the complex connections linking intelligence, learning, and language acquisition. Yet it was this very problem that troubled other testers. In making their case, they often cited answers received to specific Binet questions. Such responses illustrated the difficulties of trying to measure intelligence by gauging the use of language.

Among the questions measuring "abstract thinking" among twelve-year-olds, for instance, were several asking for definitions. "What is charity?" intelligence testers asked their subjects. "What is justice?" Responses given by delinquent adolescents frequently proved provocative. "She has no idea of what charity is, says 'Board of Healthy,'" Goddard observed of one child who spoke in broken English. "Justice, has nothing to say. . . . " The irony here was obvious to Grace Fernald, if not to Goddard. Why should such a child "define charity as anything but the 'board of health'?" she inquired. "That seems to be about the nearest thing to it that experience ever gave her," she exclaimed. "Justice! Where has it ever touched her?"[60]

Questions designed for higher mental ages often called for even more linguistic skill. For these adolescents, Binet had recommended testing words with similar sounds but different meanings. Ask the difference, testers were told, between evolution and revolution; pleasure and honor; event and prevent; pride and pretension. Perhaps no test question was more ironic when applied to street children than the one asking them to explain the difference between "poverty" and "misery."

According to practitioners, the answers which they received to such questions often reflected more linguistic complexity than they had anticipated, thus making simple right-or-wrong scoring difficult. These problems were evident in the work of one tester who had students write answers to Binet's questions, and who then published their replies – with all their ambiguities, errors, and misspellings. One student, for instance, defined justice as follows: "When a man does wrong, he is brought before a justice and asked why he did it and how he did it." Goodness, another wrote, "is when a man says 'Goodness! you're house is on fire.'" "Charity," a student replied, "is a place that begins at home." "An event is a circus," another child explained; "Prevent is to stop you from going to it."[61]

To counter this emphasis upon language, both Binet and Goddard had incorporated numerous nonverbal test items within their scales. Among the items designed for younger children, for instance, were rearranging weights or copying designs. Standardizing similar tests for older persons, however, proved much more difficult. Among Binet's tests for older adolescents, for example, was one requiring paper cutting. Yet even Goddard found this test disappointing. High school students failed it; college graduates failed it; even psychologists failed it.[62]

These failures pointed to a deeper and more fundamental problem plaguing the early testing movement: neither Binet nor Goddard had been able to standardize tests gauging the annual mental growth of "normal" adolescents. The best that Goddard could do was to follow the strategy suggested by Binet. Thus, on his 1911 Binet–Simon Revision, he offered tests for children aged three to twelve; his scale then skipped to age fifteen, and added a final category for all others called "Adult." Yet this solution was hardly satisfactory, contemporaries argued. After all, it was adolescents who were most likely to cause social problems – the group, according to reformers, who were most often in need of testing.[63]

This problem of a scarcity of tests for adolescents was compounded by another. According to the most frequent complaint found in the early testing literature, Binet's tests for older children were simply too hard. Even testing enthusiasts often found these tests poorly standardized – a problem they thought would soon be corrected. In one study, for instance, only three of sixty-eight University of Michigan seniors passed all five of Binet's tests for "adults." Even Goddard conceded that the tests for mental ages above twelve worked poorly, and recommended that they not be used. Others made similar recommendations for Binet tests above age ten. Yet these controversial tests for older children continued to appear on copies of the Binet scale disseminated throughout the country.[64]

To make matters worse, these problems in testing older children were magnified by Goddard's method of interpreting test results. In general, Goddard usually considered feebleminded any child whose mental age was three or more years below his or her chronological age. In practice, this meant that the older the adolescent, the more likely he or she would be found feebleminded, for the gap between mental age – most often discovered to be ten, eleven, or twelve – and chronological age widened accordingly. In one study of 100 delinquents, for instance, all 66 diagnosed as "feebleminded" were fourteen or older, while all 34 diagnosed as "normal" were below this age.[65]

By 1915, these serious issues concerning poor standardization and dubious test interpretation had provoked two influential American psychologists to attempt their own major revisions of intelligence testing. Both tried to save what they saw as the best features of the Binet scale while correcting for some of its most glaring weaknesses. Both also made use of more sophisticated statistical techniques – techniques which soon superseded Goddard's crude calculations and simple statements about the meaning of mental age.

The most popular of these revisions proved to be the one proposed by

Lewis Terman and his Stanford colleagues. Working largely with local schoolchildren, Terman restandardized the Binet scale, making most of his changes to its upper half. He also adopted an innovation first proposed by German psychologist William Stern: instead of simply reporting a mental age, he now divided mental age by chronological age to produce an "intelligence quotient" ("IQ"). Using this method, Terman proposed, testers could better gauge the relative standing of children of different ages, a process that would make testing more fair to older children. Published in several professional articles as well as in a 1916 volume designed for teachers entitled *Measurement of Intelligence,* Terman's "Stanford–Binet" Revision soon supplanted Goddard's as the intelligence test most frequently used in schools.[66]

By this time, however, a competing revision had already been devised by Harvard psychologist Robert Yerkes. Working with colleagues at the Boston Psychopathic Hospital, Yerkes proposed an even more radical change: abandoning the concept of "mental age" altogether. Instead, in designing his "Point Scale," Yerkes retained Binet's central idea – a scale consisting of a series of increasingly difficult tasks. To score his results, he simply assigned each task a number of points, a process that allowed testers to give partial credit for some answers. Yerkes then established norms based upon the average point scores earned by children of different ages.[67]

Such a scoring method was far more flexible, Yerkes told his psychological peers, for testers could establish norms not only for different age groups, but also for diverse segments of the population. Separate norms might be needed, he proposed, for "age, sex, language, handicap, race, and social and economic status." To illustrate this, Yerkes tried out his tests on children from two city schools representing sharply different social environments. He classified his subjects as "favored" and "unfavored," "thus escaping implications as to the relative importance of heredity and environment." The "unfavored" group, Yerkes reported, scored about 20 percent lower than the "favored." It was "obvious that to state the mental capacity of an individual from the unfavored group in terms of the norm for the favored group," he concluded, "would be extremely unfair."[68]

Both of these revisions would eventually prove profoundly influential, but in different ways. Terman's ideas about IQ were almost as easy for the public to understand as Goddard's concept of "mental age"; they would soon win his scale wide popular acceptance. Yerkes' Point Scale never achieved the same popularity, but it too proved important. For while Yerkes would soon abandon his provocative ideas about establishing separate norms for different population groups, his ideas about scoring would eventually be adopted. In particular, Yerkes' Point Scale became the prototype for many other intelligence scales, including those created in the 1930s by psychologist David Wechsler.[69]

While these two newer intelligence scales were first being introduced and debated, practitioners of all sorts continued to use the older, problem-plagued versions largely disseminated by Goddard and his followers. As a consequence, the early testing controversy really took place on two levels simultaneously. On one level, professional psychologists debated the merits and weaknesses of several versions of intelligence testing; on another, the public discussed the findings reported by hundreds of Binet testers already at work in institutions, schools, courts, and clinics.

These two controversies finally collided at the 1915 Chicago meeting of the American Psychological Association. While psychologists gathered for their annual professional meeting, several studies produced by intelligence testers caught the attention of the local press. The embarrassing publicity which ensued finally forced the psychology profession to take much more notice of what it was disseminating.

Among the most sensational was an APA paper by J. E. Wallace Wallin. Pointedly entitled "Who is Feeble-Minded?" it attacked the high rates of fee-blemindedness reported by countless testers. "During the last three or four years," this psychologist noted, thousands of "social and scholastic misfits" had been found feebleminded based on Binet testing. Using roman numerals to signify mental ages, Wallin set up his hypothesis: if "every Binet X-, XI- or XII-year old prostitute" was feebleminded, then "logical consistency demands that we so brand every successful Binet X-, XI-, or XII-year old housewife." If every low-scoring "murderer, robber or thief" was feeble-minded, then so too was every low-scoring "farmer, laborer or business man."[70]

Wallin tested his thesis in Iowa by giving Binet tests to farmers, business-men, and housewives – all "law-abiding citizens" who, when measured by social standards, could not "by any stretch of the imagination be considered feeble-minded." "Mr. A," for example, was a wealthy sixty-five-year-old farmer with only a few years of schooling who had become president of the local school board and father of nine successful children. Even so, he scored a mental age of only 10.6 on Goddard's 1911 scale.

"No doubt if a Binet tester had diagnosed this man 45 or 50 years ago," Wallin added, he would have declared him "a 'mental defective.'" Had he been a criminal, testers would have given "expert testimony under oath" that he was "unable to distinguish between right and wrong." By the "automatic standards now in common use," Wallin concluded, this man should have been institutionalized. "But is there anyone who has the temerity, in spite of the Binet 'proof,'" he asked, to maintain that he should have been "restrained from propagation because of mental deficiency (his wife is still less intelligent than he)?"

Other Iowans included a "kind mother" who had "raised and educated a

family of respectable, law-abiding and self-supporting children." Since this
fifty-nine-year-old housewife tested to a mental age below eleven, her IQ was
only .19, Wallin calculated. Had such a woman behaved immorally, she too
might have been institutionalized. "She is an imbecile, according to the intelligence quotient," he explained, "but a very desirable imbecile!"

These life experiences disproved Goddard's arguments linking low test
scores to social irresponsibility, Wallin insisted. When put to the "acid test
of fact," he argued, "the theories and hypotheses which have been the main-
stay of the Binet testers" suffered "immediate collapse." It was time, Wallin
urged his colleagues, to abandon the very concept of the "high grade moron" – the concept which lay at the center of Goddard's psychology.[71]

For psychologists attending this APA meeting, even more embarrassing
than Wallin's presentation was a study conducted by Mary Campbell, a dis-
gruntled former worker in Chicago's municipal court laboratory, who gave
intelligence tests to this city's politicians. The following day, newspapers all
over Chicago reported the results. "Who's Loony?" asked a front page headline in the *Chicago Daily Tribune,* answering "Why, Every One, Even the
Mayor." The subtitle explained the story. "Psychopathic Test Shows City
Officials Have Minds of Ten Year Olds. Dry Leader About Eight." The *Chicago Herald*'s front page was equally mocking. "Experts Assail Mentality
Test Result as Joke," it announced, adding "Hear How Binet–Simon Method
Classed Mayor and Other Officials as Morons." The tests being used in this
city's courts, these newspapers explained, were faulty and their results often
absurd, for the city electrician had failed a twelve-year test, the mayor, some
ten-year tests, and the fire marshal, a test for age eight. (One politician had
refused to take Campbell's tests, for he would not allow any woman but his
wife to "feel his head.")[72]

Not amused by either study was psychologist Samuel Kohs, the Vineland-
trained tester employed at the Chicago House of Correction. Especially up-
setting were Wallin's arguments, for this APA member, Kohs believed, was
putting in jeopardy the applied work of so many of his colleagues. Wallin's
attack was "adapted, consciously or unconsciously, to prejudice and lead the
public astray," he retorted angrily. Wallin had "magnified enormously out of
proportion a group of distinctly harmful examiners"; as a consequence, the
contributions of psychologists doing "valuable, sane, practical work" had
"shrunk correspondingly."

Such problems, once recognized, were being corrected within the psychology profession itself, Kohs insisted. After all, psychologists had begun to
abandon tests for higher mental ages, while no psychologist would have calculated an IQ on a fifty-nine-year-old woman in the way that Wallin had
suggested mockingly. "It is truly comforting to know that there are still some
Binet testers left who have preserved their sanity," he asserted defensively,

and "whose scientific perspicacity and caution have really suffered no impairment." The public, Kohs believed, could trust the good will and common sense of most testers. "As a group," he replied, "we are not as careless, as conscienceless, as wildly extreme as some may imagine. . . . "[73]

Such trust was crucial, Kohs understood, for if publicity could help to create a social market for testers, it could also help to destroy one. "Our ultimate appeal in all this work is to public opinion," he pleaded, "and we can hardly afford to sacrifice the normal progress of applied psychology" simply due to "some remediable and inconsequent defect in our procedure." Yet Wallin remained adamant in his arguments. The problem was far more serious than psychologists realized, he insisted. Not enough was being done to stop the social dangers caused by faulty testing.[74]

In trying to settle these bitter and quite public disputes which had broken out at its 1915 meeting, the APA could actually do little more than refer the matter to its standing committee on mental measurements. In the face of such conflicts between members, this professional association found itself relatively powerless. Some psychologists, however, were able to use the controversy to press the APA to respond to a related issue long bothering them. Through a series of statements introduced by Guy Whipple, they began to try to circumscribe the activities of "teacher testers" and other "amateurs" using intelligence tests. Even if the APA could not control the testing done by its members, it could at least try to control the testing done by everyone else.[75]

"Whereas, Psychological diagnosis requires thorough technical training . . . ," Whipple's resolution began,

> And whereas, There is evident a tendency to appoint for this work persons whose training in clinical psychology and acquaintance with genetic and educational psychology are inadequate:
> *Be it resolved,* That this Association discourages the use of mental tests for practical psychological diagnosis by individuals psychologically unqualified for this work.[76]

Evident in this resolution was a belated recognition by APA members of the tremendous impact that intelligence testers in general, and Goddard in particular, were having on their field. Equally evident were the narrow limits of professional consensus. By 1915, American psychologists could not yet agree on just what intelligence was, or on whether or not their tests were indeed measuring it. They could, however, reach agreement on one issue: testing intelligence, the APA now asserted, was far too important an activity to be entrusted to anyone but psychologists.

Still anxious to achieve a broader and more meaningful professional consensus, in early 1916 Yerkes conferred with Carl Seashore, one of the editors of the *Journal of Educational Psychology.* It was time to ground intelligence testing in some basic psychological principles, he insisted, while also calling professional attention to its "dangerous developments." In response, Seashore proposed using the pages of his journal for a symposium. Perhaps a reasoned intellectual exchange could suggest a means of finding consensus in ways that an annual meeting could not.[77]

"In view of the prevailing unrest in regard to the present status and future of the Binet tests," Seashore began in announcing his plan to readers, it was time for those with "first-hand knowledge" to have a "'showing of hands.'" Respondents, he proposed, should address three issues in particular: the present situation; what they were actually doing; and what they believed ought to be done. "The workers in this field crave a sort of heart to heart talk," Seashore explained, "free from formalities and generous in self-revelation." Within a month, he had received sixteen replies.[78]

Some respondents interpreted Seashore's assignment narrowly: to decide whether the profession should favor Terman's Stanford–Binet scale or Yerkes' Point Scale. Others used this opportunity to emphasize the need for new types of mental testing. "The scientist who makes a workable scale for measuring persistence of motive," vocational tester Walter Bingham replied, "will take rank with Binet. The task seems fantastic," he conceded, "but no more fantastic than did the proposal in 1905 to have a graded scale of intelligence."[79]

Goddard's reply once again emphasized the conflict between theory and practice. "I, myself, as you know, am not working on the testing of intelligence but am studying the feeble-minded," he wrote. Such a symposium might produce "interesting reading," he predicted; however, it was unlikely to generate "any great unanimity" until "more people get down to solid work."[80]

Also evident in this symposium were starkly differing views concerning the professional role of the "clinical psychologist." During this decade, psychological practitioners were still struggling to position themselves in the occupational space between medicine and pedagogy. And while Goddard had been able to establish close working relationships with both doctors and teachers, his followers were usually pulled toward one of these fields or toward the other. Each of these occupations offered a very different model of how to shape professional behavior; each also suggested a different conceptualization of the meaning of "testing," and a different understanding of how to integrate Binet testing with other forms of testing. Each thus presented "clinical psychologists" with a very different vision of their future.

For Robert Yerkes, the future direction was clear: psychologists ought to establish an expanded diagnostic role which would parallel medical practice. This conceptualization was nurtured by his own experiences in working with physicians at the Boston Psychopathic Hospital. Intelligence testing, Yerkes believed, would ultimately be only one of many assessment techniques used by psychologists to examine patients. Instead of calling their work "mental testing," he proposed that his peers begin to substitute the broader term "psychological examining."[81]

By contrast, Lewis Terman eschewed this medical model. Instead, he emphasized the need for psychologists to maintain close professional ties to educators. Thus, although Terman had supported the APA's recent resolutions, he still saw no reason why all schoolteachers could not be taught to give Binet tests. After all, this skill could be learned "in ten weeks by anyone who has intelligence enough to teach school," he insisted. The APA's claims to expertise, this prominent tester told his peers, were little more than professional snobbery, for much of the rhetoric about "psycho-clinical expertness (whatever that is)" was "hollow buncombe." There was no argument for limiting intelligence testing that did not also apply to school achievement tests – tests that teachers were being urged to administer. "No one has sought to envelop them in tabus for the glorification or personal advantage of the élite who have been initiated into their mysteries," Terman retorted. "Let us abandon pretense," he urged his colleagues, "and take the same common sense attitude toward the Binet scale."[82]

Most problematic of all in this symposium, contributors could not agree on any means of settling such disputes. Both Yerkes and Seashore still favored some sort of APA resolution. Terman, however (perhaps because of his past experiences as a normal school psychologist on the fringes of his own profession), resented any such attempt to control applied psychology from above. "The proposal to decide the matter 'by agreement' can of course not be taken seriously," he replied scornfully. "Scientific disputes were settled that way by the churchmen of the Middle Ages," he added, "but the method is a little out of date now." Progress, Terman argued, would come not from any "psychological decree" but from "demonstrable results of scientific research."[83]

To psychologists hoping to resolve such controversies within their own professional community, this symposium offered little hope. Intelligence testing, Whipple believed, was suffering not only from charlatans but also from "friends in whom eagerness exceeded exactitude." Yet his hopes for establishing an APA committee to oversee and regulate such behavior were becoming "daily more remote and difficult." Instead, he conceded wistfully, such questions would have to be settled through the "pragmatic test of 'delivering the goods.'" It was the public, and not the profession, his arguments suggested, which would ultimately decide the fate of intelligence testing.[84]

Yet other psychologists saw grave dangers in allowing such scientific controversies to be decided solely by the whims of the marketplace. Among the most critical was University of Chicago psychologist James Angell, in whose city testing had just suffered its most humiliating public episode. If many more such scandals were to occur, he feared, the fledgling field of scientific psychology would itself be seriously damaged.[85]

There was a "crying need for authoritative expression regarding the actual demonstrable achievements of the 'test' movement," Angell lamented. In making exaggerated claims, reputable psychologists had "sinned but rarely," he declared, but this was not true of "the great rabble of 'near psychologists,' fake psychologists and their kith and kin, exploiting a newspaper press." These persons had "poisoned the public mind" with "groundless" expectations. Failure to fulfill promises would lead the public to rise up in "righteous disgust with the essential fraud perpetuated upon it," thus threatening to sweep away "all that is at bottom really good and sound and wholesome in the mental test propaganda." It was time for "some frank speaking by our leaders" concerning "the real facts of the case," Angell proclaimed. Otherwise, he warned, the psychology profession would find itself battling to save a "discredited implement which an outraged public has already discarded."

What was most needed, Angell insisted, was "a campaign of conservatism" – a campaign far different from the one Goddard had been engaged in. "Let us frankly recognize and clearly state that our enterprise is still in a pioneer condition," Angell pleaded. Even with the best of distribution curves, "we cannot dogmatize with certainty regarding the meaning of our results," he emphasized. Psychologists ought to proceed more slowly, he argued. "Science as well as art is long," he reminded his peers, "and patience, that un-American virtue, is wholly indispensable."[86]

Thus, by 1917, professional psychologists were deeply embroiled in a complicated and many-sided debate over the nature and future of intelligence testing. They argued over the scientific meaning of the variable called "intelligence" and its relationship to other variables such as language skill, schooling, and social environment. They fought over the future direction of their new profession, and wondered if they should be emulating physicians or educating schoolteachers. Most frustrating of all, they struggled to find a means of resolving these problems within their own professional community, and out of the eye of the public.

While these psychologists were struggling among themselves, the need to reach some sort of professional consensus became even more pressing. By 1917, "clinical psychologists" found themselves under attack from another direction: the medical profession. Within a few short years, intelligence testing had begun to influence not only psychological practice, but the practice of medicine as well. The result would be a professional controversy of a very different sort.

Intelligence Testing and the Medical Community

Whereas Goddard's work had taken years to attract the attention of the APA leadership, his impact upon American physicians had been far more immediate. Most doctors working in institutions for the feebleminded admired God-dard enormously for injecting a research spirit into a nearly moribund field, and for raising public consciousness of their own social as well as scientific importance. Many doctors studying insanity were equally impressed with Goddard's emphasis on research. Moreover, the early efforts of Vineland's "Extension Department" blended especially well with the broader aims of the mental hygiene movement – to move mental health issues beyond institu-tions and into the community.[87]

Largely due to Goddard, Binet's tests were quickly incorporated into sev-eral important American medical texts. Translating the third French edition of J. Rogues de Fursac's *Manual of Psychiatry* in 1911, for instance, psychia-trist Aaron Rosanoff also added the Binet tests. The irony here was especially poignant: Binet's ideas were being appended by an American doctor to a French text – a testament to Goddard's skill in working collaboratively with physicians. When prominent psychiatrists William Alanson White and Smith Ely Jelliffe published *The Modern Treatment of Nervous and Mental Diseases* in 1913, they invited Goddard to write its long section on "The Educational Treatment of the Feeble-Minded," which included the Binet scale.[88]

Equally interested in intelligence testing were several physicians employed by school systems. Among the most influential was Dr. Walter Cornell, chief medical inspector for the Philadelphia public schools, who also served as a medical consultant at Vineland. An ardent proponent of the child hygiene movement (and a member of the Feeble-Minded Club), in 1911 Cornell offered a ten-day "Short Physician's Course" for medical inspectors at the Training School, which included training in how to administer Binet testing. His course drew ten doctors from New Jersey, New York, Michigan, Ohio, Virginia, and Canada. Repeating it in 1912, he enrolled another twenty physi-cians.[89]

Other doctors began using Binet's tests without any training at all. They often used them, however, in their own ways. Instead of determining a mental age, for instance, many doctors simply asked Binet's questions and consid-ered the answers while making a subjective diagnosis. Thus, while psycholo-gists battled over standardization and scoring, age norms and distribution curves, most physicians ignored such issues. Moreover, they saw no problem in omitting any Binet question that struck them as inappropriate or inconven-ient, or in inventing new ones whenever they wished.

Typical of such medical adaptations, for instance, was the "Special Modi-fication of Binet–Simon Scale" prepared by Dr. Aaron Rosanoff as part of the massive Nassau County (New York) Mental Hygiene Survey of 1916. "The following tests, selected, for the most part, from amongst those of the Binet–Simon Scale," its directions explained, required "practically no special equipment or apparatus." In other words, Rosanoff had "modified" Binet's

scale by simply omitting questions requiring weights, pictures, or other materials cumbersome for a field worker to carry around. He also added some
extra questions of his own for persons older than fifteen. "Who is the President of the United States?" this test asked. "Who was the first President?"
"Who was the civil war President? What countries are at war?"[90]

Such practices were often scorned by professional psychologists. Guy
Whipple, for instance, was especially critical of persons who felt free to modify test questions at will. One such examiner, he complained, had invented
new questions, varied them from child to child, "and finally wondered why
the Binet method was so unreliable." It was impossible to give "fool-proof
directions for the conduct by an amateur of a psychological test," Whipple
concluded. "The motto: 'Don't monkey with the method' ought to be hung
on the walls of every laboratory."[91]

To physicians, however, these practices were quite acceptable. These different approaches to intelligence testing highlighted a broader distinction between the two professions, each of which took a different approach toward
diagnosing mental abnormalities. Most scientifically trained psychologists
now regarded the mind as an entity worthy of some sort of objective measurement; by contrast, most psychiatrists were proud of their own subjective skill
in making mental diagnoses based largely on clinical experience.

During the prewar decade, psychiatrists and clinical psychologists would
increasingly find themselves working together – in institutions, clinics, and
schools. And while their new interaction might occasionally lead to sharp
clashes pitting objective against subjective methods, far more often it led
to a blending, with each profession profoundly influencing the other. Thus,
psychiatrists such as Rosanoff became increasingly intrigued with new psychological assessment techniques, while clinical psychologists such as Goddard gained increasing confidence in their own ability to make subjective
judgments based on experience.

The effects of this new blending upon medical practices were captured
especially well in the advice given to American physicians in 1914 by Dr.
E. H. Mullan of the U.S. Public Health Service. In discussing how to diagnose feeblemindedness, Mullan recommended using psychological testing.
"It might be well for the physician, after surveying a number of intellectual
tests," he proposed, "to select a certain number or to make up certain tests
of his own, say four or five. . . . " Doctors should become "thoroughly familiar with them" by practicing whenever they had the chance. "Thus, when the
feeble-minded person presents himself," Mullan concluded, "your own-made
machine will assist you in analyzing his mental condition, and proving him
to be feeble-minded." What Mullan was proposing was a means of using
modern psychological methods which did not challenge the physician's right
to make a subjective diagnosis in the end.[92]

For physicians unwilling to develop their "own-made machine," there was an even easier way to incorporate new psychological methods into their practice: they could simply hire a trained tester as an assistant. Yet this pattern soon produced an unexpected consequence which would later come to haunt doctors: it opened a side door into medical practice for nonphysicians. This opening proved especially significant for women, many of whom slowly entered the medical field through psychology. Their careers often followed a similar trajectory – from schoolteaching, to graduate work in educational psychology, to expertise in psychological testing, to careers allied to the medical profession.[93]

The early career of Leta Hollingworth, trained at Teachers College by Edward Thorndike, illustrates this process of professional transformation. "I, who had prepared myself for work in schools, found myself working in a hospital – a great surprise to me," Hollingworth confessed after being hired at New York's Clearing House for Mental Defectives. A conversation she reported upon assuming her next job – at Bellevue Hospital – shows just how crucial testing had become to her professional identity. "And what do you do?" a hospital physician asked Hollingworth in 1915. "I am a psychologist," she replied. "And what is that?" he inquired. "I give mental tests," she answered simply.[94]

Yet in assisting doctors in diagnosing mental deficiency, psychologists also began to acquire new powers – powers that in the past had been reserved only for physicians. Especially important would be their new roles in determining who should be institutionalized, and in testifying as expert witnesses concerning the mental condition of criminal defendants. Both roles suggested a new legal relationship between the psychology profession and the state – a fact reflected in a growing number of courtroom determinations made with the aid of intelligence tests.

Such practices were as yet far from uniform, for judges still had wide latitude in accepting or rejecting mental testing. In one 1916 case, for instance, a New York Supreme Court justice scoffed at intelligence testers, and refused to use their findings to commit a wayward girl to an institution for the feebleminded. "Standardizing the mind," Justice John Goff declared, "is as futile as standardizing electricity. . . . " Perhaps more revealing than Goff's verdict, however, was the response it provoked in the *New York Times*. "Scorns the Aid of Science," stated the title of an editorial criticizing the judge's decision. "As a matter of demonstrated fact," the *Times* reported, "the Binet–Simon tests, intelligently applied – it is true that they have not been so applied always – are as trustworthy as the multiplication table. Minds can be standardized," this editorial continued, "just as electricity is every day and long has been." Such judges could delay but not stop the "inevitable substitution of certainty for uncertainty," the *Times* concluded. Thus, by 1916, judges were making headlines not for using, but for refusing to use, psychological testing in the courts.[95]

Even more important than individual court rulings were new state laws. The first to recognize the new role of psychologists explicitly was a 1915 Illinois law concerning "Commitment of the Feeble-Minded." It required this state's courts to take special measures if a dependent or delinquent child was found feebleminded "on the testimony of a psychologist or of a physician." By 1917, a California law granted similar powers to "psychologists making use of standardized psychological tests" while laws in Kansas and Oregon entrusted such powers to "two qualified physicians, or one physician and one clinical psychologist." Such state laws gave American psychologists an authority unprecedented in their profession – the authority to curtail the liberties of their fellow citizens.[96]

For some psychologists, this authority simply constituted a new social and legal recognition of what they believed they had already earned: a new status as scientific diagnosticians of the mind. "The view that the diagnosis of mentality is a task for the psychologist," Lewis Terman boasted in 1916, "is now so generally accepted that it is hard for us to realize how novel and revolutionary it was a dozen years ago."[97]

Yet despite Terman's assertion, this view was not necessarily "generally accepted" by the medical profession. Instead, the new laws rankled many doctors in general, and many psychiatrists in particular. The resulting rivalry often pitted psychologists against psychiatrists in an increasingly hostile professional battle over who could best diagnose feeble minds. Such a battle would bring the strengths and weaknesses of intelligence testing to the foreground once again.

This battle grew most bitter in New York City. At the New York Medical Society's 1917 meeting, Dr. William Burgess Cornell, an institutional physician, decried the new trend of allowing psychologists to make mental diagnoses. From Vineland, he observed, testers with "more or less psychological training" had been sent into "schools, reformatories, workhouses, jails, prisons, institutions . . . and even into hospitals for the insane." Their work had led to a "loud clamor for the solution of the so-called problem of the feebleminded." Calling public attention to such matters had "undoubtedly been a progressive step," Cornell believed. However, it had also led to "bad diagnoses" and "bad statistics."[98]

The Binet scale was beginning to threaten the work of psychiatrists, Cornell told his fellow doctors. In moving from Paris to America, it had also moved "from the realm of pedagogical psychology to that of pathological psychology," he explained, "and the latter, if it is not psychiatry, approaches it with infinite closeness." In America, a "lay person" was now diagnosing abnormal mental states – an activity so novel that "testers themselves appreciate it and have called themselves clinical psychologists." Even worse, members of this "new hybrid genus" had "grown intolerant" and "declared that no one shall use the scale except the *delecti.*" Some had even tried to stop doctors from using such testing. Many of these persons "dearly love to be called 'doctor' on all occasions," Cornell reported. "Personally," he added, "I see no excuse or reason for calling any one with the Ph.D. degree 'doctor.'" It was time to call a halt to these "tendencies of the clinical psychologist to invade the work of the physician, or more properly the psychiatrist."

If things had gotten out of hand, Cornell argued, it was American physicians' own fault, for they had paid far too little attention to diagnosing feeblemindedness. "Binet was a physician," Cornell now claimed (incorrectly). In Europe, this research had been done by doctors, he insisted, but in America, "the clinical psychologist found himself alone in the field. So well entrenched has he (or usually she) become," Cornell complained pointedly, that such persons now aspired to control institutions. Were this to happen, he warned his medical colleagues, "physicians would only find an incidental use, such as signing death certificates and prescribing for stomach ache."

Cornell's comments stressed the need to protect patients from a dangerously oversimplified understanding of their condition. Feeblemindedness was "not the simple affair our psychologists would have us believe," he argued, nor was "its immutable transmission as a unit characteristic quite so frequent as Goddard would have us believe." Psychologists, he claimed, were far too unconcerned with finding causes. "The whole matter is very simple!" Cornell summarized. "The scale is applied, the child is feeble-minded – and should be segregated – nothing easier!" In such work, there was "much need of common sense."

Equally evident in Cornell's comments, however, was a concern for the well-being of his own profession. The expertise of medical superintendents was now being challenged, he argued, by psychologists (and often by women psychologists) who had found new uses for their skills. Fortunately, the "medical profession is waking up," Cornell concluded, "and is realizing that it has been losing opportunities."

Yet even as sharp a critic as Dr. Cornell had no desire to dispense with intelligence testing. After all, such tests had already proven their usefulness to medicine. Instead, he simply hoped to dispense with psychologists. There was no reason why a psychiatrist could not give intelligence tests "as part of his routine work," Cornell proposed. This would allow mental diagnosis to remain a strictly medical function. "With legal commitments and registrations coming more and more into use," he observed, "we pause and consider before placing a life-long label of such consequence to all parties concerned."[99]

Cornell's fears were obviously widely shared, for by June 1917, the New

York Psychiatric Society had agreed upon some resolutions of its own. In a widely circulated announcement, this Society published the conclusions of its committee investigating "those who have termed themselves 'clinical psychologists.'" And while praising psychologists' "earnestness and success" in making their science useful, these doctors had also "observed with much distrust" their growing tendency to deal with "abnormal mental conditions." This work had often been done independently of doctors, the only persons truly "competent to deal with questions involving the whole mental and physical life of the individual."[100]

Especially dangerous, these doctors argued, were new state laws recognizing the psychologist's "so-called expert testimony." Both involuntary commitments and criminal examinations were among the "most serious responsibilities assumed by physicians," they insisted. In a sharply worded statement, this committee recommended that "the sick, whether in mind or body," be cared for "only by those with medical training."[101]

This statement was clearly aimed at Goddard and his followers. Also feeling its sting, however, were other psychologists working in medical settings. Clinical psychology, these resolutions suggested, had begun to infringe upon the practice of medicine. Their publication in the *Psychological Bulletin* quickly drew a pointed response from that journal's editor, Shepherd Ivory Franz.[102]

A respected psychological scientist employed at the Government Hospital for the Insane in Washington, D.C., Franz began by stressing professional rapprochement. The new resolutions, he proposed, offered a chance to consider "the independence, the interdependence, the correlation and coordination, and the responsibilities of different but allied lines of work." Yet despite his conciliatory tone, Franz strongly defended his own discipline.

If several states had decided to allow courts to use psychologists as experts, he retorted, it may have been because "medical expert testimony was not satisfactory." Moreover, physicians were no less likely than psychologists to make diagnostic errors, Franz argued. Nor were they really "competent to deal with the whole mental life of an individual," as they liked to claim.

Most insulting to Franz, and most indicative of the growing movement among physicians to circumscribe psychological expertise, was the very composition of the National Committee for Mental Hygiene. As of January 1917, Franz noted, the ninety members of this influential reform organization included numerous physicians as well as "college presidents, bankers, merchants, women of wealth, social workers, professors of the social sciences," and "professors of education," but not a single psychologist.[103]

Even more significant would be the emerging rift between the National Committee for Mental Hygiene and the Vineland-created Committee on Provision for the Feeble-Minded. In the past, these organizations had been mutually supportive, with leaders from the mental hygiene movement serving as board members on the Committee on Provision. By 1917, however, medical superintendents of institutions for the feebleminded would find themselves facing increasing pressure to eschew Vineland's leadership and instead join with other doctors active in mental hygiene. The Committee on Provision would remain an influential social organization through World War I; growing opposition by the medical community, however, would eventually lead to its demise as an independent entity by 1919.[104]

According to doctors, these measures were necessary to protect the public. It was important to reduce the undue influence of any single institution, they argued, as well as to check the dangerous excesses of the testing movement. The best way to do both was to insist that all mental diagnoses remain firmly under medical control. The problems of the feebleminded, medical leaders now argued, ought to be subsumed under the broader campaign to promote mental hygiene – a campaign led by psychiatrists.

To Vineland personnel who had devoted their energies to the Committee on Provision, these recommendations suggested a bitter setback. Alexander Johnson was especially angered by this controversy, which he interpreted as a clear instance of professional jealousy. His own staff, he believed, had simply been far more successful than the medical community in making the public appreciate the importance of their cause. When the "National Mental Hygiene Committee got fairly going and discovered that the care of the feeble-minded was something more popular and essential than what they had organized to do for the insane," Johnson would later write, they "desired to enlarge their work to take in ours." These psychiatrists had stressed "the medical side," while "our officers were laymen emphasizing the educational and social aspects; this was all too much for the doctors," he concluded bitterly, "and most of them deserted us for what they considered a medical association." Thus, as Johnson saw it, his own committee's efforts at social reform had been undermined in the end by medical elitism.[105]

As such episodes illustrate, by 1917 the Vineland school's campaign on behalf of the feebleminded had begun to generate strong criticisms from American physicians. Just as they had affected the psychological community, Goddard's ideas had also forced the medical community to reexamine the meaning of feeblemindedness and the ways it was diagnosed. Like psychologists, doctors too were soon engaged in a complex effort to understand the new practice of measuring minds – a practice that increasingly affected decisions made by institutional superintendents, school medical inspectors, and hospital personnel. In their own ways, they too had begun to raise questions about who should do the testing, what their results meant, and how they ought to be integrated with other medical findings.

Equally apparent was a new professional power struggle. As Alexander

Johnson suspected, these controversies pitting testers against doctors did not merely represent a battle over ideas. In fact, during these years, Goddard's ideas about feeblemindedness and those of most medical superintendents were as yet largely indistinguishable. Long before Goddard had entered this field, institutional physicians had been promoting institutionalization, lobbying for sterilization laws, and warning the public about the social dangers of degeneracy. During this decade, the most striking characteristic distinguishing the leaders of the Vineland school from those of other institutions was still their lack of medical degrees.

Even so, something more than mere professional jealousy was indeed at stake in these emerging controversies. In both obvious and subtle ways, intelligence testers were redefining as well as expanding the social meaning of mental deficiency. Their new science now had real social consequences – for the state, for the psychology profession, for the medical profession, and most of all, for the growing number of American citizens whose lives would be affected by the determinations made by mental testers. In the decades to follow, these consequences would lead to increasing tensions between psychologically and medically trained practitioners as both groups battled to assume control over the new science of measuring minds.

Reformers, Testers, Doctors, and the Public

By 1917, both the Vineland staff and Goddard himself had succeeded in bringing the problems of the feebleminded to the public's attention in a variety of new ways. They had done so largely through a well-planned lobbying effort and a popular educational program. Central to both programs was a publicity campaign which emphasized Christian concern, social fear, and the need for state action – all presented to the public in a manner which combined scientific certainty with missionary zeal.

This campaign had been especially effective because it blended so well both with older medical ideas and with newer ideas about social reform. The movement to provide more state care for and control over the feebleminded garnered strong support from a variety of other progressive campaigns. It found a welcome home within the eugenics movement, while also earning an equally important place within movements promoting mental hygiene, social hygiene, and child welfare. It won followers among fiscal conservatives who hoped to reduce state spending on dependents and delinquents, while simultaneously gaining endorsements from social progressives who hoped to increase state funding for public health and public schooling. It also earned special support from women's organizations newly active in political affairs. The combined pressure exerted by all of these groups was apparent in a variety of new state policies. These included the opening of new institutions, the passage of sterilization laws, the growth of special classes, and the increasing legal recognition granted clinical psychologists in courtrooms.

Equally apparent was the rapid and widespread dissemination of intelligence testing. While testers and doctors argued among themselves over the merits and liabilities of psychological diagnoses, the public heard a different message: scientists had discovered a way to measure minds and diagnose feeblemindedness. Ordinary citizens could learn about intelligence testing from popular writings or public lectures, from the *New York Times* or the *Chicago Tribune,* from traveling exhibits or live demonstrations. If Goddard's goal had been to disseminate Binet testing, then he had succeeded to a degree that astonished even his strongest supporters.

Yet it was this very success that had begun to worry a growing number of professional psychologists and physicians. In response, both professions began to recognize the need to curb what each now perceived as the socially dangerous excesses of the testing movement. Neither profession would advocate an end to intelligence testing; instead, each would try to control this procedure largely by policing its own borders. Thus, by 1917, psychologists had passed a series of resolutions to keep "amateur" testers out of their field, while psychiatrists had resolved to keep psychologists out of theirs. Evident in both responses was a mixture of genuine public concern and professional self-interest. Equally apparent was a broader battle over the nature of expertise in diagnosing mental conditions – a battle which would increasingly pit psychologists against psychiatrists, and which would continue in many different forms for the rest of the twentieth century.

Despite its growing bitterness, this battle had yet to take its toll on the career of Henry Herbert Goddard. Still the nation's most accomplished scientist studying the feebleminded, he continued to prepare articles and lectures for medical audiences, teachers' meetings, social work organizations, and the public. Yet the effects of these professional power struggles were increasingly evident in the work of his colleagues. Lewis Terman, for instance, would later complain that his early efforts to promote child hygiene in the schools had been hampered because "many in the medical profession seemed to regard disease as a resource to be conserved for their financial benefit rather than as an evil to be got rid of" – a statement that suggests another appeal of the eugenics movement. His former Clark classmate and teaching colleague at the Los Angeles Normal School, psychologist Arnold Gesell, would respond to this dilemma in a different way. By the time he became Assistant Professor of Education at Yale University, he had already begun studying medicine. Gesell would complete his Yale medical degree in 1915, a fact which made it far easier for him to establish and run the "psychoclinic" that later became Yale's "Clinic of Child Development."[106]

Meanwhile, as these professional rivalries continued to simmer, the intelligence testing controversy would once again reach a new level of public visibility. By 1917, Henry Herbert Goddard had become involved in two new testing debates which would absorb the nation's attention for decades to come. The first involved him in efforts to gauge the minds of immigrants; the second, in gauging the minds of soldiers as America prepared to enter World War I. Both actions would raise new questions about psychologists' working relationships – with each other, with the medical profession, with government agencies, and with the American public. Both would also bring the intelligence testing movement far more publicity than even its most enthusiastic promoters had ever dreamed possible.

8

Psychological Work and the Nation: The Political Meaning of Intelligence

A National Agenda

On May 4, 1914, Julia Lathrop, director of the Children's Bureau, wrote a confidential letter to psychologist Henry Herbert Goddard of the Vineland Training School. It concerned a request she had received from President Woodrow Wilson. "The President has asked the Bureau of Education and the Children's Bureau to join in formulating a plan for the study of the feeble-minded in this country," she explained. In preparing her agency to undertake such work, Lathrop was quietly seeking advice from a few of the nation's experts.[1]

"First, what is your own view of the value of such a study?" she wrote Goddard. "How should it be conducted? Where should it be conducted? Am I right in thinking that your own studies and those of others into the family history of feeble-minded persons is enough to warrant our dealing with other phases of the question," she asked, or "should similar studies be made in various parts of the country?"[2]

What the nation needed, Lathrop believed, was a broad plan delineating the different responsibilities of each federal agency which found itself dealing with this social problem. Feeblemindedness, she now knew, was an issue affecting the nation's educational, medical, and welfare institutions. Responding to it would mean coordinating the activities of the Bureau of Education, the Public Health Service, and the Children's Bureau. Yet in deciding just how her own federal agency might best go about studying such a problem, Lathrop was perplexed.

In his confidential reply, Goddard reaffirmed the importance of such studies and the need for more research. "I do not think that our work on the heredity and family history of the cases is enough to warrant its being omitted from future investigation," he answered Lathrop. "No State is willing to believe that the cases within its borders are the same as we have found in New Jersey," he argued. "They have to have this demonstrated to them." If the Children's Bureau were to begin such studies, he suggested, it might start in states which still had no institutions at all for feebleminded persons.[3]

Yet in conceiving of a plan that would move beyond the state level, Goddard too was baffled. The problems of the feebleminded, he agreed, affected government agencies concerned with improving education, caring for children, reducing crime, and curbing prostitution. Even so, Goddard wondered about how scientific research might be conducted at the national level. After all, education, crime, and social welfare were all issues which fell within the province of the states.

Perhaps the federal government might open a research laboratory of its own, Goddard proposed, to which children could be brought – a plan "somewhat similar to the work that we are doing here," he added. Or maybe Lathrop could send "trained experts to gather up information" from around the country. Or perhaps she would simply have to conduct her research through questionnaires – a method he found "the least satisfactory."[4]

In the years that followed, no such plans were carried out. The president's interest notwithstanding, a coordinated federal plan for studying feeblemindedness at the national level was never articulated. Instead, the Children's Bureau would join with the Public Health Service and the Bureau of Education to produce a few local studies, including surveys of defectives, dependents, and delinquents in Delaware and the District of Columbia.[5]

Yet if the national government would fail to sponsor a broad program of scientific research to answer these social questions, it would still play a major role in advancing the new science of measuring minds. Its actions would emerge in response to two issues which clearly fell within the province of federal power: regulating immigration, and going to war. By 1918, American intelligence testers in general, and Henry Herbert Goddard in particular, would make their mark on both of these activities. Both actions would have a profound influence on the practice of intelligence testing; both would also instigate a new set of controversies concerning the broader relationship between psychological science and political power.

Most apparent in this new work was a radical expansion of the population subjected to testing. By the time World War I had ended, American intelligence testers had turned their attention from diagnosing children to classifying adults. They had also moved far beyond the population of "defectives, dependents, and delinquents," for they now considered themselves capable of gauging the minds of everyone. Psychological science, testers insisted, had much to say about the intelligence of the entire American nation, and about the racial, ethnic, and class differences found within it.

Also evident was these psychologists' expanding sense of their own political importance. During the prewar decade, Goddard and his followers had already convinced several states to employ psychological expertise in dealing with institutional commitments, school practices, and criminal sentencing. After testing immigrants and soldiers, they found even more uses for their science. Discoveries made by intelligence testers, psychologists claimed, had broad implications for understanding the rights and duties of American citizenship – indeed, for the very survival of democracy itself.

These changes would all prove significant, for in using their science to assess the minds of immigrants and soldiers, psychologists would subtly transform the nature of the controversies surrounding intelligence testing. Both actions would raise new questions about the relationships among psychological, medical, and pedagogical experts. Both would also focus new attention on the relative importance of heredity and environment. By the time World War I had ended, these controversies had begun to redefine the political meaning not only of "feeblemindedness," but also of the broader and more crucial concept, "intelligence."

In the immediate postwar years, Goddard and his fellow testers would draw ever bolder and more sweeping conclusions about the significance of their science for American politics. In the process, they would transform the political contours of the intelligence testing controversy in ways felt ever after. By the end of this decade, intelligence testers had made their psychology a part of American political discourse. They had also helped to frame a series of issues which would be fought, and refought, for the rest of the twentieth century.

Measuring the Minds of Immigrants

Perhaps no practice would raise more questions about the underlying assumptions guiding intelligence testing than the efforts to gauge the minds of immigrants. In the years between 1910 and 1918, the findings of intelligence testers began to play a role in one of the most explosive of American political controversies: immigration restriction. If some psychologists had started to question the effects of school learning, language skill, and social environment on test scores, the tests conducted on immigrants would bring these issues to the forefront. Questions concerning the minds of immigrants would also capture the attention of the public, for the same years saw a rise in concern over both the quantity and the quality of foreign immigration.[6]

In the two decades preceding 1910, more than twelve million immigrants crossed the Atlantic. Especially striking to many contemporaries were the differences between the "old" and "new" immigration. In previous decades, most émigrés to America had been northern or western Europeans, coming largely from the British Isles, Germany, and Scandinavia. After 1890, however, the majority left southern or eastern Europe. As a group, many critics argued, these newer arrivals from nations such as Italy, Russia, and the Balkans tended to be less Protestant, less educated, more impoverished, and more culturally "alien" than most of the immigrants who preceded them. All of these factors would provoke widespread public debate over the mental fitness of such persons for American citizenship.[7]

Inflaming this debate was a growing literature emphasizing the inherited "racial" traits which distinguished these non-Anglo-Saxon European populations and calling for more restrictive legislation to protect America's biological future. Among those who connected theories of racial difference with the need for immigration restriction most closely were numerous social scientists. In his 1914 study, *The Old World in the New,* for instance, sociologist Edward Ross had argued that Italians were inherently prone to crime, Slavs innately servile, and Jews prone to be crafty businessmen with a passion for Gentile girls. Even more popular was Madison Grant's 1916 text, *The Passing of the Great Race* – a work whose very title summarized its fears about the declining proportion of American "Nordics." The new immigration, Grant argued, included "an increasing number of the weak, the broken and the mentally crippled. . . . Our jails, insane asylums and almshouses are filled with this human flotsam and the whole tone of American life, social, moral, and political has been lowered and vulgarized by them. With a pathetic and fatuous belief in the efficacy of American institutions and environment to reverse or obliterate immemorial hereditary tendencies," Grant added, "these newcomers were welcomed. . . ."[8]

Many such writers found a comfortable home within the eugenics movement. Among the most ardent advocates of limiting immigration, for instance, were Charles Davenport's former Harvard classmates Robert De-Courcy Ward and Prescott F. Hall. Leaders of the Immigration Restriction League, these two New Englanders pressed Davenport to incorporate their activities within his eugenics organization. Restricting immigration ought to form a part of the broader eugenic goal of improving the hereditary stock of the nation, they proposed. "The same arguments which induce us to segregate criminals and feebleminded and thus prevent their breeding," Hall argued in 1910, "apply to excluding from our borders individuals whose multiplying here is likely to lower the average of our people." By 1913, Davenport had established a subcommittee on immigration within the American Breeders' Association, which included Hall and Ward.[9]

Yet while the writings of many of his contemporaries linked the new immigration to racial inferiority and national decline, Goddard's writings did not. In fact, immigration was rarely mentioned in Goddard's many litanies of America's social ills. Although he emanated from the same Anglo-Saxon lineage, if not the same social class, as many of these prominent restrictionists, Goddard did not share their xenophobia. He too warned frequently of the social dangers posed by biologically bad "stock"; even so, he did not argue that such threats were primarily of foreign origin.

"Of three large families of defectives . . . investigated at Vineland," Goddard declared in 1912, none contained "recent immigrants." Even the "nameless feebleminded girl" who had caused the Kallikak family so much trouble was "not referred to anywhere as being a foreigner"; instead, she may have been "one of the class that was indentured to this country in Colonial times." As such statements show, Goddard's writings reflected a consistent prejudice against the poor; still, they evidenced none of the nativism of the period.[10]

Nor were his works filled with contempt for non-Anglo-Saxon "races." Goddard was usually eager to promote new uses of Binet testing; in his own studies, however, he did not compare native-born with foreign-born children, or whites with blacks (as a few others were just beginning to do).[11] Equally absent from his publications was the socially acceptable antisemitism of the day. His Kallikak study had largely been financed by Jewish philanthropist Samuel Fels. Even more significant, in an era in which Jews commonly faced difficulty in gaining academic appointments, Goddard trained Jewish psychologists such as Samuel Kohs in his laboratory. Despite his frequent warnings about the problems caused by poor heredity, his writings never suggested that the inheritance of biological traits made some races or immigrant groups inferior to others. In his massive 1914 survey, *Feeble-Mindedness,* Goddard noted a few foreign-born children living in his institution who had "passed the customs officers," thus suggesting how "our immigration laws have failed to protect us." Even so, in summarizing his data from more than three hundred cases, he discussed crime, alcoholism, prostitution, and poverty without linking any of these issues to race, religion, ethnicity, or immigration – striking omissions in a sociological text of this era.[12]

The same attitudes were apparent in his articles on schoolchildren. In writing about special education, Goddard rarely made any mention of ethnicity or race. Even in studying the school system in which immigration had perhaps made its most profound impact – the public schools of New York City in 1911 – he paid no attention at all to ethnic, racial, or religious differences among the children he was testing. Binet testing, Goddard consistently maintained, worked equally well with any child; it was therefore unnecessary to analyze any other variables.

In fact, it was the very lack of attention to these issues that had provoked one of the many criticisms of Goddard's study from New York's Director of Ungraded Classes, Elizabeth Farrell. Had Goddard been testing "the thrifty, energetic German element in the neighborhood of Ninth Street," Farrell had asked, "or the more recent arrivals in America in the district south of Grand Street?" What was the nationality of those he tested? What was their native language? In answering such questions, Farrell declared, Goddard's school survey had supplied "not a word."[13]

Yet if Goddard had consistently refrained from making any statements about the innate inferiority of any racial, religious, or ethnic group, the same could not be said of other prominent testers. Lewis Terman, for instance, felt no such hesitation. In his studies of California schoolchildren, he had found many "dull normals" – persons who appeared normal but tested poorly. This type was "very, very common among Spanish-Indian and Mexican families of the Southwest and also among negroes," Terman reported in his influential 1916 volume, *Measurement of Intelligence*. "Their dullness seems to be racial, or at least inherent in the family stocks . . . ," he concluded. This suggested "quite forcibly that the whole question of racial differences in mental traits will have to be taken up anew and by experimental methods." Even before beginning such a study, Terman suspected what he would find.[14]

Ironically, it was Goddard and not Terman who would first have the opportunity to study such questions experimentally. He got the chance when he and Edward Johnstone were invited to visit Ellis Island. They came not because they personally favored more restrictive legislation, but because immigration officials believed their expertise might prove helpful in enforcing laws already in existence. Frustrated by frequent complaints that they were failing to keep "undesirables" out of the country, in 1910 these officials invited Goddard and Johnstone to observe their procedures and perhaps suggest ways to help the Public Health Service in "recognizing and detaining more of the mental defectives."[15]

By 1910, diagnosing mental deficiency had become a serious concern of doctors working at Ellis Island, for Congress had already passed several laws ordering them to keep persons suffering from this condition out of the country. An 1882 law, for example, excluded "lunatics," "idiots," and others likely to become public charges. In 1903, Congress added the insane, epileptics, beggars, and anarchists to this list; going even further in 1907, it specifically excluded "imbeciles, feeble-minded and persons with physical or mental defects which might affect their ability to earn a living." According to Ellis Island doctors, the decision by Congress to place the feebleminded in "the category of those whose deportation is mandatory" had been "far-reaching in importance."[16]

As Americans would soon discover, however, passing such laws would prove far easier than enforcing them. For the Public Health Service officers employed at Ellis Island, the new laws prescribed a nearly impossible task: requiring less than a dozen physicians to certify the mental fitness of as many as five thousand immigrants who might pass before them in a single day. Usually, these doctors were able to inspect the average immigrant for only a matter of seconds; the best they could do in such a situation, they argued, was to look for telltale signs which might suggest more serious physical or mental problems, and to single out these individuals for closer scrutiny over the following days.[17]

After observing these doctors at work for one day, the two Vineland visitors described themselves as "overwhelmed." Sensing few opportunities for psychological intervention, Goddard instead praised the efforts of his embattled medical colleagues. In his first article on the subject, published in 1912, he refuted arguments which blamed rising institutional commitments on immigration. "Since we have begun to recognize the appallingly large number of mental defectives among us," Goddard observed, it was "natural that many people should conclude that these defectives are foreigners and even immigrants." Yet the facts, Goddard emphasized, did not support this. Of more than three hundred feebleminded children residing at the Vineland school, for instance, only twenty-two were foreign-born.[18]

To help the Commissioner of Immigration find out the facts from other institutions, Goddard once again sent out a questionnaire. It asked superintendents to count the number of foreign-born persons within their institutions. "The Commissioner desires to know if any particular *localities abroad* are sending more defectives than others," Goddard inquired. "How many of these aliens does it seem to you ought to have been recognized by the immigration officer and sent back?" he asked pointedly.[19]

The results, he soon reported, confirmed his earlier conclusions. According to sixteen institutions responding, less than 5 percent of their populations were known to be foreign-born – "a surprisingly small number," Goddard emphasized. In all, he found "practically nothing on which to base a conclusion that any large proportion of clearly defective persons are escaping our immigration officers and entering the country." Assertions that mental defectives were more likely to emigrate Goddard found "grossly overestimated." And he knew of no statistics proving "that any one race or nationality is more inclined to mental defectiveness than another." Such data offered a rebuttal to "that rather large group of people who are inclined to think of our problem of the feeble-minded as closely related to that of immigration."[20]

By the time Goddard paid a second visit to Ellis Island in May 1912, however, his professional perspective had changed. Returning with several assistants, he found the medical situation "no longer so new and overpowering." By this time, Goddard had gained increasing confidence in his own staff's abilities to identify feebleminded persons. Perhaps his assistants could spot mental impairments among immigrants, he now proposed, in the same ways that they had made similar diagnoses of persons "in the field" – that is, by simply observing them. Asking permission to conduct an experiment, Goddard received "a ready response."

Goddard's experiment was simple: one psychologically trained assistant would select mentally defective persons, "her sole method of determining this being by her observation of them as they passed"; a second assistant in another room would then administer Binet tests (aided by an interpreter). Nine persons were chosen by the first assistant as feebleminded, and three as "control cases"; the tester found these nine "at least four to nine years backward" and the three controls normal.

Thus encouraged, Goddard returned that September to conduct a week-long study. This time his assistants tested forty-four persons selected by either physicians or the "Vineland experts." While eighteen of thirty-three chosen by physicians tested normal, nine of eleven picked by his staff proved defective. "It is thus seen that of those selected by the physicians less than half were correctly selected, while of those selected by the experts seven-eighths were rightly chosen," he bragged, employing once again the inaccurate arithmetic that pervaded all his work.

On his final day, he tried another experiment. This time, he had his assistants silently count each defective they recognized. Of 1,260 persons passing through the line, they considered 83 defective, while doctors questioned 18. Using the ratios he had previously determined (seven-eighths accuracy for "Vineland experts" and 45 percent for doctors), Goddard concluded that there were actually 72 defectives, of whom physicians had spotted 8. "On this basis then," he calculated (again incorrectly), "experts would detect at least ten times as many mental defectives. . . . " Goddard meant no disparagement of these physicians, for they "do not pretend to be experts on feeblemindedness," he now argued. "The comparison simply shows what experts can do."[21]

For Goddard, such judgments now defined the very essence of "expertise." It was "always impossible for the uninitiated to understand how the expert does his work," he explained to psychologist Robert Yerkes, who apparently still had doubts. Goddard himself wondered "how the tea tester or the wine taster can tell to a nicety what grade of article they are examining. I am inclined to believe that the whole thing is a fake," he conceded, but was "told that it is not." The same was true of mental diagnoses. "I stand on the line at Ellis Island beside the Psychiatrist and as the emigrants pass along, he says, 'there is an insane person.' I don't understand how he does it," Goddard admitted. "I come to conclude that the expert has an experience . . . whose results we must accept." Goddard's underlying argument here was clear: clinical experience produced expertise. Also evident, however, was a second argument: psychiatrists were the experts in diagnosing insanity, and psychologists (trained in institutions) in diagnosing feeblemindedness.[22]

Once Goddard had conceived of a function for Vineland personnel, his assessment of the situation changed. His assistants had discovered an "appalling percentage of defectives," he now reported, for they considered feebleminded 3¼ percent of the northern Europeans they had seen, and between 7½ and 9 percent of the southern Europeans – percentages much higher than similar estimates for the American population. Yet Goddard did not suggest restricting immigration by national origin; instead, he recommended hiring more psychologically trained inspectors. "We cannot act too promptly," Goddard concluded. "The immigration officers are ready to act as soon as an appropriation is made."[23]

Goddard's new findings quickly won him praise from the *Journal of the American Medical Association.* For years, the editors of this journal had been warning doctors of the dangers posed by the new immigration. Consequently, they had been skeptical of Goddard's earlier reports emphasizing low rates of feeblemindedness among immigrants. His new results were far more welcome, for to these editors they suggested "a strong argument against Dr. Goddard's views on the relative proportions of native-born and foreign-born defectives."[24]

Yet if the editors of the *JAMA* were pleased with Goddard's new research, doctors actually working at Ellis Island expressed much more skepticism. Especially questionable to these physicians were Goddard's "snapshot lay diagnoses" and his use of Binet tests to confirm them. "Imagine Goddard or his lay assistants designating the mentally defective merely by pointing them out as they pass and then defending their selections!" replied Dr. E. K. Sprague, an Ellis Island examiner. Such methods would invite "contempt and ridicule." Moreover, using a set of tests designed for schoolchildren on illiterate immigrants was "as sensible as to claim that with a single instrument any operation in surgery can be successfully performed."[25]

In the years that followed, Goddard's ideas in general, and ideas about Binet testing in particular, would provoke a broad range of responses among Ellis Island physicians. Reflected in these reactions was the unique situation these doctors found themselves facing. Most concurred with broader medical assessments of the day concerning the importance of their work in protecting the nation. "The most important function of the National Government from a mental hygiene standpoint," argued Dr. E. H. Mullan, an Ellis Island doctor, "is the exclusion of immigrants afflicted with insanity, imbecility, and feeble-mindedness." Yet at the same time, many of these doctors recognized the enormous power they wielded, and felt keenly the pain involved in separating families or dashing the hopes of an aspiring immigrant; consequently, they took their own legal and moral role very seriously. Such a physician "represents the country at large," declared Dr. L. L. Williams, Chief Medical Inspector of Ellis Island, "and must protect it against invasion by the unfit; but in discharging this duty," Williams added, "he must accord to the alien at least that measure of mercy which a court of justice would consider to be his right, and must give him the benefit of every reasonable doubt." To do otherwise, this doctor concluded, would "savor of tyranny."[26]

In many ways, these doctors reported, their work differed markedly from that of others charged with diagnosing the feebleminded, among them Goddard. After all, they had to make such diagnoses without knowing their patients' histories, with subjects who often spoke no English, and at a time of obvious emotional distress. As a result, Dr. Bernard Glueck concluded, the "practice of psychiatry at a port of entrance" had "very little in common with the practice of psychiatry in general." Even more crucial were the different consequences which ensued from their work. Diagnosing immigrants was a very different social enterprise than diagnosing schoolchildren, these doctors claimed, for its effects were both more devastating and more irreparable.[27]

If a tester misdiagnosed an American schoolchild, explained Dr. Williams, it "merely means that he will be wrongly classified, an error easily corrected." By contrast, an "error which results in unjustly deporting an alien from New York to Eastern Europe," he added, "is a grievous blunder and is without remedy." Dr. Sprague seconded this view. Persons diagnosing mental deficiency in institutions, schoolrooms, or courtrooms were usually "animated with the desire to benefit the unfortunate," he reported; their work thus represented "the highest form of philanthropy." At Ellis Island, however, examiners were "far from in any way benefiting one of their fellow creatures." Such a diagnosis could destroy an immigrant's life by "turning him an outcast into some European port." In making decisions "of such gravity," these doctors insisted, it was always better to err on the side of the immigrant. Immigration officers "may not certify all mental defectives," Sprague admitted, but they could at least rest assured that "those that are certified are below par mentally."[28]

Yet despite their caution and conservatism, these doctors had nonetheless been charged by the government with a serious and daunting task: diagnosing mental impairments among thousands of persons from many different countries who passed before them each day. Moreover, if doctors disagreed with Goddard's "snapshot lay diagnoses," their own methods were hardly much better, for they too had to spot mental infirmities largely with a single glance. In the years that followed, these doctors too would become increasingly interested in developing more scientific methods of assessing minds – methods, that is, which would allow them either to confirm or to overturn their initial impressions by using more reliable criteria.[29]

Toward this end, during the prewar decade a number of Ellis Island examiners began to experiment with a range of new methods for gauging mental impairment. The results they reported suggest an intricate blending of personal impressions and popular stereotypes, of objective with subjective assessment techniques, and of traditional medical practices with new psychological ideas. These experiences would also shape their responses to intelligence testing, for reflected in their reports was neither an uncritical acceptance nor a complete rejection of Binet's new ideas.[30]

Gauging the mental state of an immigrant was an extremely complex task, these doctors reported, for even the process of asking questions was fraught with special problems. Especially important in questioning immigrants, Ellis

Island doctors insisted in their writings, was the need to understand that "almost every race has its own type of reaction." Some populations, for instance, answered questions expansively, while others replied reservedly. If "an Englishman reacts to questions in the manner of an Irishman," Dr. Mullan explained, "his lack of mental balance would be suspected"; conversely, an Italian who exhibited the demeanor of a Finn might actually be "suffering with a depressive psychosis." "The English and German immigrants answer questions promptly and to the point," Dr. Knight reported, "but should they become evasive as do the Hebrews, we would be inclined to question their sanity."[31]

Like other doctors, these physicians regarded mental diagnosis as essentially a subjective skill gained through experience. Even so, no immigrant was diagnosed by only one physician; instead, Ellis Island inspectors developed an elaborate evaluation procedure that guaranteed separate assessments by at least three different doctors over several days. Moreover, they left all final decisions about deporting an immigrant to nonmedical officers.[32]

Among the techniques that these doctors adopted to help make their own assessments were numerous "mental tests." Most promising, they believed, were various "performance tests" – tests, that is, which tried to eschew the use of language as much as possible. Some of these they borrowed from other psychiatrists or psychologists; some they invented themselves. Among the nonverbal tests devised by these doctors, for instance, was the "Cube Test" invented by Dr. Howard Knox, which required subjects to move wooden blocks in particular patterns. Equally useful, they reported, were a variety of puzzles. These included the "Ship Test," a ten-part wooden puzzle which depicted a steamship – an object, these doctors surmised, that all arriving Ellis Island immigrants ought to be able to recognize.[33]

These doctors also experimented with other types of tests designed to gauge reasoning ability. Simple arithmetic problems proved surprisingly valuable, they reported, for even unschooled peasants had apparently learned to calculate sums in their daily lives. What these examiners were seeking was not a quick answer, they explained, but simply evidence of a mind in action. Thus, Dr. Mullan recorded approvingly the slow but effective method that one immigrant had used to add 15 + 15: "20 and 20, 40; take away 5, 35; take away 5, 30, *30.*"[34]

Many questions written by Binet were soon added to this eclectic battery of mental tests. Dr. Knox, for instance, incorporated many Binet items into a scale he designed for illiterates, for he asked his subjects to repeat a series of digits, follow commands, identify objects in a picture, copy shapes, and rearrange weights. He also reported some results with more difficult Binet items, including gauging reactions to the following story originally designed to test twelve-year-olds: "A man walked into the woods. He saw something

Fig. H.—SHIP TEST WITH PIECES ARRANGED FOR SUBJECT TO PLACE IN FRAME.

The Ship Test, which was first used at Ellis Island and then by army testers. From Robert Yerkes, ed., *Psychological Examining in the U.S. Army* (1921).

hanging from a tree that frightened him, and he ran back to notify the police. What did he see?" While the correct answer to this question was "a person hanged," immigrants often gave other answers. One Englishman, Dr. Knox reported, had replied in a cockney accent, "A brawnch." He was still passed, Knox explained, for doctors tried to give all immigrants "the benefit of doubt."[35]

Like many of their medical peers, these physicians did not regard any such tests as determining in themselves; instead, they simply saw them as supplying additional data to be considered by the physician, who would ultimately make a subjective judgment. No immigrant was certified as feebleminded solely because he "failed on this test or that test or because he is at a certain mental age," Mullan explained. Even if one gave a "generally poor showing," if there was "an occasional fair answer" or "some well-executed performance," the person was usually passed. "In other words," Mullan summarized, "from the mental field of the subject there has been sent forth a ray of hope." Such a ray might be nourished in a good environment, he believed, and was therefore enough to justify entry.[36]

As a result of such experiences in working with immigrants, at least some Ellis Island doctors developed an increased sensitivity toward cultural differences. Dr. Mullan, for instance, later described his experiences in surveying a prison for the Public Health Service. Of 250 black prisoners aged eleven to twenty-one, none had passed the Binet test for age eleven asking them to rearrange the following words into a sentence: "A – defends – dog – good – his – master – bravely." Their failure, this doctor hypothesized, was probably due to poor reading skills as well as "environmental conditions," for these men rarely used words such as "defends" or "bravely" in their daily vocabularies. Feeling free to improvise in changing the test, Mullan simply rewrote Binet's sentence using new words: "Eggs – supper – boys – the – for – eat – and – bacon." This time, he reported, his results were "uniformly good."[37]

As these reports show, doctors working for the U.S. Public Health Service had begun to use Binet's tests to help identify the feebleminded, both at Ellis Island and elsewhere. Yet they were hardly using them in the manner that Goddard recommended. Like many other doctors, these physicians largely ignored the concept of mental age; moreover, they too felt free to drop, add, or change questions at will. Intelligence tests, this work suggested, offered objective data which might aid the physician in making what was still a subjective diagnosis.

Even so, both the increasing national alarm over feeblemindedness and the growing body of data acquired from mental testers did play a role in influencing American immigration policy. Deportations due to this mental condition had begun to increase dramatically, from 186 in 1908 to 555 in 1913 to 1,077 in 1914. The percentage of immigrants affected was still quite small but growing sharply; between 1912 and 1914, the deportation rate for feeblemindedness increased from 19 to 88 persons excluded per 100,000 immigrants (under 0.1 percent, or slightly less than one person per thousand).[38]

Meanwhile, Goddard continued to advocate a more experimental approach toward studying the mentality of immigrants. By 1913, he had designed an extensive research project whose results would not be published for another four years. Included would be the findings from intelligence tests, performance tests, and field work. Equally evident would be the effects of contemporary debates within his own profession, for his new work would reflect a revised approach toward Binet testing.[39]

Goddard began by having his assistants administer a battery of mental tests to 165 Ellis Island immigrants, among them Jews, Hungarians, Italians, and Russians. In choosing his subjects, he avoided both the very obviously feebleminded and the "very obviously high grade intelligent immigrant." He was not trying "to determine the percentage of feeble-minded among immigrants in general," he explained, "or even of the special groups named." Even so, Goddard still believed that his results might shed light on the "great mass of 'average immigrants.'"[40]

Among the methods his assistants tried were various nonverbal "performance tests." While Ellis Island doctors often lauded this type of testing, Goddard was less impressed with these devices as measures of the mind. Tests which emphasized recognizing shapes or following mazes, he proposed, might really be gauging merely the sharpness or weakness of an immigrant's eyesight. Far better were the Binet tests, he believed, for these gauged more complex processes of mental development.[41]

By the time Goddard interpreted all his data in 1917, the impact of the previous year's rancorous testing debate among psychologists was unmistakable. Evidently in response to his critics, Goddard now decided to drop all the controversial tests for higher mental ages. Instead of using twelve as the boundary for normal mentality (as he had always done previously), he now diagnosed as feebleminded only those who failed to earn a mental age of about ten. Even so, 83 percent of the Jews, 80 percent of the Hungarians, 79 percent of the Italians, and 87 percent of the Russians he was studying still fell into the feebleminded category. These results were "so surprising and difficult of acceptance that they can hardly stand by themselves as valid," he reported. "Perhaps," Goddard now conceded, "the tests are too hard. . . . "[42]

Since the tests for Jews were conducted by a Yiddish-speaking psychologist (rather than an interpreter), he focused most on these data. The resulting analysis contained no warnings about the inferiority of non-Anglo-Saxon "stock" common in literature of the day. To the contrary, Goddard demonstrated far more sympathy for immigrants than he had ever shown his poor fellow countrymen. In the past, Goddard had been unwilling to entertain any arguments about the role of environment in influencing test scores. Binet tests, he had always insisted, were actually gauging a normal process of mental development; they therefore worked equally well with rich or poor, rural or urban, native-born or foreign-born. In studying immigrants, however, even Goddard would finally begin to consider this question far more seriously.[43]

For the very first time, Goddard decided to restandardize his scale to take into account environmental differences. Trying a new experiment, he now suggested that any question be kept only if about 75 percent of this population could pass it. To his surprise, this criterion forced him to eliminate quite a few of Binet's questions.[44]

Goddard's response to this finding was ambivalent. On one hand, he felt his new procedure was fair; on the other, he strongly suspected that his new standard was really "too low for prospective American citizens," for eliminated were many questions which still struck him as easy. By closely examining the omitted questions, Goddard too finally began to grapple with the issues already perplexing many of his peers – issues concerning the influence of schooling, environmental circumstances, and language skill on test scores.[45]

After working at Ellis Island, Goddard could understand why Binet's test requiring subjects to draw a design from memory had proven too hard for persons "who have never had a pen or a pencil in their hands." Also dropped was a test asking subjects to use three words in a sentence. Such a task was "perhaps beyond the reach of the people who have lived as most of these immigrants have," Goddard explained. "We are really asking for a sort of creation, an element of literary ability . . . ," he now saw for the first time.[46]

Harder for him to understand, however, was why most immigrants were failing a test which asked them to name any sixty words in three minutes. "How could a person live even fifteen years in any environment," he asked ingenuously, "without learning hundreds of names of which he could certainly think of 60 in three minutes?"[47]

Even more perplexing were high failure rates on other questions – failures that posed challenges to the very premises of Binet testing. Many of Binet's questions, for instance, had been chosen precisely because the answers supplied by older children differed markedly from those of younger children. In administering the same questions to immigrants, however, even this basic assumption often failed to hold true.

Among Binet's questions for children, for example, were several that asked them to define common objects. Children under the age of seven, Binet had observed, usually defined such objects only by their use: to them, a table was "to eat on" and a horse "to ride on." With nine-year-olds, however, Binet had noted a difference, for these older children supplied a diverse range of definitions that were usually "better than use"; a table, for instance, might be defined as a piece of furniture, or a wooden object with four legs.

Among immigrants, however, similar distinctions between the answers given by young and old were not nearly so evident. To most immigrants, age notwithstanding, a "table" remained simply "to eat on." To Goddard, the fact that even adults usually failed to define such terms in any way "better than use" was troubling. Could this be because "never having been to school . . . and never having been called upon to explain to anyone what a 'table' is," he proposed skeptically, the immigrant was "satisfied with the simplest expression that came to his mind even though his definition of a table did not differentiate it from a plate . . . ?"[48]

Even more puzzling, how was one to understand these adults' widespread ignorance of the date? In children, Goddard believed, a growing precision in recognizing and conceptualizing intervals of time was a sign of a developing mind. Children of six frequently could not identify the current month or year, Binet had discovered; children of nine usually could. At Ellis Island, however, even adults regularly failed such questions. "Must we again conclude that the European peasant . . . pays no attention to the passage of time?" Goddard wondered. "That the drudgery of life is so severe that he cares not whether it is January or July, whether it is 1912 or 1906?" Was it possible that an intelligent person could fail to acquire "this ordinary bit of knowledge, even though the calendar is not in general use on the continent, or is somewhat complicated as in Russia?" Goddard now asked. "If so," he added, "what an environment it must have been!"[49]

While Goddard found many of these test failures to be both inexplicable and disturbing, Ellis Island physicians offered far more sophisticated explanations of their own in confronting the same phenomena. "What are likely to be considered matters of universal knowledge may be absolutely unknown to them," one doctor wrote, "on account of the extreme limitations of their surroundings." It was "almost impossible for Americans to realize the narrowness of the lives of some of the poorer classes of the countries of Europe," he continued, for many peasants lived "a life of sordidness and hard-working monotony almost beyond belief, resulting in a mental equipment which is correspondingly limited and stunted." Many could not tell time; they knew it was Sunday only by the church bells. "The farmer of southern Italy, tilling a few acres of land and living in a hut, the bare walls of which contain only one ornament, an unframed picture of the Madonna, and the only articles of furniture of which are a bed, a table, a chair or two, a few kitchen utensils, and a little bedding," this doctor argued, "can hardly be expected to define the word 'charity.'" Such an individual could not even name the items in an ordinary American home. "His possessions, his ideas, his vocabulary, and his experiences are all extremely limited," he concluded, "and he must be judged and measured accordingly."[50]

In testing immigrants, doctors too had run into similar problems. They found it especially difficult to gauge abstract thought – a key concept for Binet testers – since many unschooled peasants were extremely reluctant to answer any question that required them to imagine a situation they themselves had never experienced. Doctors described their efforts, for instance, to use another of Binet's tests – presenting the subject with an "absurd situation" and asking for a reaction. Among these was the following: a subject "is told that a young woman's body was found in a room, cut into 18 pieces. The police say that she committed suicide. Does he think that this is likely?" Asking this very question to eleven-year-old schoolchildren, Binet had evoked laughter as well as skepticism. Asking the same thing to immigrants, however, drew markedly different reactions (including an apparent unwillingness to challenge any statement supposedly made by police). According to doctors, it was common to get responses such as "Indeed, I was not there," or "It was a great sin for her to kill herself." "Failures as bad as these by persons who have had five or six years' schooling point strongly to mental defect," these doctors agreed; however, in diagnosing the types of peasants whom they saw daily, such responses had "little significance" and should therefore not be given much weight.[51]

Despite his own personal reservations about what such findings meant, Goddard too ultimately decided to drop these and similar test questions which apparently could not be passed by most of his new subjects. Instead, he reorganized the remaining questions into a new scale which would still allow immigrants to earn a mental age of ten. Using this modified scale, he cut the percentage found feebleminded in half – to about 40 percent. "It must be admitted," he pleaded, "that this gives the immigrant the benefit of every doubt." Yet even Goddard remained skeptical, for this finding still struck him as a "startling proportion for the feeble-minded among our immigrants."[52]

"Doubtless the thought in every reader's mind is the same as in ours," he now argued, "that it is impossible that half of such a group of immigrants could be feeble-minded. . . . " Yet it was "never wise to discard a scientific result because of apparent absurdity," Goddard warned his readers. Instead, such results had to be explained. "Are these results reasonable?" he now asked.[53]

Goddard's own explanation for his "absurd" findings once again emphasized class over race. America was now getting "the poorest of each race," he reported, for recent studies showed that many new immigrants were merely laborers or servants. Of course, Goddard did not mean to "libel" any "keen, sharp, energetic worker" or to disparage the fact that "much of our population has been immigrant." He was only speaking "of 40 per cent of the immigrants and that, too, of the immigrants who come in the steerage, whereas many immigrants come second class or even first." In other words, the percentage of feebleminded immigrants had proven surprisingly high, Goddard proposed, not because of his subjects' ethnicity, but because his sample had come largely from the lower class.[54]

For Goddard, there was only one known scientific method to "establish the correctness or the error" of his conclusions: field work. The best test of all, he believed, was to see how such persons actually fared within the American environment. "Every one who makes good and becomes a useful citizen, will discredit our test," he conceded; "but every one who becomes a public charge, will confirm our diagnosis." To gather such data, in 1915 Vineland's field workers started conducting "follow-up" studies of their Ellis Island subjects. They began by paying visits to the families who had first sponsored these immigrants when they landed in America.[55]

Their findings from this field work suggest cultural differences as stark as any answers to Binet questions. Although Goddard's assistants had been able to trace the whereabouts of Deborah Kallikak's relatives, both living and dead, for nearly two centuries, they could not find these immigrants after only two years. Such persons, they now learned, often changed their names, moved frequently, and rarely knew their neighbors. Even in the few cases in which they did locate relatives, their efforts were frustrated, for whereas the garrulous country Kallikaks had welcomed the Vineland field workers and entertained them with notorious family stories, Jewish and Polish immigrants eyed them suspiciously and usually refused to disclose any information at all about their relatives. In nearly every case, these persons denied knowing the party in question, claimed never to have seen them, and offered no forwarding addresses. Goddard described a typical exchange:

> "Vat's de drouble?" was the anxious inquiry that usually met the field agent. "No trouble, my good woman, I've just come" – etc. etc. Great difficulty of comprehension on both sides generally followed with only one fact sure to be carried away – in the immigrant's struggle for adaptation in the new world there was "drouble – very much drouble."[56]

Even the most famous field worker of all, Elizabeth Kite, had been unable to gather much damaging information on these immigrants, no matter how hard she tried. Of all the Ellis Island cases she attempted to track down, she located only one. Despite several good reports about this girl's progress, Kite continued investigating until she finally found someone with something negative to say. This woman described the subject in question, an Irish domestic, as "perfectly honest, reliable and industrious, neat and good to the children." "Why then was she discharged?" Kite probed. "'Well, my daughter got angry with her for something one morning and told her it was time she went away,'" she replied. After pressing for details, Kite summarized the problem: "There was about the girl a certain obstinacy, a determination to do her own way" which "seemed incurable." To Kite, this had "directly to do with the intelligence and comes from a certain lack of power of comprehension." Apart from this, she added with evident disappointment, "I could get no history of anything bordering on what we know to be characteristic of feeblemindedness."[57]

As part of his field work, Goddard also had his investigators check New York settlement houses, schools, missions, and charity registers to see if any of the persons who tested so poorly had become dependents. Once again, none could be located. Thus, after four years of work, Goddard could announce few definitive conclusions. Like Ellis Island doctors, he too decided to give these immigrants the benefit of doubt. "The fact seems to be," he now concluded, "that a very large percentage of these immigrants make good after a fashion," notwithstanding their "relatively low mentality."[58]

Yet if this were so, how could Goddard explain his test results? Why were the 40 percent who tested feeblemindcd not failing socially? Such findings raised serious questions about Goddard's own broader arguments linking low test scores to a lack of social responsibility – questions, he recognized, that required answers.

Ironically, Goddard found his answers in his new subjects' very status as foreigners. By being "willing to do work that no one else will do," he argued, these low-scoring immigrants had earned a "relatively simple" living without competing much with more intelligent native workers. Even more critical, he hypothesized, were the efforts of ethnic communities to aid the new citizen. "They protect and care for him, partly through racial pride, partly through common humanity," Goddard reported. In other words, while the feeble-minded Kallikaks had been left to fend for themselves, thus wreaking social havoc, these immigrants were "more or less closely supervised" by caring family and friends, and fared far better.[59]

Yet this raised an even more crucial question. While Goddard believed that most of these immigrants were indeed mentally weak, he wondered about the cause. "Are these immigrants of low mentality cases of hereditary defect," Goddard now asked pointedly, "or cases of apparent mental defect by deprivation?" If the former, they still posed a threat to posterity; if the latter, then Americans need have no fears about succeeding generations.

While Goddard knew of no data to settle this "vital question," he himself believed it "far more probable that their condition is due to environment than that it is due to heredity." Their "environment has been poor" and "seems to account for the result," he decided. Thus, while he still considered many immigrants morons, Goddard saw no danger to America's biological future. Even if some "run amuck and make us trouble," he concluded, the "wise solution" was "not to exclude them all but to take care of those who are not getting along well."[60]

Such conclusions could hardly be said to support those calling for more restrictive legislation. Even so, any statement linking immigration with feeblemindedness was quickly absorbed into the anti-immigrant propaganda of the day. Perhaps for this reason, Goddard's own actions suggest some reticence about publicizing his work on immigrants. After first visiting Ellis Island in 1910, he waited two years before writing anything on this subject. In 1913, when an editor asked him to present his findings in *American Breeders' Magazine,* he replied that his studies were as yet unfinished. In his 1914 text, *Feeble-Mindedness,* he included both his own data and that of others concerning schoolchildren, delinquents, and prostitutes, but no findings at all from Ellis Island.[61]

Such reticence was apparently well founded, for when Goddard's long article entitled "Mental Tests and the Immigrant" was finally published in the

Journal of Delinquency in 1917, it was immediately summarized by an editor of the *Survey* under a very different title: "Two Immigrants Out of Five Feebleminded." "If you had gone over to Ellis Island . . . and placed your hand at random on one of the aliens waiting to be examined by government inspectors," this summary began, "you would very likely have found that your choice was feebleminded. He would probably have answered your question, "What is a horse?" by replying, "To ride on," or with some other simple reference . . . that did not differentiate it from a bicycle. He would have told you it was July if it was January. . . . "[62]

This summary soon provoked an angry letter-to-the-editor from the Council of Jewish Women's Department of Immigrant Aid. Dr. Goddard had never claimed that his sample was representative of all Ellis Island immigrants, these protesters argued. Moreover, the tests he used, they complained, were designed for American schoolchildren, not "peoples from different kinds of environment, with different languages, different education or lack of education." In addition, "Dr. Goddard himself says that even if these defectives are morons, it must be due to environment, not heredity . . . ," they added. Yet while conceding that Goddard's findings were not meant to be representative, and thus its choice of title was "unfortunate," the *Survey* stuck to its story. "Barring the title," the editor replied, "it is altogether accurate."[63]

As such conflicts suggest, the results of Goddard's efforts to use Binet testing at Ellis Island assumed different meanings to different audiences. To advocates of immigration restriction, Goddard's findings meant that scientists had now proven the mental inferiority of recent immigrants, and that new laws were needed. To the broader medical community, his articles reaffirmed the importance of the mental hygiene movement's commitment to protecting the public from insanity and imbecility. Among physicians actually working at Ellis Island, however, Goddard's studies had provoked the most complex reaction of all, for they had motivated these doctors to seek new ways to gauge immigrant minds – efforts which resulted in studies that were often more sensitive than those produced by Goddard himself.

Goddard too saw his data on immigrants as having broad implications, but of a very different sort. If low-scoring foreigners were apparently getting along well enough in the noninstitutional world, then perhaps the same was possible of low-scoring Americans – if they were supervised properly. Had he been shortsighted all along, Goddard began to wonder, in insisting upon institutionalizing this entire population? Maybe his social solutions emphasizing complete segregation had been unnecessary. Perhaps it was a "superficial view of that problem to say, we will eliminate them all as fast as we can," he proposed.[64]

Such persons might even prove useful to society, he now argued. After all, there was "an immense amount of drudgery to be done, an immense amount of work for which we do not wish to pay enough to secure more intelligent workers," he reported. Perhaps such a population could have a valuable role to play within the American economy. "May it be that possibly the moron has his place?" Goddard asked in concluding in his study of immigrants in 1917. By the end of 1918, he had his answer.[65]

Measuring the Minds of Soldiers

What finally settled such questions for Goddard, as well as for many others, was America's entry into World War I on April 6, 1917. In the ensuing nationalist fervor, war mobilization took priority over every other concern. For Progressive era industrialists, the war offered an opportunity to organize the nation's economic resources on a vast new scale. For social scientists, it meant organizing human resources in equally vast new ways.[66]

Nowhere were such strategies more needed, psychologists claimed, than in the military. Between March 1917 and November 1918, the wartime draft would transform the American army from a small body of about 200,000 professional soldiers to a massive force of 3.5 million men. To many psychologists, this transformation suggested both the necessity and the opportunity for national service. In trying to meet the needs of the military, psychological testers would markedly transform their own science, for their efforts would ultimately lead to a collaborative research project organized on a scale unprecedented in their profession.[67]

Like many of his contemporaries, Goddard quickly turned his attention toward serving the nation in its time of emergency. Within weeks, he had written an article on "The Place of Intelligence in Modern Warfare" for the *United States Naval Medical Bulletin.* "The victories of war, no less than those of peace, are frequently due to the superior intelligence of the victor," he began. Both intelligence and feeblemindedness, he asserted, were concepts deserving of serious military attention.[68]

Goddard's primary goal was to alert the military to the dangers caused by feeblemindedness. Particularly troublesome, he proposed, were an estimated 200,000 feebleminded men who might be drawn to enlist by the "spectacular features of warfare." Such men were especially likely to cause disciplinary problems, Goddard claimed, and were "notoriously intolerant of alcohol." They could also endanger many lives, he warned. A moron on sentry duty, for instance, "might be tricked into betraying the whole camp." Once again, Goddard argued for understanding rather than punishment, for officers should realize that *"mental incompetency explains the action of the unsatisfactory recruit more often than any other cause; perhaps more often than all other causes put together."*

The solution to such problems was, of course, proper diagnosis. Ideally, there ought to be "a psychological examiner at every recruiting station," Goddard declared. This was impossible, however, with so few trained psychologists. Instead, Goddard suggested another plan: psychologists could teach recruiting officers to give intelligence tests to soldiers. After all, if he could train schoolteachers to administer such tests, why not recruiters? Psychologists could then serve as "inspectors" offering instruction, encouragement, and guidance – work, he predicted, that would take "a few hours, probably a half a day at most, at any one station." Goddard never foresaw a need to test all recruits; only those who had done the "simplest kind of work, such as teamster, errand boy, etc.," or who had failed to go beyond the fifth grade might need testing.[69]

While Goddard was developing his own ideas for incorporating psychologists into the military, a far more ambitious plan was taking shape in the mind of Robert Yerkes, president of the American Psychological Association. Like Goddard, Yerkes was determined to make his discipline useful to the war effort. Yet he had absolutely no intention of teaching anyone else to administer what was quickly becoming the psychology profession's most important tool in proving its utility to the nation – tests of intelligence.

Even before the United States entered the war, Yerkes had begun to explore the possibilities for psychological involvement. On April 6, the very day war was declared, he started organizing his colleagues. "We should act at once as a professional group as well as individually," Yerkes wrote to APA Council members that evening. By the time the council met on April 21, he had already arranged for a "Psychology Committee" to join the National Research Council, the organization overseeing scientific research for the government. Yerkes also appointed members to a dozen different APA committees; their functions were to explore services ranging from studying acoustical problems to improving military recreation.[70]

Yerkes himself would chair what he considered the most important of these, the "Committee on the Psychological Examining of Recruits." Its main task would be to identify "mentally unfit" soldiers. To join him on this committee, he invited psychologists with extensive experience in using mental testing in diverse social settings: Lewis Terman and Guy Whipple, both experts in educational testing; Walter Bingham and Walter Dill Scott, the nation's foremost vocational testers; medical testers Frederic Wells (a pathological psychologist at McLean Hospital) and Thomas Haines (a psychologist and physician at the Ohio Bureau of Juvenile Research); and Henry Herbert Goddard, the country's leading expert on feeblemindedness.

In order to work effectively, Yerkes realized, these psychologists would first have to reach consensus on several issues which still divided them. Most crucially, they would have to agree upon a common means of constructing, scoring, and interpreting the results of mental tests. It was "extremely important that we psychological examiners work together," Yerkes explained to Goddard in inviting him to join this committee, for if testers continued to work independently or at cross-purposes, they would "do harm rather than good."[71]

Yerkes was not entirely successful in keeping all of these mental testers unified, for vocational tester Walter Dill Scott soon left to lead a separate "Committee on the Classification of Personnel." In the following months, Scott would develop his own set of tests for the military; these included tests to gauge proficiency in specific occupational skills, a program which soon won official recognition from the army's Adjutant General. Yet notwithstanding Scott's departure, Yerkes did succeed in convincing six of the nation's most prominent testers to work together for the first time under his leadership.[72]

This time, Yerkes was convinced, an APA committee would simply have to exert its control over intelligence testing, or the consequences could be disastrous – both for psychology and for the nation. What he planned to do, he told the National Research Council, was to eschew the "unintelligent work" done in the past by "so-called mental testers." Instead, his committee would develop new psychological tests for the military – tests that would be simpler, faster, and above all more reliable than any currently in existence. Yet such a task would take money. To fund his committee's research, Yerkes hoped to secure a government allocation of $25,000. His plans were stalled, however, when Washington refused his request.[73]

Yerkes' ideas about testing the army might have ended there, were it not for the timely intervention of Vineland's own experts on how to get things done – the Committee on Provision for the Feeble-Minded. This committee too was "vitally interested in the preparation of simplified intelligence tests," its chairman, Joseph Byers, reported. In fact, Byers too had already written to the army and navy to suggest that they make use of Binet testing – precisely the kind of independent action that Yerkes feared most. In their correspondence, Yerkes tactfully convinced Byers to aid him instead in "centralizing this work" by supporting the efforts of the APA's new committee – a committee that now included Goddard.[74]

In order to accomplish something, Byers reported to Edward Johnstone, the Vineland institution would have to become involved in war work "up to its neck." In reply, Johnstone authorized Byers to "go the limit" in offering his school's resources to the APA. What Byers could offer Yerkes' committee was free room and board at the Vineland school, full use of Goddard's psychological laboratory, and $800 toward their expenses. Yerkes quickly accepted. Thus, on May 28, 1917, Johnstone prepared his children to host summer visitors of another sort as several of the nation's most prominent psychologists left Harvard, Stanford, Carnegie Institute of Technology, and

Army testers at Vineland. Left to right, front row: Edgar Doll; H. H. Goddard; Thomas Haines; back row: Frederick Wells; Guy Whipple; Robert Yerkes; Walter Bingham; and Lewis Terman. Courtesy of the Archives of the History of American Psychology, University of Akron.

other prestigious institutions to begin working together in Goddard's laboratory at the Vineland Training School.[75]

By the time these committee members met at Vineland, Yerkes had already sketched out a plan. Their goal, as he first envisioned it, would be to develop an auxiliary role for "psychological examiners" within the army's Medical Corps – a role whose name reflected both Yerkes' experiences at the Boston Psychopathic Hospital and his future hopes for his field. Psychologists, he told the Surgeon General, would establish a battery of tests to gauge intelligence, memory, suggestibility, or other mental traits; these could then be used by the military to identify recruits suffering from "intellectual deficiency, psychopathic tendencies, nervous instability, and inadequate self-control" – persons, in other words, who might prove "peculiarly dangerous risks with respect to disaster in action, incapacity, and subsequent pension claims."[76]

Almost immediately, however, Yerkes' plan encountered strong opposition from psychiatrists (who had also organized themselves professionally as a committee within the National Research Council). In fact, the same conflicts that had begun to pit psychologists against psychiatrists within the institutional world would now be equally evident in the military. "Whether the psychiatrists will succeed in cutting us off or not is a question," Yerkes conceded in May 1917, for they were trying, he believed, to "subordinate everything psychological to their will." To resolve these professional disputes as quickly as possible, in June the National Research Council set up a "Joint Conference on Psychology and Psychiatry." Under strong pressure from physicians to narrow their aims, psychologists agreed to restrict their "examining" to the one activity in which even psychiatrists conceded their superiority and utility: intelligence testing.[77]

Ironically, while Yerkes deeply resented these constraints placed upon his profession by physicians, they would soon prove largely irrelevant. Far more important than the medical model of "psychological examining" envisioned by Yerkes would be an educational model introduced to his committee by Lewis Terman. In designing a testing program, Terman proposed, psychologists could learn more from educators than from doctors. They ought to pay less attention to what was happening in hospitals, he believed, and more to what was happening in schools.

By 1917, one of Terman's protégés, Arthur Otis, had been experimenting with transforming the Stanford–Binet scale into a written "multiple-choice" test – a type of test already being used in American schools to gauge reading skills. Such tests could be given to many persons simultaneously and then graded quickly by the use of stencils. By adopting this school strategy, Terman proposed, psychologists could do something far bolder than either Goddard or Yerkes had even dreamed: test the entire army.[78]

Following Terman's suggestion, Yerkes' committee began working on a new multiple-choice version of an intelligence test. They made rapid progress, for in less than two weeks, these psychologists had transformed Binet's oral tests for individual children into a battery of written tests for groups of adults. They spent the next two weeks trying out their new tests in training camps, and a final two weeks revising them. By July 7, 1917, the committee had in press five different versions (to prevent cheating) of their new written test, tentatively entitled Army A and later revised and renamed Army Alpha. To test illiterates, they next prepared a set of picture tests entitled Army Beta, as well as a battery of individual Performance Tests.[79]

In its final form, the Army Alpha was a stencil-scored examination which took under an hour to administer. It consisted of eight timed subtests, each containing between eight and forty questions that ranged from easy to difficult. Evident in these questions were the diverse experiences of the committee members who designed them, for while some subtests appeared to be

multiple-choice adaptations of the types of tasks suggested by Binet, others
showed the strong influence of educational or vocational testing. Included in
the Army Alpha, for instance, were tests of following directions, arithmetic
word problems, a number series completion test, an analogies test, and a test
of disarranged sentences. To gauge vocabulary, testers now asked whether
two words had the same or opposite meanings ("empty–full" asked an easy
question; "vesper–matin" a difficult one).[80]

Also added was a new "Information Test," first proposed by vocational
tester Walter Bingham. More than any other, this subtest suggested the
marked differences between testing children and testing adults. No longer
did these psychologists feel any need to discern what children of different
ages ought to know; instead, they could now ask questions covering a wide
range of information that "intelligent" adults ought to possess. Soldiers
might be asked, for instance, to identify the color of chlorine gas (green); the
Union commander at Mobile Bay (Farragut); the author of *Robinson Crusoe*
(Defoe); the occupation of Jess Willard (pugilist); the origin of silk (worm);
the product advertised by "Velvet Joe" (tobacco); or a first-class batting average (.300).[81]

Equally revealing of the culture of the day was the subtest designed to
gauge "Practical Judgment." "This is a test of common sense," its directions
began. Embedded in these questions and answers were the "commonsense"
judgments most prized by Progressive America – judgments, in other words,
that emphasized utility, efficiency, and pragmatism. "Why are doctors useful?" testers asked soldiers. "Why are cats useful animals?" "Why is agriculture valuable?" "Why is tennis good exercise?" Perhaps no multiple-choice
question captured the essence of Progressive ideology better than this one:

Why ought every man to be educated? Because
  – Roosevelt was educated
  – it makes a man more useful
  – it costs money
  – some educated people are wise

For these psychologists, in choosing between usefulness and wisdom, the correct answer was obvious.[82]

Especially striking in this committee's work was its determination to compromise over past psychological conflicts in order to produce a usable product. Perhaps nowhere was this more apparent than in its decisions about how
to report its findings. The Army Alpha would be a Point Scale – the type
of intelligence test ardently advocated by Robert Yerkes; these point scores,
however, would also be converted into "mental ages" – the type of result
strongly preferred by Lewis Terman. Even more significant, since neither
point scores nor mental ages would mean much to army officers, psycholo-

First Company of Commissioned Psychologists, School for Military Psychology, Camp Greenleaf, Georgia, 1918. Robert Yerkes is in right corner inset. From Robert Yerkes, ed., *Psychological Examining in the U.S. Army* (1921).

gists adopted a third scoring strategy: each soldier would also be awarded a letter grade from A to E, with "A" men as the highest scorers, "C" men as average, and "E" men as the lowest – a grading procedure modeled on a marking system devised by educators.[83]

With a potentially useful test in hand which borrowed far more from pedagogy than psychiatry, Yerkes' committee finally won the financial backing of the National Committee for Mental Hygiene. Sensing that such testing might prove useful to hospital personnel, this committee agreed to pay $2,500 for an extensive private trial. By July, psychologists had been dispersed to four military bases to test four thousand men. When their assessments of these men's mental abilities roughly matched assessments made by officers, support for an official army trial was assured.[84]

By August, Yerkes could report exciting results to Byers. "Within a fortnight we shall probably have forty or fifty psychologists at work," he predicted, "and if our methods prove valuable, there will doubtless be two, three, or even four times as many men engaged within two months." Yerkes himself would oversee the new enterprise. "I have accepted commission as Major attached to the staff of the Surgeon-General of the Army and shall have general direction of the psychological examining of recruits," he explained.[85]

As Yerkes' new military commission confirms, the testing program had finally won official approval from the army's medical officers. Yet despite the Surgeon General's support, the relationship between medicine and psychology remained a delicate one for the remainder of the war. The "work of the psychologist, although not strictly medical in character but instead vocational, educational and social," Yerkes argued tactfully, would be to aid medical officers by identifying persons "for whom careful psychiatric examination is obviously desirable." Despite this deference, professional rivalries continued. Psychologists, Yerkes complained bitterly, were consistently given lower military ranks than doctors; even more insulting to him, they were commissioned not in the Medical Corps but in the less prestigious Sanitary Corps.[86]

These rivalries notwithstanding, Yerkes' program made dramatic progress. By December, the army had approved a testing program for all recruits; by February, it had opened a school of military psychology at Fort Oglethorpe, Georgia, to prepare psychologists to administer the new tests; by May, its graduates were testing 200,000 men a month in camps throughout the country. By January 31, 1919, psychologists had given the Alpha and Beta tests to 1,726,966 men – a population far larger than all their previous samples combined.[87]

Many of the findings from such a large sample surprised even committee members. For the first time, intelligence testers were working with subjects who clearly represented a cross section of the American people. And while it would take these psychologists another two years to analyze and interpret all their data, some results were of immediate significance.[88]

Testers were startled, for instance, by the high rate of illiteracy they discovered among recruits. Even determining what literacy meant proved a far more complex enterprise than they had anticipated; it was crucial, however, in deciding who took the Alpha and who the Beta. In some army camps, anyone who could read English at all was given the Alpha. In others, only those who had completed the fourth grade took this test. Throughout the war, psychologists continued experimenting with methods for making such decisions quickly and consistently; none, however, proved satisfactory. (At Camp Dix, for instance, the Alpha was given to all recruits who could correctly write their name within thirty seconds and the sentence "We are in the army" within forty-five seconds.) Whatever method they used, psychological testers reached the same conclusion: American illiteracy was surprisingly high.[89]

In summarizing data gathered from more than a million and a half soldiers, Robert Yerkes reported that 25.3 percent were unable to "read and understand newspapers and write letters home" and were therefore given Beta examinations. In addition, another 5.7 percent did so poorly on the Alpha that they were retested with the Beta. Most surprising of all, over half this 31 percent were estimated to be native-born.[90]

Also striking to psychologists was the strong correlation they discovered between test scores and years of schooling. While reports varied, some army testers calculated this correlation to be as high as .81. Such a finding did not necessarily mean that these tests were really measuring education, Lewis Terman insisted. After all, he argued, it could just as easily mean the converse – intelligent persons simply stayed in school longer.[91]

The most shocking finding of all, however, concerned the low scores earned by most recruits. While many of these psychologists were hardly surprised by the poor showing made by recent immigrants or black soldiers, the low scores earned by ordinary white Americans stunned them. According to Terman, the average "mental age" of the white draft (determined by comparing Army Alpha with Stanford–Binet scores) was only 13.08 years – a score, in other words, just barely above the twelve-year mental age limit that intelligence testers had been using for years to define feeblemindedness.[92]

Such a finding forced psychologists to rethink some of their working assumptions. "It would be totally impossible to exclude all morons as that term is at present defined," Terman admitted, for this would mean eliminating from the army "47 percent of whites and 89 percent of negroes." And while critics would soon see in these test results the same problem that contemporaries had long been complaining of – the twelve-year standard was simply too high – Lewis Terman drew a very different conclusion: "Thus it appears that feeble-mindedness, as at present defined, is of much greater frequency of occurrence than had been originally supposed."[93]

To committee members, the fact that most Army Alpha scores were lower than anticipated did not in itself present a problem. Far more important, they argued, was the fact that these scores still fell roughly along a bell curve, for such a curve, these psychologists believed, offered the most convincing proof of their scientific validity. To convince the army, however, these scores would also have to be validated in other ways.

Fortunately for psychologists, their findings generally fit well with the army's own hierarchical structure and preconceptions. For instance, the group with the most years of formal education, officers (of whom nearly 75 percent had attended college), also scored the highest on the Army Alpha; by contrast, the least educated group, southern blacks (of whom nearly 20 percent reported no years of schooling at all), scored the lowest. Such findings were both credible and welcome in an army that saw educational level and social privilege as natural qualifications for choosing officers, and that was determined to keep black soldiers in segregated units assigned largely to menial tasks.[94]

These findings notwithstanding, the actual use of these test scores by the army remains hard to gauge. Some personnel officers reportedly tried to assign a fair distribution of "A," "B," "C," "D," and "E" men to each unit; there was no official policy, however, asking them to do so. Even more pointedly, although psychologists recommended that about 7,800 men be discharged

Black recruits in a segregated unit taking the army tests. From Robert Yerkes, ed., *Psychological Examining in the U.S. Army* (1921).

and 19,000 assigned to labor or development battalions because of low test scores, these recommendations were not necessarily followed. Moreover, while some officers embraced this new psychological science enthusiastically, others remained deeply skeptical about the very idea of linking poor intellectual skills to poor soldiering.[95]

Although the degree to which testers actually modified any army practices remains ambiguous, the impact of the army on the practice of testing is far clearer. Within one year, World War I changed both testers and testing in ways that would be felt for the remainder of the century. While intelligence testers may not have transformed the military, the military surely transformed intelligence testing.

Most apparent and significant was its influence in promoting unity within the psychology profession. Under the exigencies of wartime, the APA finally achieved a goal that had seemed impossibly elusive to members of the same organization only one year earlier: strict regulation over the design, administration, and scoring of intelligence tests. Before the war, Robert Yerkes and Lewis Terman had been sharp competitors; now they became collaborators (as well as lifelong friends). Moreover, courtesy of the army, a new generation of several hundred "professional psychologists" (among them a young David Wechsler) were given uniform training in the technology, methodology, and underlying assumptions of intelligence testing. Their wartime experiences would shape the practices of the American psychology profession for decades to come.[96]

Equally striking was the war's influence in adding a level of military-style discipline to the testing process itself. Nowhere was this more evident than in the official instructions for giving these tests, which were presented as a series of commands. "When I call 'attention,'" examiners stated, "stop instantly whatever you are doing and hold your pencils up. . . . Do just what you are told to do. Ask no questions. . . . " This emphasis on impersonal instructions, rigid timing, and mechanical scoring would remain recognizable fixtures of group testing for the rest of the century.[97]

Yet if psychologists' experiences with the army had maximized both unity and uniformity, they simultaneously minimized attention to diversity. During the prewar decade, many American psychologists had begun to gain at least some awareness of the influence of educational, environmental, and linguistic factors on test scores. All such gains were quickly lost, however, in the rapid pursuit of what testers now considered a more important and more immediately useful goal: distributing the entire army along a single bell curve.

For example, as early as 1911, Goddard had followed Binet's lead in revising his scale to eliminate all reading tests, for these were considered too influenced by schooling; by contrast, both reading skill and reading speed were reintroduced as crucial elements of the Army Alpha. Robert Yerkes had previously established separate norms for "favored" and "unfavored" subjects on his own point scale; there was no time to consider any such issues, however, in testing the army. Even tests for illiterates reflected this marked change in direction. A "Ship Test" depicting a steamship suggested sensitivity when it was first invented at Ellis Island; its meaning was altered, however, when the same test was adopted by army testers and used with a population that could include midwestern farmboys or southern blacks.[98]

This change in direction was closely linked to a broader change in objective. Most prewar intelligence testers, following Goddard's lead, saw themselves as fulfilling one main social function: diagnosing the feebleminded. By contrast, these testers, now largely inspired by Terman, envisioned a different role for themselves: classifying the entire army. In fact, Terman himself had always been less interested in diagnosing subnormality than in determining mental superiority; he hoped that these new tests might prove especially useful in spotting potential officers. Perhaps for this reason, the Army Alpha actually had far more in common with future college entrance examinations (for which it soon became a prototype) than with past tests for feeblemindedness. Psychologists themselves conceded as much when they advised that those who scored especially poorly be retested by older methods, which still offered better gauges of subnormality.[99]

By the time the war ended, intelligence testers had clearly changed their emphasis. Within a single year, they had moved their psychological science far beyond the population of "defectives, dependents, and delinquents" and had focused it instead on the minds of ordinary Americans. Like the draft, intelligence testers had now touched the lives of millions of "average" men.

Even Yerkes admitted this change in objective with some surprise. It had "originally seemed that psychological examining naturally belonged to the Medical Department," he reported, for its main goal had been to diagnose mental deficiency; however, wartime testing had instead convinced him that "its natural affiliation is with military personnel," for its new goal was to classify everyone. In the postwar decades, while testing would never entirely lose its medical affiliation, its identity would forever after be altered. For the rest of the century, intelligence testing would be less closely linked with hospitals than with schools, and less associated with diagnosing feeblemindedness than with classifying "normal" Americans for educational, vocational, and personnel purposes.[100]

Thus, by the time World War I had ended, American psychologists had emerged with what they saw as several dramatic gains. Working together, members of an APA committee had designed a new intelligence test – a test which now bore the imprimatur of the three most prominent American testers of the day: Yerkes, Terman, and Goddard. At the same time, they had vastly expanded their potential clientele. Finally, psychologists had gained a markedly broadened sense of their own scientific legitimacy and social utility, for their science had won official approval from the United States Army.

For APA leaders, both the successful outcome of this immense wartime collaborative experiment and the massive publicity for professional psychology generated in its wake exceeded all expectations. "Wartime publicity," Yerkes concluded, "accomplished what decades of academic research and teaching could not have equaled." To James McKeen Cattell, these tests had finally "put psychology on the map of the United States, extending in some cases beyond these limits into fairyland."[101]

Equally ecstatic was the Committee on Provision for the Feeble-Minded. It was at the Training School that Binet tests had first been developed in America, Byers boasted, and it was "again at The Training School, in its Laboratory with its Director of Research, Doctor H. H. Goddard, sitting as one of seven leading American psychologists, in 1917, that group-testing was developed and given immediate recognition as meeting a grave war-time emergency." The little laboratory at Vineland had once again proven its worth to the nation.[102]

Yet while his Vineland friends were obviously proud of his involvement, Goddard himself was cognizant of the minor role he had actually played in the work of this committee. When America entered World War I in 1917, Henry Herbert Goddard was nearly fifty-one years old. By this time, the testing movement was already being taken over a younger generation of psychologists – a generation far better trained in and reliant upon newer statistical techniques. Furthermore, in paying less attention to diagnosing the feebleminded and more to sorting the intelligent, the army testing program confirmed Goddard's waning significance and the growing influence of the movement's rapidly rising new star, Lewis Terman. Under Terman's leadership, the postwar decade would see such tests disseminated more widely than ever, especially in American schools. If the army experience represented an exciting new beginning for many testers, for Henry Herbert Goddard it signified an ending. Never again would this psychologist play any role in the creation or revision of American intelligence tests.

Responding to a request from Terman to document his part in this project for the historical record, Goddard lightheartedly disparaged his own efforts. His work was "hardly worth mentioning," he told Terman, for he "simply got behind a good man" and followed his lead. "I think I furnished a few cigars and some matches," he joked, and "faithfully served as subject in all the tests that the Committee tried upon themselves." Goddard admitted writing some multiple-choice questions that asked, for instance, "whether cows have horns because we need horn in the art to make combs etc., they add beauty and dignity to the appearance of the cow, for protection, or because our grandfathers used them for powder horns." He had also written some tests of disarranged sentences, such as "Hell to with Bill Kaiser."[103]

Although Goddard may have mocked his own contributions, he never minimized the significance of the wartime testing program. Such testing, he believed, constituted a scientific achievement of the first order. Moreover, the knowledge gained from this program, Goddard argued years later, was "probably the most valuable piece of information which mankind has ever acquired about itself."[104]

Yet just what had mankind learned? Developing a test that would distribute the entire American army – officers and enlisted men, Northerners and Southerners, whites and blacks, native-born and foreign-born – along a single bell curve was one thing. Deciding just what such a curve meant was something else. Only in the wake of the war would American psychologists begin to explore the larger social meaning of their own scientific findings.

Moreover, if the army tests had momentarily unified the psychological community, they would soon sharply polarize it. In the ensuing decade, psychologists, educators, reformers, and many others would be drawn into an increasingly intense and angry dialogue over just what the army findings really signified – both scientifically and politically. Among the first to attempt to explain their significance would be Henry Herbert Goddard.

Psychological Science and Political Philosophy

For Goddard, findings from the army tests constituted the last piece in a large scientific puzzle – a puzzle that had gradually been taking shape in his own mind ever since he had first come to Vineland in 1906. He had begun his scientific research by studying feebleminded children in his own institution. Based upon these findings, he had expanded his efforts by diagnosing subnormality within the school population. In the years that followed, he had developed his theories even further by linking low intelligence to social irresponsibility, thereby explaining the behavior of "defectives, dependents, and delinquents." Finally, he had participated in the army testing program, which had allowed him to gauge the minds of a broad sample of the entire American population. In moving from subnormality to normality, from feeblemindedness to intelligence, Goddard believed, both he and his fellow psychologists had made a number of discoveries of crucial importance to the welfare of the nation.

By 1918, Goddard had synthesized all of his ideas in a textbook. Published a year later as *Psychology of the Normal and Subnormal,* it included detailed descriptions of the brain and mind, advice about education and child rearing, and broad speculations on literature, art, philosophy, and politics. Filled with diagrams and photographs, scientific theories and personal anecdotes, this text embodied both the best and the worst of Goddard's popular style of psychological science.[105]

Surprisingly, Goddard's text largely ignored genetics (even his bibliography made no mention of Mendel, Galton, or Davenport). Instead, he stressed recent research in the field he had always loved most, physiological psychology. Crucial to complex mental processes, his text explained, were the brain's "association neurons," for they gave the mind its flexibility, its ability to connect known with unknown experiences, and thus its adaptability.[106]

"Intelligence," Goddard proposed, was "dependent upon and correlative with neuron activity." The "more elaborate and complicated the neuron pattern," he argued, "the higher the possible intelligence." By contrast, the feebleminded person adapted poorly "either because he has not inherited the necessary neurons" or because his life had been "so unusual and abnormal that he has not had the experience to bring his neurons into co-operation." In normal brains, these areas might continue to develop until about age twenty, while the feebleminded experienced an "arrested development" at some earlier age.[107]

This difference made the study of feeble minds especially valuable to scientists, Goddard insisted. For instance, most Vineland children had no fear of snakes. "From this it is evident," he concluded, "that the so-called instinctive fear of snakes is not instinctive but acquired." Educators and parents also had much to learn, he reported, including the fact that a child's happiness must come first. The most useful lessons of all, however, had come from intelligence testing.[108]

"That intelligence can be measured is no longer in doubt," Goddard stated confidently. Even if one had previously been skeptical of this new science, the army testing program had "settled the question," he declared. Yet these tests had also led to a very disturbing discovery: the low mentality of the average American.[109]

The result of this discovery, Goddard conceded, was "a division of students into two camps." The first camp, "following the hypothesis wherever it leads," he reported, "accepts the conclusion that vast numbers of people are of less intelligence than was supposed"; the second found this view "ridiculous because it makes half the human race little above the moron." Goddard put himself squarely in the first camp, "not because of a blind acceptance of the hypothesis," he explained, "but because the conclusion, surprising as it is . . . on the whole explains the facts of modern civilization more clearly than anything that has been proposed."[110]

For Goddard, the army findings simply confirmed what he had gradually come to suspect: that most of the country's problems were caused by a large class of persons of low mental power. In the past, he argued, well-meaning reformers had tried to remedy these problems by advocating either more punishment or more education. Neither approach had succeeded, he summarized, because society had failed to understand "the nature of the average man."[111]

To illustrate this misunderstanding, Goddard reproduced the famous portrait of a French peasant painted by Jean Millet in 1863, as well as the popular poem it had inspired by Edwin Markham entitled "Man with the Hoe." In 1913, Markham had produced a brief, excerpted version of Goddard's Kallikak study for a popular magazine; Goddard now tried his hand at interpreting Markham's most famous poem. "Bowed by the weight of centuries he leans / upon his hoe and gazes on the ground," this poem began. And while many others had seen in this peasant's stooping posture and uncomprehending gaze an indictment of social oppression, Goddard now saw something else: "a perfect picture of an imbecile." The poem "seems to imply that environment has made the man what he is," Goddard explained, while "our view makes the cause heredity." In thus using his science to explain the lot of poor peasants and laborers, Goddard was now openly pitting hereditary against environmental explanations.[112]

This could now be done with some confidence, he believed, because the

Reproduction of Jean Millet, *Man With the Hoe*. Goddard noted the "typical imbecile look, open mouth and low forehead." From H. H. Goddard, *Psychology of the Normal and Subnormal* (New York: Dodd, Mead, 1919), facing p. 240.

army tests had proven his case. Specifically, they had shown that most un-skilled laborers were really of low intelligence. This finding, he emphasized, had broad economic as well as social significance. The "people who are doing the drudgery are, as a rule, in their proper places," Goddard now insisted. "This fact should be recognized and they should be helped to keep their proper places, encouraged and made happy, but not promoted to work for which they are incompetent."[113]

At the same time, Goddard proposed, the army results also had clear polit-ical implications, for they placed a grave responsibility on the small minor-ity who possessed intelligence enough for leadership. Such persons "must so work for the welfare of the masses as to command their respect and affection," Goddard declared. This meant eschewing selfish aims and adopt-ing a paternal attitude toward their less fortunate countrymen. "Democracy, then," he concluded, "means that the people rule by *selecting* the wisest, most intelligent and most human to tell them what to do to be happy."[114]

Goddard made these connections between psychological science and political philosophy even more explicit later that year, for in 1919 he was paid $1,000 by Princeton University to deliver the prestigious Louis Clark Vanuxem Lectures (an honor which would also be bestowed upon Thomas Hunt Morgan, Alfred North Whitehead, and Thomas Mann, among others). Speaking on "Human Efficiency and Levels of Intelligence," his lectures offered a broad interpretation of the meaning of the army findings. In bringing his Vineland research to Princeton, Henry Herbert Goddard reached the pinnacle of his scientific career.[115]

"Stated in its boldest form," Goddard declared in his opening lecture, "our thesis is that the chief determiner of human conduct is a unitary mental process which we call intelligence," a process conditioned by a "nervous mechanism" determined by germ cells and only rarely affected by "any later influence." If this were true, then one could "restate practically all of our social problems in terms of mental level." To illustrate this, Goddard devoted one entire lecture to the ways that such testing could improve workplace efficiency, and another to its potential for reducing delinquency.[116]

He saved the broadest subject of all for his final lecture, entitled "Mental Levels and Democracy." "What about democracy," Goddard asked bluntly, for "can we hope to have a successful democracy where the average mentality is thirteen?" While many of his contemporaries had begun to doubt so, Goddard gave a slightly different response. "The fact that we here in the United States have done it for a hundred and forty years is of course an all sufficient answer," he admitted, "unless new conditions are arising. . . . "[117]

In analyzing contemporary conditions, Goddard again ignored issues that others considered most pressing. Absent from his lectures, for instance, was any mention of the eugenics movement, sterilization legislation, immigration restriction, or racial segregation. Even more striking, especially in light of publications by other army testers that would follow in the 1920s, Goddard's summary and analysis of the army results made no mention at all of the differences in scores between white and black soldiers, or between old and new immigrants.

Yet while Goddard's lectures said nothing about racial or ethnic differences discovered by army testers, class differences were another matter. Most crucially, by 1919 Goddard saw a direct link between psychological science and contemporary political controversies. Writing in the wake of the Russian Revolution, he now emphasized the significance of intelligence testing for "another problem that looms up rather large at the present time, namely socialism and especially its extreme form of Bolshevism."[118]

Findings from army intelligence tests, Goddard insisted, offered a clear scientific disproof of the arguments of socialists. "These men in their ultra altruistic and humane attitude, their desire to be fair to the workman," he wrote condescendingly, "maintain that the great inequalities in social life are wrong and unjust. For example," Goddard continued, "here is a man who says, 'I am wearing $12.00 shoes, there is a laborer who is wearing $3.00 shoes . . . I live in a home that is artistically decorated . . . there is a laborer that lives in a hovel with no carpets, no pictures and the coarsest kind of furniture.'" Pointing out such inequalities constituted a "fallacious" argument, he told his Princeton audience, for it "assumes that if you were to change places with the laborer, he would be vastly happier. . . . " "Now the fact is," Goddard retorted, "*that workman* may have a ten year intelligence while you have a twenty." Demanding the same lifestyle for him was as absurd as insisting "that every laborer should receive a graduate fellowship."[119]

"As for an equal distribution of the wealth of the world," Goddard added, "that is equally absurd," for differences in intelligence explained as well as justified differences in wealth. Once again, he argued, the low mentality of most laborers could be inferred from their behavior. "It is said that during the past year, the coal miners in certain parts of the country have earned more money than the operators," Goddard reported, "and yet today when the mines shut down for a time, those people are the first to suffer. They did not save anything. . . . " This apparent mental incapacity, he proposed, was largely responsible for the labor strife becoming ever more visible in American society. "Socialism is a beautiful theory," he concluded, "but the facts must be faced."[120]

Even so, Goddard did confess to sharing one goal with the socialists: the "desire to make all people happy." Yet this did not mean, "as socialism is too apt to claim," he added quickly, "that all people are to be treated alike. Children are not to be made happy by placing them in the same level as adults." To the contrary, there would be far less unhappiness in the world, Goddard believed, if people would simply accept the reality of fixed mental levels and adjust their social institutions accordingly. Such a realization would foster a new contentment, he predicted, as well as a new understanding of previously inexplicable behavior. "Aunt Polly's efforts were wasted," Goddard argued in describing this fictitious character's futile efforts to enforce middle-class conduct upon her lower-class nephew, "because she did not appreciate the mental level of Huckleberry Finn."[121]

Perhaps nowhere was the possibility of such contentment more in evidence than in the "Village of Happiness" itself, the Vineland institution. To Goddard, this institution now offered a microcosm of a political world. "The inmates of the Vineland Training School, imbeciles and morons, did not elect Superintendent Johnstone and his associates to rule over them," Goddard conceded; "*but they would do so if given a chance because they know that the one purpose of that group of officials is to make the children happy.*" In an equivalent way, the perfect government would arise, he predicted, if the nation's four million superior people, as estimated from the army sample, would devote themselves to "the comfort and happiness of the other ninety-six million."[122]

As both his textbook and his lectures make clear, by the war's end Goddard's ideas about the social significance of his own scientific research had once again undergone a radical broadening. He could now make claims, based upon army findings, about the minds not only of "mental defectives" but of "average workers" as well. In the process, Goddard transformed his psychological science into an explicit political philosophy – a philosophy which now offered a simple hereditarian explanation for class differences, which regarded social efforts to ameliorate them with condescension, and which was openly hostile to socialism.

A parallel transformation was also evident within his profession. In the postwar world, political views often implicit in many earlier psychological writings became explicit. Supremely confident of their science, numerous army testers would begin to make bold pronouncements concerning the significance of their data for American politics. Some would use the low test scores earned by unskilled laborers as an explanation for economic inequality, and as proof of the scientific superiority of capitalism over socialism. Others would draw equally broad conclusions from the low scores earned by black soldiers and use them to confirm the innate inferiority of this race. Still others would emphasize the low scores earned by recent immigrants from Poland, Russia, and Italy, and use them to argue for new laws restricting immigration by national origin. By the 1920s, all of these arguments would begin to fuel explosive new controversies – controversies which would focus explicit attention on both the scientific and the political meaning of "intelligence," and on the relationship between the two. In the wake of World War I, the intelligence testing debate would move from the margins to the mainstream of American political discourse. It would remain there for the rest of the century.[123]

For Goddard personally, the political implications of his Vineland experiences were even more far-reaching. A dozen years of doing "psychological work among the feeble-minded" had gradually altered his view not only of his own science, but also of the world around him. Since coming to Vineland, he had gained something more valuable than the answers to contemporary questions. He had also acquired a bold new vision of an ideal future society.

At its heart was an unusual utopian dream: the world as modeled on a benevolent institution for the feebleminded. Within such a world, psychological scientists would determine the intelligence level, and consequently the proper educational training and vocational role, of all citizens; at the same time, the "average" worker, now understood to be well-intentioned but intellectually limited, would be given more parental care and less personal autonomy. By 1918, the Training School at Vineland had come to signify something more than an institution for Goddard, something more even than a scientific laboratory. In Goddard's thinking, the "Village of Happiness" had itself become both a model and a metaphor for a future utopian state.

9

Leaving Vineland: Popularity, Notoriety, and a Place in History

Past and Future

"When the history of sociology is written," Henry Herbert Goddard prophesied in his last annual report from his New Jersey institution, "a large place will be given to the work of the Research Laboratory of the Vineland Training School for the years 1906–1918." Offering his own "brief retrospect," he showed himself more than satisfied, for his laboratory had already proven that "its value to the world is beyond price." From humble origins, it had steadily broadened its inquiries, in the process making numerous scientific "discoveries" which would significantly benefit society.[1]

If his laboratory had achieved great results, Goddard claimed, it was because it had remained true to the ideals of scientific research. "We made no attempt to predict or even guess what these results might be," he recalled; "we determined to observe, study, collect data and follow wherever these should lead." Summarizing where he had been led, Goddard enumerated the accomplishments he thought would appear in future histories. By demonstrating "the hereditary character of feeble-mindedness," his work had also suggested "the hereditary character of other mental traits." There was as well the "demonstration that feeble-mindedness is transmitted according to Mendel's law," the "discovery of the moron," the "view that each human being has a potentiality for a definite amount of intelligence (intelligence level) and beyond that point all efforts at education are useless," the "impetus given to special schools and classes," and the "intensified interest in all social problems, resulting from the recognition of intellectual levels." Lastly, Goddard cited the "starting in America of the use of mental tests, a movement which made possible the classification of an army of millions of soldiers" – an accomplishment, he argued in 1918, that might prove "one of the critical points that may win the world war for the Allies."

These achievements, Goddard boasted, had led contemporaries to compare his laboratory to that of the Mayo Brothers. His own comparisons were even bolder, for its discoveries "have made it comparable to Darwin's 'discovery' of the theory of evolution or Lister's discovery of antiseptic surgery, or

301

the 'germ theory' of disease – so significant is it for human welfare and progress." "The writer has sometimes been given undue credit for his part in the above achievements," he added more modestly. Goddard was glad to share credit with this school's philanthropists, teachers, and doctors, and to recognize above all the "unfailing optimism, wisdom, humanism and energy" of its superintendent, Edward Ransom Johnstone.

Nonetheless, at the age of fifty-two, after working for twelve years under Johnstone, Goddard decided to accept a position as head of his own agency. In 1918, he made plans to move to Columbus to replace Dr. Thomas Haines, a physician, psychologist, and fellow army tester, as director of the Ohio Bureau of Juvenile Research. "In leaving my little corner in the laboratory," he concluded in his Vineland report, "my fondest hope is that I can spread the Vineland spirit all over the State of Ohio."[2]

The decision to leave Vineland in 1918 marked a major turning point in Goddard's work and life. While the Training School would continue to play an important social role in the years after his departure, Goddard's own career began a precipitous decline. In the ensuing decades, he would leave the science of intelligence testing to others and would instead concentrate his energies in several new directions. In none of these fields, however, would Goddard achieve the prominence or prestige he had experienced as a world-renowned authority on mental deficiency. Although he continued to publish scholarly articles and popular monographs for another thirty years, never again would his science earn him the recognition, respect, or influence he had enjoyed in his dozen years at Vineland.

Far more surprising, Goddard would become a part of the "history of sociology" much sooner than he had ever anticipated, for the end of World War I also proved to be a major turning point in American science and social thought. While his research had always engendered the skepticism of a few astute critics, Goddard's psychological science had fit exceptionally well within the dominant medical, biological, and social paradigms of the prewar decade. Equally important was its congruence with broader progressive campaigns emphasizing educational efficiency, mental hygiene, and preventive approaches toward crime and poverty. Yet in the decade that followed, the opposite would prove true, for by the mid-1920s, most of Goddard's "discoveries" would be seriously challenged, significantly modified, or completely overturned by new research in medicine, biology, sociology, anthropology, education, and psychology. At the same time, the broader political implications of the intelligence testing movement, especially as expressed by army testers, would fuel an explosive and divisive new discourse which increasingly pitted scientists stressing "nature" against those emphasizing "nurture." By the end of this decade, the psychological meanings that Goddard and his fellow testers had attached to both feeblemindedness and intelligence had become intensely controversial, scientifically and politically, in entirely new ways.

Even more influential in transforming Goddard's reputation would be the scientific, economic, and political developments of the following two decades. During the 1930s and 1940s, both the Great Depression at home and the rise of fascism abroad would have an enormous impact on the meanings that American scientists attached to "nature" and "nurture." Especially traumatic in changing scientific and social thought would be World War II. In its wake, American social scientists would begin to think in new ways not only about their future, but also about their past. As a consequence, Goddard would see his own role in history rewritten once again.

Surviving until 1957, Henry Herbert Goddard lived long enough to see what several different generations of scientists and historians had to say about his laboratory, its accomplishments, and its influence upon American society. What he saw would leave him feeling increasingly isolated and depressed, self-doubting and shaken, a far different man from the optimistic and self-assured scientist who penned his own predictions in 1918. Such a future could hardly be envisioned, however, in the wake of what certainly appeared to be the most successful psychological accomplishment ever undertaken, the massive army testing project of World War I. Still basking in its afterglow, Goddard left Vineland in 1918 to begin his new career in Ohio.

A New Career

The decision to leave Vineland had not been an easy one, as evidenced by the list Goddard made for himself spelling out the pros and cons of changing positions. Moving, he noted, meant uprooting himself and spending "immense energy" starting new work, whereas at Vineland the same energy would produce "immediate scientific results." He could also expect less freedom in a state institution. On the other hand, Columbus offered the stimulation of a university environment, as well as an "unexcelled opportunity to apply what we have already learned, to the solution of a great social problem." And at Columbus he could investigate not only the "psychological aspect" of this problem but also "the physical and physiological – very dear to my heart." There was as well the financial advantage: at a salary of $7,500, Goddard would become Ohio's second-highest-paid civil servant (surpassed only by the governor). "Not that I like to consider this," he noted to himself, "but circumstances compel me to give it consideration." Even so, leaving Vineland's loving community of friends and co-workers would be painful. "I do not expect to be as happy at C. as at V.," Goddard predicted, in words which quickly proved true. "To paraphrase Shakespeare: 'Go' says judgment. 'Stay' says feelings," he confessed in his acceptance letter. "In such a case one probably ought to follow judgment and burn the bridges."[3]

The Columbus offer certainly seemed too promising to pass up. In 1918, the state of Ohio was a leader in implementing progressive reforms. Among these was its Bureau of Juvenile Research, an institution which Goddard had helped create four years earlier as a consultant to Democratic governor James Cox. Its primary purpose was to recommend appropriate placements for juveniles who came before the courts – either in state institutions, industrial schools, their own home, or new homes. Simultaneously, the bureau had a scientific role: to gather data on this population which would be used to prevent delinquency. "The idea," a Columbus newspaper story explained, "is to study the causes of juvenile delinquency and check them at the root rather than to punish their effects." Its headline announced Goddard's arrival in May 1918: "Ohio Gets Expert at Big Pay to Better Children."[4]

By that September, Goddard had hired a staff of psychologists, physicians, and clerical workers to assist him. Most influential was psychologist Florence Mateer, a talented researcher who had worked at Vineland before earning a Clark Ph.D., and who had recently served as Dr. Walter Fernald's assistant at Waverley. By 1920, the Bureau of Juvenile Research had examined 3,578 individuals, most of them "problem children" referred by courts, schools, social workers, hospitals, or parents. Of these, most important were 236 "observation cases" – delinquents facing charges ranging from unmanageability to murder – who spent from two days to several months in residence at Goddard's bureau.[5]

At the time that Goddard began his new work, the field of forensic psychology was largely dominated by the writings of his contemporary, psychiatrist William Healy. In his many studies of delinquency, among them *The Individual Delinquent* (1915), *Pathological Lying, Accusation, and Swindling* (1915), and *Mental Conflicts and Misconduct* (1917), Healy had increasingly begun to incorporate psychoanalytic insights along with his statistics. The influence of newer psychiatric theories was soon evident in Goddard's new work as well, for he quickly turned his attention from the "subnormal" to the "abnormal" child mind.[6]

To prevent delinquency, Goddard argued, psychologists would need to gauge not only the quantity of mind but its quality. For abnormal mental functioning, he used a popular medical label as broad and sweeping in its implications as "feebleminded" had been a decade earlier: "psychopathic." "Psychopathic means *mentally suffering* – having a diseased brain," Goddard explained. In children, this disease manifested itself in a wide range of symptoms, including nervousness, violent tempers, chronic lying, sexual perversions, and general disobedience. And whereas parents often saw such children as "bad," Goddard insisted that they were really sick. "These cases need to be studied, treated and *cured* of their mental disturbance – if this is possible," he maintained.[7]

In seeking to cure such cases, Goddard tried a variety of approaches. A first step was to provide free medical and dental care at his clinic. Goddard also tried to modify behavior; eschewing all forms of corporal punishment, he instead implemented many techniques that Johnstone had developed at Vineland, such as rewarding good behavior with special privileges, and offering much kindly and fatherly advice.[8]

Equally important, Goddard believed, was a scientific system of classification. In the following months, he and Mateer worked diligently to standardize their psychological, medical, legal, and educational recordkeeping. Delinquents were subjected to a battery of tests, including intelligence tests, performance tests, and tests of school subjects such as spelling, arithmetic, and geography. Using all of their results, they hoped to develop a more precise means of distinguishing feebleminded, psychopathic, and normal adolescent minds.[9]

It was this search for precision, they believed, that would make their work a science. While most psychiatrists felt comfortable using subjective diagnostic criteria, psychologists such as Goddard and Mateer continued to press for more objective standards. The psychiatric literature, Mateer lamented, offered "no standardization of procedure." Instead, what she and Goddard sought most was a test that would clearly distinguish the normal from the psychopathic, just as Binet testing distinguished the normal from the subnormal. Most promising, they argued, was the Kent–Rosanoff Association Test, which offered a quantified assessment of unusual word associations.[10]

While Mateer assumed responsibility for gathering precise scientific data, Goddard attended to an equally important problem: changing public attitudes. By 1921, he had published *Juvenile Delinquency,* a small monograph which explained his bureau to the public. "With the list of physical diseases and defects . . . with the mental weaknesses . . . with the environmental conditions which we all know have surrounded these children, how can we hold them responsible?" he asked. "How can we regard them as criminal?" Courts should see themselves as "child savers, *not* as child punishers," he insisted.[11]

Goddard also answered the charge that he was merely trying to "save criminals from the consequence of their crimes" – in short, that he was too soft on crime. He was proposing no "maudlin sentiment of soft sympathy or misplaced kindness," he declared, but "cold scientific rational treatment." And if such treatment proved "less cruel" by using a method that made "both officers and victims more human and less vicious, that develops whatever is good in this misdemeanants, that cares for them in a way that even makes them happy or that sends them back to society feeling that they have been justly treated and with the desire to act socially rather than anti-socially," he proposed, "will we complain of the method?"[12]

Most striking in this monograph was its unbounded optimism. Society was finally on the right track, Goddard asserted, for scientific study would surely find solutions for pressing social problems. Crime had increased because "we have made no effort to understand these children," he told his readers. "Pessimism? No, the foundation of optimism. All this can be changed." Goddard's unswerving faith in the power of social science was especially evident in his conclusion. "There is no longer any need for hit or miss guesswork. Scientific handling is entirely within reach," he reassured the public. "Juvenile delinquency can be largely eradicated."[13]

Yet Goddard's high hopes for his new agency were soon dashed by internal dissent. By 1920, daily functioning within this clinic was being increasingly undermined by a bitter feud pitting Mateer against Dr. Gertrude Transeau, its chief physician. Lacking the administrative skill that Johnstone had wielded so invisibly, Goddard found himself "literally worn out with the petty bickerings." Both women were deeply dedicated, he told a friend, but neither could subordinate "personal pride and ambitions to the welfare of the work. Consequently, if some one does not speak to them with the proper deference or in just a tone that they prefer," he complained, "they run to me and tell me they will not stand it." On April 4, 1921, this rivalry erupted into a public scandal when Transeau convinced ten other women employees to join her in resigning, as a protest against Mateer's power. Two weeks later, Mateer offered her resignation as well.[14]

To Goddard, this embarrassing episode was largely the fault of cantankerous women who simply could not get along. Yet personal animosities notwithstanding, also involved were deeper professional rivalries, for the challenge that physician Transeau posed to psychologist Mateer was being echoed in clinics throughout the country. In fact, by the early 1920s, the whole new specialty of clinical psychology found itself embattled. Facing growing medical antagonism, as well as stiff competition from new "Child Guidance" clinics run by psychiatrists, clinical psychologists struggled to maintain the status they had acquired a decade earlier. By the end of the decade, the trend was clear: instead of running clinics, psychologists were increasingly relegated to more circumscribed roles in clinics run by physicians. Their frustration was captured in the title of J. E. Wallace Wallin's bitter 1929 article, "Shall We Continue to Train Psychologists for Second String Jobs?" Despite such protests, the winners and losers in this professional power struggle were soon unmistakable, for by 1934 there would be 87 psychological clinics in America, compared with 755 clinics run by psychiatrists.[15]

Whatever its deeper causes, in the politically charged climate of Ohio's state capital in 1921, the mass resignations of twelve employees from a single agency prompted a spate of rumors and an official inquiry. And while Goddard himself was fully exonerated of any wrongdoing, the incident afforded the state's newly elected Republican legislature an opportunity to make stark cuts in his bureau's funding. In its new budget, the bureau's appropriations for staff salaries were cut from $40,000 to $25,000, and Goddard's salary from $7,500 to $4,000 – a figure more in line with that paid to medical superintendents of other state institutions. Humiliated and disillusioned by the entire episode as well as by this state's intricate politics, Goddard too offered his resignation. The following year, he was replaced by a physician.[16]

Fortunately, Goddard was quickly able to land a new job in the city of Columbus – as professor in the newly established Department of Abnormal and Clinical Psychology at Ohio State University. He would remain at Ohio State for the next sixteen years, earning the fond affection and enduring loyalty of several generations of future psychologists. Yet while Goddard's new academic post was clearly prestigious, it also signified a defeat of sorts, for his return to the classroom brought to an end his unusual sixteen-year career in nonacademic laboratory psychology. In blending scientific research with social reform, Goddard had hoped to forge a new public role for his profession outside the university. In the case of his own career, this experiment had now come to a close.[17]

Despite his career change, Goddard's work at the bureau continued to influence his scholarly and popular writing for years to come. Perhaps no experiences proved more provocative than his meetings with Bernice R., identified in Goddard's writings by the pseudonym "Norma." Goddard first met Norma, an attractive and intelligent nineteen-year-old, when she was brought to his bureau on the evening of September 22, 1921. By the following morning, however, they had to be reintroduced, for the same woman now insisted that she was a four-year-old child named "Polly." In encountering "Norma–Polly," Goddard got the chance to study one of the most intriguing psychological issues of the day: "multiple personality." Over the following months, he used his skills as a trained hypnotist to try to make these two personalities (as well as a third, which appeared later) conscious of each other's existence.[18]

While similar cases had been described by other psychologists, among them William James and Morton Prince, by the 1920s they were still rare enough to capture media attention. Newspapers around the country were soon chronicling Goddard's progress in resolving this "Duel of Dual Personality." "Columbus Scientist Is Killing Baby so Girl Can Inhabit Body Alone in Strange Case of Double Personality," stated a typical headline.[19]

Several years after leaving the bureau, Goddard decided to turn his notes from this case into a scientific article; by 1927, he had also published a popular monograph on the same subject. Its purpose, he explained, was to show the public how children became "mentally and morally malformed as the result of mistreatment by their parents or teachers." Most damaging of all was "the idea of badness." Equally evident, however, was a second purpose, which brought Goddard back to subjects explored decades earlier in his dissertation on faith healing. Too many persons, he argued, still accepted supernatural explanations for strange behavior, for the publicity surrounding this case had brought him dozens of letters suggesting treatments from Christian Science to clerical exorcisms. His desire to answer such claims is evident in the provocative theological question he used for his title: *Two Souls in One Body?*[20]

The public had no need for supernatural explanations, Goddard wrote reassuringly, for psychological scientists now understood and could cure such behavior. He himself was "familiar with the views of the world's greatest experts in these cases – Charcot, Janet, Prince, Sidis, Freud, Meyer, Southard, to mention a few." Like the Freudians of his day, Goddard dismissed as fantasy the one charge that both of Bernice's personalities had insisted upon – that they had been sexually molested by their father. Clearly uncomfortable with even discussing such a subject, in his scientific article he described it only in Latin phrases. "The vita sexualis was manifested through a hallucinosis incestus patris," Goddard reported. In his popular monograph, he omitted any mention at all of this charge, even as a "hallucination."[21]

Bernice's hysteria, Goddard instead concluded, was the natural result of traumatic life experiences (including the sudden death of her twin sister, and later deaths of both parents) imposed upon an exhausted nervous system. In many ways, his treatment resembled the "rest cure" advocated by nineteenth-century doctors such as S. Weir Mitchell (whom he quoted at length in an Appendix) far more than any "talking cure" promoted by followers of Freud, for his notes show Bernice following a tedious regimen of taking naps and avoiding overstimulation. Yet while Goddard ended his story with the confident assertion that Bernice's condition was steadily improving, the real case was far more complex, for this young woman could still barely function, and had to be sent to a state institution.[22]

If Goddard had hoped that the sensational subject matter of this monograph would bring him popular success of the kind he had experienced with the Kallikaks, then he was surely disappointed. "Somehow the book has had no recognition as yet," he told a friend more than a year after its publication, for "very few people have ever heard of it. I think perhaps the title was unfortunate," he concluded.[23]

Although Bernice R. soon receded from media attention, the same was not true of Goddard himself. During this decade, the story of the Kallikaks continued to be told and retold in materials circulated from New York to Montana, while the word "moron" penetrated ever more deeply into the

American vernacular. Despite his new research interests, Goddard continued testifying in court cases involving potentially feebleminded defendants, always arguing for the defense and against the death penalty. He also continued to travel around the country to address teacher associations, social workers, civic organizations, church groups, and women's clubs, often giving speeches that were syndicated in local newspapers.[24]

Goddard's lectures usually offered his audiences a mixture of scientific theory and homespun advice concerning poor heredity, poor environment, differential schooling, and kindly child-rearing. "'Bad Boy' Will Be Unknown in School of Future, Psychologist Addressing St. Paul Teachers Says," read a typical headline summarizing one such speech. "Kids are Like Blueberries," reported a more cryptic headline in the *Omaha World-Herald.* It was as foolish "to expect all children to take the same kind of school work," this newspaper stated in explaining this lecture given by "America's Freud," "as it would be to try to grow blueberries on Nebraska alkaline soil."[25]

Also reflected in these lectures was another project which increasingly absorbed his attention. In 1922, Goddard had been approached by a Cleveland women's group interested in supporting public classes for gifted children. Their leader, philanthropist Roberta Bole, implored Goddard to serve as their scientific advisor. For the next six years, he spent two days each month visiting these classrooms. By 1928, he had published *School Training of Gifted Children,* a detailed description of the Cleveland program and a glowing paean to a new kind of education.[26]

What Goddard found was the nation's "most thoroughgoing and progressive experiment" in educating gifted children, a program as open and unstructured as anything advocated by his contemporary John Dewey. Whereas most programs for the gifted emphasized rapid promotion through the grades, Cleveland instead offered a parallel curriculum emphasizing "enrichment." Especially appealing to Goddard was its lack of traditional order, for in these classrooms he saw "no formality, no regularity, no silence . . . no straight rows of seats . . . no repeated commands to 'sit up straight.'" Instead, children worked independently or in groups seated at small tables, surrounded by a piano, a typewriter, an aquarium, and growing flowers. They chose their own subjects for study, wrote poetry and plays, paid frequent visits to museums, factories, and businesses throughout the city, and even put on their own circus.[27]

This educational experiment, Goddard argued, offered a model not only for other gifted classes but for what might be accomplished in normal classrooms. All that was really needed, he proposed in words reminiscent of G. Stanley Hall, was "a little more ingenuity on the part of the teacher, a little more faith in the child, a little more willingness to adapt the school to the child rather than to attempt to adapt the child to the school."[28]

By the mid-1920s, Goddard had become a sharp critic of the "fundamental

Children in the Cleveland gifted classes put on a circus. From H. H. Goddard, *School Training of Gifted Children* (Yonkers-on-Hudson, N.Y.: World Book Company, 1927), p. 13.

errors" evident in ordinary schoolrooms. "Discipline is the most important word in the American teacher's vocabulary," he complained. And while this word ought to mean "to make a disciple of, a follower," in schools it meant "to keep order – policemanship." Even worse, "the idea of unquestioning obedience is dear to the hearts of many parents and teachers. For slaves it is good," he argued, but for "the children of free men" it was "pernicious." "O Education!" Goddard lamented. "What crimes are committed in thy name!"[29]

In contrast, Goddard now offered his own pedagogical philosophy. Teachers should "teach less and educate more" by letting children experience "the joy of making an original discovery." Moreover, too many useless subjects were being taught. "The number of things that we still teach for no other reason, than that the person taught may teach them to somebody else, is amazing!" he concluded. Colleges too needed a drastic change. "Instead of requiring class attendance, grades, readings, examinations, we should say, 'It is all up to you. If you are ever to move or act on your own responsibility . . . now is the time to begin,'" he proposed. "Education through happiness and pleasure," Goddard concluded, "not through punishment, unhappiness, fear and pain."[30]

By the end of this decade, Goddard's expertise in this area had been recognized nationally, for in 1930 he was invited to participate in the White House Conference on Child Health and Protection. And whereas Edward Johnstone attended the same conference as an expert on educating the feebleminded, Goddard now chaired the subcommittee on educating the gifted.[31]

Thus, in the years between 1918 and 1930, Goddard had begun to study juvenile delinquency, multiple personality, and gifted education, and had published a monograph explaining each to the public. Ironically, none of these accomplishments would prove very relevant to his scientific reputation, for during these same years he would also witness a stunning reassessment of his accomplishments of a decade earlier. During the 1920s, the new psychological meanings that Goddard and his fellow testers had attached to feeblemindedness and intelligence would come to the forefront of an explosive national debate. In the process, nearly all of Goddard's research from the prewar decade would be called into question.

Intelligence, Feeblemindedness, Kallikaks, and Morons Become Controversial

Although Goddard had left the field of intelligence testing after the war, most of his fellow army testers had not. To the contrary, the invention of group testing opened vast new possibilities for expanding the role of psychological expertise, especially in the schools. By the early 1920s, both the findings and the methods promoted by intelligence testers had become increasingly influential, and increasingly controversial. Spurring the most controversy was the 1921 publication of *Psychological Examining in the United States Army,* a nearly 900-page volume edited by Robert Yerkes, and filled with tables, graphs, and analyses. Almost immediately, these results were interpreted for the public by a wide range of commentators.[32]

Most frequently emphasized in the literature of the 1920s was the low mental age of the average American and its significance for American society. This finding proved especially useful to cultural critics intent on distinguishing the "classes" from the "masses," for it easily explained the enormous popularity of Hollywood films and tabloid newspapers. The American mass public, these critics now reported, simply contained far too many morons. In the face of such reports, some journalists defiantly announced their refusal to write for "Mr. and Mrs. Moron and the Little Morons"; others, however, now openly advocated such a strategy. Advertising executives were particularly likely to quote the army results to their copywriters. "What we are really saying," one explained, "is the great bulk of people are stupid."[33]

Also impressed by these army findings were a variety of political commentators, who now stressed the dangers of low intelligence to American democracy. America, one commentator summarized, contained 45 million people with "no sense but a majority of votes," 25 million with "a little sense," another 25 million with "fair to middling sense" who "haven't much but what there is, is good," and only 4 million with "the thing we call 'brains.'" "What Is the Matter with America" proclaimed the title of an article by popular Kansas journalist William Allen White. His answer was clear: "The moron majority."[34]

Especially emphasized in many popular publications were the low scores earned by recent immigrants and black soldiers. Both findings quickly found their way into a growing body of nativist and racist literature arguing for "Nordic" supremacy, an idea already popularized in Madison Grant's influential 1916 text, *The Passing of the Great Race*. "What is this curious thing called Race?" asked Charles W. Gould in his 1922 anti-immigration diatribe, *America: A Family Matter*. "Suppose we had imported two hundred million Russians, Poles, Syrians, South Italians, Greeks, Negroes," he proposed, and then "exploited the land down to its rock foundations, what would it have profited us? Lost in these foreign millions, America would be no more." Army data helped prove his case, he added, for they "showed the score of the conscripts little influenced by schooling. Man must first breed before he can educate intelligence," Gould concluded. Even more alarmist was Lothrop Stoddard's 1922 treatise, *The Revolt Against Civilization,* which warned of the dangers looming from "racial impoverishment" and the threat to civilized life posed by the "Under-man," and which also included more than a dozen pages on the army findings.[35]

Yet if many writers simply accepted the conclusions of army testers, many others began to question them. Among the sharpest of skeptics was the most influential journalist of the early twentieth century, Walter Lippmann. In a series of articles in *The New Republic,* Lippmann presented a sophisticated explication of the underlying assumptions, statistical errors, and unproven conclusions asserted by army testers. In the process, he also brought this scientific controversy to a new level of political visibility.[36]

"A startling bit of news has recently been unearthed and is now being retailed by the credulous to the gullible," Lippmann began. Most startling was the finding that the average "mental age" of Americans was below fourteen, the intelligence of an immature child instead of an adult. Such a statement was nonsense, he reasoned, for the "average adult intelligence cannot be less than the average adult intelligence." To argue otherwise, Lippmann maintained, was like finding "that the average mile was three-quarters of a mile long."

Lippmann also understood the reason for such an absurd conclusion: the army findings (based on results from nearly 2 million men) were being gauged against norms established earlier for the Stanford–Binet scale (based on results from a few hundred California schoolchildren). What had actually happened, Lippmann concluded, was that the army findings "had knocked the Stanford–Binet measure of adult intelligence into a cocked hat." Yet since testers refused to admit this, Lippmann had begun to suspect "that the real promise and value of the investigation which Binet started is in danger of gross perversion by muddle-headed and prejudiced men."[37]

Measuring intelligence, Lippmann explained, was not like using the "accepted standard foot and standard pound," for the very concept was an "exceedingly complicated notion which nobody has as yet succeeded in defining." In a "general way," he proposed, it might be considered "the capacity to deal successfully with the problems that confront human beings." Yet in specifying "what those problems are, or what you mean by 'dealing' with them, or by 'success,'" he added, "you will soon lose yourself in a fog of controversy."

Yet Lippmann was not rejecting all intelligence testing. To the contrary, he was simply insisting that testers be clear about what they had, and had not, done. Rather than a "measure," an intelligence test was really "an instrument for classifying." "The army was interested in discovering officers and in eliminating the feeble-minded," he summarized. "It had no time to waste, and so it adopted a rough test. . . . In that, it succeeded on the whole very well." What the army had not done, however, was "measure the intelligence of the American nation, and only very loose-minded writers imagine that it did."[38]

These loose claims had led Lippmann to fear abuses, for too many educators would "stop when they have classified and forget that their duty is to educate." Even worse, in the hands of "blundering or prejudiced men," intelligence testing could easily become "an engine of cruelty" used to stamp "a permanent sense of inferiority upon the seal of a child." This was likely to happen, he believed, because most testers were committed to "a dogma which must lead to just such abuse. They claim not only that they are really measuring intelligence, but that intelligence is innate, hereditary, and predetermined."[39]

On the question of heredity, Lippmann recommended keeping "an open mind," for even in sensational cases such as Goddard's Kallikak study, the proof was "not conclusive." Even if intelligence were inherited, psychologists had yet to prove that their test "*reveals and measures*" it alone. After all, Terman had ignored all child learning before age four. "He cannot simply lump together the net result of natural endowment and infantile education and ascribe it to the germplasm," Lippmann insisted. "In doing just that, he is obeying the will to believe, not the methods of science."[40]

Yet this suggested an even deeper problem, for "behind the will to believe,"

Lippmann argued, one could usually find "the will to power," and one "did not have to read far in the literature of mental testing to discover it." Testers were selling themselves as scientists who could predict a child's future, a power so "intoxicating" that it had proven "too strong for the ordinary critical defenses of the scientific methods," he concluded. "With the help of a subtle statistical illusion . . . ," he believed, "self-deception as the preliminary to public deception is almost automatic." The claims of these testers would soon pass "into that limbo where phrenology and palmistry and characterology and the other Babu sciences are to be found," Lippmann predicted. Yet this fate could be avoided if psychologists would admit that they had developed not "measurements of intelligence" but "simply a somewhat more abstract kind of examination," thus saving themselves "the humiliation of having furnished doped evidence to the exponents of the New Snobbery."[41]

While Goddard ignored Lippmann's attack, Lewis Terman was infuriated. In a long reply, he garnered all the sarcasm he could muster. Mr. Lippmann had now taken his place alongside William Jennings Bryan, the opponent of evolution, he began, in attempting to annihilate that "group of pseudoscientists known as 'intelligence testers.'" And what had testers done to merit this fate? They had simply declared that the average man was "not a particularly intellectual animal," that some men were "much stupider than others," and that the "offspring of socially, economically, and professionally successful parents have better mental endowment, on the average," than those of "janitors, hod carriers, and switch tenders."[42]

If Americans had been fooled, Terman noted mockingly, so had the rest of the world. "For example, the innocent-minded Germans are being shamefully taken in at this very moment," he proposed, for Germany's new republican government was using testing in the schools to break up the old Prussian caste system. "If the German people don't wake up," Terman wrote in words whose real irony would only be apparent a decade later, "they will soon find themselves in the grip of a super-Junker caste that will out-Junker anything Prussia ever turned loose."

Mr. Lippmann, he concluded in a clear political allusion, "has been seeing red." And while Terman now resented having to explain anything so technical to a layman, he felt compelled to address several charges in detail. He also responded to Lippmann's fears that tests could be used as an "engine of cruelty." "Very true," Terman admitted, "but they simply aren't. That is one of the recognized rules of the game," he wrote reassuringly.[43]

Neither Terman's sarcasm nor his reassurances quieted Lippmann, who quickly penned a reply. Mr. Terman had "felt impelled to pause in those labors, which he modestly compares with Darwin's," he began, to answer the "contemptible creature who had challenged his dogmas." Lippmann was denying not the influence of heredity, but "Mr. Terman's unproved claim that he had isolated the hereditary factor. Mr. Terman's logical abilities are so primitive," he added, "that he finds this point impossible to grasp." Finally, Lippmann answered Terman's charge that he had an "emotional complex" about this issue. "Well, I have," he conceded. "I admit it. I hate the impudence of a claim that in fifty minutes you can judge and classify a human being's predestined fitness in life," he retorted. "I hate the pretentiousness . . . the abuse of scientific method . . . the sense of superiority which it creates and the sense of inferiority which it imposes." Thus, while Lippmann saw for the testing movement a "considerable future," he also feared it would become "the happy hunting ground of quacks and snobs" if men like Terman were allowed to lead it for much longer.[44]

These fears were quickly realized, for works by psychologists using army results to confirm popular prejudices soon appeared. *Is America Safe for Democracy?* asked a book by Harvard psychologist William McDougall which promoted eugenics and disparaged immigration. Far more detailed was *A Study of American Intelligence,* by Princeton psychologist and former army tester Carl Brigham. Among the foreign-born, Brigham reported, army testers had found a strong correlation between intelligence scores and length of residence in the United States. In other words, recent arrivals from Italy, Russia, or Poland had scored far lower than those arriving decades earlier from England, Germany, or Norway. To Brigham, this suggested a dangerous decline in immigrant intelligence; it could best be explained by analyzing the percentage of Nordic, Alpine, or Mediterranean blood found in each country of origin. "In a very definite way," he concluded, "the results which we obtain by interpreting the army data by means of the race hypothesis support Mr. Madison Grant's thesis of the superiority of the Nordic type." Writing a glowing forward to Brigham's book, Robert Yerkes too warned Americans not to "ignore the menace of race deterioration or the evident relations of immigration to national progress and welfare." By the early 1920s, Yerkes was offering his scientific expertise to congressmen intent upon passing what would become the Immigration Restriction Act of 1924. Although the impact of this new psychological expertise is hard to gauge in a political climate already strongly predisposed toward restricting immigration by country of origin, it nonetheless demonstrates Yerkes' eagerness to make his profession's findings a part of the political process.[45]

Even more consistent with widespread popular prejudices were the testers' findings about the innate inferiority of black soldiers. While a handful of studies comparing white with nonwhite children had been done in the prewar decade, the easy availability of new group tests as well as the conclusions drawn by army testers quickly spawned a new and growing literature. By the 1920s, this issue drew increasing attention. The fact that the army's northern blacks had far outscored southern blacks proved little about education or

environment, many white commentators concluded; instead, it simply suggested that the smarter members of this race had migrated northward in
search of better opportunities.[46]

To black intellectuals, such arguments were exasperating but hardly surprising, considering the racial climate of the day. They had to be challenged,
college educator Horace Mann Bond insisted, for they were giving "the professional race-hatred agitator a semblance of scientific justification for his
mouthings." Reviewing the army results in *Opportunity: A Journal of Negro
Life,* he focused on differences not only between races but also between states,
for blacks from Illinois, New York, and Ohio had outscored whites from
Mississippi, Kentucky, and Arkansas. If Brigham was correct, Bond argued,
then whites from New York and Massachusetts, states with large numbers of
supposedly inferior immigrants, should not have so outscored whites from
Georgia and South Carolina, "with the purest racial stock of the so-called
Nordic branch now existent in America." Such results could only be explained in two ways, he reasoned logically: either southern whites were naturally of low intelligence, or the Army Alpha was really measuring education
and environment. "Instead of furnishing material for the racial propagandists and agitators," Bond concluded, testers ought to be emphasizing "the
sad deficiency of opportunity which is the lot of every child, white or black,
whose misfortune it is to be born and reared in a community backward and
reactionary in cultural and educational avenues of expression." Black sociologist Charles Johnson produced an equally sharp critical response; even so,
he was far less sanguine about challenging ideas about race with logic. "It is
so absurdly easy to prove almost anything where there exists a *will to believe,*"
Johnson concluded in reviewing the army findings, "that the elaborate ges-
tures of scientific thoroughness at times seem grotesquely out of place."[47]

These close connections between theories of native intelligence and theories of racial supremacy brought even more controversy to the testing debate.
During the 1920s, intelligence testers would increasingly find their arguments
about the meaning of "nature" challenged by a new generation of anthropol-
ogists, many trained by Franz Boas, who brought a new sophistication to the
meaning of "nurture" through their studies of "culture," and who began to
question the very concept of "race." To Boas, the army findings suggested
differences in cultural assimilation, not racial inheritance. "I believe all our
best psychologists recognize clearly that there is no proof that the intelligence
tests give us an actual insight into the biologically determined functioning of
the mind," he asserted confidently in 1926.[48]

As Boas' assertion suggests, interpretations of the army results had also
begun to divide American psychologists. In fact, far more contentious than
the challenges posed by other disciplines would be the debates that now broke
out within the psychology profession itself. In the years following the war,

American psychologists would become increasingly polarized into two camps: those who accepted the conclusions of army testers, and those who opposed them. The intensity of their arguments would gradually escalate as the century progressed.

Among the earliest and most passionate of opponents to the army testers was William Bagley. As one of the founding editors of the *Journal of Educational Psychology,* Bagley had closely followed the prewar intelligence testing debate with no sense of alarm; in the postwar era, however, he saw the same movement as fraught with "social dangers of so serious and far-reaching a character as to cause the gravest concern." By disparaging the low IQ of ordinary Americans, testers were now attacking the very heart of democracy – faith in the abilities of the "common man." Equally dangerous, he declared, were their claims to have measured "innate" intelligence unaffected by education. Finally, Bagley despised the "pro-Nordic propaganda" which preached "that the long-headed blonds are beyond doubt the Chosen People and should proceed forthwith to the full enjoyment of their heritage – the earth." By 1922, Bagley had begun to publish a series of powerful articles attacking the political preconceptions, faulty logic, and poor science of his fellow psychologists.[49]

He would not attack his colleagues so openly, Bagley explained, were he not convinced that at stake was "a great ideal – an ideal that has already cost more in the terms of human striving and suffering and sacrifice than anything else in this world of ours." This ideal included a respect for the average man, and a belief in his potential to improve himself and his world. "What, may I ask, would have been the effect of the anti-slavery agitation if the hypothesis of an unmodifiable 'general intelligence' had been current at that time?" Bagley wrote. "What would be the case of the universal franchise? Indeed, why not . . . the divine right of kings if only these doctrines could be tempered with a little Mendelism?" Nor was Bagley offended by claims that he was defending "sentiment" instead of "science," for he was "devoutly grateful that 'sentiment' had an opportunity to work a few miracles before modern psychology 'discovered' these innate, racial, and unchangeable differences in native ability." Even so, Bagley feared for the future, for his hereditarian colleagues were endorsing solutions "openly inhumane and blatantly anti-democratic," and which were likely to lead to "an upheaval beside which the late war would look like an afternoon tea."[50]

Disturbed but undeterred by the vehement opposition his arguments were now engendering, both from public commentators and from his own profession, Terman continued to press forward. Yet by the middle of this decade, even he recognized that the controversy over intelligence testing had assumed a new political meaning. "Rightly or wrongly, some have felt that educational democracy is at stake . . . ," he conceded. "Catch phrases" such as "democracy and the I.Q." had become "as charged with emotion as were once such slogans as 'states rights,' 'abolition,' or 'taxation without representation.'" And while Terman still hoped that these debates could be settled by "a research undertaking" and not "a spectacular combat of educational gladiators," his attempts to do so left him increasingly frustrated. By the end of this decade, intelligence testing had become a crucial component of a much larger and far more contentious nature–nurture debate, a debate which could no longer be resolved by simple statements from testers about "the facts."[51]

Thus, in the post–World War I decade, American intelligence testing had begun a new chapter in its history. During these years, group testing became ever more widely disseminated, ever more hotly contested, and ever more closely linked to both old and new forms of elitism and racism. By the 1920s, the American founder of this movement, Henry Herbert Goddard, no longer played any central role in these debates; even so, the fact that the movement he started had become both so educationally influential and so politically controversial would leave its mark on Goddard's scientific reputation.

If in the postwar era the meaning of "intelligence" had provoked new controversies, the same was true for the meaning of "feeblemindedness." For students of this condition, the army tests were equally problematic, for if they proved that many ordinary Americans were really morons, they also proved the converse: many morons were really quite ordinary. This finding was especially significant for the numerous social scientists who had followed Goddard in discovering high rates of feeblemindedness within America's defective, dependent, and delinquent populations. Among these was former army tester Carl Murchison, who now reexamined the relationship between feeblemindedness and crime.[52]

"After hearing the guards in a certain penitentiary describe in condescending terms their ideas on criminals," this psychologist reported in his 1926 study, *Criminal Intelligence,* Murchison decided to test the intelligence not only of these criminals but also of their guards. The criminals, he discovered, scored "just 75 per cent higher" than their guards. "The only reason the guards continued to live," he added, "was because the architects of that prison had done their job well."[53]

To Murchison, the army results meant that scientists should throw out more than a decade of research blaming crime on feeblemindedness. The "pre-war prevailing opinion that criminality and feeble-mindedness are closely related," he concluded, "was certainly not built upon a solid foundation of collected facts." Yet the "progenitors of the theory are not to be condemned," he added in referring to Goddard and his followers, for there had been "no existing norms of general intelligence in the civil population." Drawing new conclusions, Murchison rejected not only Goddard's correlations but also what he called his "maternalistic fallacy" – the idea that it is "unethical to punish severely those criminals who are young, or feeble-minded, or insane." Instead, he advocated abolishing juries, parole, and the system giving special protection to the mentally ill, and increasing the use of the death penalty.[54]

While Murchison believed that psychological definitions of feeblemindedness had led to too much social leniency, others saw too much harshness. Too many persons, critics began to claim, were being institutionalized unnecessarily, for the army tests showed that nearly a million Americans of low intelligence were apparently functioning well enough in the outside world. Such arguments also had an economic component, for they offered welcome news to hard-pressed state legislatures unwilling to allocate ever more funds for institutions. If feeblemindedness was really as common as the army tests showed, then the solution Goddard and his followers had advocated for this social problem – complete institutionalization – was indeed hopeless, as well as pointless.[55]

Those seeking other solutions found a leader in Dr. Charles Bernstein, superintendent of New York's Rome institution. Unlike nearly all other superintendents of his day, Bernstein had long opposed increasing institutionalization, and had instead been developing his own innovative strategies for placing persons with mental ages as low as five in supervised work situations around the city. The feebleminded, he argued, deserved a "world test" to see how well they could really do if provided with special training and extra supervision. In the postwar climate, his views received a new hearing.[56]

Among those whose thinking on this issue changed most radically was Superintendent Walter Fernald. Since 1916, Fernald had been conducting a follow-up study of persons taken out of his Waverley institution, often by family members acting against his express recommendations. He had expected to find such persons causing social havoc; he was stunned, however, to find the converse, for many were leading "useful and blameless lives." Most of these people were doing remarkably well in the outside world, he concluded. Only after the war was Fernald willing to publish his results.[57]

In 1924, Fernald confessed his own dramatic change of heart in a presidential address to the American Association for the Study of the Feebleminded. The most significant advance of the last thirty years, Fernald asserted, had been Goddard's "inspired recognition of the vast significance of Binet's theory and technique." The "discovery of the *concept of mental age*," he continued, "did more to explain feeble-mindedness, to simplify its diagnosis and to furnish accurate data for training and education, than all the previous study and research from the time of Seguin." Yet Fernald now also saw liabilities in this work, for the "intelligence factor" had proven so impressive that it had overshadowed all other factors. And while he praised Goddard's heredity research as "another great advance," he now believed the resulting "black charts" had overemphasized the negative while ignoring the positive. "At this time no paper on mental defect failed to emphasize strongly the criminal and antisocial tendencies of the feeble-minded," he stated, "with little reference to the non-criminal defectives."[58]

Such work had led, Fernald argued, to "what I like to call the '*Legend of the Feeble-minded.*'" This "legend" suggested that "almost all" such persons were "of the highly hereditary class," and "almost invariably immoral . . . antisocial, vicious and criminal" – in short, "highly dangerous people roaming up and down the earth seeking whom they might destroy." It had made the feebleminded person "an object of horror and aversion," an "Ishmaelite" who ought to be "ostracized, sterilized and segregated for his natural life at public expense."

Fortunately, "much water has run over the dam since that period of pessimism, say 1911 or 1912," Fernald reported. Older studies (including his own) had been "far too sweeping," he argued, for he now recommended less institutionalization, more out-patient clinics, and, above all, more optimism. "It is possible," Fernald concluded, "that in the past we have a bit *overstressed* the protection of the public welfare and *understressed* the obligation of society to its less endowed members, to their great disadvantage." Thus, by 1924, Fernald was already describing, and distancing himself from, what a later generation of historians would label the "myth of the menace of the feeble-minded."[59]

While Fernald had begun to place Goddard's "epoch-making" studies within a historical framework, even more striking was a 1922 dissertation by sociologist Stanley Davies. Published by the Mental Hygiene Association a year later as *Social Control of the Feebleminded,* Davies focused on social attitudes in the past and present. In the process, he produced the first historical account which conceptualized changing social policies toward the feebleminded as something other than a continuous march of progress.[60]

This history, Davies argued, could be divided into four distinct (albeit radically unequal) eras. The first era ran from ancient times to the advent of Seguin's work in the early nineteenth century. The second, which he labeled the "period of physiological education," lasted until the century's end. This was followed by the "alarmist period, roughly dating from 1900 to 1915," in which feeblemindedness was portrayed as "the mother of crime, pauperism and degeneracy," and finally the "modern period," which instead emphasized extra-institutional solutions. Goddard's two most influential books, *The Kallikak Family* and *Feeble-Mindedness,* had clearly been produced at the height of the "alarmist period."[61]

Like many others, Davies too believed that the army tests proved the need for a new understanding of this mental condition. "To continue to use a definition of mental deficiency which would include a half or even a fourth

of the population," he proposed, "is about as ridiculous as to frame a definition that would include all the population." Perhaps, he added, "we have been
too ready in the past to classify as mental defectives and commit to institutions, individuals of merely low average intelligence."[62]

While Goddard's ideas about the meaning of feeblemindedness were being
challenged by superintendents and sociologists, they were also being reconsidered by a new generation of geneticists. Most influential were the followers
of Thomas Hunt Morgan, whose studies of the fruit fly suggested the complexities involved in inheriting even simple traits, and the folly of using the
same theories to explain complex human behavior. Characteristics which
demonstrated a pattern of continuous variation within a population (such as
height), geneticists now argued, suggested not one but many genes in action,
a theory which contradicted Goddard's explanation of low intelligence as a
simple "unit character." In fact, the very concept of the "unit character" as
the product of a single gene unaffected in any way by its environment, Herbert Jennings reported in 1925, was "an illustration of the adage that a little
knowledge is a dangerous thing. The doctrine is dead – though as yet, like
the decapitated turtle, it is not sensible of it," he declared, for while geneticists
had rejected this doctrine years earlier, it continued to appear in eugenic
arguments. "Neither eye color, nor tallness, nor feeblemindedness, nor any
other characteristic, is a unit character," Jennings insisted, and it would be
better for science if this expression were to disappear altogether.[63]

Also troubling many geneticists were the increasingly close connections
between their science and the new racism. In response, a few, such as Morgan, quietly withdrew from eugenics organizations, while a few others, such
as Raymond Pearl, began to attack this movement in the popular press. Eugenics had recently "fallen in some degree into disrepute," Pearl explained,
due to the "ill-advised zeal" of "devotees" who explained "poverty, insanity,
crime, prostitution, cancer, etc." with "simple and utterly hypothetical Mendelian mechanisms." Such works had left eugenics "a mingled mess of ill-
grounded and uncritical sociology, economics, anthropology, and politics,
full of emotional appeals to class and race prejudices, solemnly put forth as
science, and unfortunately accepted as such by the general public." Goddard
himself was still rarely mentioned in these attacks; even so, since his scientific
research played such a central role in bolstering eugenic arguments, attacks
on the legitimacy of eugenics became indirect attacks on him.[64]

Of all the new challenges to Goddard's science, none proved more direct,
or more destructive, than the writings of psychiatrist Abraham Myerson.
Since 1917, Myerson too had been writing stinging reviews attacking the
"pseudoscience" of eugenics. In his 1925 volume, *The Inheritance of Mental
Diseases,* he reanalyzed the statistics gathered from numerous older studies
purporting to prove the inheritance of a range of physical and mental condi-

tions. And while he also criticized the heredity studies produced by physicians such as Martin Barr, his analysis of Goddard's Kallikak study proved especially devastating.[65]

Attacking the "appalling" research methods used by eugenic field workers, Myerson once again insisted upon medical supremacy in diagnosing disease. Perhaps it was unnecessary to have "laboratories, blood tests, clinical examinations, and to take four years in medical school plus hospital experience," he proposed, if a woman such as Elizabeth Kite could "as a result of a dozen or two lectures make all kinds of medical, surgical, and psychiatric diagnoses in an interview." Myerson himself felt "shame in the presence of the work done by the field worker in this case," he argued with "due humility," for he frequently had to reverse his first impressions. "Judge how superior the field workers trained by Dr. Goddard were!" he exclaimed. Not only does "their 'first glance,' tell them that a person is feeble-minded, but they even know . . . that 'a nameless girl' living over a hundred years before . . . is feeble-minded. They *know* this," he emphasized, "and Dr. Goddard acting on this superior female intuition, founds an important theory of feeble-mindedness, and draws sweeping generalizations, with a fine moral undertone, from their work."

Goddard's entire argument, Myerson concluded, rested upon his field worker's "surmises as to the mental and physical state of the dead and the quick." The Kallikak story contained "all the dramatic flavor of the missionary spirit" wishing to "awaken into vigilance the threatened normals." Such a work could not be taken seriously as science, Myerson insisted. "In ethics two wrongs do not make a right," he argued, "and in science a thousand instances of guess work, intuition, snap judgments and hearsay will do good neither to the Mendelian theory nor to eugenics."[66]

Myerson's sharp analysis and biting wit permanently punctured both the eugenic and the Christian logic that had sustained Goddard's monograph for more than a decade. In the following years, psychologists, psychiatrists, sociologists, geneticists, and educators would begin a critical reexamination of the Kallikak study. Under such scrutiny, Goddard's arguments quickly withered.

Thus, within a few short years, Goddard's research had come under numerous attacks from multiple directions. In less than a decade, his work had become historical, a part of the past increasingly rejected by leading scientists working in the present. Even worse, by the mid-1920s, his sociology had been labeled "legend," his reform "alarm," and his psychological science "pseudoscience."

Surprised and stung by this new criticism of his research methods in general, and of his Kallikak study in particular, Goddard reexamined his evidence. Hoping to answer Myerson, he once again contacted his Kallikaks field worker, Elizabeth Kite. "Did we ever know the real name of the mother of the bad line in the Kallikak story?" he now inquired for the first time. "One or two people" who were "opposed to the idea of the heredity of feeblemindedness," he told Kite, "have attempted to discredit the story . . . among other things stating that it is absurd to attempt to declare that this girl was feebleminded when so little is known of her that we do not even know her name. I should like to turn the tables on them if it is possible," he added, "by stating that we did know her name and that the calling her 'the nameless feebleminded girl,' was in accordance with our policy of disguising *all* names."[67]

Kite, however, could supply little useful information. The "situation," she replied, "is not so bad as would appear. Whether it was 'Moll' or 'Kate' or 'Jane' I cannot be certain," she conceded, "for as you say and yourself thoroughly understand I was after something infinitely more characteristic than *a name*. . . . " The "*real* thing I was after *I did get,*" Kite reported, "namely a distinct and vital contact with the memory which her particular type of character left upon the people who could remember that far back." Unfortunately, with the persons interviewed long scattered or dead, it was now impossible to retrieve this crucial bit of information. "What her maiden name was it never entered my head to ask, for which of course I am duly sorry now . . . ," she wrote.[68]

Instead, Kite offered Goddard more anecdotes.

You will recall that I questioned *one man who remembered her* and her great-grand-daughter. . . . It was the latter who in a sudden burst of confidence made the startling revelation "Ye see his mother had him before she wuz married." Thus having opened the way she went on without restraint – "Steven, as lived twenty miles from here wuz his half-brother an' looked so like my Grand-father that eff they'd been dressed alike you'd a knowed they was brothers. . . ." That was all I could stand for one day! I was far too excited to trust myself to ask any more questions but left as soon as possible and simply flew down the mountain road, my head bared to the cool spring breezes in an effort to maintain my mental balance.[69]

Whatever appeal such literate storytelling may have had for Goddard and his readers in 1912, it proved of no use in his new battle to defend his book as a reputable work of science.

In the long run, the most damage to Goddard's reputation came not from those who attacked his work but from those who continued to use it. While leading geneticists, psychiatrists, sociologists, and psychologists had begun to reject his theories, many others still believed them. Goddard's arguments about the feebleminded still constituted a central theme in the literature, lectures, and exhibits being produced by the eugenics movement. During the 1920s, the leaders of this movement expanded their efforts to shape public opinion and to influence legislation.[70]

Perhaps the most directly traceable use of the Kallikak study was in a 1924 legal case which reached the Supreme Court in 1927 as *Buck v. Bell,* and which became the test case for legalizing involuntary sterilization. At issue was the State of Virginia's right to sterilize Carrie Buck, an eighteen-year-old inmate of the State Colony for Epileptics and Feeble-Minded with a sup- posed mental age of nine years. Joining the state in arguing for compulsory sterilization were Charles Davenport's eugenics associate, Harry Laughlin, and his Cold Spring Harbor field worker, Arthur Estabrook – both active lobbyists for sterilization legislation.[71]

While Goddard had long believed that sterilization was an appropriate and beneficial response in individual cases, and thus always supported its legalization, he never advocated it as a widespread solution to the problem of feeblemindedness, as others were arguing. Sterilization, he wrote a few years later, "never was a distinctively Vineland idea. Vineland never opposed, yet never ardently espoused, that cause."[72]

Even so, the trial transcript demonstrates the centrality of Goddard's re- search in bolstering the arguments of far more ardent advocates, for entered into the record were both his 1911 article on the "Heredity of Feeble- Mindedness" and a slightly distorted version of "the Callicac [*sic*] case." "Old man John Callicac in 1775 had an illegitimate child by a feeble-minded woman" resulting in hundreds of feebleminded descendants, a Virginia insti- tutional superintendent recounted to the court. "That is a report that was generally published throughout most of the books on heredity," he ex- plained.[73]

Based upon this report, as well as the reports of eugenic field workers, the state contended that Carrie Buck, her mother, and her seven-month-old daughter were all feebleminded. Writing for the majority, Supreme Court Justice Oliver Wendell Holmes expressed the long-held and still popular be- lief in theories about degeneracy which linked defectives, dependents, and delinquents. "It is better for all the world," Holmes contended, "if instead of waiting to execute degenerate offspring for crime, or to let them starve for their imbecility, society can prevent those who are manifestly unfit from con- tinuing their kind. . . . Three generations of imbeciles are enough."[74]

Yet if, as Holmes' comments show, contemporary scientific debates had failed to effect a radical change in public opinion, they did affect Goddard. Like those of other scientists (among them army tester Carl Brigham, who by 1930 would recant all of his earlier conclusions linking low IQ to racial differences), Goddard's own ideas about who ought to be designated "feeble- minded" were changing. Although he still regarded the army tests as a re- markable scientific achievement, by the end of this decade Goddard too saw that they had also proven something unexpected: the concept of "mental age" no longer meant anything definitive. It was time, Goddard now believed, to reconsider the meaning of the very term that had made him most famous: the moron.[75]

In an article for *Scientific Monthly* entitled "Who Is a Moron?" Goddard faced this challenge squarely. He had invented this term for "specific scientific use," but it had apparently "filled a long-felt want in the public mind," he wrote, for the word "moron" was now being used in "polite conversation" as well as "popular literature . . . from newspapers to novels and poetry." It was also used "indiscriminately" to refer to "anybody who is a little bit dull in intelligence, or even, as some one has expressed it, to any one who does not agree with you." What exactly, its inventor asked, did this term mean?[76]

"There was a time to be sure," Goddard now admitted, "when we rather thoughtlessly concluded that all people who measured twelve years or less on the Binet–Simon scale were feebleminded." Scientists had "begun to discover our error" before 1917, but the army tests had proven this point unmistakably. So many persons tested below twelve, Goddard now argued in words that sharply contradicted his own earlier conclusions of 1918, that "to call them all feebleminded was an absurdity of the highest degree."

What the army results had really shown, he now argued, was that some persons who tested below twelve "*are* morons, but the great mass of that group *are not* morons." Goddard also now saw just what this meant – that the scientific precision he had so yearned for had been overturned. "And now must our nakedness be exposed!" he exclaimed. "In this year of grace, nineteen hundred and twenty-six, after three quarters of a century of dealing with the problem and at least a quarter of a century of intensive study of it," he conceded, "we are still limited to a definition of feeble-mindedness that is unscientific and unsatisfactory." In this sense, Binet's efforts had failed, for practitioners were once again reduced to defining feeblemindedness by using subjective judgments based upon multiple criteria – psychological, biological, and most slippery of all but most important, sociological. "Such a definition is not scientific," Goddard complained, "because it is not definite."

If feeblemindedness was now being defined largely by the social ability to get along in the world, then other changes were also in order. In the past, "we have always said that feeblemindedness was incurable," Goddard reported. "We were evidently in error," he now admitted, for even if the mental level could not be raised, if a person could be trained, he was no longer "incurable." This was "happening all the time," he reported. "It has been happening for years, but we did not know it." Goddard praised Bernstein and others for proving this point. "*We are curing some feebleminded in all our well managed institutions* – if you choose to put it that way," he concluded. However one put it, it was "a fact of tremendous significance."

Goddard drew out this significance for his readers. "Now that we have learned the facts," he proposed, "the solution is easy." Most needed was special education, especially in industrial and manual training. Moreover, he saw no dangers in allowing such persons to reproduce. "Why should we be afraid of their having children and bringing up a family like themselves?" he asked. "The problem of the moron," Goddard concluded, "is a problem of education."[77]

He repeated these themes two years later, for in 1928 Goddard returned to New Jersey to address once again the American Association for the Study of the Feebleminded, the same organization before which he had first defined the term "moron" eighteen years earlier. Once more, he frankly conceded his own mistakes. It was time to change "several of our time-honored concepts," he stated. First, morons were *"not incurable,"* and second, they "do not generally need to be segregated in institutions." Finally, he found the "eugenic aspect" of this problem to be "probably negligible."[78]

"In view of these facts," he continued, "honesty and fairness compel us to raise the question: Is it possible that during all these years we placed the limit of feeblemindedness too high? Is the real limit 7 years instead of 12?" If this were so, then the entire category of persons that Goddard had called "morons" ought not be classified as among the feebleminded. Morons, Goddard now argued, "are *not* hopeless and incurable mental defectives, but merely the lowest group of the body politic, requiring special attention and special methods in their education and training."

Such a concession was indeed painful for the man who had "discovered" the moron in the first place. Yet Goddard now believed that most morons could be educated and thus cured. "This may surprise you, but frankly when I see what has been made out of the moron by a system of education, which as a rule is only *half right,*" he explained, and if one could "add to this a social order that would literally give every man a chance," he would be "perfectly sure of the result." "I assume that most of you, like myself, will find it difficult to admit," he conceded, having "worked too long under the old concept." "As for myself," Goddard added, "I think I have gone over to the enemy. . . . " Thus, by the late 1920s, Goddard too had publicly changed his mind. Both in speeches and in print, he now candidly admitted that some of his most influential conclusions concerning the feebleminded had been in error.[79]

Utopian Dreams and Nazi Nightmares

Such admissions, however, did not mean that Goddard had repudiated all of his earlier work. Both *The Kallikak Family* and *Feeble-Mindedness* were scientific studies made in good faith, he still insisted, and he remained convinced of the essential correctness of many of their findings. Moreover, by 1929, he was still a member of the American Eugenics Society's Advisory Council and of Ohio's Committee on the Legalization of Sterilization.[80]

Yet Goddard's views on eugenics differed markedly from those of this movement's American leaders. Most striking, his 1920s publications contained none of the nativism and racism which were increasingly linked with this movement. For instance, while many of the Cleveland children he had studied were Jewish, Goddard made no comments about their ethnicity or their religion; moreover, his book on the gifted ended with illustrations produced by a five-year-old "Japanese genius" he had met while teaching summer school in Hawaii. While in Hawaii, Goddard had also been impressed by the efforts of a group of educational, religious, and civic leaders to find common ground among this territory's Anglo-Saxon, Chinese, Filipino, Hawaiian, Japanese, Korean, and Portuguese communities, and intrigued with a small pamphlet they produced outlining an "Interracial Standard of Moral Conduct and Social Ethics." Responding to Stoddard's popular diatribe, *The Rising Tide of Color*, he sketched out his own "idle thoughts" on "The Rising Tide of Character." "*Madison Grant* and *Lothrop Stoddard* have tried to show that the progress of the world in civilization is threatened by the 'passing' of the Nordic Race or the dangerous increase of the yellow man, the brown man and the black man. This is an error," Goddard scribbled. "The difference in humans is *intellectual* not *racial*. The distribution of Intelligence in the different races is probably the same." The Hawaiian situation proved that environmental change "tends to eradicate racial differences. The best people of all races must work together," he concluded. "Narrow nationalism must give way to a broad humanism."[81]

Goddard's increasing emphasis on improving education also put him at odds with many eugenicists. "Anti-vivisectionists, anti-evolutionists, and the Ku Klux Klan," he wrote in a 1925 article, were "all evidences of the failure of education." Invited by the editor of *Eugenics* to submit an article in 1930, he produced a short essay on "The Child's Inheritance and What Can Be Done With It," which emphasized the difference between biological and social inheritance. This was "a live question" since "in the past we have been radically wrong in our education because we did not understand this fundamental difference," he argued. With "the right kind of teacher, and the right kind of school system," Goddard stated, "crime and insanity could be eliminated." By this time, he conceded, some readers had "thrown down the magazine with the remark, 'Oh, he is one of those fellows with a panacea; his is *education*.'" Yet to Goddard, such logic was "inescapable," for most undesirable social traits were products of "the unconscious education which we have often mistaken for the hereditary traits." Not surprisingly, *Eugenics* chose not to publish Goddard's essay. "Of course I suspected that a journal that published articles about the inheritance of a boat-building tendency would find my article a little hard to swallow," he told a friend.[82]

Ever an enthusiastic traveler, Goddard's views were also influenced by a grand tour taken with his wife in 1931 (sponsored by his wealthy Cleveland

Goddard riding a camel on his Mediterranean trip, 1931. Courtesy of the Archives of the History of American Psychology, University of Akron.

patron, Roberta Bole, both as a gift and as a means of studying education for the gifted in other countries). This trip took him around the Mediterranean and throughout Europe. In the Holy Land, he visited the school established by Maine's Quaker missionaries, Eli and Sybil Jones. In Rome, he saw the Catacombs. "It looked very familiar to me," he later recalled, "for it was very much like a Quaker meeting except that the people were dead and not simply sleeping." Most surprising was his visit to the Soviet Union, which in 1928 had made Goddard an honorary member of the Leningrad Scientific-Medical Pedological Society. Both Leningrad and Moscow, he discovered, were "perfectly normal cities" with impressive school systems. Americans had been getting "a great deal of misinformation," he concluded, "much of it downright prejudice and apparently made up to suit the politics of the narrator."[83]

Goddard's own politics were also being transformed. Slowest to change were his attitudes toward the poor, even during the Great Depression. Problems "in unemployment and the consequent poverty and starvation," he argued in 1931, were largely "due to the fact that the great mass of these people have not had the intelligence and foresight to save some of their earnings, when they had employment at good wages, in anticipation of just such difficulties." Yet once again, Goddard was not simply defending laissez-faire capitalism. "*It is perfectly clear to those who understand this situation,*" he emphasized, "*that half of the world must take care of the other half.*"[84]

By 1934, Goddard had modified his views, for he had become an ardent admirer of Franklin Delano Roosevelt and a strong supporter of his policies. He now advocated minimum wage laws, shorter working hours, state-sponsored medical care, national relief, and even a redistribution of wealth. "As a nation," he told a newspaper interviewer, "we must put our resources in a pot and apportion them to our people." Americans should ignore the "loud cries" of "hired propagandists who have a stake in the old order," he argued. "Such sheer capitalistic claptrap is the cry that a redistribution of wealth would wholly destroy leadership and initiative," Goddard added, "that the principle of liberty is involved in every restriction on unbridled use of money, that every man who raises his voice above a whisper belongs to a mysterious group known as Communists." Although America was "not ready for the collectivism of Russia, and may never be," it could still learn much from the Russian example in furthering its own evolution, and avoiding revolution.[85]

While Goddard's political views during the 1930s reflected change, more consistent was his scientific idealism. In fact, Goddard had always pinned his hopes for the future on scientists rather than politicians, and he continued to do so. Especially revealing was a speech of 1931 which dealt once again with the subject of feeblemindedness. Invited back to Vineland as its guest of honor for a ceremony commemorating the twenty-fifth anniversary of the founding of his laboratory, Goddard described in detail his own utopian vision. "At this point in preparing this address," he jokingly told his audience, "I fell asleep, and had a dream (Johnstone taught me to dream)." In his dream, he saw Vineland twenty-five years in the future – in the faraway year 1956.[86]

"Yes, this is the Vineland Laboratory," his dream guide explained. "It is a national institution now, supported by a syndicate of intelligent financiers, who have resolved that the problem of low intelligence and its relation to world welfare and social efficiency *shall be solved*." At the heart of their solution was research, for the institution consisted of a series of buildings with working scientists.

The first building Goddard saw was for "physical, physiological and biochemical examination." "Yes, there was a biochemical department in the early days," his guide explained, "but it was just a little ahead of its time" and was now making great strides. A second building contained "educational research." For the "different grades of mentality, we are trying out the value of different educational materials, such as pictures, models, oral descriptions and the like." A third building housed psychological research addressed to other types of questions. "What makes a child happy? How does fear arise, and how is it eliminated?" In a fourth building, sociological researchers sought to determine the grades of intelligence that "will be useful in a civilization becoming ever more complicated. It is now pretty well recognized that we cannot make masters out of morons," his guide remarked.

Finally, Goddard saw the "biological building," whose "central aim is how to develop a better human being." Surrounding it were small homes for feebleminded couples, the subjects of an experiment. "They are living thoroughly normal lives. . . . The medical staff keeps in close touch with them and, whenever a pregnancy occurs, special treatments are begun to see if anything can be done to produce a better child than would naturally come from two feeble-minded parents. Special types of food, glandular treatment, special exercises as stimulants of various kinds, are given. . . . The result? It is too soon to know." Still, "we shall at least have the most perfect records ever obtained of the pre-natal and post-natal influences on children." Within this dream utopia, money was never an obstacle. "We pay any salary necessary to get the man we want," his informant explained. "Should we spare any expense? No. A million dollars, a hundred million dollars, a billion or two, is not too much to pay for information that will lead to the cure, prevention, elimination or understanding of the feeble-minded."

In the 1930s, as in the 1910s, Goddard still saw society as on the verge of eradicating feeblemindedness, with all its attendant social consequences. "We solved the problem of smallpox. . . . We solved the problem of malaria. . . . And we shall solve the problems of the social adjustments of the entire world," he predicted. "Just as the mosquito was the key to the malaria problem, so are the feeble-minded the key to the great social problems of the present."

In Goddard's version of a brave new world, social as well as genetic engineering would produce both a better society and a more intelligent man – a man who, for instance, would be smart enough to avoid wars. "The World War reflected an appalling lapse of intelligence," Goddard argued. "Either there was not sufficient intelligence to prevent it or we had so badly organized ourselves that our intelligence could not function in such a situation."[87]

Goddard remained completely unconscious both of his own deep class biases and of the dangerous potential for political totalitarianism within his technocratic, eugenic utopia. His contemporaries, however, were soon to learn otherwise. By the end of this decade, the idea of breeding in "superior" men and breeding out "inferior" had taken on a frightening reality in the hands of Nazi eugenicists. And while the Nazis burned other books, they printed a new 1933 edition of *Die Familie Kallikak,* shortly after passing a sweeping new sterilization law.[88]

Its liabilities notwithstanding, Goddard had certainly never intended his Kallikak study to prove Anglo-Saxon racial superiority. Moreover, while he may have shared some common stereotypes of his day, including the belief that Jews (and Scots) were cheap, or smart, he never included such ideas in his published works. Whatever its flaws and biases, Goddard's Kallikak study expressed no antisemitism.[89]

Even so, this book's warnings against mixing "good blood" with "bad" took on a new connotation in a society that would soon pass the Nuremberg decrees. Within this context, the public health metaphors that Goddard and his contemporaries frequently employed – metaphors that compared the feebleminded to "mosquitos" or other pests – took on a new literalness. Within the Third Reich, these same metaphors would be used to justify the extermination of entire populations deemed "parasitic." In Nazi propaganda, both fears of the feebleminded and faith in eugenic solutions would be repeatedly promulgated as central tenets. By the late 1930s, Nazi Germany had implemented not only a vast sterilization program but also a secret policy of "euthanasia" which targeted institutional populations, including the insane and feebleminded, as the first to be exterminated by poison gas.[90]

All of these actions added a new urgency to American efforts to distinguish the Nazi version of eugenics from legitimate genetics. They also intensified the campaign to drive the Kallikaks from the realm of science. Goddard's basic Mendelian logic was faulty, Amram Scheinfeld argued in his 1939 book, *You and Heredity,* for if feeblemindedness was a recessive gene, as Goddard implied, then Martin Kallikak, Jr., must have inherited it from *both* parents. More importantly, he explained, the idea that feeblemindedness emanated from a single gene had long since been abandoned by serious geneticists like Thomas Hunt Morgan.[91]

By 1940, psychologist Knight Dunlap had pronounced Goddard's book scientifically dead. The "Kallikak phantasy has been laughed out of psychology," he declared in *Scientific Monthly.* Yet Dunlap still worried that even in "books written by psychologists who ought to know better, the Kallikaks skulk in the corners of the pages, and leap out upon unwary students." The family's fame "began with an anecdote perpetrated with incredible innocence by an eminent expert on 'intelligence,' and repeated with astonishing solemnity by many after him," he reported. Dunlap then pointed out more logical flaws. "The promoter of the legend inferred that the girl was feeble-minded because she had feeble-minded descendants," he summarized. "This procedure, of assuming the conclusion in the premises from which it is presumably drawn, is called by the logicians, 'Begging the question.'" Such arguments, and the new political climate, were clearly having an impact, for although *The Kallikak Family* had gone through twelve American printings since 1912, Macmillan would publish it for the last time in 1939.[92]

In 1942, at the age of seventy-six, Goddard surprised his critics by publishing a short defense of the Kallikak study in *Science.* "For a decade the data were accepted apparently without question," he observed, probably because there were "enough people who were familiar with the details to explain how the study was made." Over time, however, "the inevitable happened and writers appeared who did not know, who obviously had not read the originals, and who therefore thought they detected certain flaws in the techniques which did not exist."[93]

Goddard's strongest argument in his own defense was that his critics had made several factual errors in describing his research. Scheinfeld, for instance, had claimed that Goddard had begun the Kallikak study in 1898, before Mendelian genetics had even been rediscovered. Copying and then compounding this error, two other biologists had insisted that not only Mendelism but Binet testing as well "did not even exist when the study was made." Such mistakes were easy for Goddard to refute.[94]

Much more problematic, however, were the deeper questions raised by Myerson and others concerning diagnoses of persons alive and dead. Dr. Myerson "argues that because he can not correctly diagnose feeble-mindedness, nobody can," Goddard retorted. His field workers, however, had been "carefully trained" for this task, for they had spent "weeks and months in the institution."

Goddard's central defense thus rested once again upon institutional expertise. "It is well known that superintendents of such institutions quickly learn, and when a new arrival appears they not only know whether he is a fit subject for their institution or is normal and does not belong there," he insisted, "but they also know his *grade*." Goddard was certainly correct in arguing that institutional physicians had long made such claims in defending their own subjective diagnoses. Such a defense was bitterly ironic, however, for it was exactly this problem of subjective diagnoses, and the unscientific arrogance that sustained them, that had motivated Binet to construct tests to measure intelligence in the first place.

Impossible to contest was the charge that chagrined Goddard more than any other: that he had judged the mentality of a woman whose name he had not even known. In defending himself, he now decided to advance the argument he had recently proposed to Kite. "She is nameless to the reader only," Goddard claimed, even though Kite had informed him otherwise. "We had her name; and not only her name but her history," he lied.[95]

After so many years of silence on the subject, Goddard's self-defense drew an enthusiastic response from many of his old friends. "You have been very patient to have waited so long before answering your critics," Lewis Terman wrote. "In my opinion the Kallikak study will stand as one of the most notable studies of the kind ever made." Carl Seashore was also "very glad that you struck back against those who treated your valuable early investigations lightly." Goddard had taken "a definite stand on issues in which your name will stand as a classic," he added. "So, I guess you are a Yankee," Vineland teacher Helen Hill responded. "I think you have those gentlemen, again as a New Englander would say, 'Where the wool is short.'" Charles Davenport's reply, however, suggested a darker undertone. "I wonder why it is that people with such names as Abraham Myerson and Amram Scheinfeld should think it necessary to attack so much of the work on heredity?" he asked.[96]

Yet despite his friends' enthusiasm, Goddard's arguments did little to convince his opponents of his book's scientific merit. Two years later, Scheinfeld published a reply entitled "The Kallikaks After Thirty Years." Aware that he had wounded an older scientist, Scheinfeld's response was kind on several counts. He had no desire to criticize Goddard personally, he noted. After all, it was "unfair to be too critical" of work done "in the infancy of genetics." Nor was Goddard "responsible for the misuses of his study." Moreover, Scheinfeld credited Goddard for his "brilliant pioneering work, and for his many contributions to the study of mental defectives." Even the Kallikak study, he graciously conceded, had "proved the inspiration for much useful research by others." Nonetheless, "it would ill serve the cause of science," Scheinfeld insisted, to claim that such findings "stand unshaken after all these years."[97]

The "ghosts of the notorious Kallikaks have not easily been exorcised," Scheinfeld lamented; yet scientists could not let the case rest, because "today the Kallikak study has acquired new implications." Indeed, by 1944, debates such as this one had assumed a haunting relevance and a far more serious import. "What should interest us now," Scheinfeld noted in phrasing the question that would dominate historical scholarship thereafter, "is why, in view of the easily apparent flaws. . . . and its rejection in authoritative circles for many years, it has continued to be given such strong credence. . . . "

The answer, he concluded, was simple: "there are persons everywhere who relish the thought that some groups, races, classes or strains (always including the ones to which they themselves belong) are born to be superior and dominant, and that other groups are destined by nature to be inferior and subordinate." This concept permitted "those on top to smugly keep their place, while relieving them of the necessity of doing very much for those at the bottom, except, perhaps, to suggest that they be prevented from reproducing." Such views had done "great damage to the eugenic movement." "No one in possession of the facts can doubt the existence of pathological genes in human germplasm," Scheinfeld conceded; far more dangerous, however, was "the unjustified extension of pseudogenetic principles into sociology." It was just such principles, and the hatreds they engendered, that had "helped to bring on the present war."[98]

If Goddard ever saw any connection between his own work and ideological developments in Nazi Germany, he never wrote of it. Although he was by then in his seventies, World War II did touch Goddard personally in at least one way. Among those he had met on his 1908 European tour was Dr. Salomon Krenberger, editor of a small journal studying handicapped children, who later became director of Vienna's Jewish Institute for the Deaf and Dumb. Over the following two decades, these two men had maintained a warm correspondence. (On one visit to Vienna, Goddard had even been taken to meet Krenberger's friend Sigmund Freud). Although Krenberger died before Hitler came to power, in 1938 his daughter Selina contacted her father's former American colleague, Henry Herbert Goddard, to seek his help in escaping Nazi Austria.[99]

In response, Goddard took immediate action. He first contacted Jewish philanthropist Samuel Fels, who agreed to aid Miss Krenberger in finding work in America. Even more helpful was Goddard's Jewish former student, Samuel Kohs, who had by then left the field of clinical psychology to work for the National Coordinating Committee for Aid to Refugees. Like "the thousands proscribed by Hitler," Goddard wrote Kohs, Miss Krenberger "wants to leave Austria-Germany. In desperation, she has written me. . . . I would be glad to help her," he added, "if I could and if I knew how." Responding the same day, Kohs sent Goddard information on sponsoring a refugee; he also arranged for a law firm with expertise in handling such cases to help Goddard complete the necessary affidavits and financial transactions.[100]

"I am so glad about your help and so happy with my affidavits," Selina Krenberger wrote Goddard in March 1939. Ironically, Goddard's fame even among the Nazis was proving useful. "You are so well-known and esteemed everywhere. . . . ," she wrote. "Your name alone opens possibilities." Yet in the months that followed, Krenberger's letters became ever more desperate as plans for her departure suffered frustrating but typical bureaucratic delays.[101]

Not until October 26, 1939, did Selina Krenberger have good news to report to her sponsor. "With great joy and thankfullness I received the letter from the Italian line about the booking of the ship ticket and your wire," she finally wrote. Although she still had much to do, she hoped to sail from Italy on December 6, 1939, "arriving as very doubtful 'Christmas gift' in U.S.A." The nightmare of Nazi Austria, Krenberger added, had left her physically and emotionally drained. "Sometimes I am so tired of heart, soul and body," she wrote, "that I fear to loose [sic] my strength and never come to see you and your country." Goddard's help had restored her hope. "With this ticket," she thanked him, "you gave me back the right of self-destination." Yet despite Goddard's best efforts, Krenberger's hopes for a new life were destroyed, for she was unable to escape from Austria. Her letters ended abruptly in 1939; by 1941, she had died in a Jewish hospital in Vienna.[102]

As World War II worsened, Goddard himself became increasingly introspective, as well as increasingly convinced of the need for all people to reject narrow nationalisms and to promote pacifism. Wars were avoidable, he believed, but this one had come about because Americans had been "asleep at the switch" and had "allowed a relatively few isolationists to lob[b]y us out of our moral duty of joining the League of Nations." American support would have made the League a success, Goddard asserted, "and Hitler would not be."[103]

Despite his growing pacifism, Goddard saw no dilemma in fighting Hitler, for this was "not a war any more than it could be properly called a war if the dangerous inmates of the State Hospital got out and the State Militia was called to corral them." Even a Quaker need have no "conscientious scruples against offering his service to bring this crazy man to terms." The future "looks dark," Goddard wrote a friend in 1943, but he felt "as sure of victory as I do of my dinner today." To explain why he knew that Hitler would lose, he quoted Victor Hugo's words about why Napoleon had lost the battle of Waterloo.

> Was it possible for Napoleon to win the battle? We answer in the negative. Why? On account of Wellington, on account of Blucher? No; on account of God. . . . It was time for this man to fall; his excessive weight in human destiny disturbed the balance. . . . The moment had arrived for the incorruptible supreme equity to reflect. . . . Streaming blood, over-crowded graveyards, mothers in tears, are formidable pleaders. When the earth is suffering from an excessive burden, there are mysterious groans from the shadow, which the abyss hears.

Does Hugo's account also "fit the present situation?" Goddard asked. "There can be no doubt about it," he concluded.[104]

A Life in Retrospect

In the 1930s and 1940s, Goddard watched as the nation was plunged into depression and world war. During these same decades, his personal life suffered a series of losses from which he would never recover. Most devastating was the death of his wife, Emma, in October 1936, following a long illness. After forty-seven years of marriage to a "perfect mate," Goddard was so desolate that his friends in both Ohio and New Jersey began to worry among themselves.[105]

"Our good, kind and gentle friend and colleague, Goddard, has been passing through the dismal shadows known only to those who have had similar experiences," wrote George Arps, his Ohio State colleague, to Vineland psychologist Edgar Doll. Goddard's isolation was compounded two years later when, upon reaching the age of seventy-two, he had to retire (unwillingly) from Ohio State University. This loss of his work, mixed with the increasingly frequent and often mocking attacks on his past accomplishments, left him depressed and filled with self-doubt. To his Ohio contemporaries, Goddard now became a reclusive figure who lived alone in a deteriorating part of the city, befriended mainly by a few devoted graduate students.[106]

These changes in a personality once known for its enthusiasm and optimism were especially striking to his former Vineland co-workers. After meeting Goddard in 1939, Kohs was haunted by a vivid impression of loneliness.

Goddard in Columbus, Ohio, c. 1940. Courtesy of the Archives of the History of American Psychology, University of Akron.

Doll described him as "rather gravely depressed and dispirited." "Dr. Goddard is not infirm," Helen Hill concluded. "He drives all over the country, visits, and other things," she stated. "He has in some way lost faith in himself." Neither the many warm tributes Goddard received in recognition of his seventy-fifth birthday in 1941 nor the honorary law degree awarded by Ohio State University in 1943 did much to alter his prolonged sense of depression.[107]

By the mid-1940s, Goddard had instead turned for solace inward, toward a study of religion. Most satisfying were the ideas of an obscure theological writer named Cephas Guillet, author of *The Forgotten Gospel* (1940) and a "Plan for a Peaceful World" (1943). In corresponding with Guillet and others, Goddard began to explore his own religious background.[108]

Although raised "a Quaker of the so-called Orthodox variety," Goddard had long considered himself "unorthodox." "In all my teaching and study, I have been seeking a philosophy and a religion that was not contradicted by science and the common everyday experiences of men of average intelligence," he explained. "I have never thrown away my religion; I have tried to revamp it so that I could live it." Christian ideals had proven especially useful in working with children, for he had soon learned that one could not teach "a child whom one had offended," but with "a child who trusted, it was easy to work wonders."[109]

Goddard had also come to have no faith in "the God of the old Testament and little if any more the God of the new. And least of all, the God of the church," he continued. "I should not think much of myself if I asked people to compliment me, to continually praise me," he argued. "How can I respect a God that avowedly 'seeketh such to worship him?'"

Seeking a conception of God "worthy of my poor adoration," he had accepted "the Biblical phrase, 'God is a spirit.' Not a Ghost. . . . A spirit of the Universe." His psychology suggested another biblical phrase, "God is love," as well as its converse. "'Love is God.' That is Jesus' teaching. That is my creed," Goddard stated. "Where love is manifest, there is, at least a little, of God." And for himself, he concluded, that was enough.[110]

As for reconciling religion with science, Goddard liked best the words of Albert Einstein, a scientist who had no inhibitions about "using the word God." "That deeply emotional conviction of the presence of a superior reasoning power, which is revealed in the incomprehensible universe," Einstein had written, "forms my idea of God." For Goddard, such an idea "more than takes the place of 'The old-time religion which was goodenough for father, but is NOT goodenough for me.'"[111]

Like his Quaker forebears, Goddard also believed that his faith required a social commitment. In the postwar years, he committed himself to various causes, sending not only money but also letters containing advice or expressing "a 'concern' – as we Quakers say." Among the groups he supported was the Emergency Committee of Atomic Scientists, the group led by Einstein to oppose the use of nuclear weapons. He could not support the American Committee on United Europe, he informed its chairman, William Donovan, for although its name sounded promising, this group proposed sending more troops to Western Europe. "I have spent some time in Russia," he explained in the midst of the Cold War, and was "thoroughly convinced that the only trouble with Stalin and his supporters, is: They are overcome by fear. They know that they are unpopular" and believe "they will sooner or later be attacked." What was most needed, Goddard argued, was trust and friendship. "We have sent them millions of money – that was good," he advised. "But every squad of soldiers that we send destroys the friendship that the money was producing." If only Roosevelt were still alive, he lamented, things would be better, for "he was HUMAN. He knew how to get along with people who did not believe as he did . . . he was friendly with Stalin to the last." For Goddard, politics and religion had become increasingly intertwined. "We are told that 'A soft answer turneth away wrath,' and WE KNOW THAT IS TRUE, but we haven't the courage (or the intelligence) to act upon it," he told Donovan. "By this time – if you have read this far – you have thrown my letter in the fire and characterized me as 'One of those impractical idealists.'"[112]

During these years, Goddard also continued to write a few articles for academic journals, including several on a topic still perplexing him: "What Is Intelligence?" After summarizing numerous definitions, he presented his own. Crucial to intelligence, he argued, was the ability to learn from experience. "It will be asked: 'Is not intelligence inherited?'" he stated. "Much unnecessary argument has been wasted on this topic," he answered, for one inherited not intelligence but a brain, its organ. "We know too little about the genes that account for the growth and functioning of the brain cells," Goddard concluded; "yet we know enough to safely conclude that the possibilities of variation are great." A child may inherit a brain "average, above average, or below average in its capacity to retain impressions (memory)." Yet intelligence meant more than this, for it also involved the range and quality of one's experiences, either personal or vicarious, as well as habits of mind (often acquired in school) which allowed one to make connections between them. "Intelligence," he concluded, "is the degree of availability of one's experiences for the solution of immediate problems and the anticipation of future ones."[113]

In a Christmas letter written when he was nearly eighty, Goddard cited this definition and then used it to assess his own life. While most of his experiences were "common to teachers," a few were "out of the ordinary." "I have sailed the oceans: crossed the continent by air: visited Europe – with brief excursions into Asia and Africa: I have climbed Mont Blanc, the Matterhorn, Jungfrau and Finsteraarhorn," he reported. Whether he had acquired "enough intelligence to prophesy, you must judge," he wrote. "I have solved some of life's problems – though not always wisely. I have made many mistakes," he added, "because I did not have sufficient intelligence to see the end from the beginning."[114]

Among those receiving this letter was Goddard's most famous pupil ever, Deborah Kallikak. "Did I tell you that our Deborah was so thrilled to get one of your Christmas messages?" wrote Helen Reeves, a Vineland teacher, who then quoted Deborah's reaction. "'The nicest thing about it Miss Reeves,' she says, 'is that he thought I have the brains to understand it which of course I do.'" Such a statement, Reeves believed, summed up Goddard's unusual talents as a writer. "A definition of intelligence that a moron can understand – that's an achievement!"[115]

Reeves was not the only contemporary to recognize Goddard's skills as an accessible writer. Equally impressed was Dr. George Crane, the money-making author of *Psychology Applied,* who also produced a newspaper column and a daily radio show dispensing psychological advice. "I find the public hungry for practical psychology," Crane wrote Goddard in 1945. And since he believed that Goddard still possessed "the unusual faculty of writing in an interesting manner," he urged him to try his hand at more popular materials.[116]

Why not "turn your keen mind to writing a few booklets on vital current topics," Crane suggested, "such as 'FACTS ABOUT VOCATIONAL GUIDANCE' or 'METHODS FOR PEPPING UP SUNDAY SCHOOL ATTENDANCE' or 'CLINICAL PSYCHOLOGY FOR FACTORY FOREMEN'?" After all, one educator had claimed that 70 percent of American children were not enrolled in Sunday schools. "Why?" Crane asked. "Apparently because nobody has thus far written a snappy little manual. . . . You are the man to do it," Crane told Goddard. He might even include a chapter for clergymen "listing simple magic tricks that can be employed to good effect as moral lessons," as well as some "self-rating charts," he added, "for the public likes to analyze itself." "Very few of our American psychologists can approach you in motivating people," Crane concluded. "You have a distinctive talent. And your prestige and name are powerful."[117]

At a time when Goddard's "prestige and name" were fast losing scientific credibility, Crane's flattery, as well as his ideas about making money, evidently came as a welcome surprise. His entreaties earned him an author. In fact, Goddard already had a manuscript in hand for Crane to publish. He had written it years earlier, he explained, in response to a colleague's request that he address the question, "How can I save my child from becoming a criminal?" And even though Goddard believed that writing such a book would be like preaching "a WONDERFUL sermon" but "the people who ought to hear it never go to church," he had still completed it in his retirement years.[118]

He had even submitted the manuscript for review to Harper and Brothers in 1943. It had been rejected, however, based on an anonymous reader's report from another psychologist. Goddard's book, this report stated, contained "a hodge-podge of historical and anecdotal material." "The author has read widely, has thought some, has a good feeling about children," the reader contended, "but lacks understanding." The book's problems, this reviewer argued, were epitomized in its title. "Anyone who would suggest titles like 'How Can I Save My Child From Becoming A Criminal'" seemed to lack "a fundamental awareness of what is involved in the bringing up of children and in the world we live in," this report concluded.[119]

Crane, however, had a different reaction. "Mail me the manuscript at once!" he replied. Not surprisingly, he quickly agreed to publish it after deciding with Goddard upon an even more timely title – *How to Rear Children in the Atomic Age.* Although Goddard's book largely consisted of admonitions telling parents to raise their children with love and kindness, his preface stressed its relevance to world conditions. "We have men of vision who understand the new world, the atomic age, but we do not have enough of them," he argued. "Eventually we must have men who were born and bred since 1945: men who will be unhampered by the old disproved traditions. That means that we must start with the children and give them better care, better bringing-up, better schooling." By 1948, Goddard's book was being marketed along with *Dr. Crane's Radio Talks,* a work that included essays on "How to Lose 10 Pounds in 10 Days" and "Logical Proof of God," as well as a 100-point "Test for a Good Wife." Under Crane's promotional eye, Goddard's final book sold more than a thousand copies.[120]

Goddard had no objections to Crane's popular sales techniques; he was humiliated, however, by a statement printed in one of his book's advertising circulars. "Educators recognize Dr. Goddard as one of the 'titans' in the development of modern scientific psychology," it stated. "He ranks with such giants as William James, Walter Dill Scott, Alfred Binet and similar outstanding pioneers."[121]

"That is so ridiculous as to be nauseous," Goddard wrote Crane in 1946, as soon as he saw it. "It could be stamped as a downright lie, were it not that the term "Educators" may include a couple of people who do not know the difference between geniuses and a plodding, commonplace psychologist, who occasionally hits upon an idea worth repeating." Such a statement, "if seen by anyone who knows me," Goddard admitted, "would hurt me more than all my sins."[122]

If this were true, then Goddard must have been equally embarrassed by the hyperbolic remarks Crane published in his "Tribute" introducing the book, for he compared Goddard's "contributions to scientific progress" to those of Luther Burbank, Thomas Edison, the Wright brothers, and Henry Ford, as well as psychological "titans" like William James, E. B. Titchener, Walter Dill Scott, James McKeen Cattell, Carl Seashore, John B. Watson, and Lewis Terman. Crane's tone and message can be gauged by the grandiose biblical quotation which supplied the opening words to his Tribute: "'. . . and there were giants in those days who inhabited the land.'"[123]

Meanwhile, Goddard's opinion of himself continued to plummet. His last book, with its many autobiographical allusions, emphasized the need to learn from mistakes. And while these remarks were ostensibly about training children, they suggest as well a struggle to come to terms with mistakes of a more adult variety, for the book included a series of quotations concerning failures and mistakes from Disraeli, Huxley, Gladstone, Plutarch, Pope, and Confucius, among others. "The little I have seen of the world," he quoted Longfellow, "teaches me to look upon the errors of others in sorrow, not in anger." "Honest error," he quoted Chesterfield, "is to be pitied, not ridiculed."[124]

Whatever Goddard thought of his own errors, he did receive favorable recognition from his scientific peers on one final occasion before his death. In 1946, two years before his last book appeared, the University of Pennsylvania held a ceremony in conjunction with the American Psychological Association to commemorate the fiftieth anniversary of the founding of Witmer's Psychological Clinic. Goddard received an honorary degree in recognition of his "definitive pioneer work in the clinical approach to the problems of feeblemindedness – its detection, causes, and consequences – and in the training of feebleminded children." The honor, Goddard wrote to Penn's president, "comes as a complete surprise and leaves me all but speechless." Goddard accepted "with feelings of humility and gratitude which defy expression in words." This time, he was undoubtedly in distinguished company, for the other recipients were John Dewey, E. G. Boring, Wolfgang Kohler, Lewis Terman, and Robert Woodworth.[125]

Yet such honors did little to change his increasingly critical sense of himself. The more Goddard reflected upon his own life and work, the more convinced he became that his accomplishments were largely due to a series of lucky accidents. In 1947, he made plans to move from Columbus to California to live with his wife's younger sister, Mrs. Mabel Whiting, and her daughter Alice. "Dr. Goddard Moves to California; Weighs Title for Autobiography," stated his final interview with a Columbus newspaper. "With a smile," this interviewer reported, Goddard "admitted considering the title, 'As Luck Would Have It.'" "And therein lies the story," the reporter added, "of how the New England-born Quaker-reared authority on abnormal psychology found himself in a new field shortly after the turn of the century."[126]

Meanwhile, Goddard's letters to former Quaker schoolmates, Vineland coworkers, Ohio State students, and many others became increasingly autobiographical, with his reflections on the failures of his own education and early upbringing growing increasingly bitter. Especially important were his letters to his former student, psychologist Robert Fischer. Punctuating their correspondence were Fischer's repeated expressions of deep affection and sincere admiration and Goddard's self-deprecating replies. Goddard could not "escape from the 'Sainthood' into which I have put you by the mere excuse that I am ignorant of the facts," Fischer wrote him. "As the number one person on my list you must expect the love and respect that I am going to continue sending your way." Receiving such letters, Goddard admitted, made him feel like a bear that had fallen into a barrel of honey, for they poured forth so much sweetness. "Of course I like it," he added, "especially as I have never had much of it. And with my confirmed conviction that I deserve none of it, it is not likely to do me much damage." Yet Fischer's compliments also offered "the strongest argument" for writing his autobiography. "I see now that I must write it," he explained, "to show how much I have been over-estimated in certain quarters."[127]

Meanwhile, Goddard continued working on what was to be his autobiography, "As Luck Would Have It," by gathering together old correspondence, making lists of summer vacations spent hiking and camping, and jotting down random remembrances. Yet his increasingly faulty memory made the completion of such a book impossible. By the late 1940s, Goddard had agreed to let Fischer tell his story for him. "I want to write a decent and an intelligent biography – not eulogistic, but factual. . . . ," Fischer explained. After all, even Leonardo da Vinci, the greatest scientist of his era, was "imperfect and fallible." Toward this end, he urged Goddard to share his early correspondence with "the men that counted in the field in your time."[128]

Yet Goddard could supply Fischer with few such letters, he claimed, probably because such men "realized that I *didn't count*." In explaining why, Goddard again reviewed his early life. "I never wrote to them and so they never wrote to me. You see, my father died when I was nine years old and my mother was a Quaker preacher who 'felt called' to travel and visit Friends meetings. . . ," he added. "If I could have gone with her, I might have learned some of the social graces. As it was I was placed in a boarding school. . . . Nobody knew me or cared a whit whether I lived or died."[129]

Goddard even began to reassess his own temperament. While his contemporaries consistently praised him for his gentle manner and modest demeanor, he now saw these same traits in a new light. Haunting him was a comment made many years earlier by Mrs. Bole. "I think you will grow up," she had joked on one occasion, "but I am sure you will never grow old." "It was said in fun," Goddard recalled. "But it 'opened my eyes'" to the realization "that I had *not grown up*." This explained why he felt "ALWAYS inferior" to others, a trait his friends had mistakenly attributed to modesty. His own temperament, he now believed, was more like that of a "half grown boy" than a "full grown man" – a fact which might also explain his long-standing interest in child development. After all, Goddard reported, he himself had never become the kind of male who inspired jealousy in other men. "Possibly I would have accomplished more if I had been more aggressive," he proposed.

This also explained why so many persons professed such a fondness for him. "Everybody loves a child – if he is fairly well behaved," he concluded.[130]

Yet while Goddard had begun to reanalyze his early life and work, others were doing the same. And while he could see no underlying explanation for his own scientific accomplishments besides luck, they could. By the late 1940s, Goddard would gain a first glimpse at the way that post–World War II historians would remember and interpret his science. It came in 1948, in a letter from a psychology graduate student, Nicholas Pastore.

"I have prepared the first draft of a doctorate . . . ," Pastore began, "dealing with the following question: 'Is there a relationship between the outlook of scientists on nature–nurture problems and their outlook on social, political, and economic issues? (That is, does the hereditarian assume a conservative orientation and the environmentalist a liberal orientation?)'" To answer this question, Pastore had produced brief sketches summarizing the views of twenty-four scientists prominent in "nature–nurture discussions." For those still alive, among them Henry Herbert Goddard and Lewis Terman, he sent copies for comment.[131]

"Goddard, a student of psychology under Hall at Clark University, was long interested in questions pertaining to the growing child," Pastore's summary began. Citing numerous quotations from *The Kallikak Family, Feeble-Mindedness,* and other books and articles, all but two written no later than 1920, Pastore documented Goddard's early beliefs that intelligence tests measured only "inborn capacity," that feeblemindedness functioned as a "Mendelian recessive," and that it constituted a social "menace" which led to crime, prostitution, pauperism, and disease. Such ideas "led Goddard to emphasize heredity and to minimize the role of environment," Pastore argued. Goddard had also believed in "the unmodifiability of that which has been inherited," Pastore continued, and as a result he had "urged 'rough and ready' methods for dealing with the criminal." "The modern view, now accepted by Goddard," he conceded, "does not place the same stress upon the ineducability of the feeble-minded."

To document Goddard's politics, Pastore quoted this psychologist's post–World War I writings, which stressed the low intelligence of the average man, the link between mental differences and class differences, and the import of the army findings for American democracy. Goddard had a "tendency to introduce other than strictly scientific issues into his work and to express them in a rather popular and dramatic style," Pastore noted. Even so, his works had an "essential practical character," for they were aimed at influencing policies, and were "much quoted by eugenists in support of their views on the importance of heredity." Such writing "suggests that Goddard would oppose plans for most social reforms," he concluded. "In brief," Pastore's summary ended, "Goddard's views on heredity place him in the category of hereditarians. Correspondingly, his attitude toward social and political issues makes him conservative."[132]

Surprised and disturbed to read what he called "Pastore's 'guess' at Goddard's philosophy," Goddard took seriously the task of answering it. Yet it was hard to challenge such a summary, for Pastore had included many quotations from Goddard's most famous publications. In reply, Goddard defended some statements and explained others. The problem, he argued, was that Pastore was interpreting his 1910s writings within a 1940s context, a context in which the nature–nurture debate had assumed new meanings. "In other words," Goddard wrote, "you are discussing a problem of recent origin, by studying writings that were produced before there was any such problem." For instance, "Goddard was NOT led to 'emphasize heredity and deemphasize environment,'" he wrote of himself. "He was studying heredity and had no inclination to deemphasize environment, because in those days environment was not being considered." "Had you been writing in 1910–1920, you would never have written that," Goddard insisted, for the context would have been different. As for Pastore's larger argument, Goddard put his response in a postscript. "P.S. I am not a Hereditarian any more than I am an environmentalist. We are what we are by heredity AND environment," he noted. As to his politics, he added another postscript. "I have always been anything but conservative," he explained. "Now in my old age I am becoming somewhat conservative."[133]

Yet the more Goddard thought about Pastore's thesis, the more worried he became. "I am in trouble!" Goddard wrote Fischer. "The enclosed letter sounds innocent enough the first time you read it," he continued. "BUT! Upon second thought, what is an Hereditarian or an Environmentalist?" he added. "In my innocence (soft for 'stupidity') I thought that scientists had reached the conclusion that the sensible position is a truce. We are indebted to both. . . ," he stated. "Then again: Do we '*assume* an orientation'?" he asked, "and what exactly is 'A Liberal Orientation?'" Goddard had answered the letter, he told Fischer, "and now, childlike I am scared, at what I have done. Please read the document and tell me if it is something that I *ought* to understand," he requested. "NOW DONT THINK THIS IS A JOKE," he added, for it was "SERIOUS!"[134]

Not surprisingly, Fischer was quick to reassure Goddard that such a position could hardly be taken seriously. "To pick out snatches of a man's writing over a period of a professional lifetime, and then to impute to the man certain attitudes and beliefs," he added, "is not only valueless from a research standpoint – it is even harmful. But then you see, Dr. Goddard," Fischer added, "I am a young man. . . . You, having the temperance of experience behind you, probably in your most agitated moment would still encourage this chap."[135]

Fischer's reply alleviated most of Goddard's fears about Pastore's argument. "I think he is probably Jewish," he confided privately to Fischer, " – and not as bright as they come." Such a statement from Goddard was striking; perhaps he now felt that the Jews did blame him for what had happened during the war. "I should like to see his finished thesis," he added. "(I wonder if he will send me one.)"[136]

Goddard wouldn't need to see the thesis, for a year later Pastore's book was published as *The Nature–Nurture Controversy*. In documenting how scientific arguments about nature and nurture had long been mixed with extrascientific concerns, contemporary politics, social biases, and personal prejudices, this study quickly became an early classic in the emerging field known as the sociology of science. Pastore's findings, wrote his advisor, Goodwin Watson, shed light "not only on the heat and bitterness of the nature-nurture controversy, but also on the functioning of the man of knowledge" for they questioned "whether even these very superior men had learned how to free their research from the limitations of their social frame of reference." In the post-war world, Pastore's thesis would also serve a more explicit historiographic function: in linking hereditarians with conservatism and environmentalists with liberalism, it established the larger political framework within which the history of intelligence testing would be discussed ever after.[137]

Yet in the case of this movement's two earliest American promoters, such a framework was problematic. And while Goddard's response was ineffectual in convincing Pastore to modify his text, Lewis Terman's was not. Equally disturbed at his own depiction (and his own quotations), he too had complained to Pastore (and to Watson) about being taken out of context. "How on earth did you ever get the idea that I would defer any effort toward social betterment. . . ?" Terman asked. "I have never said or written anything of the sort."[138]

While Goddard's response had emphasized past science, Terman stressed present politics, for he too had since become an avid New Dealer and a strong supporter of the welfare state. "I still strongly suspect the existence of race differences," he admitted, "but I am now inclined to think that they may be less than I formerly believed them to be." Moreover, he found the Nazi racial pronouncements "beneath contempt," and saw America's dismal failure to ensure the civil rights of minorities as "a national disgrace." In short, Terman saw his own politics as considerably left of center, a fact confirmed by his financial support for the Loyalists in the Spanish Civil War, his vehement opposition to European fascism and American isolationism in the 1930s, and his strong public denunciations of McCarthyism and ongoing efforts to rally the Stanford faculty to fight it.[139]

Lewis Terman, Pastore finally concluded, represented "a contradiction to the hypothesis," for he was a consistent hereditarian who now held "strong liberal views." Had Pastore considered Henry Herbert Goddard's writings or politics past 1920, he would have had to classify him also as a "contradiction to the hypothesis," for he too held "strong liberal views" and had even become far less of a "hereditarian."[140]

However Goddard felt about the way that contemporary psychologists, or future historians, would remember him, by 1949 he was too frail to do much about it. By this time, his writings, while still surprisingly sharp, had become increasingly nostalgic, his memories more defensive, his letters more disjointed. His annual Christmas messages to friends were now filled with miscellaneous bits of poetry, humor, quotations, and observations. "'The 'osophies and 'ologies / Appealing to the mind / Can never take the place of love / And trying to be kind,'" he quoted from Edith Lowell in one such letter. "We want United Nations and Peace," he added; "'One flag, one Land, one heart, One hand, one Nation, evermore!'" "This is a *very* STRANGE Christmas letter (I am afraid *you* will want to use stronger adjectives than 'very strange')," he confessed in his 1951 message.[141]

As one of the most long-lived members of the generation of reformers born in the post–Civil War decade, he also wrote many tributes and memorials to friends who had passed away. "I am now 'The last leaf,'" he explained, "dried and withered." When in 1952 the Feeble-Minded Club gathered at Vineland to celebrate the fiftieth anniversary of the fateful first meeting between Edward Johnstone, Earl Barnes, and Henry Herbert Goddard, they sent Goddard a group picture. The Feeble-Minded Club, Goddard wrote in considering this group's history, had begun less as a "club" than as "a situation; a condition in which, at the start three men without plan or purpose, found themselves. I am the only man living who was 'in at the beginning,'" he now noted, "and hence who knows all the secrets."[142]

Whatever secrets Goddard possessed would soon be lost, however, both to himself and to history. Increasingly suffering from senility (or from what may have been the early stages of Alzheimer's disease), his memory grew noticeably weaker year by year. In one Christmas letter, he asked his friends to start using full names, for he was having trouble remembering even persons "whom I would have said I could never forget." That kind of forgetting "indicates that, as the telegraph linesmen say, 'the line is down,'" he joked.[143]

A physiological psychologist to the last, he described this process with clinical detachment. His "association fibers" were, "if not actually diseased, at least not functioning properly," he explained. A 1952 letter told even more. "*Right now* I am reading Hoover's autobiography," he wrote. "I began it, because the name 'Herbert Hoover' was familiar. . . . But I found myself asking myself 'Who was he? Where did I know him?' Later I found myself saying 'Wasn't he President?'" Too ashamed to ask anyone such a question, he continued reading until the author "alluded to something 'when I was President.'"[144]

By the time he was in his late eighties, Goddard's correspondence had begun to dwindle; it finally stopped altogether. He spent his last years living quietly at home, cared for by his niece. At the time of his death in 1957, he was still famous enough to merit an obituary in the *New York Times.* "Dr. Henry Goddard, Psychologist, Dies; Author of 'The Kallikak Family' Was 90," its headline stated. In its opening sentence, this article cited what contemporaries still saw as this psychologist's most memorable accomplishment. "Dr. Henry Herbert Goddard, an American psychologist credited with introducing the term moron into the English language," it began, "died Wednesday. . . . "[145]

To his friends, the passing of this once great but now sad figure was bittersweet. "Our very much beloved Dr. Goddard died Ju[ne] 18, 1957, at his home in Santa Barbara," Vineland schoolteacher Alice Morrison Nash informed psychologist J. E. Wallace Wallin, a sometimes sharp critic who had remained a lifelong friend. The body had been cremated, and an urn with ashes sent back to Vineland, where Nash arranged for it to be buried alongside his wife under a single small headstone. Located nearby was the grave of Edward Johnstone. "God Made Him So," stated the poem on Johnstone's monument, "and Deeds of Week-day Holiness Fell from Him Gentle as the Snow. . . . " At Johnstone's side, in death as in life, now lay Goddard. Gathering at the gravesite, his Vineland friends held a small ceremony. "It was a simple but lovely Farewell Service to a truly great man," Nash wrote. She was pleased, for it was just the kind of event that "would have pleased Dr. Goddard," she explained, "had he arranged this, his last visit to Vineland."[146]

Epilogue: Psychological Legacies, Historical Lessons, and Luck

"Of what use is a measure of intelligence?" asked its inventor, Alfred Binet, in 1908. Binet himself could think of many answers "in dreaming of a future where the social sphere would be better organized than ours." Within an "ideal city" of the future, such tests could prove useful in many ways, he proposed. Doctors could use them to differentiate the normal from the subnormal, Binet argued, and to gauge different degrees of subnormality with some uniformity. Schoolteachers could use them to decide who belonged in a special class; more broadly, they might even revise the curriculum so that what was taught at school might better match the natural evolution of the child's mind. Courts too could find them helpful in considering the mind of the accused and assessing "penal responsibility." Binet even suggested the "very great utility to humanity that would result" from giving these tests to "young recruits before enlisting them."[1]

Such ideas were only dreams, Binet added wistfully, for he and his co-worker, Theodore Simon, were "far from flattering ourselves that we have inaugurated a reform. Reforms in France do not succeed except through politics," he explained, "and we cannot readily imagine a secretary of state busying himself with a question of this kind. What is taught to children at school!" Binet exclaimed. "As though legislators could become interested in that!"[2]

Alfred Binet's death in 1911 would prevent him from seeing just how interested future legislators would become in such questions, or how profoundly his own psychological legacy would affect the twentieth century. Within the single decade between 1908 and 1918, all of Binet's predictions had come true, for intelligence testing had been used in medical clinics, public schools, courtrooms, and the army – not in France, but in the United States. And all of these uses were connected to the activities of Henry Herbert Goddard.

In using intelligence tests to redefine "feeblemindedness," Goddard helped make Binet's name world-renowned and his scale the most widely disseminated psychological product yet invented. During the same decade, Goddard himself also became world famous as a disseminator of the biological ideas of Gregor Mendel, especially as explained by leaders of the eugenics movement. By fusing together Binet's psychology, Mendel's biology, and his own

348

sociology, he redefined "intelligence" as a biologically inherited, socially unchangeable, and easily measured entity with profound political implications.
In the process, he also helped spark what would become one of the most
explosive scientific controversies of the century.

If Binet died too soon to experience fully either the fame or the controversy
generated by his invention, Goddard lived long enough to see his own popularity gradually transformed into notoriety, and his science greeted with ridicule instead of respect. In the decades following World War I, numerous
American testers found even more uses for Binet's invention, especially in
offering scientific confirmation for nativist, racist, and antisocialist political
programs. As eugenic science became more suspect, so too did Goddard's
writings. By the 1920s, the concept of the "moron" had begun to lose its
scientific legitimacy, although its popular use increased. By 1936, this word
had been included in H. L. Mencken's study *The American Language*. By the
1940s, "little moron" jokes had become a familiar part of American folk
humor.[3]

If contemporary scientific, social, and political events were beginning to
change the reception granted to Goddard's earlier writings, they also influenced his own thinking. By the 1920s, Goddard candidly admitted that he
had made many errors in his earlier scientific studies. His later writings
placed far greater emphasis upon school improvements, environmental reform, and better child-rearing techniques. Yet despite these changes, Goddard never fully understood why the generation that experienced World War
II found his early science so dangerous, or why his Kallikak study became
so controversial.

Goddard's death in 1957 did not end the controversies which swirled
around him during the last decades of his life, for the Kallikaks outlived him.
As late as 1961, their story was still being taught as science in a textbook
written by Columbia University psychologist Henry Garrett. A native Virgin-
ian and the only academic psychologist to testify against school desegrega-
tion in the cases which reached the Supreme Court as *Brown v. Board of
Education* in 1954, Garrett feared that integration would lead to social mixture which would in turn "breed down" the white race. Goddard's own writings had never espoused white supremacy; even so, the usefulness of his science in bolstering racist arguments of this type continued to cast a pall over
his reputation in the decades after his death.[4]

Such uses of the Kallikak study by scientists were becoming increasingly
rare, however, for in the post-World War II world, it became much harder to
take such stories seriously. Like the "moron," the Kallikaks too would leave
science and enter the world of fiction and folklore, blending easily into other
stereotypical portrayals of troublesome poor white families. By the century's
end, the term "Kallikaks" had become a part of American language and

legend, a word with a vaguely disreputable history whose very sound suggested something funny.

Its new meaning was illustrated in the summer of 1977, when a pilot episode for a new sitcom appeared on American television. Entitled simply *The Kallikaks,* "as in the rollicking, fun loving Jukes and Kallikak clans," one newspaper explained, it featured the comic antics of a colorful family of poor whites, its characters a kind of mixture between the lazy schemers of *Tobacco Road* and the appealing innocents of *The Beverly Hillbillies.* "I think the show is going to be a scream," one of its stars predicted. (She was wrong, for it only lasted five episodes.) This family's name inspired not only low comedy but more sophisticated satire as well, for in 1987 *The New Yorker* featured a large cartoon entitled "The Jukes and Kallikaks Today." Its four panels portrayed "Ed Jukes," who had by now become a prominent real estate developer, "Amanda Kallikak," an attorney for Pastoral Nuclear Waste Manufacturers, Inc., "Lance Kallikak," a hot young graffiti artist, and "Judy Ann Jukes," the newest queen of workout videos.[5]

If the very words "moron" and "Kallikaks" had largely become jokes, the same was not true of Goddard himself. By the 1970s, his writings as well as those of other early testers were being rediscovered by a generation newly sensitized to the social damage caused by racism and nativism, and to the potential dangers of the scientific combination now called "sociobiology." In succeeding decades, Henry Herbert Goddard would again earn a small place in American social history, usually as a eugenicist.[6]

His reappearance was linked to the reemergence of the heredity-environment debate in its most polarized and bitter form yet. In 1969 it was reignited by an article in the *Harvard Educational Review* entitled "How Much Can We Boost I.Q. and Scholastic Achievement?" Its author, psychologist Arthur Jensen, used the past research of intelligence testers to challenge contemporary social programs which provided compensatory education. Jensen's article provoked a storm of controversy and more than 120 replies. Once again, the meaning and measurement of IQ had become a part of American political discourse, this time in arguments over antipoverty programs, racial equality, and affirmative action. By the century's end, such controversies would become familiar, for at least once a decade, the IQ debate would explode anew. This process was demonstrated most recently by the massive media attention and fierce academic criticism that greeted Richard Herrnstein and Charles Murray's 1994 study, *The Bell Curve,* a book that once again drew sweeping political conclusions about the meaning of class and race based on IQ scores.[7]

The strong parallels between political arguments produced by some hereditarian theorists in the latter half of the twentieth century and those used in its opening decades were hardly lost on modern critics. Among the works to

"The Jukes and Kallikaks Today." From *The New Yorker* 63 (June 8, 1987): 25. Drawing by R. Chast; © 1987, The New Yorker Magazine, Inc.

emphasize this connection most explicitly was psychologist Leon Kamin's 1974 study, *The Science and Politics of I.Q.* In analyzing Jensen's arguments, Kamin also examined the racist, nativist, and class-biased statements produced by Lewis Terman, Robert Yerkes, and Henry Herbert Goddard. Equally questionable as science, he argued, were the highly suspicious statistics generated by England's most famous early IQ tester, Sir Cyril Burt. Kamin's suspicions soon triggered a broader inquiry by British journalists as well as psychologists, which led to an even more disturbing finding: investigators could not prove that Burt's later studies had ever been conducted. "Pioneer of IQ Faked his Research Findings," announced a sensational front-page headline of the London *Sunday Times* in October 1976, in what quickly became one of the most highly publicized and hotly contested scientific fraud controversies of this century.[8]

The claim that hereditarian thinkers of the past had injected their arguments not only with personal biases but also with fraudulent data, presumably to enhance their political arguments, added a new dimension to the already virulent IQ debate, and a new means of explaining its history. In 1981, Goddard too was accused of fraud. The charge appeared in Stephen Jay Gould's popular study of past social science biases, *The Mismeasure of Man*. It grew from the observation that several photographs in the Kallikak study had been retouched, apparently with ink; over the years, the photos themselves faded, while their ink markings became more visible (see photographs on pp. 177–178). "It is now clear," Gould argued, "that all the photos of noninstitutionalized kakos were phonied by inserting heavy dark lines to give eyes and mouths their diabolical appearance." His theory was seconded by James H. Wallace, Jr., director of Photographic Services for the Smithsonian Institution. "The harshness clearly gives the appearance of dark, staring features, sometimes evilness, and sometimes mental retardation," Wallace concluded. "It would be difficult to understand why any of this retouching was done were it not to give the viewer a false impression of the characteristics of those depicted." Henry Herbert Goddard, Gould concluded, was guilty not only of producing a biased study but of "conscious skullduggery" as well – a charge which was soon repeated in other historical texts.[9]

Despite all the flaws in Goddard's Kallikak study, this charge of conscious photographic fraud is not convincing. After all, Goddard had never wanted to prove that the feebleminded looked "evil" to the average reader. To the contrary, he had always insisted upon the opposite: that girls such as Deborah Kallikak (and presumably her biological cousins as well) were usually quite ordinary-looking to the untrained observer, and often surprisingly attractive, and therefore all the more dangerous to society. What Goddard called the "discovery of the moron" was a discovery precisely for this reason. Even in photographing "criminal imbeciles," he had portrayed his subjects respectfully (see photographs on pp. 213 and 216). In supplying photographs for public exhibitions, he had been equally careful to avoid suggesting that the feebleminded could be recognized by facial features alone.

For instance, in preparing materials for a eugenics display at the New York Child Welfare Exhibit of 1910, Goddard had recommended that Charles Davenport not use the face of an "*evidently imbecile* child" because it would "defeat the very thing that we are struggling very hard to overcome in the popular mind, and that is the idea that the defective children show their defect in their faces." Instead, he suggested showing "a *fine looking, normal appearing* boy or girl and lay the emphasis on the fact that they are really feebleminded." In light of such attitudes and actions, it is extremely improbable that Goddard doctored his own Kallikak photographs two years later to make feebleminded persons look "diabolical."[10]

In 1987, psychologist Raymond Fancher challenged Gould's theory by offering a much better explanation for the ink markings in this book. Photographs in early-twentieth-century publications were routinely doctored in precisely this crude manner, he discovered, often by editors in ways unknown to authors. This was done because facial features in some photographs seemed to wash out when reproduced. Such an explanation seems likely in this case, for the ink markings are visible on the outdoor snapshots taken by field workers, and not on the better-quality institutional photographs of Deborah Kallikak taken by Goddard himself.[11]

There was a particular irony in the sudden rise and equally quick fall of this historical claim of Kallikak fraud. After all, one hardly needed to prove fraud to discredit the Kallikak study. To the contrary, it was Goddard's very scientific candor which had long provided the most damaging evidence of all. With complete faith in his methods as well as his findings, Goddard had presented all his evidence to the public – including patient records, verbatim field reports, and frank admissions about what he did and did not know about this family. It was this very frankness which would make Goddard's own biases so transparent to later generations, and which would, thirty years later, lead him to lie in attempting to defend his study – by claiming that he had really known the name of the person his text identified as a "nameless feebleminded girl."[12]

Even without the charge of fraud, Goddard's writings in general, and his Kallikak study in particular, have again returned to histories of science – but hardly in the ways that he himself prophesied nearly eighty years ago. By the second half of the century, they were largely being used to demonstrate the ways in which science can go very wrong. In fact, few works can match *The Kallikak Family* in illustrating the potential for injecting personal biases into purportedly objective studies, or the naiveté of offering simple scientific explanations for complex social problems. Goddard had written the Kallikak study as a didactic story, a parable warning his contemporaries about the social dangers resulting from "bad blood"; this book's own history has now become an even more potent cautionary tale, an object lesson warning today's students about the far greater dangers emanating from "bad science."[13]

These historical lessons serve an important and necessary social function. They are extremely useful in reminding modern audiences that social science has never been a value-free enterprise, and in fostering a healthy skepticism toward the claims of experts. Moreover, because these writings make the underlying connections between science and politics so visible, they are particularly relevant to contemporary controversies.

Yet at the same time, this very relevance has proven an impediment to a fuller understanding of psychology's past. Especially considering the intense media coverage focused on the heredity–environment debate, it has become increasingly difficult to see the history of IQ testing in anything but the light of current controversies. As a result, this history is most often framed within simplified, polarized, and largely static dichotomies – a story of hereditarians and environmentalists, conservatives and liberals, good science and bad science.

Such frameworks cannot fully capture the rich texture and multidimensional nature of the earliest testing controversies. In considering Henry Herbert Goddard's ideas about intelligence, the label "hereditarian" is in one sense perfectly accurate and in another highly inadequate; calling Goddard a "conservative" is more problematic still. Neither label really does justice to the scientific and political complexities of the pre–World War I decade, or to the ways in which Goddard's views changed in the decades that followed.

Even more important, most likely to be missed by such dichotomized accounts are the ways in which "good science" and "bad science" were deeply and thoroughly intertwined in the work of an earlier era. Using all the tools known to them, and working within the social constraints of their day, Goddard and his contemporaries struggled to make scientific sense out of the diverse phenomena presented to them by doctors, teachers, parents, and children. The science they produced was filled with instances of impressive creativity and acts of egregious oversight, close casework and sweeping theorizing, sharp insight and stunning blindness. Only when all of these instances are pieced together can we recapture the complex history of psychological science. Only then can we see the creation of such a science as a fluid social process, a process making progress in fits and starts, a process whose trajectory was as often circular as linear.

This circularity in the history of psychology will be evident, for instance, if we consider past efforts to conceptualize just one issue which captured the attention of these scientists: the connections between low intelligence and alcoholism. Nineteenth-century physicians believed they saw such a connection, and often blamed feeblemindedness on drinking – by the father as well the mother. Goddard too saw a connection, but he concluded that in most cases, feeblemindedness was cause and alcoholism its consequence. By the mid-twentieth century, most physicians dismissed this connection as simply another legacy of the narrow-minded moralism of the past. Only in recent decades has it been rediscovered once again in the new literature on "fetal alcohol syndrome," a syndrome whose victims are frequently described as especially prone to fail in school or wind up in jail – language that bears an eerie similarity to some of Goddard's own case descriptions.[14]

Equally important in recapturing this history is the need to frame the social issues of the day in less dichotomized ways. Here too Goddard's actions are best understood not as simply good or bad, but as complex responses to new and difficult questions. For instance, today's educators have overwhelmingly rejected most of Goddard's ideas about segregated special education in favor of "mainstreaming," a concept that respects the rights of all students to be educated within the "least restrictive environment." Even so, many questions which Goddard and his contemporaries first raised are still being discussed today, for the special class itself has not disappeared. Still meriting debate are questions concerning who belongs in such a class, what criteria should be used to make such decisions, just how similar or how distinctive the mission and curriculum of such classes should be, and how to balance the benefits gained by specialized attention against the losses associated with labeling and stigmatization. Even if most of Goddard's answers are no longer accepted, the questions he helped to shape live on – and still defy easy solution.[15]

If the educational issues suggested by Goddard's science are still complicated and contentious, so too are the legal issues. In 1989, a question which Goddard had hoped to see before the Supreme Court finally reached it in the case of *Penry v. Lynaugh, Director, Texas Department of Corrections.* At issue was the pending execution of Johnny Paul Penry, convicted of a brutal rape and murder in 1979. Before the trial, a clinical psychologist diagnosed this twenty-two-year-old as having an IQ of 54 and the "mental age" of a "6 1/2 year old kid." In a 5 to 4 decision, the Supreme Court ruled that the execution of criminals diagnosed as mentally retarded did not in itself constitute "cruel and unusual punishment"; however, it also decided that juries must be allowed to consider such evidence in determining sentencing, and therefore granted Penry a new trial. In writing their opinions, several justices noted some of the complex legal problems associated with the concept of mental age. Problematic or not, the reason for its use in such trials is clear: "mental age," in 1989 as in 1914, was still an idea easily grasped by any jury of laymen.[16]

The histories of these issues affecting American medical, educational, and legal policies, all linked to the history of intelligence testing, have largely been overshadowed by the bigger battle pitting hereditarians against environmentalists. Ironically, this battle has also obscured some of the intelligence testing movement's broader legacies. Broadest of all is the way Binet's innovation dramatically transformed the scientific study of the child mind. Perhaps the best way to assess the magnitude of this transformation is to compare the science of "child study" at the end of the nineteenth century with its position at the end of the twentieth.

In the 1890s, the very idea of studying children scientifically was still a novel, unstructured, and highly controversial enterprise. By the 1990s, such science had evolved into a massive academic industry in its own right, and although its scientific findings were still often controversial, its social legitimacy was by now beyond question. The idea that the mind of the child could, and should, fall within the province of scientifically trained psychologists had become a familiar component of modern American life, as intrinsic a part of the broader twentieth-century "psychological society" as the Freudian concept of therapy. In explaining how this idea gained such widespread modern acceptance, historians have underestimated both the influence and the complexity of G. Stanley Hall's model of child study, Alfred Binet's innovation in mental measurement, and Henry Herbert Goddard's success in fusing the two and disseminating them.

At Clark University, Hall took American psychology in an applied direction by encouraging his graduate students to work collaboratively with both Adolf Meyer of the Worcester State Hospital for the Insane and E. H. Russell of the Worcester Normal School. His efforts helped to establish a "new psychology" bounded by psychiatry on one side and pedagogy on the other. At the same time, through his popular Child-Study societies, Hall created an institutional form which introduced, legitimated, and promoted the Darwinian study of the child mind to teachers and parents. This form was soon copied across the nation and around the world.

Hall's model of institutional collaboration found its strongest intellectual expression in the work of Child-Study enthusiast Alfred Binet. Children's minds, Binet proposed, were related to and yet distinguishable from both their bodies and their schooling. Seeking to measure them, he blended medical, pedagogical, and psychological techniques to create a new concept of age scaling and a new form of mental testing. This type of testing quickly proved useful to both doctors and teachers, while still establishing psychology as a distinct scientific entity.

While Binet devised a new type of mental testing, it was Goddard who first figured out how to introduce it to society. Working outside the university, he showed his contemporaries how to use psychological tests to address practical questions being asked in clinics, classrooms, and courtrooms. With his acute sensitivity toward how such institutions worked, his intuitive understanding of popular audiences, and his personal missionary zeal, Goddard succeeded in making psychological testing an accepted part of American life.

This success, however, was in large part shaped by Goddard's willingness to simplify Binet's ideas, to overlook the most complex and difficult issues raised by his own science, and to ignore the many astute objections and warnings of his critics. Both feeblemindedness and intelligence, he insisted, were relatively simple concepts which could be readily understood by the public as well as the profession. It was this very simplicity that would make Goddard's science so immensely popular, and so easy to disseminate; it was this same simplicity, however, that would also make his science so readily adaptable to the broader theories of nativists and racists, both during his own lifetime and in the decades following his death. Ever after, intelligence testing would be associated with simplistic genetic theories purporting to prove the biological inferiority of entire races, ethnic groups or social classes.

If one of Goddard's most enduring legacies was a dangerously oversimplified popular conception of the meaning of intelligence, another was the social role he carved out for his contemporaries. In the prewar decade, Goddard gradually expanded the parameters of his profession by helping to institutionalize the new quasi-medical, quasi-pedagogical diagnostic entity which Lightner Witmer had already christened "clinical psychology." It was intelligence testing that first made the scientific study of the child seem viable and visible and important to society, and that first established a new social role for psychologists as diagnosticians of the mind. It was intelligence testing that made a new "clinical psychology" seem a genuine possibility with something different and valuable to offer the public.

Goddard's immense influence in helping to establish a new social niche for his professional peers is evident in the work of American psychologists whose early careers reflected the issues he raised and the occupational doors he helped open: Leta Hollingworth at Bellevue Hospital; J. E. Wallace Wallin at the St. Louis Board of Education's Psycho-Educational Clinic; Samuel Kohs at the Chicago Municipal Court; Arnold Gesell at Yale's Clinic of Child Development. In the years that followed, clinical psychologists would move far beyond intelligence testing in offering the public the benefits of their expertise. Even so, their initial acceptance within American hospitals, schoolrooms, courtrooms, and clinics owes much to the early testing movement and its leader, Henry Herbert Goddard.

These broad and diverse legacies to psychology raise intriguing historical questions – questions not only about what happened, but also about what might have happened had circumstances been different. What might have happened, for instance, had Binet lived just one year longer – just long enough to read and review Goddard's Kallikak study? Or what might have occurred had America not entered World War I in 1917, thus suddenly unifying a deeply divided testing community and abruptly ending much of its internal questioning about the roles played by learning, language, and social class? Most provocative of all, what might intelligence testing, and thus American psychology, be like had Binet's earliest American proponent been someone other than Henry Herbert Goddard?

Such a question suggests the deeper relationship between Goddard's science and Goddard himself. It is hardly new, for historians have long seen a connection between the lives of testers and the types of tests they produced and disseminated, a fact emphasized in studies stressing the political motives of early IQ testers. Yet these discussions of motivation have usually been framed within contexts which are both too narrow and too dichotomized.

Implicit in much recent writing, for instance, is a potentially dangerous assumption: that bad science is usually the product of bad motives or, more broadly, of bad character. Since the Kallikak study is so bad, such reasoning goes, then its author must have been at best a bumbling fool, or at worst a Nazi sympathizer.[17]

In making such assumptions, modern historical assessments are sharply at odds with the testimony of contemporaries. In the years before 1918, Goddard's ideas often engendered strong critical reactions, but not even his most fervent adversaries questioned his character, or his motives. To the contrary, in describing Goddard himself, the accounts of nearly all who knew him – colleagues, co-workers, schoolteachers, students, parents, and the children he drew to him – were deeply respectful and overwhelmingly positive.

Henry Herbert Goddard, stated psychiatrist William Healy, a frequent critic, was "the dearest man on earth." Goddard's Ohio State dean, George Arps, found him "afflicted, or shall I say possessed, with an enormous amount of human tenderness" and was "suspicious that Henry Herbert never lost his original divinity." Even his clerical assistants often came to love him. "I have been working at one place or another ever since I have been in the eighth grade – factory, cafeteria, libraries, offices, schools," one woman wrote, " – and never have I found any director or employer quite so consistently considerate as yourself."[18]

Other letters also cast doubt on narrow interpretations of Goddard's motives. "I wonder whether you will even remember me," stated one such letter from a Dayton attorney in 1947, "but in the year 1920 I was representing a defendant. . . . " It was this lawyer's first murder case, and his client, a soldier, had killed a pawnbroker. "My defense was that Jones was of 9 year mentality and I asked you to come to Dayton, which you did. . . . " Jones was found guilty, but due to Goddard's testimony, the jury recommended mercy; the governor later commuted his sentence, and Jones was sent home. After twenty-seven years, this lawyer still remembered the trial vividly, and wanted Goddard to know "how grateful I have always been for your study of this case at that time."[19]

Most loyal of all throughout Goddard's lifetime were his students. One thanked him for giving her "an ideal toward which I can ever strive, in my own teaching. And the greatest of all the lessons you have given is that of kindness." Others credited him with personally inspiring them in their life's work. "The greatest tribute that has ever been paid to the Woods Schools was your visit," wrote Mollie Woods, a Vineland summer school graduate who founded her own special schools. Were it not for Goddard, wrote Martin Reymert, director of the Moose Foundation's Laboratory for Child Research, "I would not have had the intense scholarly pleasure of establishing this Laboratory which I have now directed for sixteen years. If I had not been placed here, I might even be dead," he told Goddard, "since I had firmly in mind to return to Norway and with my temperament, I am sure the gestapo would not have let me live very long." Jewish psychologist Samuel Kohs regretted that his own psychological career had been cut short by "narrow prejudices"; even so, he regarded Goddard as "a sort of godfather" who had helped him more than anyone else. Goddard was "a real person, of most lovable character," Kohs wrote, and "one of the finest and warmest personalities I know." Even his most famous pupil of all, "Deborah Kallikak," loved Goddard and named her favorite cat "Henry" after "a dear, wonderful friend who wrote a book."[20]

Explaining Goddard's character and motives in the midst of a changing world proved an especially difficult challenge for several of his students. Among those who tried was his former assistant, Edgar Doll, who later became director of psychological research at the Vineland school. Asked to write Goddard's obituary for *Science* in 1957, Doll vacillated between a tribute and a defense. Dr. Goddard had come to Vineland because he "felt called to a lifework which was to bring hope to a world where humility and a sense of belonging were to guide his fruitful efforts," he reported. By establishing a psychological laboratory devoted solely to studying the feebleminded, he had made many important discoveries, including his findings on the Kallikak family. "This is not the place to moralize or to defend," Doll conceded, but he still felt it "grossly improper to laugh this epochal investigation out of court as unjustly and as disparagingly as later 'students' speciously did." All he could do against this "tide of revolt or even revulsion" was to "let the disparagement run its ungenerous course." Despite these attacks, Doll still admired Goddard enormously. "This writer is grateful to Goddard as person, teacher, and many varieties of hero," Doll confessed. "In a real sense," he concluded defensively, Goddard "belongs with and to the moron and the Kallikaks – their best friend and most benevolent proponent."[21]

Equally determined to explain her former teacher to a new generation was Marie Skodak Crissey, a psychologist who did her undergraduate work with Goddard at Ohio State and finished her graduate work at the University of Iowa's Child Welfare Research Station, a project famous for its efforts to raise children's IQs through environmental improvement. Notwithstanding his reputation, Goddard had never been a "rigid hereditarian" of the modern sort represented by Jensen, Crissey insisted in a 1979 oral history. What he really was, she added, was much harder to capture "because there aren't many people now in the world like him; he was an old-time gentleman." Explaining Goddard's social vision was equally difficult. "Have you been to New England?" she asked her student interviewer. "Were you interested in history?" A man like Goddard could only be explained within an older New England tradition of "noblesse oblige," she argued, with its strong sense of social responsibility toward "the inadequate, the ineffective, and the needy."[22]

Yet it is the very nature of such "benevolence," such "noblesse oblige," that has been most seriously and deeply challenged in the post–World War II era. In the second half of the twentieth century, American society would be reshaped by a variety of new struggles aimed at redressing long-standing power inequities. These would include attacks on the very concepts of imperialism, racism, class differences, and sexual discrimination – concepts often justified in the past as forms of kindly caretaking. Such struggles would lead to increasingly critical scrutiny of all forms of paternalism (and even of its Progressive era analogue, "maternalism"). In the wake of such scrutiny, the very phrase "benevolent paternalism" has almost become an oxymoron. Also transformed has been the connotation of the term used to describe the very model of benevolence that Goddard embraced most wholeheartedly, "missionary."[23]

These transformations have influenced not only contemporary life, but perspectives on the past as well. "Put succinctly," historian David Rothman summarized, by the 1970s "a claim once considered to be of the most virtuous sort, the claim to be acting benevolently, had now become – to understate the point – suspect; if the last refuge of the scoundrel was once patriotism," he stated, "it now appeared to be the activity of 'doing good' for others, acting in the best interest of someone else." To many modern critics, "benevolence" had "devolved into mischievousness," Rothman concluded, or more specifically, "the exercise of power in disguise."[24]

This new emphasis on power relationships has been most influential in revising the historical literature dealing with institutions. Past institutional personnel, many modern accounts suggest, were actually less concerned with "social benevolence" than with "social control" – the need to control the behavior of the unruly poor, troublesome minorities, or sexually promiscuous women. In Goddard's case, there is certainly plenty of evidence to support such a theory, for much of his rhetoric suggests a desire to control the behavior of paupers, prostitutes, or others who failed to conform to contemporary standards of middle-class morality.[25]

Yet while adding a much-needed corrective to older portrayals of institutional personnel as self-abnegating saints, this new emphasis has unfortunately led to yet another set of polarized opposites. Thus, in many modern accounts, reformers must be either practitioners of social benevolence or promoters of social control, challengers of the status quo or its defenders, saints or scoundrels.[26]

Here, too, a more fruitful historical framework would be one which asks different questions and which allows for more complex answers. The early history of the testing movement is clearly a story about renegotiating power relationships. This included the powers of individuals, the powers of social classes, and the powers of professions. Superseding and influencing all three, however, was a new conceptualization of the powers of the state. Like the eugenics movement, the early intelligence testing movement took shape within the context of broader international efforts to reconceptualize state policies toward the health, education, and welfare of citizens. These efforts took different forms in different countries. In the United States, they led to the state's growing powers to mandate compulsory education for all children, to enforce public health measures, and to act as a parent in court cases involving juveniles. Within this broader framework, it is easy to see why some of Goddard's ideas found receptive audiences in Progressive America in the 1910s, Leninist Russia in the 1920s, and Nazi Germany in the 1930s. This diverse political acceptance raises deeper and more difficult questions concerning the role of psychological expertise within the modern welfare state – questions about the interrelationships between state benevolence and malevolence, about social responsibility as well as social control, about the state's obligations as well as its powers.[27]

Perhaps most complex of all, and most important in assessing Goddard's legacy, is the impact of his form of benevolence upon the population he was ostensibly trying to serve. In expanding his profession's potential, visibility, and power, Goddard also brought the problems of the "feebleminded child" to America's attention in an entirely new way. He did so largely by emphasizing the social dangers posed by those he labeled "morons." In the process, he confirmed many of the public's darkest fears about persons who seemed different. And while Goddard did not create these fears, his science surely exacerbated them. In the process, he left the mentally retarded population with a legacy of suspicion, social isolation, and ostracism that would take decades to begin to undo. Only in the context of the modern civil rights movement would these Americans find a means of fighting back to establish their own rights to self-determination, social integration, and personal dignity. It is this legacy that makes Goddard's writings seem most antithetical to the dramatic social developments of the past few decades promoting deinstitutionalization, mainstreaming, and social inclusion. In this sense, the Vineland Training School of today would barely be recognizable to Goddard, for by 1997 this institution had placed all of its residents in small group homes located throughout southern New Jersey.[28]

In other ways, however, Goddard anticipated future trends, for he fought hard to transform the field of "psycho-asthenics" from a moribund and marginalized caretaking enterprise into a modern research-oriented science. Goddard remained true to his promise of vastly increased educational, medical, biological, sociological, and psychological research, even within his own small institution. He insisted that special pedagogy develop its own methods and receive state support in public schools; he challenged physicians and geneticists to pay more attention to finding a scientific explanation for birth defects, and to do something about them. And while Goddard's own research might have offered little help, the work of those who followed him dramatically altered the science of his day. Ironically, it was this new research that gradually overturned nearly all of Goddard's own findings, made his own methodological flaws so visible, and finally destroyed his own scientific reputation.

Perhaps Henry Herbert Goddard's legacies to psychology, to American society, and to the population he was attempting to serve can only be expressed within a history framed not in dichotomies, but in ironies. In America, intelligence testing was embedded within a broader pedagogical reform movement which inspired an entire generation of special educators while predicting that their efforts would have only limited success; which markedly expanded medical research while simultaneously suggesting the futility of such labors in the face of genetic determinants; which forced the state to assume more responsibility for caring for the physically and mentally handicapped, while also overseeing their segregation and exclusion from the rights of citizenship; which spoke frequently and ominously of the "menace" of the feebleminded while still repeatedly warning, "Inasmuch as ye have done it unto one of the least of these. . . . "

This sense of historical irony is only deepened if we compare Goddard's fate to the strange and diverse fates of persons whose lives intersected, at least for a time, with his, as well as with some of the larger forces shaping the twentieth-century world. For instance, in modern historical accounts, Henry Herbert Goddard (1866–1957) is most frequently associated with Charles Davenport (1866–1944), the leader of the American eugenics movement, and his co-worker, Harry Laughlin (1880–1943), an avid promoter of immigration restriction and sterilization legislation who gratefully accepted an honorary degree from Heidelberg University in Nazi Germany in 1936. A different picture emerges, however, if Goddard is considered alongside his fellow Vinelander, Alexander Johnson (1847–1941), field secretary of the "Committee on Provision for the Feeble-Minded." Although this organization too is now largely remembered as a disseminator of eugenic propaganda, Johnson's autobiography tells a more complicated story. Appropriately entitled *Adventures in Social Welfare,* its chapters include "Adventures with the Red Cross," "Adventures in Nutrition," and "Five Years Adventuring in Propaganda for the Feeble-Minded" – "adventures" which suggest a personal involvement in both eugenic and "euthenic" campaigns.[29]

This picture becomes even more complex if we add in the later history of Kallikak field worker Elizabeth Kite (1864–1954). In 1932, Kite became archivist of the American Catholic Historical Association; a year later, France honored her for her historical studies of French contributions to the American Revolution. Kite also continued traveling the world, gaining audiences with two popes and paying an admiring visit to Mahatma Gandhi's school in India. And finally, there is Goddard's boyhood schoolmate and early teaching colleague, Rufus Matthew Jones (1863–1948), who later became an internationally respected Quaker mystical philosopher, religious historian, and author of more than fifty books, among them *Social Law in the Spiritual World* (1904) and *Rethinking Missions* (1932). In 1917, Jones became chairman of the newly formed American Friends Service Committee, an organization which trained conscientious objectors to work in Europe during both world wars, and which was awarded the Nobel Peace Prize in 1947. In 1938, he visited Nazi Germany in a futile attempt to win the release of Germany's Jews.[30]

Perhaps all that these diverse yet intersecting lives prove is that Goddard was right in the end – that many a lifetime is really about luck. In his own case, he insisted, this was certainly true. After all, it was luck that led Goddard to see a book written by Rufus Jones and to write to him, thus receiving a job offer in Maine; luck that brought him to hear a Maine lecture by G. Stanley Hall, and to decide to study psychology at Clark University; luck that drew him to a New Jersey Child-Study meeting featuring Earl Barnes and Edward Johnstone, and to help start the "Feeble-Minded Club"; luck that led him to accept a new career at Vineland; luck that brought him to Europe, where his Child-Study contacts showed him Binet's studies; luck that made him world famous ever after. "I never wooed lady Luck," he concluded, "but she has done more for me than all my conscious prayers have done."[31]

Yet even Goddard could not fully accept such an explanation. Too much of a believer in both psychological science and an ultimate divine order to explain human behavior in such random terms, he struggled with the implications of the title he had chosen for his autobiography, "As Luck Would Have It." Perhaps luck, he finally suggested, was really "another name for our *unconscious* prayers."[32]

Even with all its lucky twists and turns, its scientific and social reversals, and its political inconsistencies, Goddard's long lifetime still shows evidence of an underlying pattern reflecting direction and purpose, whether conscious or not. Most striking and most consistent in his writings is its distinctive mixture of positivist science and Christian mission, a mixture also evident in the writings of other members of this transitional generation of American social scientists. Like Goddard, many of these writers were the children of ministers and missionaries; they too grew up in a world in which intellectual authority was slowly but surely shifting from faith to science. For such persons, knowledge could no longer rest on dogma; it now required scientific proof. Although eager to embrace the liberating intellectual freedom promised by modern science, Goddard tried to do so without sacrificing all of the values he had acquired from his Quaker upbringing, many of which he still cherished. As a result, much of the social science he produced suggested a deeper process of reconciliation, a process that posed difficult challenges of its own: how to reground social morality in a new foundation of scientific certainty; how to harness human passion and spirit to promote self-sacrifice and altruism within a Darwinian universe; how to organize a bolder, broader, and more effective community of "concern"; how to let a little child lead them.

Henry Herbert Goddard's psychological messages, like those of his mentor, G. Stanley Hall, still bore the marks of their simultaneous rejection of and indebtedness to the theological ideas of the preceding century – ideas still powerful and meaningful not only to Goddard, but also to the audiences who heard him. Behind his continual quest for science's "facts and laws," Goddard's language suggests a quest of another kind, a quest that connects him both to his discipline's intellectual past and to his own family's missionary heritage.

This connection is best illustrated by the words chosen by Joseph Byers, head of the Committee on Provision for the Feeble-Minded, to describe the work done at Vineland. "The first recorded question asked by man of his Maker," Byers remarked in a passage describing Goddard's laboratory, "was 'am I my brother's keeper?'" The answer, Byers affirmed, "was a plain and unequivocal 'Yes.'" Perhaps the deepest ironies of Goddard's life, and of the early American intelligence testing movement which he inspired and led, are encapsulated in the complex and multiple meanings – positive and negative, intellectual and institutional, social and moral – attached to the word "keeper." In convincing his fellow scientists to undertake "psychological work among the feebleminded," Henry Herbert Goddard also undertook a broader mission to discover just what kind of Christian "keeper" the secular new psychologist might become. It was this quest, with all its potential and all its dangers, that most profoundly influenced twentieth-century American life, and that continues today.[33]

Abbreviations Used in the Notes

AHAP	Archives of the History of American Psychology
AMO	*Proceedings of the Association of Medical Officers of American Institutions for Idiotic and Feeble-Minded Persons*
AP	*American Psychologist*
AR	*Annual Report,* Vineland Training School
CB	Records of the Children's Bureau
CBD	Charles Benedict Davenport
ERJ	Edward Ransom Johnstone
FM	*Feeble-Mindedness: Its Causes and Consequences*
GP	Henry Herbert Goddard Papers
HHG	Henry Herbert Goddard
JAMA	*Journal of the American Medical Association*
JEP	*Journal of Educational Psychology*
JHBS	*Journal of the History of the Behavioral Sciences*
JPA	*Journal of Psycho-Asthenics*
KF	*The Kallikak Family: A Study in the Heredity of Feeble-Mindedness*
NCCC	*Proceedings of the National Conference of Charities and Correction*
NEA	*Proceedings of the National Educational Association*
NYT	*New York Times*
PB	*Psychological Bulletin*
PC	*Psychological Clinic*
PS	*Pedagogical Seminary*
PSJ	*Pennsylvania School Journal*
TSB	*Training School Bulletin*

Notes

Introduction

1 HHG to George Crane, June 6, 1947, File M36, GP.

2 Robert Fischer to HHG, October 17, 1950, File M33, GP.

3 HHG to Robert Fischer, October 23, 1950, File M33, GP.

4 James McKeen Cattell, as cited in Franz Samelson, "Putting Psychology on the Map: Ideology and Intelligence Testing," in Allan R. Buss, ed., *Psychology in Its Social Context* (New York: Irvington, 1979), p. 106.

5 Lee Cronbach, as cited in Orville G. Brim, Jr., Richard S. Crutchfield, and Wayne H. Holtzman, *Intelligence: Perspectives 1965. The Terman–Otis Memorial Lectures* (New York: Harcourt Brace & World, 1966), p. vii. The literature on the history of the IQ debate is enormous. For diverse overviews, see Leon Kamin, *The Science and Politics of I.Q.* (Potomac, Md.: Erlbaum, 1974); Cronbach, "Five Decades of Public Controversy over Mental Testing," *AP* 30 (January 1975): 1–14; Paul Houts, ed., *The Myth of Measurability* (New York: Hart, 1977); Jeffrey M. Blum, *Pseudoscience and Mental Ability: The Origins and Fallacies of the IQ Controversy* (New York: Monthly Review Press, 1978); Paula Fass, "The IQ: A Cultural and Historical Framework," *American Journal of Education* 88 (August 1980): 431–458; Hamilton Cravens, "The Wandering IQ: American Culture and Mental Testing," *Human Development* 28 (1985): 113–130; Raymond Fancher, *The Intelligence Men: Makers of the IQ Controversy* (New York: Norton, 1985); Michael Sokal, ed., *Psychological Testing and American Society, 1890–1930* (New Brunswick, N.J.: Rutgers University Press, 1987); Mark Snyderman and Stanley Rothman, *The IQ Controversy, the Media, and Public Policy* (New Brunswick, N.J.: Transaction Press, 1988); and JoAnne Brown, *The Definition of a Profession: The Authority of Metaphor in the History of Intelligence Testing, 1890–1930* (Princeton: Princeton University Press, 1992). Documents relating to this controversy are reprinted in N.J. Block and Gerald Dworkin, eds., *The IQ Controversy: Critical Readings* (New York: Pantheon, 1976); Clarence Karier, ed., *Shaping the American Educational State: 1900 to the Present* (New York: Free Press, 1975); and Russell Jacoby and Naomi Glauberman, eds., *The Bell Curve Debate: History, Documents, Opinions* (New York: Times Books, 1995). Dissertations offering overviews include Thomas Weinland, "A History of the I.Q. in America, 1890–1940" (Ph.D. diss., Columbia University, 1970); Russell Marks, "Testers, Trackers, and Trustees: The Ideology of the Intelligence Testing Movement in America, 1900–1954" (Ph.D. diss., University of Illinois, 1972); David Gersh, "The Development and Use of IQ Tests in the United States from 1900 to 1930" (Ph.D. diss., State University of New York at Stony Brook, 1981); and John Carson, "Talents, Intelligence, and the Con-

structions of Human Difference in France and America, 1750–1920" (Ph.D. diss., Princeton University, 1994). The most detailed studies focus on World War I (see Chapter 8).

6 May Seagoe, *Terman and the Gifted* (Los Altos, Calif.: W. Kaufmann, 1975); Henry Minton, *Lewis M. Terman, Pioneer in Psychological Testing* (New York: New York University Press, 1988); Paul Chapman, *Schools as Sorters: Lewis M. Terman, Applied Psychology, and the Intelligence Testing Movement, 1890–1930* (New York: New York University Press, 1988). There are no biographies of Goddard; for a brief sketch, see Peter Tyor, "Henry Herbert Goddard," *Dictionary of American Biography* Supplement 6 (1956–1960): 240–241. J. David Smith examines Goddard's most famous book in *Minds Made Feeble: The Myth and Legacy of the Kallikaks* (Rockville, Md.: Aspen Systems Corporation, 1985).

7 Stephen Jay Gould, *The Mismeasure of Man* (New York: Norton, 1981), p. 160. For other works analyzing testing as part of the broader heredity–environment controversy, see Mark Haller, *Eugenics: Hereditarian Attitudes in American Thought* (New Brunswick, N.J.: Rutgers University Press, 1963); Donald Pickens, *Eugenics and the Progressives* (Nashville, Tenn.: Vanderbilt Press, 1968); Hamilton Cravens, *The Triumph of Evolution: American Scientists and the Heredity–Environment Controversy, 1900–1941* (Philadelphia: University of Pennsylvania Press, 1978); Allan Chase, *The Legacy of Malthus: The Social Costs of the New Scientific Racism* (New York: Knopf, 1977); R. C. Lewontin, Steven Rose, and Leon Kamin, *Not in Our Genes: Biology, Ideology, and Human Nature* (New York: Pantheon, 1984); Daniel J. Kevles, *In the Name of Eugenics: Genetics and the Uses of Human Heredity* (New York: Knopf, 1985); and Carl Degler, *In Search of Human Nature: The Decline and Revival of Darwinism in American Social Thought* (New York: Oxford, 1991).

8 Knight Dunlap, "Antidotes for Superstitions Concerning Human Heredity," *Scientific Monthly* 51:3 (September 1940): 221. Smith, *Minds Made Feeble.*

9 Seymour Sarason and John Doris, *Psychological Problems in Mental Deficiency* (New York: Harper & Row, 1969), p. 264 (paragraph italicized in original).

10 Nicholas Pastore, *The Nature-Nurture Controversy* (New York: King's Crown Press, 1949), p. 2.

11 On cultural biases, see Block and Dworkin, eds., *IQ Controversy;* Chase, *Legacy of Malthus;* Kamin, *Science and Politics;* and Gould, *Mismeasure of Man.*

12 Lewis Terman to Nicholas Pastore, March 4, 1948, Terman Papers. Relevant portions are reproduced in Pastore, *Nature-Nurture Controversy,* pp. 94–95.

13 Pastore noted two exceptions: Terman, whom he called a liberal hereditarian, and John B. Watson, a conservative environmentalist. *Nature-Nurture Controversy,* p. 176.

14 Draft of letter from HHG to Nicholas Pastore, April 2, 1948, File M32, GP.

15 Although Goddard's response has never been analyzed, Terman's has; see Minton, *Lewis Terman,* pp. 234–242, and "Lewis M. Terman and Mental Testing: In Search of the Democratic Ideal," in Sokal, ed., *Psychological Testing,* pp. 95–112.

16 Cronbach, "Five Decades of Public Controversy," p. 9. This vocabulary is confusing in another sense, for the same words often meant both degrees of impairment and the more general condition. "Idiot" generally meant the most severely impaired cases; "imbecile," moderate cases; and various words including "feebleminded," the mildest cases. Yet each of these words could also mean all three groups. Edouard Seguin's text *Idiocy, and Its Treatment by the Physiological Method* (New York: Wood, 1866) discussed all three. Legally, "imbecile" was often the generic term. Goddard invented the word "moron" for the mildest cases precisely because "feebleminded" was used both for this group and generically. Yet he too changed his usage in different contexts; for instance, Goddard called one book *The Criminal Imbecile* (following legal usage), although these criminals were "morons." ("Feebleminded" was sometimes two hyphenated words. I use one word, unless it is cited differently in a quotation.)

17 Stanley Davies, *Social Control of the Feebleminded: A Study of Social Programs and Attitudes in Relation to the Problems of Mental Deficiency* (New York: National Committee for Mental Hygiene, 1923); *Social Control of the Mentally Deficient* (New York: Thomas Crowell, 1930); *The Mentally Retarded in Society* (New York: Columbia University Press, 1959). By 1996, both the *American Journal on Mental Retardation* and *Mental Retardation* recommended using "people first language." "Authors should use language that emphasizes the humanity of people with mental retardation," stated the "Editorial Policy" of *Mental Retardation;* they should avoid nouns such as "subjects" or "informants" and instead use "people, participants, students, children, and adults." For current terminology, see also *Mental Retardation: Definition, Classification, and Systems of Supports* (Washington, D.C.: American Association on Mental Retardation, 1992).

18 On American eugenics, see Haller, *Eugenics;* Pickens, *Eugenics and the Progressives;* Kevles, *Name of Eugenics;* Edward Larson, *Sex, Race, and Science: Eugenics in the Deep South* (Baltimore: Johns Hopkins University Press, 1995); and Garland Allen, "The Misuse of Biological Hierarchies: The American Eugenics Movement, 1900–1941," *History and Philosophy of the Life Sciences* 5 (1984): 105–128. For overviews of the international literature, see Mark Adams, ed., *The Wellborn Science: Eugenics in Germany, France, Brazil, and Russia* (New York: Oxford University Press, 1990); and Diane Paul, *Controlling Human Heredity: 1865 to the Present* (Atlantic Highlands, N.J.: Humanities Press, 1995).

19 Kevles, *Name of Eugenics,* p. 51. Loren Graham, "Science and Values: The Eugenics Movement in Germany and Russia in the 1920s," *American Historical Review* 82 (1977): 1133–1164; Michael Freeden, "Eugenics and Progressive Thought: A Study in Ideological Affinity," *Historical Journal* 22:3 (1979): 645–671; and Diane Paul, "Eugenics and the Left," *Journal of the History of Ideas* 45:4 (1984): 567–590.

20 Daniel T. Rodgers, "In Search of Progressivism," *Reviews in American History* 10 (December 1982): 113–132. See also Clyde Griffen, "The Progressive Ethos," in Stanley Coben and Lorman Ratner, eds., *The Development of an American Culture* (Englewood Cliffs, N.J.: Prentice-Hall, 1970), pp. 120–149; John D. Buenker, John C. Burnham, and Robert M. Crunden, *Progressivism* (Cambridge, Mass.: Schenkman, 1977); and Buenker and Nicholas C. Burckel, eds., *Progressive Reform: A Guide to Information Sources* (Detroit: Gale Research, 1980).

21 Robert Wiebe, *The Search for Order, 1877–1920* (New York: Hill & Wang, 1967); John Burnham, "Psychiatry, Psychology, and the Progressive Movement," *American Quarterly* 12 (1960): 457–465. On the professionalization of psychology, see Thomas Camfield, "The Professionalization of American Psychology," *JHBS* 9 (1973): 66–75; Loren Baritz, *The Servants of Power: A History of the Use of Social Science in American Industry* (Middletown, Conn.: Wesleyan University Press, 1960); John Reisman, *The Development of Clinical Psychology* (New York: Appleton-Century-Crofts, 1966); Dorothy Ross, *G. Stanley Hall: The Psychologist as Prophet* (Chicago: University of

Chicago Press, 1972); Donald Napoli, *Architects of Adjustment: The History of the Psychological Profession in the United States* (Port Washington, N.Y.: Kennikat Press, 1981); John O'Donnell, *The Origins of Behaviorism: American Psychology, 1870–1920* (New York: New York University Press, 1985); Samelson, "Putting Psychology on the Map"; and JoAnne Brown, *Definition of a Profession.*

22 Samuel Haber, *Efficiency and Uplift: Scientific Management in the Progressive Era, 1890–1920* (Chicago: University of Chicago Press, 1964), p. ix. See also Raymond Callahan, *Education and the Cult of Efficiency* (Chicago: University of Chicago Press, 1962); Brown, *Definition of a Profession;* and Rodgers, "In Search of Progressivism."

23 On testing's role in medical debates, see Leila Zenderland, "The Debate over Diagnosis: Henry Herbert Goddard and the Medical Acceptance of Intelligence Testing," in Sokal, ed., *Psychological Testing,* pp. 46–74. While these controversies have largely been ignored by historians of testing, they have interested scholars concerned with defining mental retardation. See J. Clausen, "Mental Deficiency – Development of a Concept," *American Journal of Mental Deficiency* 71 (March 1967): 727–745; Robert Haskell, "Mental Deficiency Over a Hundred Years," *American Journal of Psychiatry* 100 (1944): 107–118; Seymour Sarason and John Doris, *Educational Handicap, Public Policy, and Social History: A Broadened Perspective on Mental Retardation* (New York: Free Press, 1979); and Jane Mercer, *Labeling the Mentally Retarded: Clinical and Social System Perspectives on Mental Retardation* (Berkeley: University of California Press, 1973).

24 For case studies connecting testing and tracking, see Chapman, *Schools as Sorters,* and Judith Raftery, "Missing the Mark: Intelligence Testing in Los Angeles Public Schools, 1922–32," *History of Education Quarterly* 28 (Spring 1988): 73–93. For broader analyses of this relationship, see Clarence Karier, "Testing for Order and Control in the Corporate Liberal State," in Block and Dworkin, eds., *IQ Controversy,* pp. 339–373; Joel Spring, *Education and the Rise of the Corporate State* (Boston: Beacon, 1972); Samuel Bowles and Herbert Gintis, *Schooling in Capitalist America: Educational Reform and the Contradictions of Economic Life* (New York: Basic Books, 1976); Sarason and Doris, *Educational Handicap;* Harvey Kantor and David Tyack, eds., *Work, Youth, and Schooling: Historical Perspectives on Vocationalism in American Education* (Stanford, Calif.: Stanford University Press, 1982); and Paula Fass, *Outside In: Minorities and the Transformation of American Education* (New York: Oxford, 1989).

25 *Penry v. Lynaugh, Director, Texas Department of Corrections,* 492 U.S. 302.

26 Leon Kamin, "Heredity, Intelligence, Politics, and Psychology: II," in Block and Dworkin, eds., *IQ Controversy,* p. 375. This theme is strongly emphasized in most studies of the eugenics movement.

27 Most of the critical historical literature focuses on asylums for the insane; for an overview of changing attitudes, see Mark S. Micale and Roy Porter, eds., *Discovering the History of Psychiatry* (New York: Oxford University Press, 1994). On institutions for the feebleminded, see Peter L. Tyor and Leland V. Bell, *Caring for the Retarded in America: A History* (Westport, Conn.: Greenwood Press, 1984); R. C. Scheerenberger, *A History of Mental Retardation* (Baltimore: Brookes, 1983); James W. Trent, Jr. *Inventing the Feeble Mind: A History of Mental Retardation in the United States* (Berkeley: University of California Press, 1994); Steven Noll, *Feeble-Minded in Our Midst: Institutions for the Mentally Retarded in the South, 1900–1940* (Chapel Hill: University of North Carolina Press, 1995); and Philip M. Ferguson, *Abandoned to*

Their Fate: Social Policy and Practice Toward Severely Retarded People in America, 1820–1920 (Philadelphia: Temple University Press, 1994). For a valuable critique of recent literature, see also Gerald Grob, "Rediscovering Asylums: The Unhistorical History of the Mental Hospital," in Morris Vogel and Charles Rosenberg, eds., *The Therapeutic Revolution: Essays in the Social History of American Medicine* (Philadelphia: University of Pennsylvania Press, 1979), pp. 135–157.

Chapter 1

1 "Lillian Capell [Goddard's secretary] – Interviewed by Peter Tyor, July, 23, 1976," GP. Even today, Vassalboro conducts annual town meetings.

2 Pliny Earl Goddard, Compiler, "Genealogy of the Descendants of Charles Wheeler Goddard of Dartmouth, Massachusetts," 1966, Robert Hutchings Goddard Papers, Clark University Archives. "Goddard, Henry Herbert," *Biographical Encyclopedia of the World,* 3rd ed. (New York: Institute for Research in Biography, n.d.), File M43, GP. "Record of Births, Deaths, and Burials Began the Yr. 12th of the 9th Mo. 1787, Vassalboro Monthly Meeting," New England Yearly Meeting Records, Rhode Island Historical Society, Microfilm at Friends Historical Library, Swarthmore College. ("Herbert Henry Goddard" is the name in church records.) HHG, Notes for "As Luck Would Have It," GP.

3 Alma Pierce Robbins, *History of Vassalborough, Maine, 1771–1971* (Vassalboro, Me.: Vassalboro Historical Society [c. 1971]).

4 U.S. Manuscript Census for Vassalboro Township, 1850, 1860, and 1870, Vassalboro Historical Society.

5 HHG to James Ritter, May 1, 1952, File M33; HHG, Notes for "As Luck Would Have It," GP.

6 Ibid.; HHG to Ritter, May 1, 1952.

7 William Penn, *Primitive Christianity Revived in the Faith and Practice of the People Called Quakers* (London: T. Sowle, 1696; rpt. Philadelphia: Henry Longstreth, 1877), p. 9.

8 Richard Bauman, *Let Your Words Be Few: Symbolism of Speaking and Silence Among Seventeenth Century Quakers* (Cambridge: Cambridge University Press, 1983).

9 Charles Gilpin, ed., *Journal of the Life and Gospel Labours of David Sands. . .* (London: Charles Gilpin, 1848), p. iv. Ralph D. Greene, "The Development of Quaker Communities in Central Maine, 1770–1800" (M.A. thesis, University of Maine at Orono, 1975). In *Kennebec: Cradel of Americans* (New York: Farrar & Rinehart, 1937), Robert Tristam Coffin states that "*Uncle Tom's Cabin* is the classic of the Kennebec country. And Mrs. Harriet Beecher Stowe is the prophet" (p. 185).

10 Thomas Hamm, *The Transformation of American Quakerism: Orthodox Friends, 1800–1907* (Bloomington: Indiana University Press, 1988); David LeShana, *Quakers in California: The Effects of Nineteenth Century Revivalism on Western Quakerism* (Newberg, Ore.: Barclay Press, 1969), pp. 29–45.

11 Excerpt from *Waterville Mail,* August 15, 1873, in Raymond Manson, "Clippings from *The Eastern Mail,* 1847–1863, and *The Waterville Mail,* 1863–1874," Vassalboro Historical Society. "Charles M. Bailey" in Manson and Elsia Holway Burleigh, *First Seventy Years of Oak Grove Seminary,* Vassalboro Historical Society, 1965; Alice Frost

Lord, "Quakers as They Persist Today in Maine are Rooted in a Few Scattered Hamlets," *Lewiston [Maine] Journal* (May 8, 1937), clipping in Depot 2, GP.

12 Rufus M. Jones, *Eli and Sybil Jones, Their Life and Work* (Philadelphia: Porter & Coates, 1889); Jones, *A Boy's Religion from Memory* (Philadelphia: Ferris & Leach, 1902), p. 21. Elizabeth Gray Vining, *Friend of Life: The Biography of Rufus M. Jones* (Philadelphia: Lippincott, 1958). Jones published fifty-four books, including autobiographical accounts such as *Finding the Trail of Life* (New York: Macmillan, 1926) and *A Small Town Boy* (New York: Macmillan, 1941). Jones and Goddard may have been related. "My mother was a Hoxie and her mother was a Goddard," Jones notes in *Finding the Trail,* p. 19.

13 Rufus M. Jones, *The Society of Friends in Kennebec County, Maine* (New York: H. W. Blake, 1892), p. 10. Excerpt from the *Waterville Mail,* July 3, 1874, in Manson, "Clippings."

14 HHG to Ritter, May 1, 1952, File M33, GP. Jones, *Finding the Trail,* pp. 99–100.

15 Jones, *Society of Friends,* pp. 10–11. Sarah Goddard's travels can be traced by following the certificates issued by the local meetings. See, for example, Vassalboro Monthly Meeting, vol. 6, p. 396, New England Records.

16 Jones, *Finding the Trail,* p. 25; and *Small Town Boy,* p. 89.

17 Edward Cook, Goddard's principal, is quoted in I. Kern Moyse, "Goddard and His Work," undated MS [c. 1915], p. 1, File 465, Yerkes Papers. HHG to [Allen] Clement, April 16, 1948, File M35.1, GP; HHG, Notes for "As Luck Would Have It," File M31, GP.

18 Jones, *Finding the Trail,* p. 132; HHG, Notes to "As Luck Would Have It," File M31, GP.

19 Jones, *The Trail of Life in College* (New York: Macmillan, 1929), pp. 30, 34; Philip C. Garrett, ed., *A History of Haverford College for the First Sixty Years of Its Existence* (Philadelphia: Porter & Coates, 1892), pp. 63–64.

20 Isaac Sharpless, *Story of a Small College* (Philadelphia: John C. Winston, 1918), p. 67.

21 *Biographical Catalog of the Matriculates of Haverford College, 1833–1922* (Philadelphia: Alumni Association, 1922), p. 245. HHG to Clement, April 16, 1948. "The Alumni Prize Orations," *The Haverfordian* 7 (1886): 159–160, Haverford College Archives.

22 Jones, *Trail of Life,* pp. 31–32. Jones, *Finding the Trail,* pp. 136–137. HHG to Frank Cartland, January 20, 1947, File M35.1; HHG to George Crane, June 6, 1947, File M36, GP.

23 HHG to Clement, April 16, 1948.

24 Ibid.

25 Ibid. HHG to Crane, June 6, 1947.

26 HHG to Crane, June 6, 1947. HHG to Clement, April 16, 1948.

27 HHG to Clement, April 16, 1948. HHG to Ritter, May 1, 1952.

28 "Hankes Report Form for Strong Vocational Interest Test – Men" (Palo Alto, Calif.: Stanford University, 1946), File M32, GP.

29 Sarah Goddard married Jehu Newlin on April 30, 1884. See *The Newlin Family: Ancestors and Descendants of John and Mary Pyle Newlin* (Greensboro, N.C.: Algie I. Newlin, 1965), pp. 215, 222, 227. Their travels are recounted in Jones, *Society of Friends,* p. 11. Jehu and Sarah Newlin, "Albums – 1887–1891," Haverford College Quaker Collection.

30 Elizabeth R. Rutter, *The Unfailing Guide,* 3rd ed. (London: Friends Tract Association, n.d.), File M35, GP.

31 Sarah W. Newlin to "My Precious Children," n.d. [c. 1886], GP.

32 Jehu Newlin to HHG, May 27, 1887, File M35.2, GP. Promisory notes written by Jehu Newlin to HHG, January 12, 1895, and August 5, 1895, File M33, GP.

33 Jones, *Finding the Trail,* pp. 47, 147–148.

34 HHG to Ritter, May 1, 1952.

35 Ibid. HHG to Dr. Henley, September 3, 1938, File M33, GP. H. E. Burtt and S. L. Pressey, "Henry Herbert Goddard: 1866–1957," *American Journal of Psychology* 70 (December 1957): 657. Ken Rappoport, *The Trojans: A Story of Southern California Football* (Huntsville, Ala.: Strode, 1974), p. 18.

36 Wedding announcement, August 7, 1889, File M33, GP. HHG, "Memorandum of Summers of HHG," File M33, GP. Emily Stogdill, Oral History, interviewed by John Popplestone, November 25, 1966, pp. 7–8, AHAP. Moyse, "Goddard and His Work," pp. 1–2. Herbert to Emmie Goddard, n.d. [c. 1929], File M33, GP. HHG, Notes for "As Luck Would Have It"; HHG, "Why was I a child of chance?" File M37, GP. Herbert to Emmie Goddard, "Saturday Morn, 1916," File M37, GP.

37 *Catalogue of Damascus Academy* (Salem, Ohio: Rukenbrod, 1890), p. 3, File M35.2, GP.

38 HHG to Robert Fischer, October 23, 1950, File M33, GP. Jones' book was *Eli and Sybil Jones, Their Life and Work.*

39 *Catalogue of Oak Grove Seminary and Bailey Institute, Vassalboro, Me.* (n.p., 1895–1896), File M35.2, GP.

40 E. E. Dudley to Sarah W. [Newlin], May 25, 1896, File M35.2, GP. HHG, "In the Beginning," *Understanding the Child* 3 (April 1933): 2. Even while at Damascus Academy, Goddard had written to Hall about becoming a fellow in Clark's mathematics department. HHG to G. Stanley Hall, November 11, 1889, Hall Papers.

41 Dorothy Ross, *G. Stanley Hall: The Psychologist as Prophet* (Chicago: University of Chicago Press, 1972), pp. 281–282.

42 G. Stanley Hall, "Recent Advances in Child Study," *NEA* (1908): 948.

43 Ross, *Hall;* John O'Donnell, *The Origins of Behaviorism: American Psychology, 1870–1920* (New York: New York University Press, 1985), pp. 110–128.

44 Ross, *Hall,* pp. 211–212.

45 E. G. Boring, *A History of Experimental Psychology* (New York: Appleton-Century-Crofts, 1957), pp. 203–209; Laurence Veysey, *The Emergence of the American University* (Chicago: University of Chicago Press, 1965), pp. 22–25; Earl Barnes, "The Present and Future of Child Study in America," in Barnes, ed., *Studies in Education* (Philadelphia: Earl Barnes, 1902), 2: 363. Darwin's "Biographical Sketch of an Infant" (1877) is cited in *Journal of Social Science* (February 1882), pp. 6–8.

46 See, for instance, G. Stanley Hall, "Child-Study: The Basis of Exact Education," *Forum* 16 (1893): 429–441. On Child Study, see Ross, *Hall,* pp. 124–133, 279–308; Edward A. Krug, *The Shaping of the American High School, 1880–1920* (Madison: University of Wisconsin Press, 1969), pp. 107–112; James Hendricks, "The Child-Study Movement in American Education, 1880–1910: A Quest for Educational Reform Through a Scientific Study of the Child" (Ph.D. diss., Indiana University 1968); Leila Zenderland, "Education, Evangelism, and the Origins of Clinical Psychology: The Child-Study Legacy," *JHBS* (April 1988): 152–165; and Sheldon White, "Child Study at Clark University: 1894–1904," *JHBS* 26 (April 1990): 131–150. *Catalogue of Oak Grove,* p. 17.

47 HHG to Ritter, May 1, 1952.

48 G. Stanley Hall, "Research the Vital Spirit of Teaching," *Forum* 17 (July 1894): 558. On Clark, see Ross, *Hall,* pp. 186–204; O'Donnell, *Origins,* pp. 161–166; Veysey, *American University,* pp. 165–170, 384–386; William Koelsch, *Clark University, 1887–1987: A Narrative History* (Worcester, Mass.: Clark University Press, 1987); and Michael Sokal, "G. Stanley Hall and the Institutional Character of Psychology at Clark, 1889–1920," *JHBS* 26 (April 1990): 114–124.

49 HHG to Ritter, May 1, 1952; HHG to G. Stanley Hall, October 19, 1901, Hall Papers.

50 Hall, "Research the Vital Spirit of Teaching," p. 568.

51 Ross, *Hall,* pp. 290–292. Questionnaires (called "syllabi") are in G. Stanley Hall, *Topical Syllabi,* vol. 1 (1894–1899) (Worcester, Mass.: Clark University, 1899), and vol. 2 (1899–1906) (Worcester, Mass.: Clark University, 1906), Hall Papers.

52 HHG to Crane, June 6, 1947.

53 Jones, *Finding the Trail,* pp. 138–139. Elizabeth Flower and Murray G. Murphey, *A History of Philosophy in America* (New York: Putnam, 1977), 2: 517–563; James R. Moore, *The Post-Darwinian Controversies: A Study of the Protestant Struggle to Come to Terms with Darwin in Great Britain and America, 1870–1900* (Cambridge: Cambridge University Press, 1979); Robert Richards, *Darwin and The Emergence of Evolutionary Theories of Mind and Behavior* (Chicago: University of Chicago Press, 1987).

54 G. Stanley Hall, *Life and Confessions of a Psychologist* (1923), as cited in O'Donnell, *Origins,* p. 159. The tragedy concerned Hall's wife and daughter, who were accidentally asphyxiated in their home. On Hall's religion, see Ross, *Hall,* pp. 207–208; Clarence Karier, *Scientists of the Mind: Intellectual Founders of Modern Psychology* (Urbana: University of Illinois Press, 1986), pp. 159–190; and Hendrika Vande Kemp, "G. Stanley Hall and the Clark School of Religious Psychology," *AP* 47 (February 1992): 290–298.

55 Hall, "Modern Methods in the Study of the Soul," *Christian Register* 75 (February 1896): 131–133.

56 Ibid.

57 Hall, "Modern Methods," p. 131. HHG to Robert Fischer, November 13, 1948, File 614, GP.

58 HHG to Fischer, November 13, 1948. HHG, "A New Brain Microtome," *Journal of Comparative Neurology* 10 (1900): 209–213. This device was used at Westboro State Hospital until at least 1938. B. W. St. Claire, Mico Instrument Company, to HHG, December 21, 1938, File M43, GP.

59 HHG, "Psychological Work Among the Feeble-Minded," *JPA* 12 (September 1907): 29–30. For Meyer's work at Clark, see Ross, *Hall,* pp. 381–382; and Gerald N. Grob, *The State and the Mentally Ill: A History of Worcester State Hospital in Massachusetts, 1830–1920* (Chapel Hill: University of North Carolina Press, 1966), pp. 279–316.

60 Goddard's dissertation was published as "The Effects of Mind on Body as Evidenced in Faith Cures," *American Journal of Psychology* 10 (1899): 431–502.

61 Rufus M. Jones, "Foreword" in Henry J. Cadbury, ed., *George Fox's "Book of Miracles"* (Cambridge: Cambridge University Press, 1948), p. ix. Donald Meyer, *The Positive Thinkers: Religion as Pop Psychology from Mary Baker Eddy to Oral Roberts* (New York: Pantheon, 1965); Gail Thain Parker, *Mind Cure in New England From the Civil War to World War One* (Hanover, N.H.: University Press of New England, 1973); Robert Fuller, *Mesmerism and the American Cure of Souls* (Philadelphia: University of Pennsylvania Press, 1982); Nathan G. Hale, Jr., *Freud and the Americans: The Beginnings of Psychoanalysis in the United States, 1876–1917* (New York: Oxford University Press, 1971), pp. 228–49; and Catherine Albanese, *Nature Religion in America: From the Algonkian Indians to the New Age* (Chicago: University of Chicago Press, 1990), pp. 105–152.

62 George Beard, *American Nervousness: Its Causes and Consequences* (New York: G. P. Putnam's Sons, 1881); John S. and Robin M. Haller, *The Physician and Sexuality in Victorian America* (Urbana: University of Illinois Press, 1974), pp. 3–43; Charles Rosenberg, "George Beard and American Nervousness," in *No Other Gods: On Science and American Social Thought* (Baltimore: Johns Hopkins University Press, 1976), pp. 98–108; Barbara Sicherman, "The Paradox of Prudence: Mental Health in the Gilded Age," in Andrew Scull, ed., *Madhouses, Mad-Doctors, and Madmen: The Social History of Psychiatry in the Victorian Era* (Philadelphia: University of Pennsylvania Press, 1981), pp. 218–240; and F. G. Gosling, *Before Freud: Neurasthenia and the American Medical Community, 1870–1910* (Urbana: University of Illinois Press, 1987).

63 Ross, *Hall,* pp. 149–51. Theodore R. Sarbin, "Attempts to Understand Hypnotic Phenomena," in Leo Postman, ed., *Psychology in the Making: Histories of Selected Research Problems* (New York: Knopf, 1963), pp. 745–785. Henri F. Ellenberger, *The Discovery of the Unconscious: The History and Evolution of Dynamic Psychiatry* (New York: Basic Books, 1970).

64 HHG, "Faith Cures," p. 431.

65 On Quimby, see Parker, *Mind Cure,* pp. 3–5, 134–137; Fuller, *Mesmerism,* pp. 118–36; Meyer, *Positive Thinkers,* pp. 33–37, 88–90; and Albanese, *Nature Religion,* pp. 107–115.

66 Parker, *Mind Cure,* p. 3.

67 HHG, "Faith Cures," pp. 446–447.

68 Ibid., p. 447.

69 Ibid., pp. 437–444, 483–484.

70 Robert Peel, *Mary Baker Eddy: The Years of Discovery* (New York: Holt, Rinehart & Winston, 1966); Martin Gardner, *The Healing Revelations of Mary Baker Eddy: The Rise and Fall of Christian Science* (Buffalo: Prometheus, 1993) and Parker, *Mind Cure,* pp. 109–129.

71 HHG, "Faith Cures," pp. 432–437, 449.

72 Ibid., p. 432.

73 HHG, "The Psychology of Health and Disease," Topical Syllabi No. VII (December 1896) and No. XIV (May 18, 1897), in Hall, *Topical Syllabi.*

74 HHG, "Are Drugs Unnecessary to the Cure of Disease?" *The Hypnotic Magazine* (March 1897): 155–158.

75 HHG, "Faith Cures," pp. 442–444, 468–471.

76 Ibid., pp. 458–464.

77 Ibid., pp. 471–478.

78 Ibid., p. 481.

79 Ibid., pp. 480–481. Richards, *Darwin;* Sarbin, "Attempts to Understand," pp. 763–767.

80 HHG, "Faith Cures," pp. 480–481.

81 Ibid.

82 William James, *The Varieties of Religious Experience: A Study in Human Nature* (1902; rpt. New York: Macmillan, 1961), pp. 91–92n. While praising the thesis, James questioned Goddard's claims for traditional Christianity. "The ideas of Christian churches are not efficacious in the therapeutic direction to-day. . . ," he noted; in explaining why, "the mere blank waving of the word 'suggestion' as if it were a banner gives us no light." Goddard "concludes by saying that 'Religion [and by this he seems to mean our popular Christianity] has in it all there is in mental therapeutics. . . .' And this in spite of the actual fact that the popular Christianity does absolutely *nothing,* or did nothing until mind-cure came to the rescue" (p. 103).

83 HHG, "Faith Cures," p. 499.

84 Ibid., p. 432.

85 "Herbert Henry W. Goddard" and his wife joined the Uxbridge, Massachusetts, meeting. "Vassalboro Monthly Meeting," p. 82. Although these photos are lost, a letter mentions them. Edith to HHG, December 5, 1937, File M37, GP. HHG, "Faith Cures," p. 431. The quotation is from *All's Well That Ends Well,* I.i.

86 HHG, "Faith Cures," pp. 500–501.

87 Ibid.

88 Ibid.

89 Ibid., p. 482. For Spencer's discussion of altruism, see *Principles of Psychology* (New York: Appleton, 1887), 2: 607–626; and Richards, *Darwin,* pp. 309–313.

90 HHG, "Faith Cures," p. 483.

91 Ibid., p. 484.

92 HHG, "Mysterious Influences," p. 1, File M42, GP. Goddard entered this in an essay contest at Ohio State University.

93 Ibid., pp. 10–11.

94 Goddard eventually climbed mountains in Europe, Canada, and the American West, and witnessed one fatal accident. See "Facts concerning the death of Dr. Frank B. Wynn on Siyeh Mt., Glacier National Park, Montana, Thursday July 27, 1922," File M33, GP.

Chapter 2

1 HHG to James Ritter, May 11, 1952, File M33, GP; G. Stanley Hall to Dr. F. W. Ellis, March 29, 1899, Hall Papers. Hall used these phrases in a letter of recommendation.

2 Dorothy Ross, *G. Stanley Hall: The Psychologist as Prophet* (Chicago: University of Chicago Press, 1972), pp. 103–133, 211–212, 279–308.

3 *Catalogue to Clark University Summer School of Psychology, Biology, Pedagogy and Anthropology* (1899): 11–12, Clark University Archives.

4 HHG to G. Stanley Hall, October 19, 1901, Hall Papers.

5 G. Stanley Hall, "Editorial," *PS* 3 (1894): 188.

6 G. Stanley Hall, *The Contents of Children's Minds on Entering School* (New York: E. L. Kellogg, 1893); Sara Wiltse, "A Preliminary Sketch of the History of Child Study in America," *PS* 3 (1894): 192. G. Stanley Hall, *Topical Syllabi,* vol. 1 (1894–1899) (Worcester, Mass.: Clark University, 1899), and vol. 2 (1899–1906) (Worcester, Mass.: Clark University, 1906), Hall Papers. In 1895–1896 alone, Hall received 60,000 returns from about 800 individual "investigators." See Ross, *Hall,* pp. 128–129, 290–293.

7 "Catalogue and Circular of the State Normal School, Worcester," 1890, p. 23, as cited in William H. Burnham, "The Observation of Children at the Worcester Normal School," *PS* 1 (1891): 219–224; E. Harlow Russell, "The Study of Children at the State Normal School, Worcester, Mass.," *PS* 2 (1892): 343–357. Hall, "The Methods, Status, and Prospects of the Child-Study of To-Day," *Transactions of the Illinois Society for Child Study* 2 (May 1896): 184; Hall, "Universities and the Training of Professors," *Forum* 17 (May 1894): 300.

8 Wiltse, "Preliminary Sketch," p. 199. HHG, "Child Study for Pennsylvania Teachers," *PSJ* 49 (September 1900): 127. For overviews of Child Study, see Chapter 1, note 46.

9 Earl Barnes, "Child Research and Social Progress," in Edgar Doll, ed., *Twenty-Five Years* (Vineland, N.J.: The Training School, 1932), p. 49. Wiltse, "Preliminary Sketch," pp. 195–198. Barnes worked for the Society for the Extension of University Teaching in London and Philadelphia. "Earl Barnes" in John Ohles, ed., *Biographical Dictionary of American Educators* (Westport, Conn.: Greenwood Press, 1978), p. 90; and "Professor Earl Barnes," *TSB* 32 (June 1935): 61.

10 "Editorial," *Educational Review* 15 (April 1898): 412. HHG, "Child Study for Pennsylvania Teachers," *PSJ* 49 (September 1900): 127. "The Passing of Child Study," *School Journal* (December 21, 1901): 664–665; Edward Krug, *The Shaping of the American High School, 1880–1920* (Madison: University of Wisconsin Press, 1969), p. 115; James Hendricks, "The Child-Study Movement in American Education, 1880–1910: A Quest for Educational Reform Through a Scientific Study of the Child" (Ph.D. diss., Indiana University, 1968), pp. 270–277.

11 William T. Harris charged "vivisection"; see Krug, *High School,* p. 111. "Passing of Child Study," pp. 664–665.

12 Hugo Munsterberg, "Psychology and Education," *Educational Review* 16 (September 1898): 106.

13 Earl Barnes, "The Present and Future of Child Study in America," in Barnes, ed., *Studies in Education* (Philadelphia: Earl Barnes, 1902), 2:371.

14 Ibid., p. 372. On educational rhetoric, see Michael Katz, *The Irony of Early School Reform: Educational Innovation in Mid-Nineteenth Century Massachusetts* (Cambridge, Mass.: Harvard University Press, 1968), pp. 115–160. See also Ross, *Hall,* pp. 114–115.

15 Paul Mattingly, *The Classless Profession: American Schoolmen in the Nineteenth Century* (New York: New York University Press, 1975), pp. xvii, 142. Geraldine Joncich, *The Sane Positivist: A Biography of Edward L. Thorndike* (Middletown, Conn.: Wesleyan University Press, 1968), p. 555.

16 G. Stanley Hall, "Child Study," *NEA* (1894): 173–179.

17 Ibid.; Adelaide E. Wykoff, "Discussion," *NEA* (1894): 185.

18 Ross, *Hall,* p. 344. Earl Barnes, "Child Study. General Conclusions" in Barnes, ed., *Studies in Education* (Stanford, Calif.: Earl Barnes, 1897) 1:364. HHG, "Child Study for Pennsylvania Teachers," p. 129. Barnes, "Present and Future," p. 371.

19 *Program of Friends' Summer School of Religious History,* File 37, GP; *Biographical Catalog of the Matriculates of Haverford College, 1833–1922* (Philadelphia: Haverford Alumni Association, 1922), p. 245.

20 HHG, "Child Study for Pennsylvania Teachers," p. 127. Goddard is quoting educator L. H. Galbraeth.

21 Munsterberg, "Psychology and Education," pp. 115, 111.

22 Russell quoted this phrase from Darwin's autobiography in "Study of Children," p. 352.

23 Earl Barnes, "The Study of Children on the Pacific Coast," *NEA* (1895): 902; A. C. Ellis and G. Stanley Hall, "A Study of Dolls," *PS* 4 (1896): 129–175.

24 J. Mark Baldwin, "Child Study," *Psychological Review* 5 (1898): 218–220.

25 William James, *Talks to Teachers on Psychology . . .* (May 1899), in Frederick Burkhardt, ed., *The Works of William James* (Cambridge, Mass.: Harvard University Press, 1983), pp. 17–18.

26 Munsterberg, "Psychology and Education," pp. 115, 113.

27 Wiltse, "Preliminary Sketch," pp. 211–212. Barnes, "Methods of Studying Children," *Studies in Education,* 1:5–6.

28 Edward Thorndike, "What Is a Psychical Fact?" *Psychological Review* 5 (November 1898): 645–650. On his debate with Munsterberg, see Joncich, *Sane Positivist,* pp. 157–160.

29 HHG, "Child Study for Pennsylvania Teachers," pp. 127–128. Hall, "Child-Study: The Basis of Exact Education," *Forum* 16 (December 1893): 431. Hall credits these priorities to Russell. See also Ross, *Hall,* p. 289.

30 HHG, "Child Study for Pennsylvania Teachers," pp. 128–129.

31 Ibid., pp. 129–130.

32 Letter "To the Teachers of Pennsylvania" in HHG, "Child Study: Special Circular from Child Study Department," *PSJ* 49 (November 1900): 227.

33 HHG to G. Stanley Hall, January 14, 1901, Hall Papers. The biblical reference, "Come over to Macedonia and help us," is from Acts 16:9.

34 For Hall's involvement with health issues, see Ross, *Hall,* pp. 289–297. HHG, "Child Study: Special Circular," pp. 227–228.

35 HHG, "Child Study: Special Circular," p. 229.

36 Earl Barnes, "Children's Ideals," *PS* 7 (April 1900): 3–12; HHG, "Negative Ideals," in Barnes, ed., *Studies in Education,* 2:392–398.

37 HHG, "Negative Ideals," p. 392.

38 Ibid., p. 398.

39 Ibid.

40 E. Meumann to HHG, February 16, 1904, File M33, GP. Edward Johnstone, "Report of the Superintendent," *AR* (1906): 28. HHG, "An den Lehrer oder die Lehrerin," GP. HHG, "Ideals of a Group of German Children," *PS* 13 (1906): 208–220.

41 HHG, "Ideals," pp. 211–215.

42 Ibid., pp. 213–214.

43 Ibid.

44 Ibid., p. 216.

45 Ibid., pp. 218–219.

46 Ibid., pp. 210–211.

47 *Historical Statistics of the United States: Colonial Times to 1970, Part I* (Washington, D.C.: U.S. Bureau of the Census, 1975), pp. 22–37.

48 Ibid., pp. 412–432. Krug, *High School,* pp. 169–170. "New York City," *Journal of Education* (June 8, 1911): 631.

49 Hall, "Universities and the Training of Professors," pp. 306–307.

50 Leonard Ayres, "The Training of the Mentally and Physically Unfortunate," in "Developments of the Past Decade," *NEA* (1911): 244.

51 Ibid., pp. 242–244. See also Luther Halsey Gulick and Leonard Ayres, *Medical Inspection of Schools* (New York: Charities Publication Committee, 1908), and Ayres, compiler, *Medical Inspection Legislation* (New York: Russell Sage Foundation, 1911).

52 HHG, "Psychological Work Among the Feeble-Minded," *JPA* 12 (September 1907): 18.

53 "Preliminary Report of the Committee on Provision for Exceptional Children in the Public Schools," *NEA* (1908): 354. Before 1900, elementary classes exceeded 50 in Chicago and Boston. Krug, *High School,* p. 96. New York City's 1910 average was 43, but over 300 classes had 60 or more. *The First Fifty Years: A Brief Review of Progress, 1898–1948* (New York: City Board of Education, 1948), p. 50.

54 See, for example, appeals by Dorothea Dix, Samuel Gridley Howe, and others in Marvin Rosen, Gerald R. Clark, and Marvin S. Kivits, eds., *The History of Mental Retardation, Collected Papers,* vol. 1 (Baltimore: University Park Press, 1976).

55 Leo Kanner, *History of the Care and Study of the Mentally Retarded* (Springfield, Ill.: Charles Thomas, 1974), pp. 114–115; Kanner tries to distinguish "special" from "disciplinary" classes. Guy M. Whipple, "Special Classes," in Paul Monroe, ed., *A Cyclopedia of Education* (New York: Macmillan, 1911), 5:384–386.

56 Teachers of the deaf first contacted the NEA about organizing their own department in 1895. A year later, the NEA organized a "Department of Education for the

Deaf and Dumb, the Blind, and the Feeble-Minded." See Alexander Graham Bell, "President's Address," *NEA* (1902): 828–829. On deaf education (and Bell's controversial role), see John Vickrey Van Cleve and Barry A. Crouch, *A Place of Their Own: Creating the Deaf Community in America* (Washington, D.C.: Gallaudet University Press, 1989); and Harlan Lane, *When the Mind Hears: A History of the Deaf* (New York: Random House, 1984).

57 "Edward Ransom Johnstone" in John F. Ohles, *Biographical Dictionary of American Educators* (Westport, Conn.: Greenwood, 1978), 2:719–720; "Honoring of Edward Ransom Johnstone, 1870–1946," *TSB* 44 (May 1947): 32–58.

58 ERJ, "The Institution as a Laboratory," and HHG, "Anniversary Address," in Doll, ed., *Twenty-Five Years,* pp. 3–4, 55–56. "Program for Special Meeting," New Jersey Association for the Study of Children and Youth, March 31, 1900, File M37, GP. See also "Child Study. Vineland Teachers' Institute, Jan. 14, 1899," Johnstone Papers.

59 ERJ, "Institution as a Laboratory," pp. 14–15; Kathrine Regan McCaffrey, "Founders of the Training School at Vineland, New Jersey: S. Olin Garrison, Alexander Johnson, Edward R. Johnstone" (Ph.D. diss., Teachers College, Columbia University, 1965), pp. 69–74, 179. By 1906, while most state institutional populations ranged from 700 to 1200, Johnstone tried to limit Vineland to under 500. McCaffrey, "Founders," p. 179.

60 Edouard Seguin, *Idiocy, and Its Treatment by the Physiological Method* (New York: Wood, 1866); HHG, "Defectives, Schools for," in Monroe, ed., *Cyclopedia of Education,* 2:276–277. Goddard is quoting Seguin, p. 68.

61 McCaffrey, "Founders." Goddard, "Letchworth Village," unpublished report, 1948, p. 4, File M37, GP.

62 S. Olin Garrison to Edward Johnston [*sic*], October 29, 1897, ERJ Papers. McCaffrey, "Founders," pp. 165–168.

63 HHG, "Anniversary Address," in Doll, ed., *Twenty-Five Years,* p. 56.

64 Ibid.

65 Ibid.; Edgar Doll, "Foreword," *Twenty-Five Years,* p. xix; HHG, "Child Study," *PSJ* 54 (September 1905): p. 124.

66 ERJ, "Discussion" following David Lincoln, "Special Classes for Feeble-Minded Children in the Boston Public Schools," *JPA* 7 (June 1903): 93.

67 "Paidological Staff," *AR* (1906): 4.

68 ERJ, "Institution as a Laboratory," pp. 7–8. "Child Study," *PSJ* 54 (September 1905): 120–125. Teachers also turned to other institutions for training; see Martin Barr, "Discussion," following Lincoln, "Special Classes," p. 90.

69 Goddard's comments are cited in "Child Study" (1905), pp. 120–122.

70 Ibid.

71 Ibid.

72 Ibid.

73 Ibid.

74 G. Stanley Hall, "Experimental Psychology in Education," *Addresses and Proceedings of the International Congress of Education* (Chicago, 1893), p. 717.

75 ERJ, "Functions of the Special Class," *NEA* (1908): 1116.

76 "Child Study" (1905): 122, 124–125.

77 Hall, "Experimental Psychology in Education," p. 717. Barnes, "Present and Future," p. 366.

78 Barnes, "Present and Future," p. 371.

79 Ibid., p. 372.

80 ERJ, "Institution as a Laboratory," p. 8.

81 "Professor Earl Barnes," *TSB* 32 (June 1935): 61; Doll, "Foreword," p. xix.

82 ERJ, "Institution as a Laboratory," pp. 8–9.

83 George Morris Philips to HHG, May 3, 1904, File 35.2, GP. The final episode which led to Goddard's leaving West Chester is unclear. A late-in-life account argues that in 1905 he complained to the Superintendent of Schools that both his students and representatives from the State Board of Examiners were cheating on teacher certification tests. As a consequence, he claims, he lost his job, but the principal agreed to a final year. He never even told his wife about this incident. HHG, Notes to "As Luck Would Have It," GP.

84 The only comparable nonacademic position may have been held by A.R.T. Wylie, a psychologist and pharmacist at the Faribault institution. A few psychologists, however, were employed in psychiatric clinics. See John A. Popplestone and Marion White McPherson, "Pioneer Psychology Laboratories in Clinical Settings," in Josef Brozek, ed., *Explorations in the History of Psychology in the United States* (Cranbury, N.J.: Associated University Presses, 1984), pp. 196–272.

85 ERJ, "Report of the Superintendent," *AR* (1906): 26.

86 Ibid., p. 29.

87 Ibid.

88 HHG, "In the Beginning," *Understanding the Child* 3 (April 1933): 2.

89 Draft, HHG to Nicholas Pastore, April 2, 1948, File M32, GP.

90 HHG, "Psychological Work Among the Feeble-Minded," *JPA* 12 (September 1907): 18.

91 Thomas Carlyle, *Sartor Resartus* (1831); rpr. as vol. 1, *Works of Thomas Carlyle . . .* (London: Chapman & Hall, 1896), pp. 166–168. These words are said by "Herr Teufelsdrockh," an authority on the Philosophy of Clothes, although commentators have interpreted them as Carlyle's views. Mary Agnes Best, *Rebel Saints* (New York: Harcourt Brace, 1925), p. 25, sees them as Carlyle's response to Lord Macauley's criticism of Fox.

92 HHG, "Psychological Work," p. 18.

Chapter 3

1 HHG, September 17 and September 24 entries, "First Year at Vineland, Sept. 17, 1906–June 24, 1907," File M43, GP. HHG, "The Research Work," Supplement 1, *TSB* 46 (December 1907): 4–8.

2 HHG, "In the Beginning," *Understanding the Child* 3 (April 1933): 2. G. Stanley Hall to HHG, February 9, 1906; Adolf Meyer to HHG, February 8, 1906; E. C. Sanford to HHG, February 9, 1906, File M33, GP.

3 Hall to HHG, February 9, 1906; Sanford to HHG, February 9, 1906.

4 Meyer to HHG, February 8, 1906.

5 Both S. Olin Garrison and E. R. Johnstone were very self-conscious about not being doctors. ERJ to S. Olin Garrison, undated [c. June 1897], and S. Olin Garrison to E. R. Johnston [*sic*], October 29, 1897, Johnstone Papers.

6 Physicians used "psycho-asthenics," probably to parallel neurasthenics, since there was "no universal, or even very general use that implies a knowledge of that condition which is termed idiocy or feeble-mindedness." "Announcement," *JPA* 1 (September 1896): 34.

7 HHG, "Psychological Work Among the Feeble-Minded," *JPA* 12 (September 1907): 18–30.

8 Henry Swinburne, *A briefe treatise of testaments and last willes . . .* (1590), as cited in Albert Deutsch, *The Mentally Ill in America: A History of their Care and Treatment from Colonial Times* (New York: Columbia University Press, 1937, 1949), p. 333.

9 Leo Kanner notes that in the most complete early index, Heinrich Laehr's *Die Literatur der Psychiatrie, Neurologie und Psychologie von 1459 bis 1799* (1899), there is "not one, however faint, allusion to mental deficiency, except for the evidence of sporadic interest in cretinism toward the end of the Middle Ages." *History of the Care and Study of the Mentally Retarded* (Springfield, Ill.: Charles Thomas, 1964, 1974), pp. 7–8. See also R. C. Scheerenberger, *History of Mental Retardation* (Baltimore: Brooks, 1983), pp. 3–50.

10 Kanner, *History,* pp. 12–17, 35–38. On Itard, see Harlan Lane, *The Wild Boy of Aveyron* (Cambridge, Mass.: Harvard University Press, 1976). Edouard Seguin, *Idiocy, and Its Treatment by the Physiological Method* (New York: Wood, 1866), p. 457.

11 The Association of Medical Officers later became the American Association for the Study of the Feeble-Minded, the American Association on Mental Deficiency, and the American Association on Mental Retardation; see Kanner, *History,* pp. 78–82; and G. E. Milligan, "History of the American Association on Mental Deficiency," *American Journal of Mental Deficiency* 66 (1961): 357–369.

12 J. Clausen, "Mental Deficiency – Development of a Concept," *American Journal of Mental Deficiency* 71 (March 1967): 727–745; Robert Haskell, "Mental Deficiency over a Hundred Years," *American Journal of Psychiatry* 100 (1944): 107–118. Case descriptions are in Seguin's appendix to *Idiocy* (first edition only); Isaac Kerlin, *The Mind Unveiled* (Philadelphia: Hunt, 1858); William Ireland, *On Idiocy and Imbecility* (London: Churchill, 1877); and J. Langdon Down, *On Some of the Mental Affections of Childhood and Youth* (London: Churchill, 1887); Goddard described more than three hundred Vineland cases in *FM* (New York: Macmillan, 1914).

13 Martin Barr, *Mental Defectives: Their History, Treatment, and Training* (Philadelphia: Blakiston, 1904), pp. 78–90; quotation p. 78.

14 HHG, "Suggestions for a Prognostical Classification of Mental Defectives," *JPA* 14 (September and December 1909/March and June 1910): 48.

15 U.S. Bureau of the Census, *Special Reports: Insane and Feeble-Minded in Hospitals and Institutions, 1904* (Washington, D.C.: GPO, 1906), p. 205.

16 Ibid., p. 212. For the religious implications of "stigmata," see Steven Gelb, "The Beast in Man: Degenerationism and Mental Retardation, 1900–1920," *Mental Retardation* 33 (February 1995): 1–9.

17 For parents' reports, see Frederick Wines, "Report on the Defective, Dependent, and Delinquent Classes of the Population of the United States, as returned at the Tenth Census (June 1, 1880)," House Miscellaneous Documents, 47th Cong., 2d sess., 1882–1883, vol. 13, pt. 21:240–241. HHG, *FM,* pp. 518–521, 271.

18 Among conditions recognized before 1900 were tuberosclerosis and amaurotic family idiocy [Tay-Sachs disease]; see Kanner, *History,* pp. 87–109. "Mongolians" [persons with Down syndrome] were included in Down's larger "physiognomical" or "ethnological" classification system, as were Ethiopians, Malays, Americans [Native Americans], and Caucasians. J. Langdon Down, "Observations on an Ethnic Classification of Idiots," *London Clinical Lectures and Hospital Reports* 3 (1866): 259–262. Steven Gelb, "Typological Madness: A Brief History of Ethnological Idiocy," paper presented to Annual Meeting of Cheiron (International Society for the History of the Behavioral and Social Sciences), 1995.

19 HHG, *FM,* pp. 428, 52. For these and other "moral imbeciles," see Barr, *Mental Defectives,* pp. 264–281.

20 G. E. Johnson, "Contribution to the Psychology and Pedagogy of Feeble-Minded Children," *JPA* 2 (December 1897): 70–71; HHG, *FM,* pp. 385–386.

21 The "historical cook" is first cited in the *Edinburgh Review* (July 1865) and collected by Seguin in *Idiocy,* p. 444. HHG, *FM,* pp. 423–426, includes drawings by "Frank H."

22 HHG, *FM,* pp. 124–126.

23 Isaac Kerlin, "The Organization of Establishments for the Idiotic and Imbecile Classes," *AMO* (1877), p. 20.

24 Edouard Seguin, *New Facts and Remarks Concerning Idiocy,* Lecture before the New York Medical Journal Association, October 15, 1869 (New York: Wood, 1870), pp. 8–9.

25 Draft of letter from HHG to Nicholas Pastore, no date [1948], File M32 (Ephemeras), GP.

26 HHG, "The Grading of Backward Children," *TSB* 5 (November 1908): 12.

27 Various classification systems are described in Kanner, *History,* and Scheerenberger, *History of Mental Retardation.* See also Clausen, "Mental Deficiency – Development of a Concept." For a contemporary analysis, see Jane Mercer, *Labeling the Mentally Retarded: Clinical and Social System Perspectives on Mental Retardation* (Berkeley: University of California Press, 1973); and Mercer, "Who Is Normal? Two Perspectives on Mild Mental Retardation," in E. Gartly Jaco, ed., *Patients, Physicians and Illness: A Sourcebook in Behavioral Science and Health* (New York: Free Press, 1972), pp. 56–75. The categorization of classification systems used in this chapter is my own.

28 Ireland, *On Idiocy, especially in its Physical Aspects* (Edinburgh: Oliver & Boyd, Tweeddale Court, 1874), pp. 1–34, quotation p. 6; see also Ireland, *The Mental Affections of Children: Idiocy, Imbecility and Insanity* (Philadelphia: Blakiston, 1900), p. 39.

29 Isaac Kerlin, "A Clinical Lecture on Idiocy and Imbecility," *Medical and Surgical Reporter* 46 (May 27, 1882): 534. Ireland, *On Idiocy,* p. 28. A. C. Rogers, "On the *Ascribed* Causation of Idiocy as Illustrated in Reports to the Iowa Institution for Feeble Minded Children," *AMO* (1884): 296–301.

30 Kerlin, "Clinical Lecture," p. 534. Hervey Wilbur, "The Relation of Speech or Language to Idiocy," *AMO* (1878): 66–80, quotation p. 74.

31 Jean Esquirol, *Des Maladies Mentales,* 2:340, cited in Alfred Binet and Theodore Simon, *The Development of Intelligence in Children,* trans. Elizabeth Kite (Baltimore: Williams & Wilkins, 1916), pp. 15–17. See also G. E. Johnson, "Contribution to the Psychology and Pedagogy of Feeble-Minded Children," *JPA* 1 (March 1897): 91. Another version is cited in Barr, *Mental Defectives,* p. 79.

32 Wilbur, "The Relation of Speech," pp. 67, 76–79.

33 Ibid., p. 75.

34 J. C. Bucknell and D. H. Tuke, *Manual of Psychological Medicine,* 4th ed. (Philadelphia: Lindsay, Blakiston, 1879), p. 152; Barr, *Mental Defectives,* p. 80.

35 Barr, *Mental Defectives,* p. 90.

36 A. F. Tredgold, *Mental Deficiency* (New York: Wood, 1908; rpt. 1922), pp. 92–96. A version is cited in the "Discussion" following HHG, "Suggestions for a Prognostical Classification," p. 53.

37 HHG, "Psychological Work," pp. 18–30.

38 HHG, "Report of the Research Department," *AR* (1907): 38.

39 HHG, "Psychological Work," p. 19.

40 Ibid., pp. 19, 27–28.

41 Ibid, p. 19.

42 Ibid., p. 20.

43 Ibid.

44 Ibid., pp. 22–23.

45 Ibid., pp. 23–24.

46 Ibid., p. 24.

47 Ibid., pp. 20–22.

48 Ibid., pp. 29–30. On Goddard's attempts to use Meyer's "case method," see HHG, "The Research Work," p. 3. See also Adolf Meyer to HHG, January 29, 1907, File M35, GP.

49 Seguin, *New Facts and Remarks,* pp. 18–19, 25. On psychophysical measurements, see Gail Hornstein, "Quantifying Psychological Phenomena: Debates, Dilemmas, and Implications," in Jill Morawski, *The Rise of Experimentation in American Psychology* (New Haven, Conn.: Yale University Press, 1988), pp. 1–34, esp. pp. 3–8.

50 Hervey Wilbur, "The Classifications of Idiocy," *AMO* (1877): 29–35, esp. pp. 34–35.

51 Shepherd Ivory Franz as cited in HHG, "Psychological Work," p. 29. For Franz's work, see John A. Popplestone and Marion White McPherson, "Pioneer Psychology Laboratories in Clinical Settings," in Josef Brozek, ed., *Explorations in the History of Psychology in the United States* (Cranbury, N.J.: Associated University Presses, 1984), pp. 216–222.

52 James McKeen Cattell, "Mental Tests and Measurements," *Mind* 15 (1890): 373–381; Cattell and Livingston Farrand, "Physical and Mental Measurements of the Students of Columbia University," *Psychological Review* 3 (1896): 618–648. Michael

Sokal, "James McKeen Cattell and the Failure of Anthropometric Mental Testing, 1890–1901," in William R. Woodward and Mitchell G. Ash, eds., *The Problematic Science: Psychology in Nineteenth-Century Thought* (New York: Praeger, 1982), pp. 322–345. For the broader concepts shaping psychophysics, see also Hornstein, "Quantifying Psychological Phenomena."

53 Cattell and Farrand, "Physical and Mental Measurements," p. 648.

54 By 1909, Goddard had collected information on the height and weight of about 8,000 institutional cases; see "From the Report of the Research Department," *TSB* 6 (June 1909): 169–171. For examples of earlier anthropometric and psychometric research, see Allen Gilbert, "Researches upon School Children and College Students," *University of Iowa Studies in Psychology* 1 (1897): 1–39; and A.R.T. Wylie, "Report of Committee on Psychological Research," *JPA* 6 (September 1901): 21–26 and "Contribution to the Study of the Growth of the Feeble-Minded in Height and Weight," *JPA* 8 (September 1903): 1–7. See also James Allen Young, "Height, Weight, and Health: Anthropometric Study of Human Growth in Nineteenth-Century American Medicine," *Bulletin of the History of Medicine* 53 (1979): 214–243; and James B. Gilbert, "Anthropometrics in the U.S. Bureau of Education: The Case of Arthur MacDonald's 'Laboratory,'" *History of Education Quarterly* 16 (Summer 1977): 169–195. HHG, September 24 entry, "First Year." For the instruments in Goddard's laboratory, see Popplestone and McPherson, "Pioneer Psychology Laboratories," pp. 238–243.

55 HHG, "Research Work," p. 8; HHG, "Psychological Work," p. 26.

56 HHG, November 27 entry, "First Year" (emphasis in original).

57 Meyer to HHG, February 8, 1906.

58 Lightner Witmer, "The Psychological Clinic," *Old Penn* 8 (1909): 100. See also Lightner Witmer, "Psychological Diagnosis and the Psychonomic Orientation of Analytic Science: An Epitome," *PC* 16 (1925), reprinted in Robert A. Brotemarkle, ed., *Clinical Psychology: Studies in Honor of Lightner Witmer . . .* (Philadelphia: University of Pennsylvania Press, 1931), pp. 388–390; and Samuel W. Fernberger, "The History of the Psychological Clinic," pp. 10–36 in the same volume; Popplestone and McPherson, "Pioneer Psychological Laboratories," pp. 235–238; and John O'Donnell, "The Clinical Psychology of Lightner Witmer: A Case Study of Institutional Innovation and Intellectual Change," *JHBS* 15 (1979): 3–17.

59 Naomi Norsworthy, *Psychology of Mentally Deficient Children,* Archives of Psychology, no. 1, November 1906 (New York: Science Press, 1906).

60 Ibid., pp. 4–9, quotations pp. 7, 9. On Norsworthy, see Popplestone and McPherson, "Pioneer Psychology Laboratories," pp. 253–255.

61 HHG, "Research Work," p. 6.

62 "Remarks by Francis Galton, F.R.S.," p. 380, following Cattell, "Mental Tests and Measurements." For a discussion of this issue, see Read D. Tuddenham, "The Nature and Measurement of Intelligence," in Leo Postman, ed., *Psychology in the Making: Histories of Selected Research Problems* (New York: Knopf, 1963), p. 477.

63 Michael Sokal, "James McKeen Cattell and the Failure," p. 338.

64 HHG to "Dear Friends (or 'would be' Friends)," March 6, 1952, File M35.2, GP.

65 HHG, European Diary, File M33.1, GP. HHG, "Two Months Among the European Institutions," *TSB* 5 (July 1908): 11–16.

66 Binet and Simon's 1905 articles, "Sur la necessité d'établir un diagnostic scientifique des états inférieurs de l'intelligence," "Méthodes nouvelles pour le diagnostic du niveau intellectuel des anormaux," and "Applications des méthodes nouvelles au diagnostic du niveau intellectuel chez des enfants normaux et anormaux d'hospice et d'école primaire," appeared in *L'Année Psychologique* 11 (1905): 163–190, 191–244, and 245–336. Goddard had them translated by Elizabeth Kite and reprinted with 1908 and 1911 articles in *The Development of Intelligence in Children* (Baltimore: Williams & Wilkins, 1916). Further citations are to this volume. Kite translated other Binet articles in *The Intelligence of the Feeble-Minded* (Baltimore: Williams & Wilkins, 1916). For Ovide Decroly's work with Binet's scale, see "Les tests de Binet et Simon pour la Mesure de l'intelligence," *Archives de Psychologie* 6 (1908): 27–130.

67 HHG to George Crane, June 6, 1947, File M36, GP.

68 Theta Wolf, *Alfred Binet* (Chicago: University of Chicago Press, 1973).

69 HHG, March 31, 1908 entry, European Diary. HHG, "Two Months," p. 15.

70 Many factors account for Binet's inability to secure a prestigious appointment as chair of psychology. According to Wolf, he had not attended the "right" schools, had not studied philosophy, did not know German, came from a well-to-do family, and had no visible socialist proclivities. Of equal significance, he was described as "difficult, dominant – perhaps even domineering" in the laboratory. Between 1901 and 1904, Binet lost three potential positions to other candidates. See Wolf, *Alfred Binet,* pp. 22–28.

71 Ibid., esp. pp. 283–326. Binet was president of La Société from 1902 until his death in 1911; he was so central to its work that it was renamed La Société Alfred Binet in 1917. On La Société's members, see pp. 284–288.

72 HHG, "Child Study for Pennsylvania Teachers," *PSJ* 49 (September 1900): 127; HHG, Letter "To the Teachers of Pennsylvania" in "Child Study: Special Circular from Child Study Department" *PSJ* 49 (November 1900): 227. Alfred Binet, "Un livre récent de William James sur l'education: *Causeries pédagogiques." Bulletin de la Société libre pour l'étude psychologique de l'enfant,* no. 48 (1908): 167; and "Le passé et l'avenir de notre Société," *Bulletin de la Société . . . ,* no. 19 (1904): 548–549, as translated in Wolf, *Alfred Binet,* pp. 289, 311. The French teacher is François Zuza, cited in Wolf, p. 285.

73 Binet and Simon, *Development of Intelligence,* pp. 9–10.

74 Wolf, *Alfred Binet,* pp. 1–3, 19.

75 Binet and Simon, *Development of Intelligence,* p. 89.

76 Ibid., pp. 23–24.

77 Ibid., p. 14.

78 Ibid., p. 22.

79 Ibid., p. 24.

80 Alfred Binet, "La mesure en psychologie individuelle," *Revue philosophique,* 2nd sem. (1898): 122–123, as cited in Tuddenham, "The Nature and Measurement of Intelligence," p. 487. Binet and Simon, *Development of Intelligence,* p. 40.

81 Binet and Simon, *Development of Intelligence,* pp. 165–166.

82 Ibid., pp. 45–181.

83 Ibid. See also Alfred Binet and Theodore Simon, "Méthodes nouvelles pour diagnostiquer l'idiotie, l'imbécillité et la débilité mentale," *Atti del V Congresso internazionale di psicologia* (Rome: Forziani, 1906), pp. 507–510.

84 HHG, "The Binet and Simon Tests of Intellectual Capacity," *TSB* 5 (December 1908): 3–9. Binet and Simon, *Development of Intelligence,* pp. 182–273, quotations pp. 195, 224, 228. While Binet used the phrase "mental level," Goddard used "mental age."

85 HHG, "In the Beginning," p. 3.

86 HHG, "Introduction" to Binet and Simon, *Development of Intelligence,* p. 5. See also "In the Beginning," pp. 3–4.

87 HHG, "Suggestions for a Prognostical Classification," p. 48.

88 Ibid.

89 Ibid, pp. 49, 52.

90 Dr. Bernstein in "Discussion," following HHG, "Suggestions for a Prognostical Classification," p. 53.

91 Dr. Keating in ibid., p. 54.

92 Circular Letter from Walter Fernald to the Committee on Classification, April 23, 1910, in "Report of Committee on Classification of Feeble-Minded," *JPA* 15 (September and December 1910): 63.

93 HHG to Fernald, April 29th, 1910, in ibid., pp. 64–67.

94 HHG, "In the Beginning," p. 5.

95 HHG, "Four Hundred Feeble-Minded Children Classified by the Binet Method," *JPA* 15 (September and December 1910): 17–30. See also Edmund B. Huey, "Retardation and the Mental Examination of Retarded Children," *JPA* 15 (September and December 1910): 31–43; and F. Kuhlmann, "Binet and Simon's System for Measuring the Intelligence of Children," *JPA* 15 (March and June 1911): 76–92. Psychologist A.R.T. Wylie was also on the classification committee.

96 HHG, "Four Hundred Feeble-Minded Children," pp. 18–19.

97 Ibid., p. 19.

98 Ibid.

99 Fernald quoted in HHG, "In the Beginning," pp. 3–4.

100 "Report of Committee on Classification," pp. 61–67, quotation p. 61. "Minutes of the Association," *JPA* 15 (March and June 1911): 130–135; HHG, "In the Beginning," pp. 2–6.

101 HHG to Editorial Department, *Century Dictionary,* October 14, 1914, File M615, GP.

102 A. C. Rogers, "Editorial: The New Classification (Tentative) of the Feeble-Minded," *JPA* 15 (September and December 1910): 70.

103 Ibid., pp. 68–71.

104 Ibid., pp. 68–69.

105 Ibid., pp. 70–71.

Chapter 4

1 Francis Burke Brandt, "The State in its Relation to the Defective Child," *NEA* (1901): 877. Club members are listed as Vineland's "Paidological Staff" in *AR* (1906): 4.

2 HHG, "What Can the Public School Do for Subnormal Children?" *NEA* (1910): 914–915.

3 Carroll Pearse, "Public Schools for the Exceptional Child," *NEA* (1909): 873–874; Lucius Button, "The Care of Our Feeble-Minded School Children," *TSB* 7 (February 1911): 323.

4 "Preliminary Report of the Committee on Provision for Exceptional Children in the Public Schools," *NEA* (1908): 346; "Survey of Educational Progress, 1900–1910," *U.S. Commissioner of Education Report* (1911), 1:19; Guy M. Whipple, "Special Classes," in Paul Monroe, ed., *A Cyclopedia of Education* (New York: Macmillan, 1911), 5:384–386. Whipple reports that by 1911, 94 American cities had classes for "defectives," while 207 had remedial classes for "backward" children.

5 Olive Jones, "The Systematic Care of the Exceptional Child," *NEA* (1909): 347.

6 Mary Pogue, "Concerning our Limitations in Educating Mentally Deficient Children," *NEA* (1905): 895–896.

7 Maximilian Groszmann, "What Consideration Should be Given to Subnormal Pupils?" *NEA* (1910): 163; David Lincoln, "Special Classes for Feeble-Minded Children in the Boston Public Schools," *JPA* 7 (June 1903): 84–85.

8 ERJ, "The Functions of the Special Class," *NEA* (1908): 1116. HHG, "Report of the Research Department," *AR* (1909): 43.

9 ERJ, "Functions of the Special Class," p. 1116.

10 Lightner Witmer, "Clinical Records," *PC* 9 (1915): 1. For a comparison of Witmer's approach to Goddard's, see John O'Donnell, "The Clinical Psychology of Lightner Witmer: A Case Study of Institutional Innovation and Intellectual Change," *JHBS* 15 (1979): 3–17, esp. pp. 7–8.

11 ERJ, "The Institution as a Laboratory," in Doll, ed., *Twenty-Five Years* (Vineland, N.J.: The Training School, 1932), pp. 1–15.

12 HHG, "What Can the Public School Do," pp. 916–917.

13 ERJ, *Dear Robinson: Some Letters on Getting Along with Folks* (Vineland, N.J.: n.p., 1923), pp. 60–61.

14 Ibid.

15 ERJ, "What the Summer School Taught Us," *TSB,* no. 46 (December 1907): 1; HHG, "Report of the Research Department," *AR* (1909): 37; ERJ, "Report of the Superintendent," *AR* (1909): 22; [ERJ], "The Winter Training Class for Teachers," *TSB* 8 (May 1911): 46; [ERJ], "Summer School," *TSB* 8 (September 1911): 76–77.

16 ERJ, "Report of the Superintendent," *AR* (1909): 25–26. Edouard Seguin, *Idiocy, and Its Treatment by the Physiological Method* (New York: Wood, 1866).

17 ERJ, "Foreword," *Directory of Graduates, The Training School* (Vineland, N.J.: n.p., n.d. [c. 1924]), File M43, GP.

18 HHG, "Report of the Research Department," *AR* (1908): 41; HHG, "Report of the Research Department," *AR* (1909): 40.

19 HHG, "What Can the Public School Do," pp. 914–917.

20 HHG, "Psychological Work Among the Feeble-Minded," *JPA* 12 (September 1907): 23. HHG, "What Can the Public School Do," pp. 914, 916.

21 HHG, "What Can the Public School Do," pp. 915–917.

22 Ibid., p. 917.

23 ERJ, "Some Aides in Administration," *JPA* 12 (1908): 88. Joseph Byers, *The Village of Happiness: The Story of the Training School* (Vineland, N.J.: Smith, 1934).

24 Mary C. Breen, "Echoes from the Summer School: The Class of 1912," *TSB* 9 (September 1912): 74–75.

25 Harriet E. Monks, "Class Song – 1912," *TSB* 9 (September 1912): 75. This song had several more verses (emphases in original).

26 S. I. Schermerhorn, "From the Field: Richmond, Va.," *TSB* 9 (April 1912): 26–27.

27 "Miss Gleason" in "Special Class Items," *TSB* 7 (June 1910): 236; C.M.C., "Grand Rapids" in "Alumni Department," *TSB* 7 (November 1910): 289; "Newton Centre, Mass." in "Special Class Department," *TSB* 8 (May 1911): 48.

28 [ERJ], "What the Summer School Taught Us," p. 1.

29 Ibid.

30 Ibid.

31 "Alumni Department," *TSB* 7 (November 1910): 286.

32 C.M.C., "Grand Rapids," in ibid., p. 289; "Morrisville, Pa." in "Special Class Department," *TSB* 7 (February 1911): 332; "Toledo, Ohio" in "Special Class Department," *TSB* 8 (March 1911): 16.

33 "Following our lead," Johnstone wrote, "summer schools for this purpose have been established at the New York University, the University of Washington, the University of Pittsburgh, the University of Pennsylvania, the University of California, Harvard University, Columbia University, and the State Normal College of Greeley, Colo." "Report of the Superintendent," *AR* (1913): 23–24. Witmer's work was certainly influential as well, especially in starting the University of Pennsylvania's summer program.

34 *Directory of Graduates.* On the summer school's influence, see Kathrine Regan McCaffrey, "Founders of the Training School at Vineland, New Jersey: S. Olin Garrison, Alexander Johnson, Edward R. Johnstone" (Ph.D. diss., Teachers College, Columbia University, 1965), pp. 180–223.

35 ERJ, "Report of the Superintendent," (1909): 18; "Seattle, Wash." in "Special Class Department," *TSB* 8 (November 1911): 111.

36 HHG, "What Can the Public School Do," p. 913.

37 James H. Van Sickle, Lightner Witmer, and Leonard P. Ayres, *Provision for Exceptional Children in Public Schools,* U.S. Bureau of Education Bulletin, 1911, no. 14 (Washington, D.C.: GPO, 1911), p. 17.

38 New York City, Department of Education, *Sixth Annual Report of the City Superintendent of Schools* (1904), p. 47. Oliver P. Cornman, "The Retardation of the Pupils of Five City School Systems," *PC* 1 (January 15, 1908): 247. Only first-graders aged eight or older were considered "overage," since in many school systems, compulsory education began at age seven. If those only one year behind were included, New York in 1904 would have recorded 71 percent overage. James Bryan, "A Method for Determining the Extent and Causes of Retardation in a City School System," *PC* 1 (April 15, 1907): 41–52.

39 Bryan, "Method for Determining the Extent," p. 47; Cornman, "Retardation," p. 249; Leonard Ayres, *Laggards in Our Schools: A Study of Retardation and Elimination in City School Systems* (New York: Charities Publications Committee, 1909), p. 3.

40 For a related discussion of the need to make the curriculum match child development, see John Dewey, *The Child and the Curriculum* (Chicago: University of Chicago Press, 1902).

41 Ayres documented these problems. For instance, in Camden, only 17 of every 100 children who started the elementary course completed all eight grades. Ayres, *Laggards,* p. 4.

42 Cornman, "Retardation," pp. 245–250.

43 Ayres, *Laggards,* pp. 3, 220.

44 Van Sickle, Witmer, and Ayres, *Provision for Exceptional Children,* pp. 26–28.

45 Ayres, *Laggards.* The "laggard" profile was developed by comparing "promoted and nonpromoted pupils." See Van Sickle, Witmer, and Ayres, *Provision for Exceptional Children,* pp. 28–29.

46 Ibid., p. 29.

47 Ibid., p. 13; HHG, "What Can the Public School Do," pp. 912–913.

48 Van Sickle, Witmer, and Ayres, *Provision for Exceptional Children,* p. 17.

49 Ibid., pp. 23–30.

50 Ayres, *Laggards,* p. 219; Cornman, "Retardation," p. 257. For an analogous philosophy about transforming pedagogy, see Geraldine Joncich, *The Sane Positivist: A Biography of Edward Thorndike* (Middletown, Conn.: Wesleyan University Press, 1968), pp. 282–311.

51 HHG, "What Can the Public School Do," p. 919.

52 HHG, Notes to "As Luck Would Have It," GP.

53 Ibid.

54 HHG, "Two Thousand Normal Children Measured by the Binet Measuring Scale of Intelligence," *PS* 18 (June 1911): 232–259. This paper and his previous Vineland Binet study "aroused the schoolmen of America"; HHG, Notes to "As Luck Would Have It," GP.

55 HHG, "Two Thousand Normal Children," pp. 232–236. Binet's 1908 sample included 203 children. Although Goddard claimed he tested about 2,000 children, his tests for seventh- through twelfth-graders proved unsatisfactory. His actual results comparing age with mental age concerned 1,536 children (not 1,547, as claimed), while his most complete data on age, mental age, and grade involve 1,290 children. Here I am citing the numbers Goddard used to make his points in his article, not the actual numbers evident from his published data.

56 Ibid., pp. 234–237.

57 Estimates of children at least two years advanced are in Ayres, *Laggards,* pp. 73–88 (which found from 1 to 5 percent) and Van Sickle, Witmer, and Ayres, *Provision for Exceptional Children,* p. 18 (4 percent).

58 In Goddard's New Jersey district, 23 percent in the first six grades were at least two years "overage," while 51 percent were "overage" by one or more years. These figures are based on the chart comparing age and grade for 1,290 children in Table IV, Section H, in HHG, "Two Thousand Normal Children," p. 245.

59 Medical estimates varied. The English Royal Commission of 1904 found from .110 to .468 percent feebleminded, with the higher percentages in rural areas. See A. F. Tredgold, *Mental Deficiency* (New York: Wood, 1908; rpt. 1922), pp. 11–12. Since the most serious cases would not be in public schools, Goddard's 3 percent school estimate would seem even higher. In his New Jersey district, 1.7 percent in the first six grades were four years behind; see Table IV, Section H, in HHG, "Two Thousand Normal Children," p. 245.

60 HHG, "Two Thousand Normal Children," pp. 236, 250.

61 Ibid., pp. 241-247.

62 Ibid., pp. 241-242. Again, I am citing the figures Goddard used, although children whose mental age matched grade in school actually constituted 45.2 percent, not 43.2 percent, as he stated.

63 Ibid., p. 241.

64 Ibid., pp. 246–249.

65 Ibid., pp. 242, 249.

66 Ibid., pp. 247, 250.

67 Ibid.

68 Ibid., pp. 251–254. One test involved cutting out shapes of paper. Goddard tried it not only on thirteen-year-olds but also on twenty-five "teachers, superintendents of schools, psychologists, mathematicians and biologists" – probably the Feeble-Minded Club; of these, only six passed it.

69 Ibid., pp. 254–258.

70 HHG, "Two Thousand Children Tested by the Binet Measuring Scale for Intelligence," *NEA* (1911): 870–878.

71 Chapter 234 of New Jersey Public Law (1911). For the earlier version of this bill, see HHG, "Two Thousand Children," p. 870. The following year, a new law reimbursed school boards $500 for each teacher of the blind, deaf, or feebleminded (Chapter 141, Laws of 1912). These laws, however, affected mainly large cities, for smaller districts rarely identified enough children to start special classes. See Joseph Vincent Summers, Jr., "The Development of Public School Special Education through Legislation in the State of New Jersey from 1911–1968" (Ph.D. diss., Temple University 1969), pp. 81–86.

72 HHG, "Two Thousand Children," p. 870. This change had evidently taken place between the time that Goddard delivered his paper and the publication of the NEA's proceedings, for he added this information as a footnote to his address.

73 James McKeen Cattell, for instance, was equally ignorant of statistical methods. "Cattell was mathematically illiterate – his addition and subtraction were often inaccurate," according to his biographer, Michael Sokal. Cattell had Clark Wissler, a graduate student, calculate the correlation coefficients for his mental measurements. See Sokal, "James McKeen Cattell and the Failure of Anthropometric Mental Testing, 1890–1901," in William R. Woodward and Mitchell Ash, eds., *The Problematic Science: Psychology in Nineteenth-Century Thought* (New York: Praeger, 1982), p. 337. Theta Wolf reports similar arithmetic errors in one of Binet's early articles; see *Alfred Binet* (Chicago: University of Chicago Press, 1973), p. 98.

74 Goddard's article is filled with addition errors. Children testing within one year of their age constituted 76%, not 78% as he reported; backward children constituted 16%, not 15%, and the gifted 5%, not 4%. The number whose grade matched their mental age was 584, not 558 (45.2%, not 43.2% as he reported).

75 The *Journal of Educational Psychology* was edited by William Bagley of the University of Illinois, J. Carleton Bell of the Brooklyn Training School, Carl Seashore of the University of Iowa, and Guy Whipple of Cornell. They hoped to be "middlemen" linking education and psychology through their "middle-magazine." "Editorial," *JEP* 1 (January 1910): 1.

76 Leonard Ayres, "The Binet-Simon Measuring Scale for Intelligence: Some Criticisms and Suggestions," *PC* 5 (November 15, 1911): 189–196. For answers to each point raised by Ayres, see F. Kuhlmann, "A Reply to Dr. L. P. Ayres' Criticism of the Binet and Simon System for Measuring the Intelligence of Children," *JPA* 16 (1911): 58–67.

77 Ayres, "Binet-Simon Measuring Scale," pp. 189–190.

78 Ibid., pp. 190–193.

79 Ibid., pp. 194–196.

80 Ibid.

81 Clara Schmitt, "The Binet–Simon Tests of Mental Ability. Discussion and Criticism," *PS* 19 (June 1912): 186–200. William Healy, *The Individual Delinquent* (Boston: Little, Brown, 1915), cites tests used at his clinic.

82 Schmitt, "Binet-Simon Tests," pp. 192–197.

83 Ibid., pp. 190–192.

84 Lewis M. Terman, "Trails to Psychology," in Carl Murchison, ed., *History of Psychology in Autobiography* (Worcester, Mass.: Clark University Press, 1932), 2: 297–331; May Seagoe, *Terman and the Gifted* (Los Angeles: W. Kaufmann, 1975); Henry L. Minton, *Lewis M. Terman, Pioneer in Psychological Testing* (New York: New York University Press, 1988); and Paul Chapman, *Schools as Sorters: Lewis M. Terman, Applied Psychology, and the Intelligence Testing Movement, 1890–1930* (New York: New York University Press, 1988).

85 Lewis M. Terman, "The Binet–Simon Scale for Measuring Intelligence: Impressions Gained by its Application upon Four Hundred Non-selected Children," *PC* 5 (December 15, 1911): 199–206.

86 Lewis M. Terman and H. G. Childs, "A Tentative Revision and Extension of the Binet–Simon Measuring Scale of Intelligence," *JEP,* in three parts: 3 (February 1912): 61–74; 3 (April 1912): 198–208; and 3 (May 1912): 277–289. Quotations pp. 70–71.

87 Terman, "The Binet–Simon Scale for Measuring Intelligence," p. 204.

88 J. Carleton Bell, "Abstracts and Reviews. Recent Literature on the Binet Tests," *JEP* 3 (January 1912): 101–110. Quotations pp. 102, 107–108.

89 Ibid., p. 110.

90 Alfred Binet and Theodore Simon, *The Development of Intelligence in Children,* trans. Elizabeth Kite (Baltimore: Williams & Wilkins, 1916), pp. 253, 258–259.

91 Ibid., pp. 253–254.

92 Ibid., pp. 254, 259.

93 Ibid., p. 240.

94 Ibid., p. 259.

95 HHG, "Two Thousand Normal Children," pp. 237–238.

96 William Healy to HHG, April 27, 1912, File M614, GP. The actual number of New York pupils was 770,243; surveyors usually worked with an estimated figure of 750,000. See Paul Hanus, *Adventuring in Education* (Cambridge, Mass.: Harvard University Press, 1937), p. 178.

97 Paul Hanus, "Introduction," *Report on Educational Aspects of the Public School System of the City of New York to the Committee on School Inquiry of the Board of Estimate and Apportionment* [hereafter *Educational Aspects*], Part II (New York: City of New York, 1911–1912). On surveys, see Jesse Sears, *The School Survey* (Boston: Houghton Mifflin, 1925); and Martin Balmer, Kevin Bales, and Kathryn Kish Sklar, eds., *The Social Survey in Historical Perspective, 1880–1940* (New York: Cambridge University Press, 1991).

98 Hanus, *Adventuring,* pp. 175–193. Hanus describes the political intrigues of School Inquiry chairman John Purroy Mitchel (then Board of Alderman president) and a Bureau of Municipal Research director he calls "X" to use his survey to "discredit the Board of Education and the administration of the city superintendent, thus supplying X and his bureau with material which Mitchel might use in forwarding his ambition to become mayor of New York." (Mitchel later did become the city's mayor.) Hanus, p. 182.

99 New York City began educating handicapped children in public schools in 1899, when it allowed the Association for the Aid of Crippled Children to start a class. Hospital classes for tubercular children were begun in 1904; five years later, this city tried "open air" classes for those considered susceptible to this disease. Speech instruction for stutterers was begun in 1907. In 1908 it opened a School for the Deaf, and the next year, six classes for children with serious vision problems. *The First Fifty Years: A Brief Review of Progress, 1898–1948* (New York: City Board of Education, 1948), pp. 21–48.

100 HHG, "Elementary Schools, Section E. – Ungraded Classes," *Educational Aspects,* Part II, Subdivision I, pp. 361–381. "Isabella Thompson Smart," in John William Leonard, ed., *Woman's Who's Who of America* (New York: American Commonwealth Co., c. 1914), p. 753.

101 This city had 126 ungraded classes in 1911; Goddard visited 125. By 1912, there were there were 131. HHG, "Ungraded Classes."

102 Ibid., pp. 363–369, 376–380.

103 Ibid., pp. 362–363.

104 Ibid.

105 Ibid. pp. 362–363, 369–373.

106 Ibid., pp. 369–373.

107 William Maxwell, "The School Inquiry Reports," in *Annual Report of the City Superintendent of Schools* 15 (1912–1913): 169. Maxwell saw the survey as a political attack on his administration; Hanus agreed, for he believed the politician he called "X" "cherished a personal animosity against the city superintendent of schools (Dr. William H. Maxwell, in many respects the best superintendent New York City has ever had), and he wanted to 'get' Maxwell." Hanus, *Adventuring,* p. 186.

108 Farrell's reply was published in three parts in *PC:* "A Study of the School Inquiry Report on Ungraded Classes," 8 (April 15, 1914): 29–47; 8 (May 15. 1914): 57–74; and 8 (June 15, 1914): 99–106 [hereafter "School Inquiry Report," pts. 1, 2, and 3 respectively].

109 Ibid., pt. 3, pp. 99–101. Farrell is apparently quoting notes from her visit to the Massachusetts School for the Feebleminded at Waverly.

110 Ibid., pt. 2, p. 63.

111 HHG, "Ungraded Classes," p. 369.

112 Farrell cites Witmer in "School Inquiry Report," pt. 1, p. 44. Witmer also demonstrated his support for Farrell by publishing her long response in his journal, *PC.*

113 On methods used by Farrell and Smart to diagnose children for ungraded classes, see Mary Sutton Macy, M.D., "The Subnormal Child in New York City Schools," *JEP* 1 (March 1910): 132–144. Macy's analysis is very close to Goddard's, for she estimates that by 1905, 7,000 children were entitled to special education, but city programs reached only 361, or 6 percent. Goddard may have been influenced by Macy, for she was a Lecturer in Physiology and Pathology in Education at New York University, where he also lectured.

114 Isabelle Thompson Smart, M.D., "Some Urgent Needs for Advancement in the Education of Mentally Defective Children," *NEA* (1908): 1145. Smart's diagnosis of one of Goddard's cases is in Farrell, "School Inquiry Report," pt. 1, pp. 35–38.

115 Farrell cites Fernald and Barr in "School Inquiry Report," pt. 1, pp. 43–44.

116 Ibid., pt. 3, p. 106.

117 Ibid., pt. 3, pp. 105–106.

118 Minutes to Board of Education meetings determining the procedure for dealing with the School Inquiry Reports are in *Journal of the Board of Education* (1912): 1380–1381, 1407–1408, and (1913): 248, 464–467, and 576–577. "Ungraded Classes – Report of the Special Committee Appointed to Consider the Report on 'Ungraded Classes' Submitted to the Committee on School Inquiry of the Board of Estimate and Apportionment by Dr. H. H. Goddard," Board of Education, City of New York, Document No. 2 (1914), Depot Box 5, GP, and Document No. 7 (1914), New York City Board of Education Records, Teachers College.

119 "Ungraded Classes," Document No. 7 (1914). The committee also suggested that "a circular of information be prepared explaining the most obvious signs of mental deficiency for the use of teachers and principals, to aid them in discovering defective children." This report was presented to the Board of Education on March 11 and amended and adopted on May 27, 1914.

120 Obvious once again in retrospect are numerous mathematical mistakes. In New Jersey, Goddard had found not 2 percent but 3 percent feebleminded – a figure that should have led him to estimate not 15,000 but 21,500 feebleminded students, thus adding even more incredulity to his New York report. Moreover, in New Jersey the feebleminded were those four or more years behind, and in New York, three or more years. Even Farrell missed these important inconsistencies in a study which saw such numbers as so meaningful. Other evidence from New Jersey would also have aided Farrell, for Goddard found rates of feeblemindedness higher among rural than among urban children – a finding confirmed by medical estimates. Farrell quotes these medical estimates in explaining why New York differed from Goddard's New Jersey school district; she could have quoted Goddard as well.

121 *NYT,* February 8, 1913, p. 10. The *Times* reported many of the controversies surrounding the Hanus survey results.

122 "North Dakota" in "Alumni Department," *TSB* 7 (November 1910): 289; "Washington, D.C." in "Special Class Department," *TSB* 8 (November 1911): 111.

123 "Special Class Department," ibid., p. 112; "Newark" in "Special Class Department," *TSB* 8 (March 1911): 15.

124 "Orange, N.J.," in "Progress of the Work for Defectives," *TSB* 9 (March 1912): 15–16; "Newton Centre, Mass." in "Special Class Department," *TSB* 8 (May 1911): 47.

125 Goddard's central role in structuring these curricula is cited by Farrell in "School Inquiry Report," pt. 2, p. 64. *New York University Summer School Bulletin* (1912): 2.

126 "Arnold Gesell," in Edwin Boring et. al., eds., *History of Psychology in Autobiography* (Worcester, Mass.: Clark University Press, 1952), 4:127–128. Gesell says he worked at New York University from 1911 to 1915, although summer school bulletins first mention this program in 1912.

127 Farrell, "School Inquiry Report," pt. 2, p. 64. *New York University Summer School Bulletin* (1913): 50–53.

128 Clara Harrison Town, "The Binet-Simon Scale and the Psychologist," *PC* 5 (January 15, 1912): 239. On Town, see John A. Popplestone and Marion White McPherson, "Pioneer Psychology Laboratories in Clinical Settings," in Josef Brozek, ed., *Explorations in the History of Psychology in the United States* (Cranbury, N.J.: Associated University Presses, 1984), pp. 256–257.

129 Wallin often asked Goddard to recommend him for positions, even while criticizing Goddard's work in print. He lost one position, he wrote Goddard, because he was reputedly "a difficult man to get along with." While his writings were significant in questioning the overstatements of other testers, Wallin largely remained an isolated maverick. Wallin's letters to Goddard are in File M615, GP. See also Wallin, *Odyssey of a Psychologist* (Wilmington, Del.: J. E. Wallace Wallin, 1955); and Popplestone and McPherson, "Pioneer Psychological Laboratories," pp. 245–248.

130 J. E. Wallace Wallin, "Danger Signals in Clinical and Applied Psychology," *JEP* 3 (April 1912): 224–226. Farrell cites Wallin in "School Inquiry Report," pt. 1, p. 43. See also Wallin, *The Functions of the Psychological Clinic* (New York: Wood, 1913); and "Experimental Studies of Mental Defectives. A Critique of the Binet–Simon Tests," *Educational Psychology Monographs* No. 7 (1912).

131 Wallin, *Odyssey,* pp. 118–119. Allan Chase, *The Legacy of Malthus: The Social Costs of the New Scientific Racism* (New York: Knopf, 1977), pp. 238–242.

132 G. M. W[hipple], "The Amateur and the Binet–Simon Tests," *JEP* 3 (January 1912): 118–119.

133 Lewis M. Terman, "Concerning Psycho-Clinical Expertness," *TSB* 11 (March 1914): 10. For a similar defense, see Edgar Doll's review of Wallin's book, *The Functions of the Psychological Clinic,* in *TSB* 11 (March 1914): 10–12.

134 HHG, "The Binet Tests and the Inexperienced Teacher," *TSB* 10 (March 1913): 9–11.

135 Ibid., p. 9.

136 Whipple, "The Use of Mental Tests in the School," *Fifteenth Yearbook of the National Society for the Study of Education* (Bloomington, Ill.: Public School Publishing, 1916), pp. 153–154; his list of cities using testing in 1914 is from Wallin. HHG, "Introduction" to Binet and Simon, *Development of Intelligence,* p. 6. "It will seem an exaggeration to some to say that the world is talking of the Binet–Simon Scale," Goddard wrote, "but consider that the Vineland Laboratory alone, has without effort or advertisement distributed to date 22,000 copies of the pamphlet describing the tests, and 88,000 record blanks."

Chapter 5

1 HHG, "Psychological Work Among the Feeble-Minded," *JPA* 12 (September 1907): 23; HHG, *FM* (New York: Macmillan, 1914), pp. 19–20.

2 HHG, *School Training of Defective Children* (Yonkers-on-Hudson, N.Y.: World Book, 1914). Many New York survey reports were published in this World Book series.

3 HHG, *FM;* HHG, *KF* (New York: Macmillan, 1912).

4 J. David Smith, *Minds Made Feeble: The Myth and Legacy of the Kallikaks* (Rockville, Md.: Aspen Systems Corporation, 1985); Seymour Sarason and John Doris, *Psychological Problems in Mental Deficiency* (New York: Harper & Row, 1969), pp. 256–274; and Stephen Jay Gould, *Mismeasure of Man* (New York: Norton, 1981), pp. 158–174.

5 This passage is found in Exodus 20:5 and Deuteronomy 5:9.

6 Samuel Gridley Howe, "On the Causes of Idiocy," in Marvin Rosen, Gerald R. Clark, and Marvin Kivitz, eds., *The History of Mental Retardation, Collected Papers* (Baltimore: University Park Press, 1976), 1:31–60.

7 Ibid.

8 Edouard Seguin, *New Facts and Remarks Concerning Idiocy* (New York: Wood, 1870), p. 39.

9 Howe, "On the Causes of Idiocy," 33.

10 Peter Tyor, "Segregation or Surgery: The Mentally Retarded in America, 1850–1920" (Ph.D. diss., Northwestern University, 1972), pp. 91–97.

11 Frederick H. Wines, "Report on the Defective, Dependent, and Delinquent Classes of the Population of the United States . . . (June 1, 1880)," *House Miscellaneous Documents,* 47th Cong., 2d sess., 1882–1883, vol. 13, pt. 21:179–298.

12 Daniel Pick, *Faces of Degeneration: A European Disorder, c. 1848–1918* (Cambridge: Cambridge University Press, 1989); J. E. Chamberlin and Sander L. Gilman, eds., *Degeneration: The Dark Side of Progress* (New York: Columbia University Press, 1985); Charles Rosenberg, "The Bitter Fruit: Heredity, Disease, and Social Thought," in *No Other Gods: On Science and American Social Thought* (Baltimore: Johns Hopkins University Press, 1976), p. 43; Steven Gelb, "Degeneracy Theory, Eugenics, and Family Studies," *JHBS* 26 (1990): 242–246; and Gelb, "The Beast in Man: Degenerationism and Mental Retardation, 1900–1920," *Mental Retardation* 33 (February 1995): 1–9.

13 Isaac Kerlin, "Provision for Idiotic and Feeble-Minded Children," *NCCC* 11 (1884): 258.

14 Kerlin, "Report of the Standing Committee: Provision for Idiots," *NCCC* 12 (1885): 174.

15 Ibid. Versions of this biblical quotation describing life in heaven can be found in Matthew 22:30, Mark 12:25, and Luke 20:35.

16 Richard Dugdale, *The Jukes: A Study in Crime, Pauperism, Disease, and Heredity* (New York: G. P. Putnam's Sons, 1877). For a summary, see Mark Haller, *Eugenics: Hereditarian Attitudes in American Thought* (New Brunswick, N.J.: Rutgers University Press, 1963), pp. 21-25. For an analysis of such studies as a sociological genre, see Nicole Hahn Rafter, *White Trash: The Eugenic Family Studies, 1877–1919* (Boston: Northeastern University Press, 1988).

17 Haller, *Eugenics,* pp. 8–57; Richard Hofstadter, *Social Darwinism in American Thought* (New York: Braziller, 1955); George W. Stocking, *Race, Culture, and Evolution: Essays in the History of Anthropology* (New York: Free Press, 1968); Hamilton Cravens, *The Triumph of Evolution: American Scientists and the Heredity–Environment Controversy, 1900–1941* (Philadelphia: University of Pennsylvania Press, 1978); Robert Bannister, *Social Darwinism: Science and Myth in Anglo-American Social Thought* (Philadelphia: Temple University Press, 1979); Robert Richards, *Darwin and the Emergence of Evolutionary Theories of Mind and Behavior* (Chicago: University of Chicago Press, 1987); and Carl Degler, *In Search of Human Nature: The Decline and Revival of Darwinism in American Social Thought* (New York: Oxford, 1991), pp. 1–55.

18 Pick, *Faces of Degeneration,* pp. 109–152; Haller, *Eugenics,* pp. 40– 47; Gould, *Mismeasure of Man,* pp. 123–145; and Arthur Fink, *Causes of Crime: Biological Theories in the United States, 1800–1915* (Philadelphia: University of Pennsylvania Press, 1938).

19 G. Stanley Hall, "The New Psychology as a Basis of Education," *Forum* 17 (August 1894): 716–717.

20 On Galton's eugenics, see Haller, *Eugenics,* pp. 8–14; Daniel J. Kevles, *In the Name of Eugenics: Genetics and the Uses of Human Heredity* (New York: Knopf, 1985), pp. 3–19; Raymond Fancher, *The Intelligence Men: Makers of the IQ Controversy* (New York: Norton, 1985), pp. 18– 40; Theodore Porter, *The Rise of Statistical Thinking, 1820–1900* (Princeton: Princeton University Press, 1986), pp. 130–146; Derek Forrest, *Francis Galton: The Life and Work of a Victorian Genius* (London: Elek, 1974); Ruth Schwartz Cowan, *Sir Francis Galton and the Study of Heredity in the Nineteenth Century* (New York: Garland, 1984); and Diane Paul, *Controlling Human Heredity: 1865 to the Present* (Atlantic Highlands, N.J.: Humanities Press, 1995), pp. 30–36.

21 Francis Galton, *Hereditary Genius: An Inquiry into its Laws and Consequences* (London: Macmillan, 1869; rpt. 1950), p. 12.

22 These surgical techniques were vasectomy and salpingectomy.

23 Martin Barr, "Moral Paranoia," *AMO* (1895): 530–531.

24 On sterilization, see Tyor, "Segregation or Surgery," pp. 178–231; Philip R. Reilly, *The Surgical Solution: A History of Involuntary Sterilization in the United States* (Baltimore: Johns Hopkins University Press, 1991); Edward Larson, *Sex, Race, and Science: Eugenics in the Deep South* (Baltimore: Johns Hopkins University Press, 1995); James W. Trent, Jr., *Inventing the Feeble Mind: A History of Mental Retardation in the United States* (Berkeley: University of California Press, 1994), pp. 184–202; Allan Chase, *The Legacy of Malthus: The Social Costs of the New Scientific Racism* (New York: Knopf, 1977), pp. 12–23; and Rudolph J. Vecoli, "Sterilization: A Progressive Measure?" *Wisconsin Magazine of History* 43 (Spring 1960): 190–202. HHG, *KF,* p. 109. "The New Jersey Sterilization Law," *TSB* 8 (January 1912): 135–136.

25 Charles Rosenberg, "The Social Environment of Scientific Innovation: Factors in the Development of Genetics in the United States," in *No Other Gods,* p. 203. See also Peter Bowler, *The Non-Darwinian Revolution: Reinterpreting a Historical Myth* (Baltimore: Johns Hopkins University Press University Press, 1988).

26 Martin Barr, *Mental Defectives: Their History, Treatment, and Training* (Philadelphia: Blakiston, 1904), pp. 95–97, 108.

27 Ibid., pp. 106, 96; Barr, "Moral Paranoia," p. 530.

28 Barr, "Some Studies in Heredity," *JPA* 1 (September 1896): 2. On medical hereditarianism, see Rosenberg, "Bitter Fruit," pp. 25–53.

29 Studies of causation are cited in G. E. Johnson, "Contribution to the Psychology and Pedagogy of Feeble-Minded Children," *JPA* 1 (December 1896): 96–102; and Barr, "Some Studies in Heredity," p. 5.

30 ERJ, "Some Reasons for Mental Deficiency," *TSB* Supplement, No. 1-46 (December 1907): 13–18, quotations p. 14.

31 Ibid., pp. 13–17.

32 My summary of ideas about heredity as a "force" is largely based on Ruth Schwartz Cowan's analysis; see "Francis Galton's Contribution to Genetics," *Journal of the History of Biology* 5 (Fall 1972): 389–412. "Naturalists spoke of the 'force' of inheritance or the 'power' of inheritance or the 'principle' of inheritance – just as they might speak of the 'force' of gravity, the 'power' of electric attraction or the 'principle' of inertia," she writes (p. 399). The medical literature confirms Cowan's arguments. Barr, "Some Studies in Heredity," p. 1.

33 HHG, "Psychological Work," p. 23.

34 For the development of experimental biology and genetics, see Cravens, *Triumph of Evolution,* pp. 15–55; Jane Maienschein, *Transforming Traditions in American Biology, 1880–1915* (Baltimore: Johns Hopkins University Press, 1991); Kenneth Ludmerer, *Genetics and American Society: A Historical Appraisal* (Baltimore: Johns Hopkins University Press, 1972); and William Provine, *The Origins of Theoretical Population Genetics* (Chicago: University of Chicago Press, 1971). On medicine and genetics, see Ludmerer, pp. 63–73.

35 CBD to Edward Johnston [*sic*], March 9, 1909, Davenport Papers.

36 HHG to CBD, March 15, 1909, Davenport Papers.

37 Wines, "Report on the Defective, Dependent, and Delinquent Classes," pp. 240–241. A. C. Rogers, "On the *Ascribed* Causation of Idiocy, As Illustrated in Reports to the Iowa Institution for Feeble Minded Children," *AMO* (1884): 296–301.

38 ERJ, "Some Reasons for Mental Deficiency," p. 14.

39 HHG to CBD, March 15, 1909, Davenport Papers.

40 Ibid.

41 HHG to J. S. Woodward [President of the Board of Trustees of the Carnegie Institution], June 1, 1908, File M33 AA2, GP. HHG to CBD, March 15, 1909, Davenport Papers.

42 CBD to HHG, March 18, 1909, and HHG to CBD, April 19, 1909, Davenport Papers.

43 On Davenport, see E. Carleton MacDowell, "Charles Benedict Davenport, 1866–1944," *Bios* 17 (1946): 3–50; Charles Rosenberg, "Charles Benedict Davenport and the Irony of American Eugenics," in *No Other Gods,* pp. 89–97; and Kevles, *Name of Eugenics,* pp. 41-56.

44 HHG to CBD, May 5, 1909, and CBD to HHG, May 7, 1909, Davenport Papers. R. C. Punnett, *Mendelism* (New York: Wilshire, 1909 [American Edition]).

45 Cowan, "Francis Galton's Contribution," pp. 409–411. See also Porter, *Rise of Statistical Thinking,* pp. 279–286.

46 Cowan, "Francis Galton's Contribution," p. 407.

47 Ibid., pp. 408–412.

48 William Bateson, "Problems of Heredity as a Subject for Horticultural Investigation," *Journal of the Royal Horticultural Society* 25 (1900), as cited in Cowan, "Francis Galton's Contribution," p. 408.

49 For an overview of the impact of Mendelian ideas, see Peter Bowler, *The Mendelian Revolution: The Emergence of Hereditarian Concepts in Modern Science and Society* (Baltimore: Johns Hopkins University Press, 1989).

50 CBD, "Fit and Unfit Matings," *TSB* 7 (October 1910): 258–262. Biologist Edwin Grant Conklin spoke of the "Mendelian Law of Alternative Inheritance" as consisting of the Principle of Unit Characters, the Principle of Dominance, and the Principle of Segregation. See Conklin, *Heredity and Environment in the Development of Men* (Princeton: Princeton University Press, 1915, rpt. 1919), pp. 97–98.

51 "Mendelism [*sic*] Inheritance in Man (From Punnett's Mendelism)," *TSB* 8 (September 1911): 78–80, from R. C. Punnett, *Mendelism,* 3rd ed. (New York: Macmillan, 1911).

52 [HHG], "Review Punnett's Mendelism," *TSB* 8 (September 1911): 80. On the need for "pedigree studies," see Ludmerer, *Genetics and American Society,* pp. 55–62.

53 CBD to HHG, July 7, 1909, Davenport Papers. Although both claimed they were unsure if feeblemindedness was a "unit character," this hypothesis guided all their research.

54 Mendel is cited in Donald MacKenzie, *Statistics in Britain, 1865–1930: The Social Construction of Scientific Knowledge* (Edinburgh: Edinburgh University Press, 1981), p. 123.

55 On biometrics, see Porter, *Rise of Statistical Thinking*, pp. 270–314; on the biometrician-Mendelian controversy, see Provine, *Origins of Theoretical Population Genetics*, pp. 25–89; Ludmerer, *Genetics and American Society*, pp. 45–62; and MacKenzie, *Statistics in Britain*, pp. 120–152. HHG, *KF*, p. 110.

56 MacKenzie, *Statistics in Britain*, p. 123. My analysis of the biometrician–Mendelian controversy's impact on Goddard's work is based on MacKenzie's broader analysis of this debate, and particularly on his discussion of the significance of "Green Peas, Yellow Peas, and Greenish-Yellow Peas," pp. 120–129.

57 Bateson as cited in ibid., p. 126. MacKenzie sees in this controversy a contest over professional competencies, for some biometricians had little training as naturalists, while Mendelians were statistically ignorant. Ludmerer's research confirms this. According to Lionel Penrose, "Bateson was so totally unmathematical that in connection with gene frequencies he had difficulty distinguishing 2x from x^2." Ludmerer, *Genetics and American Society*, p. 45. For a different view, see Robert Olby, "The Dimensions of Scientific Controversy: The Biometric-Mendelian Debate," *British Journal of the History of Science* 22 (1989): 299–320.

58 Goddard and Davenport worked out the symbols for these heredity charts. See, for example, CBD to HHG, May 7, 1909, and CBD to HHG, July 21, 1909, Davenport Papers.

59 Between 1911 and 1924, Davenport trained 258 "eugenics field workers." See Garland Allen, "The Eugenics Record Office, Cold Spring Harbor, 1910–1940," *Osiris* 2nd Ser. 2 (1986): 225–264; Haller, *Eugenics*, pp. 63–68; and Kevles, *Name of Eugenics*, pp. 54–56.

60 Besides Vineland, Davenport also had field workers studying the insane in Matawan, New Jersey; delinquents at the Juvenile Psychopathic Institute in Chicago; albinos in Massachusetts; and the Amish in Pennsylvania, among other groups. Kevles, *Name of Eugenics*, p. 55. On the field work program, see CBD, H. H. Laughlin, David Weeks, ERJ, and HHG, *The Study of Human Heredity* (Cold Spring Harbor, N.Y.: Eugenics Record Office, Bulletin No. 2, 1911).

61 On social work during this era, see Roy Lubove, *The Professional Altruist: The Emergence of Social Work as a Career, 1880–1930* (New York: Atheneum, 1973); Regina Kunzel, *Fallen Women, Problem Girls: Unmarried Mothers and the Professionalization of Social Work, 1890–1945* (New Haven, Conn.: Yale University Press, 1993); and Elizabeth Lunbeck, *The Psychiatric Persuasion: Knowledge, Gender, and Power in Modern America* (Princeton: Princeton University Press, 1994), pp. 34–45.

62 CBD et. al., *Study of Human Heredity*, p. 9. Elizabeth Kite, "Method and Aim of Field Work at the Vineland Training School," *TSB* 9 (October 1912):81–87.

63 "Notes – Elizabeth S. Kite," Box 3, and "Something about the Author" from "Autobiography: A 'Conversion Story,'" Box 1, Kite Papers. Smith, *Minds Made Feeble*, pp. 50–51. See also "Elizabeth S. Kite," in John William Leonard, ed., *Woman's Who's Who of America* (New York: American Commonwealth, 1914), p. 462.

64 Kite's translations are in Alfred Binet and Theodore Simon, *The Development of Intelligence in Children* (Baltimore: Williams & Wilkins, 1916) and *The Intelligence of the Feeble-Minded* (Baltimore: Williams & Wilkins, 1916). Elizabeth Kite, "A Visit to Algiers," *The Friend: A Religious and Literary Journal* (Fifth Mo. 13 [May 13], 1905): 355.

65 Kite, "Method and Aim of Field Work," pp. 85–86; Kite to HHG, December 13, 1928, File M35.1, GP.

66 Kite, "Method and Aim of Field Work," p. 84.

67 Ibid.

68 Ibid.

69 Ibid., p. 86.

70 Galton, *Hereditary Genius,* pp. 33, 43.

71 HHG, *KF,* pp. 14–15.

72 Galton, *Hereditary Genius,* pp. 37–38. Galton, "The History of Twins as a Criterion of the Relative Powers of Nature and Nurture," *Journal of the Anthropological Institute of Great Britain and Ireland* 5 (1876): 391–406; Fancher, *Intelligence Men,* pp. 33–34.

73 HHG, *KF,* pp. 51–54.

74 Ibid.

75 HHG, "Heredity of Feeble-Mindedness," *American Breeders' Magazine* 1 (Third Quarter, 1910): 176.

76 On the ABA, see Barbara Kimmelman, "The American Breeders' Association: Genetics and Eugenics in an Agricultural Context, 1903–1913," *Social Studies of Science* 13 (May 1983): 163–204.

77 CBD to HHG, May 7, 1909, Davenport Papers. On the committee's membership, see CBD to HHG, May 24, 1909; HHG to CBD, May 27, 1909; CBD to HHG, July 9, 1909; and HHG to CBD, July 18, 1909, Davenport Papers. Members are listed on letterhead of the ABA's "Sub-committee on Heredity of the Feeble-Minded." A. C. Rogers, "Modern Studies in Heredity," *JPA* 14 (September and December 1909/ March and June 1910): 118.

78 HHG to CBD, May 27, 1909, Davenport Papers.

79 Haller, *Eugenics,* pp. 62–65, 31–33. On Bell's eugenics, see also Harlan Lane, *When the Mind Hears: A History of the Deaf* (New York: Random House, 1984), pp. 340–375.

80 CBD to HHG, July 9, 1909, Davenport Papers.

81 This discussion of Goddard's presentation is based on the version later published as "Heredity of Feeble-Mindedness," pp. 165–178.

82 Ibid., pp. 175–176.

83 CBD to HHG, April 2, 1910, Davenport Papers.

84 HHG to CBD, April 13, 1910, Davenport Papers.

85 Ibid.

86 Ibid.

87 CBD to HHG, April 15, 1910, Davenport Papers.

88 Ibid. Rosenberg considers Davenport's role as an "ambassador to the laity from the world of science" in "Charles B. Davenport and the Irony of American Eugenics," p. 97.

89 HHG, "Heredity of Feeble-Mindedness," appeared in *American Breeders' Magazine* 1 (1910): 165–178; and in *Annual Report of the American Breeders' Association* 6 (1911): 103–116. CBD, "Fit and Unfit Matings."

90 The appeal of eugenics to groups interested in initiating a more open sexual dialogue between men and women is discussed in Kevles, *Name of Eugenics,* pp. 24–26.

91 On the appeal of eugenics to nativists, racists, and imperialists, see Haller, *Eugenics,* pp. 58–94; Kevles, *Name of Eugenics,* pp. 57–112; Ludmerer, *Genetics and American Society,* pp. 1–43; Chase, *Legacy of Malthus,* pp. 97–137; and Garland Allen, "The Misuse of Biological Hierarchies: The American Eugenics Movement, 1900–1940," *History and Philosophy of the Life Sciences* 5 (1984): 105–128.

92 H. H. Laughlin, "Report of the Committee to Study and to Report on the Best Practical Means of Cutting Off the Defective Germ-Plasm in the American Population. L. The Scope of the Commitee's Work," Bulletin No. 10A (Cold Spring Harbor, N.Y.: Eugenics Record Office, 1914).

93 HHG, *Proceedings of the Ninth Annual Meeting of the New Jersey State Conference of Charities and Correction* (1910): 93–103; quotation p. 95.

94 Ibid., pp. 98–100. For an analysis of how degeneration and eugenic theories merged in family studies, see Gelb, "Degeneracy Theory, Eugenics, and Family Studies."

95 HHG, *Proceedings,* p. 102.

96 Ibid., p. 103.

97 HHG, *KF,* p. vii.

98 Ibid., pp. vii-ix.

99 Ibid., pp. ix-x.

100 Ibid., pp. 1–2.

101 Ibid., p. 2.

102 Ibid., pp. 2–9.

103 Ibid., pp. 10–11, 7.

104 Ibid., p. 12.

105 Ibid.

106 Ibid., pp. 13–16.

107 Ibid., pp. 16–17.

108 Ibid., p. 18.

109 Ibid., pp. 18, ix.

110 Ibid., pp. 18–19.

111 Ibid.

112 Ibid., pp. 29, 98–99.

113 Ibid., pp. 29–30, 97.

114 Ibid., pp. 50–51, 68.

115 Ibid., pp. 52–53, 68–69.

116 Ibid., pp. 33–49.

117 Ibid., pp. 70–71; Goddard's faith in Kite's work is evidenced by his inscription in her copy of this book: "To Elizabeth S. Kite, – without whose indefatigable labor the material in this book would never have been brought to light; and without whose skill and excellent judgment would not have been worth publishing, even when collected." "In Memoriam: Elizabeth S. Kite," *TSB* 50 (January 1954): 201.

118 Kite's field report is reprinted in HHG, *KF,* pp. 71–73.

119 Ibid., pp. 20–29, 73–93.

120 Ibid., pp. 83, 88.

121 Ibid., pp. 94–95.

122 This book's photographs have become controversial. In *Mismeasure of Man,* p. 171, Gould argued that these photos were probably "phonied" by Goddard to make the feebleminded look menacing or evil. For replies challenging Gould, see Raymond Fancher, "Henry Goddard and the Kallikak Family Photographs: 'Conscious Skulduggery' or 'Whig History'?" *AP* 42 (June 1987): 585–590; Sigrid S. Glenn and Janet Ellis, "Do the Kallikaks Look 'Menacing' or 'Retarded'?" *AP* 43 (September 1988): 742–743; Michael J. Kral, "More on Goddard and the Kallikak Family Photographs," *AP* 43 (September 1988): 745–746; and Leila Zenderland, "On Interpreting Photographs, Faces, and the Past," *AP* 43 (September 1988): 743–744. This issue is discussed in the Epilogue.

123 HHG, *KF,* p. 51.

124 Surprisingly, many family studies emphasized Anglo-Saxon rural families. For an astute analysis of these studies as a genre, see Rafter, ed., *White Trash,* pp. 1–31.

125 Ibid., p. 52. The term "Euthenics" was coined by Ellen Richards, whose book *Euthenics, the Science of Controllable Environment . . .* (1910) advocated a good food supply, fresh air, and the alleviation of poverty to diminish the "social and mental inequalities among men" (p. 23). Davenport rightly saw this as the opposite of his own philosophy stressing "the importance of blood." CBD, "Euthenics and Eugenics," *Popular Science Monthly* 77 (January 1911): pp. 16–17.

126 Elizabeth Kite, "Unto the Third Generation," *Survey* (September 28, 1912): 789–791. This article as well as Kite, "Two Brothers," *Survey* (March 2, 1912): 1861-1864, obviously describe the "Kallikak" family, although Kite only identifies them by this name in one footnote. "Elizabeth S. Kite," in Leonard, ed., *Woman's Who's Who* p. 462.

127 *KF,* pp. 50, 102–103.

128 Ibid., p. 103.

129 Ibid.

130 Ibid., p. 101 (emphasis in original).

131 Ibid. Goddard was clearly wrong, for in 1939 the Nazis began the systematic extermination of mentally retarded persons. See Hugh Gregory Gallagher, *By Trust Betrayed: Patients, Physicians, and the Licence to Kill in the Third Reich* (New York: Holt, 1990); and Michael Burleigh, *Death and Deliverance: "Euthanasia" in Germany, c. 1900–1945* (Cambridge: Cambridge University Press, 1995).

132 HHG, *KF,* pp. 102, 11–12.

133 Ibid., p. 104–105, 11.

134 Ibid., p. 105. Goddard estimated that 1 of 300 Americans was probably feebleminded. Using the 1910 census, he estimated that 307,185 needed institutionalization. The actual institutionalized population numbered 23,856. See "Estimated Number of

Feeble-Minded in the United States," *TSB* 8 (March 1911): 12–13. Later school studies used much higher estimates of about 2 percent.

135 HHG, *KF,* pp. 105–106.

136 Ibid., pp. 106–109.

137 Ibid., p. 109.

138 Ibid., pp. 109–111.

139 Ibid., pp. 114–115.

140 Ibid., pp. 113, 117.

141 A. C. Rogers, "Reviews and Notices: *The Kallikak Family. A Study in the Heredity of Feeble-Mindedness,*" *JPA* 17 (December 1912): 83–84. Review of *The Kallikak
Family, Medical Record* 83 (January 18, 1913): 126.

142 Review of *The Kallikak Family, American Journal of Psychology* 24 (1913): 290–
291; [James McKeen Cattell], "The Kallikak Family," *Popular Science Monthly* 82
(April 1913): 415–416. Although anonymous, I am assuming this review was by Cattell, the editor, because he made the very same argument in another article in this
journal. See Cattell, "Families of American Men of Science," *Popular Science Monthly*
86 (May 1915): 504–515, esp. p. 513.

143 "A Unique Study in Social Heredity," *Dial* 53 (October 1, 1912): 247; and Review
of *The Kallikak Family, Independent* 73 (October 3, 1912): 794.

144 Edwin Markham, "Book of the Month: The Kallikak Family," *Hearst's Magazine*
23 (February 1913): 329–331.

145 Alice Kauser [agent for Joseph Medill Patterson] to HHG, March 1, 1913; HHG
to Kauser, March 7, 1913 and March 11, 1913; Kauser to HHG, March 12, 1913;
Bleecker Van Wagenen to HHG, March 26, 1913, and April 2, 1913, File M615, GP.
Alice Kauser was a play broker who brought the works of numerous modern writers,
among them Henrik Ibsen, to Broadway.

146 HHG, *KF,* p. 100.

Chapter 6

1 HHG, *Proceedings of the Ninth Annual Meeting of the New Jersey State Conference
of Charities and Correction* (1910): 93–103, quotation pp. 93–94. Elizabeth Kite, *Research Work in New Jersey* (Trenton: New Jersey Department of Charities and Correction, 1913), p. 4.

2 HHG, *New Jersey State Conference,* p. 95.

3 Nicole Hahn Rafter, ed., *White Trash: The Eugenic Family Studies, 1877–1919* (Boston: Northeastern University Press, 1988), includes studies blending degeneration
with eugenic theories. See Steven Gelb, "Degeneracy Theory, Eugenics, and Family
Studies," *JHBS* 26 (1990): 242–246; and Gelb, "The Beast in Man: Degenerationism
and Mental Retardation, 1900–1920," *Mental Retardation* 33 (February 1995): 1–9.

4 On public health campaigns, see, for example, Charles Rosenberg, *The Cholera
Years: The United States in 1832, 1849, and 1866* (Chicago: University of Chicago
Press, 1962); and Barbara Gutmann Rosenkrantz, *Public Health and the State:
Changing Views in Massachusetts, 1842–1936* (Cambridge, Mass.: Harvard University
Press, 1972).

5 HHG, "The Prevention of Feeble-Mindedness," *Transactions of the Fifteenth International Congress on Hygiene and Demography* (Washington, D.C.: GPO, 1913), 3:463.

6 HHG, *FM* (New York: Macmillan, 1914).

7 Ibid., pp. 55, 266, 436–437.

8 Ibid., pp. 23–26.

9 Ibid., pp. 48–49.

10 Ibid., pp. 36, 209, 52.

11 HHG to Jane Griffiths, November 5, 1912, File M614, GP.

12 HHG, *FM,* pp. 47, 436. For cases with one feebleminded relative, see pp. 125–126, 139.

13 Ibid., p. 55n, 261.

14 Ibid., p. 443.

15 Ibid., pp. 448–460.

16 Ibid., pp. 460–465.

17 HHG, *FM,* pp. 539–557, 435.

18 On poverty, see James Patterson, *America's Struggle Against Poverty, 1900–1980* (Cambridge, Mass.: Harvard University Press, 1981), pp. 1–34; and Michael Katz, *Poverty and Policy in American History* (New York: Academic Press, 1983).

19 HHG, "Infant Mortality in Relation to the Hereditary Effects of Mental Deficiency," *Virginia Medical Semi-Monthly* 17 (1912–1913): 534–537, quotation pp. 535–536.

20 Robert Hunter, *Poverty* (New York: Macmillan, 1904), pp. 11–12, 62–64.

21 HHG, *FM,* p. 15.

22 Ibid., pp. 15–16; Hunter, *Poverty,* p. 69.

23 HHG, *FM,* p. 16.

24 Ibid., p. 17.

25 Walter Hines Page, "The Hookworm and Civilization," *World's Work* 24 (1912): 509. See also John Ettling, *The Germ of Laziness: Rockefeller Philanthropy and Public Health in the New South* (Cambridge, Mass.: Harvard University Press, 1981).

26 HHG, *FM,* pp. 16–17.

27 Anne Moore, *The Feeble-Minded in New York* (New York: State Charities Aid Association, 1911), p. 3.

28 Ibid, p. 11.

29 Ibid., p. 23.

30 Ibid., pp. 23, 28, 30, 20.

31 Ibid., pp. 15–18.

32 A. F. Tredgold, *Mental Deficiency* (New York: Wood, 1908; rpt. 1922), p. 93.

33 HHG, *FM,* p. 14; HHG, *KF* (New York: Macmillan, 1912), p. 64.

34 Most antithetical to Goddard's views on poverty were those of the socialists. For very different views of a contemporary, see, for instance, Kathryn Kish Sklar, *Florence Kelley and the Nation's Work,* vol. 1 (New Haven, Conn.: Yale University Press, 1995).

35 HHG, *FM,* p. 58.

36 Ibid., pp. 58–71.

37 Ibid., p. 67.

38 Ibid., p. 29.

39 See John McPhee, *The Pine Barrens* (New York: Farrar, Straus & Giroux, 1967).

40 HHG, *KF,* pp. 65–66.

41 Ibid.

42 Ibid., p. 70; HHG, *FM,* p. 11.

43 HHG, *KF,* pp. 70–71.

44 Ibid., p. 70.

45 HHG, *FM,* p. 5.

46 Ibid., pp. 2–3, 17.

47 Ibid., p. 2.

48 Ibid., p. 3.

49 Ibid.

50 On juvenile justice, see Bernard Wishy, *The Child and the Republic: The Dawn of Modern American Child Nurture* (Philadelphia: University of Pennsylvania Press, 1968); Anthony Platt, *The Child Savers: The Invention of Delinquency* (Chicago: University of Chicago Press, 1969); Steven Schlossman, *Love and the American Delinquent: The Theory and Practice of "Progressive" Juvenile Justice, 1825–1920* (Chicago: University of Chicago Press, 1977); Harold Finestone, *Victims of Change: Juvenile Delinquents in American Society* (Westport, Conn.: Greenwood Press, 1976); David Rothman, *Conscience and Convenience: The Asylum and Its Alternatives in Progressive America* (Boston: Little, Brown, 1980); and Mary Odem, *Delinquent Daughters: Protecting Adolescent Female Sexuality in the United States, 1885–1920* (Chapel Hill: University of North Carolina Press, 1995). On *parens patriae,* see Ellen Ryerson, *The Best-Laid Plans: America's Juvenile Court Experiment* (New York: Hill & Wang, 1978).

51 William Healy, *The Individual Delinquent* (Boston: Little, Brown, 1915). HHG, "The Responsibility of Children in the Juvenile Court," *Journal of Criminal Law and Criminology* 3 (September 1912): 365–375.

52 On "moral imbecility" and "moral insanity," see Arthur Fink, *Causes of Crime: Biological Theories in the United States, 1800–1915* (Philadelphia: University of Pennsylvania Press, 1938), pp. 211–239; Steven Gelb, "'Not Simply Bad and Incorrigible': Science, Morality, and Intellectual Deficiency," *History of Education Quarterly* 29 (Fall 1989): 359–379; Charles Rosenberg, *The Trial of the Assassin Guiteau: Psychiatry and Law in the Gilded Age* (Chicago: University of Chicago Press, 1968); and Janet Tighe, "A Question of Responsibility: The Development of American Forensic Psychiatry, 1838–1930" (Ph.D. diss., University of Pennsylvania, 1983).

53 Isaac Kerlin, "The Moral Imbecile," *NCCC* (May 1890): 244–250; Kerlin, "Juvenile Insanity," *AMO* (1879): 86–94; Martin Barr, "Moral Paranoia," *AMO* (1895): 522–531; and Barr, *Mental Defectives: Their History, Treatment, and Training* (Philadelphia: Blakiston, 1904), pp. 268–281.

54 HHG and Helen F. Hill, "Feeble-Mindedness and Criminality," *TSB* 8 (March 1911): 6.

55 HHG, "Estimated Number of Feeble Minded Persons in State Reformatories and Industrial Schools," *TSB* 9 (March 1912): 8–10.

56 Ibid.

57 Ibid.

58 HHG, "Responsibility of Children in the Juvenile Court," p. 371.

59 Ibid.

60 HHG, *FM,* pp. 9–10.

61 HHG and Hill, "Feeblemindedness and Criminality," pp. 3–6.

62 Ibid.

63 HHG and Helen F. Hill, "Delinquent Girls Tested by the Binet Scale," *TSB* 8 (June 1911): 50–56, quotations p. 50.

64 Ibid., pp. 50–51.

65 Ibid., pp. 51–55.

66 Ibid., p. 52.

67 Ibid., pp. 51, 54.

68 Ibid., p. 53.

69 Ibid., pp. 51, 54. Mrs. E. Garfield Gifford and HHG, "Defective Children in the Juvenile Court," *TSB* 8 (January 1912): 132–135.

70 HHG, *FM,* p. 13. See also Mark Thomas Connelly, *The Response to Prostitution in the Progressive Era* (Chapel Hill: University of North Carolina Press, 1980), and Ruth Rosen, *The Lost Sisterhood: Prostitution in America, 1900–1918* (Baltimore: Johns Hopkins University Press, 1982).

71 HHG, *FM,* p. 14. HHG, "Feeble-Mindedness a Source of Prostitution," *Vigilance* (April 1913): 3–11, quotation p. 3. On feeblemindedness and prostitution, see Connelly, *Response to Prostitution,* pp. 41–44; Rosen, *Lost Sisterhood,* pp. 21–23; Estelle Freedman, *Their Sisters' Keepers: Women's Prison Reform in America, 1830–1930* (Ann Arbor: University of Michigan Press, 1981), pp. 116–121; and Nicole Hahn Rafter, *Partial Justice: Women, Prisons, and Social Control* (New Brunswick, N.J.: Transaction, 1990), pp. 64–74.

72 HHG, "Feeble-Mindedness a Source of Prostitution," p. 3.

73 HHG, *FM,* pp. 13–14.

74 HHG, "Feeble-Mindedness a Source of Prostitution," pp. 3–4.

75 Ibid., p. 5; HHG, *FM,* p. 99.

76 HHG, "Feeble-Mindedness a Source of Prostitution," p. 6. On women's reformatories, see Freedman, *Their Sisters' Keepers,* and Rafter, *Partial Justice;* see also Peter Tyor, "'Denied the Power to Choose the Good': Sexuality and Mental Defect in American Medical Practice, 1850–1920," *Journal of Social History* 10 (Summer 1977): 472–489; Steven Schlossman and Stephanie Wallach, "The Crime of Precocious Sexuality: Female Juvenile Delinquency in the Progressive Era," *Harvard Educational Review* 48 (February 1978): 65–94; Regina Kunzel, *Fallen Women, Problem Girls: Unmarried Mothers and the Professionalization of Social Work, 1890–1945* (New Haven, Conn.: Yale University Press, 1993); and Odem, *Delinquent Daughters.* See also Elizabeth Lunbeck's discussion of "hypersexuality" in *The Psychiatric Persuasion: Knowl-*

edge, Gender, and Power in Modern America (Princeton: Princeton University Press, 1994), pp. 185–208.

77 Ellen Fitzpatrick, *Endless Crusade: Women Social Scientists and Progressive Reform* (New York: Oxford University Press, 1990), p. 92. Katharine Bement Davis, "Feeble-Minded Women in Reformatory Institutions," *Survey* 27 (March 2, 1912): 1850.

78 Davis, "Feeble-Minded Women," pp. 1849–1851. Weidensall came to see the Binet tests as not reliable with this population; see *The Mentality of the Criminal Woman* (Baltimore: Warwick & York, 1916). On Davis' work, see Fitzpatrick, *Endless Crusade,* pp. 92–129; and Rafter, *Partial Justice,* pp. 69–74. On testing female criminals, see Freedman, *Their Sisters' Keepers,* pp. 116–121.

79 Dr. Louise Morrow and Olga Bridgeman, "Delinquent Girls Tested by the Binet Scale," *TSB* 9 (May 1912): 33–36; HHG, *FM,* pp. 14–15.

80 HHG, "Feeble-Mindedness a Source of Prostitution," pp. 6, 9–10.

81 Ibid., pp. 4, 7.

82 Ibid., p. 11.

83 Thacher is quoted in ibid., pp. 7–9.

84 Ibid.

85 Ibid., p. 8, 10.

86 Ibid., p. 10.

87 HHG, *FM,* pp. 13–14.

88 Ibid., p. 15.

89 *Report of the Commission for the Investigation of the White Slave Traffic, So Called* (Boston: Wright & Potter, 1914).

90 Ibid., pp. 29–30. These findings were added in a footnote in HHG, *FM,* pp. 15–16. Binet tests, however, were not the sole determinant of feeblemindedness for Fernald; instead, they formed part of a broader medical diagnosis. Among 135 persons labeled "normal," 17 had a mental age of twelve, 71 of eleven; 32 of ten; and 4 of nine (while 11 were not tested). Thus, some persons marked "feebleminded" had the same mental age as others marked "normal."

91 *Report of the Commission,* p. 29.

92 HHG, "Feeble-Mindedness a Source of Prostitution," p. 11.

93 Matthew Hale, Jr., *Human Science and Social Order: Hugo Munsterberg and the Origins of Applied Psychology* (Philadelphia: Temple University Press, 1980), pp. 111–121. The Munich "expert witness" was Albert Schrenck-Notzing.

94 Ibid., pp. 111–118. On the murder of former Idaho governor Frank Steunenberg, the confession of Harry Orchard, and the trial of "Big Bill" Haywood (defended by Clarence Darrow), see also Kevin Tierney, *Darrow: A Biography* (New York: Crowell, 1979), pp. 206–226; and Melvyn Dubofsky, *We Shall Be All: A History of the Industrial Workers of the World* (New York: New York Times Book Co., 1969; rpt. 1973), pp. 96–105.

95 Hale, *Human Science,* pp. 115–118.

96 Hugo Munsterberg, *On the Witness Stand: Essays on Psychology and Crime* (New York: Doubleday, 1908), pp. 234, 240–241.

97 HHG, *KF,* p. 54; HHG, "Relation of Feeble-Mindedness to Crime," *Bulletin of the American Academy of Medicine* 15 (April 1914): 105–112, quotation p. 108.

98 HHG, "Relation of Feeble-Mindedness," p. 108.

99 The Gianini case is in HHG, *The Criminal Imbecile: An Analysis of Three Remarkable Murder Cases* (New York: Macmillan, 1915), pp. 1–41, quotation p. 1.

100 "Gianini Confessed He Killed Teacher," *NYT* (May 15, 1914), p. 10.

101 In addition to Goddard's summary, facts about this case are from Steven Gelb's study of the trial transcript. See Gelb, "The Strange Case of Jean Gianini: The 1914 Crime and Trial of an Adolescent Teacher-Killer," paper presented to History of Education Society, 1989; and "Sentenced in Sorrow: The Role of Asylum in the Jean Gianini Murder Defence," *Health and Place* 3 (1997): 123–129.

102 HHG, *Criminal Imbecile,* pp. 11, 14.

103 Ibid., p. 15.

104 Ibid., pp. 15, 31.

105 Ibid., pp. 21–22.

106 Ibid., pp. 32–33.

107 Gelb, "Strange Case."

108 Ibid.

109 HHG, *Criminal Imbecile,* pp. 34–41, 1–2.

110 Ibid., p. 2.

111 The Pennington case is in ibid., pp. 42–63.

112 Ibid., pp. 46–47, 52–53.

113 P. M. Kerr, "The Mental Status of Roland P.," *Alienist and Neurologist* 36 (May 1915): 131–154, quotation p. 136.

114 Ibid., p. 136.

115 Ibid., p. 137.

116 HHG, *Criminal Imbecile,* p. 53.

117 Kerr, "Mental Status," pp. 139–140.

118 Ibid.

119 Ibid., p. 144.

120 Ibid., pp. 144–146. Prosecuters claimed Roland's cousin was put in Elwyn merely to create sympathy for Roland.

121 HHG, *Criminal Imbecile,* pp. 87–88.

122 Kerr, "Mental Status," p. 131, 152.

123 HHG to Elizabeth Whitacre, March 2, 1916, File M918, GP. HHG, *Criminal Imbecile,* pp. 88–89.

124 HHG, *Criminal Imbecile,* pp. 65–82. Hoyt Landon Warner, *Progressivism in Ohio, 1897–1917,* pp. 405–406, 419–420; Thomas H. Haines, "The Ohio Plan for the Study of Delinquency," *Popular Science Monthly* 86 (June 1915): 576–580; and Hamilton Cravens, "Applied Science and Public Policy: The Ohio Bureau of Juvenile Research and the Problem of Juvenile Delinquency, 1913–1930," in Michael Sokal, ed., *Psycho-*

logical Testing and American Society, 1890–1930 (New Brunswick, N.J.: Rutgers University Press, 1987), pp. 158–194.

125 Judge Olson is cited in Sarah Schaar, "The Institute as I Know It," in Agnes A. Sharp, *A Dynamic Era of Court Psychiatry, 1914–1944* (Chicago: Psychiatric Institute of the Municipal Court of Chicago, n.d. [1944]), pp. 12–13.

126 HHG, *Criminal Imbecile.* The third case concerned Fred Tronson of Portland, Oregon, convicted of second-degree murder in 1914. See pp. 65–82.

127 HHG, *FM,* pp 474–492, quotations pp. 481, 474, 492.

128 Ibid., p. 10–13.

129 Ibid., p. 18.

130 Ibid., p. 19.

131 L. W. Crafts, "Bibliography of Feeble-Mindedness in Relation to Juvenile Delinquency," *Journal of Delinquency* 1 (September 1916): 195–208; Fink, *Causes of Crime,* pp. 211–239.

Chapter 7

1 "Editorials," *Survey* (October 17, 1914): 73.

2 *The United States Children's Bureau, 1912–1972* (New York: Arno, 1974). On women reformers and progressive politics, see Ellen Fitzpatrick, *Endless Crusade: Women Social Scientists and Progressive Reform* (New York: Oxford University Press, 1990); Theda Skocpol, *Protecting Soldiers and Mothers: The Political Origins of Social Policy in the United States* (Cambridge, Mass.: Harvard University Press, 1992); Molly Ladd-Taylor, *Mother-Work: Women, Child Welfare, and the State, 1890–1930* (Urbana: University of Illinois Press, 1993); Sonya Michel and Seth Koven, "Womanly Duties: Maternalist Politics and the Origins of the Welfare State in France, Great Britain and the United States, 1880–1920," *American Historical Review* 95 (October 1990): 1076–1108; and Linda Gordon and Theda Skocpol, "Gender, State and Society: A Debate," *Contention* 2 (Spring 1993): 139–189.

3 On the child-saving movement, see Anthony Platt, *The Child Savers: The Invention of Delinquency* (Chicago: University of Chicago Press, 1969); and Susan Tiffin, *In Whose Best Interest? Child Welfare Reform in the Progressive Era* (Westport, Conn.: Greenwood, 1982); see also Ronald D. Cohen, "Child-Saving and Progressivism, 1885–1915," and Hamilton Cravens, "Child-Saving in the Age of Professionalism, 1915–1930," in Joseph M. Hawes and N. Ray Hiner, eds., *American Childhood: A Research Guide and Historical Handbook* (Westport, Conn.: Greenwood Press, 1985), pp. 273–309 and 415–488. On women social scientists, see Rosalind Rosenberg, *Beyond Separate Spheres: Intellectual Roots of Modern Feminism* (New Haven, Conn.: Yale University Press, 1982); Fitzpatrick, *Endless Crusade;* and Elizabeth Scarborough and Laurel Furumoto, *Untold Lives: The First Generation of American Women Psychologists* (New York: Columbia University Press, 1987).

4 On mental hygiene, see Gerald Grob, *Mental Illness and American Society, 1875–1940* (Princeton: Princeton University Press, 1983), pp. 144–178; David Rothman, *Conscience and Convenience: The Asylum and Its Alternatives in Progressive America* (Boston: Little, Brown, 1980); Theresa Richardson, *Century of the Child: The Mental Hygiene Movement and Social Policy in the United States and Canada* (Albany: State University of New York Press, 1994); and Christine Mary Shea, "The Ideology of

Mental Health and the Emergence of the Therapeutic Liberal State: The American Mental Hygiene Movement, 1900–1930" (Ph.D. diss., University of Illinois at Urbana-Champaign, 1980). On social hygiene, see Charles Walter Clarke, *Taboo: The Story of the Pioneers of Social Hygiene* (Washington, D.C.: Public Affairs Press, 1961); James F. Gardner, "Microbes and Morality: The Social Hygiene Crusade in New York City, 1892–1917" (Ph.D. diss., Indiana University, 1973); and David Pivar's work-in-progress, "Citizens and Pariahs: The Ironies of Social Hygiene Reform in America, 1900–1921." Barbara Rosenkrantz notes the change from "public health" to "public hygiene" in *Public Health and the State: Changing Views in Massachusetts, 1842–1936* (Cambridge, Mass.: Harvard University Press, 1972), esp. pp. 132–134. Arnold Gesell, "The University in Relation to the Problems of Mental Deficiency and Child Hygiene," *Transactions of the Fourth International Congress on School Hygiene* (Buffalo: Courier Company, 1914), 5:617. On changing the "Department of Child Study" to "Department of Child Hygiene," see Percy E. Davidson, "Secretary's Minutes," *NEA* (1911): 870.

5 Lewis M. Terman, "Trails to Psychology," in Carl Murchison, ed., *History of Psychology in Autobiography* (Worcester, Mass.: Clark University Press, 1932), 2:324. See also Lewis Terman, "Professional Training for Child Hygiene," *Popular Science Monthly* (March 1912): 289–297.

6 HHG, "Research in School Hygiene, in the Light of Experiences in an Institution for the Feeble Minded. Abstract," in *Lectures and Addresses Delivered Before the Departments of Psychology and Pedagogy in Celebration of the Twentieth Anniversary of the Opening of Clark University* (Worcester, Mass.: Clark University, 1910), pp. 51–53. Goddard's address was to the Department of Pedagogy.

7 ERJ, "Report of the Superintendent, *AR* (1912): 20; ERJ, "Report of the Superintendent," *AR* (1914): 22.

8 Walter Cornell, "Division of Medical Research," in HHG, "Report of the Research Director," *AR* (1910): 39.

9 Goddard's laboratory is described in "Notes and News," *PB* 12 (October 15, 1915): 403. Gesell, "University in Relation," p. 615. See, for example, Drs. Charles L. Dana, William N. Berkeley, Walter S. Cornell, and HHG, "Functions of the Pineal Gland with Report of Feeding Experiments," *Medical Record* 83 (May 10, 1913): 835–847.

10 ERJ, "Report of the Superintendent," *AR* (1914): 22.

11 Both options are discussed in HHG, "Sterilization and Segregation," *Bulletin of the American Academy of Medicine* 13 (August 1912), reprinted as *Publications of the Russell Sage Foundation, Pamphlet no. 12* (New York: Department of Child Helping, 1913); and HHG, "The Elimination of Feeble-Mindedness," *Annals of the American Academy of Political and Social Science* 37 (January–June, 1911): 505–516. For overviews of the sterilization debates, see Chapter 5, note 24, in the present volume.

12 HHG, "Sterilization and Segregation," pp. 7–8; HHG, "Elimination of Feeblemindedness," pp. 514–516. HHG, *KF,* pp. 107–109. Stevenson Smith, Madge W. Wilkinson and Lovisa C. Wagoner, "A Summary of the Laws of the Several States . . . ," *Bulletin of the University of Washington* No. 82 (May 1914), pp. 24–26; and J. H. Landman, *Human Sterilization: The History of the Sexual Sterilization Movement* (New York: Macmillan, 1932), pp. 64–65, 289. By 1932, no sterilizations had been performed in New Jersey institutions. Goddard cites his objections to types of provisions found in the New Jersey law in "Sterilization and Segregation," pp. 9–10.

13 HHG, "Sterilization and Segregation," pp. 5–7; HHG, "The Basis for State Policy: Social Investigation and Prevention," *Survey* (March 2, 1912): 1852–1856.

14 Kathrine Regan McCaffrey, "Founders of the Training School at Vineland, New Jersey: S. Olin Garrison, Alexander Johnson, Edward R. Johnstone" (Ph.D. diss., Teachers College, Columbia University, 1965), p. 273.

15 Joseph Byers, *The Village of Happiness: The Story of the Training School* (Vineland, N.J.: Smith, 1934), pp. 53–65; James D. Eadline, *Training School at Vineland, New Jersy* (Vineland, N.J.: The Training School, 1963); McCaffrey, "Founders," pp. 273–285.

16 HHG, "Sterilization and Segregation," p. 5.

17 Byers, *Village*, p. 76.

18 Ibid., pp. 76–77.

19 Letters from parents are in Department of Public Health and Charities of Philadelphia, "The Feeble-Minded World," Bulletin No. 4 (1911): 10–13. At this time, Vineland had about four hundred residents, with two hundred on a waiting list. See also Philip M. Ferguson, *Abandoned to Their Fate: Social Policy and Practice Toward Severely Retarded People in America, 1820–1920* (Philadelphia: Temple University Press, 1994), pp. 143–144.

20 Byers, *Village*, p. 77.

21 Ibid., pp. 76–80; Alexander Johnson, *Adventures in Social Welfare* (Fort Wayne, Ind.: Alexander Johnson, 1923), pp. 389–417; and ERJ, "The Extension of the Care of the Feeble-Minded," *JPA* 19 (September 1914): 3–15.

22 Byers, *Village*, pp. 80–82; On the South, see Johnson, *Adventures*, pp. 399–407; and Johnson, "Care of the Degenerate," in *The Call of the New South* (Nashville, Tenn.: Southern Sociological Conference Press, 1912), p. 171. See also Steven Noll, *Feeble-Minded in Our Midst: Institutions for the Mentally Retarded in the South, 1900–1940* (Chapel Hill: University of North Carolina Press, 1995); and Edward Larson, *Sex, Race, and Science: Eugenics in the Deep South* (Baltimore: Johns Hopkins University Press, 1995), pp. 57–58.

23 One pamphlet illustrating the Kallikak story is captioned "There are Many Girls in Montana Needing the State's Protection – 1919." File M614, GP (see illustration on p. 230).

24 HHG to Mrs. Fiske, Gaiety Theatre, New York City, March 2, 1915; Fiske to HHG, March 23, 1915. File M615, GP.

25 Edgar Doll, "Three Phases of the Fifteenth International Congress on Hygiene and Demography," *JEP* 4 (January 1913): 41–44. The American Federation for Sex Hygiene sponsored the sex hygiene exhibit, which also included the eugenics exhibit. The mental hygiene exhibit included displays from Vineland and Faribault. HHG, "The Prevention of Feeble-Mindedness," *Transactions of the Fifteenth International Congress on Hygiene and Demography* (Washington, D.C.: GPO, 1913), 3:463–468; and *Fifteenth International Congress on Hygiene and Demograpy* (Washington, D.C.: GPO, 1913), vol. 1: "Some Lessons and Suggestions from the Exhibition."

26 Byers, *Village*, p. 80. On popularizing eugenics, see Daniel Kevles, *In the Name of Eugenics: Genetics and the Uses of Human Heredity* (New York: Knopf, 1985), pp. 57–69.

27 On the national Committee on Provision, see Byers, *Village,* pp. 80–90; Johnson, *Adventures,* pp. 389–417; Mark Haller, *Eugenics: Hereditarian Attitudes in American Thought* (New Brunswick, N.J.: Rutgers University Press, 1963), pp. 125–128; and James W. Trent, Jr., *Inventing the Feeble Mind: A History of Mental Retardation in the United States* (Berkeley: University of California Press, 1994), pp. 172–181. Robert D. Dripps, "A Review of the Campaign in Pennsylvania," *JPA* 22 (March and June 1918): 162–172.

28 Byers, *Village,* pp. 84–85.

29 HHG, "The Menace of Mental Deficiency from the Standpoint of Heredity," *Boston Medical and Surgical Journal* 175 (1916): 271; HHG, "Prevention of Feeble-Mindedness," p. 464; HHG, "The Possibilities of Research as Applied to the Prevention of Feeble-Mindedness," *NCCC* (1915): 307.

30 The *New York Times* covered Goddard's school survey (see Chapter 4), Anne Moore's survey (see Chapter 6), and numerous other events, including the Shrank case.

31 Carlos F. MacDonald, William Mabon, and Max G. Schlapp, "Alienists on Schrank," *NYT,* October 16, 1912, p. 5.

32 "Defective Children Will be Classified," *NYT,* October 16, 1912, p. 5; "Defectives and How to Treat Them," *NYT,* October 17, 1912, p. 10; "Endow New Work Among Defectives," *NYT,* December 13, 1912, p. 8. On the Clearing House for Mental Defectives, see Max Schlapp and Leta Stetter Hollingworth, "An Economic and Social Study of Feeble-Minded Women," *Medical Record* 85 (June 6, 1914): 1025–1028; and Schlapp and Alice Paulsen, "Report on 10,000 Cases from the Clearing House for Mental Defectives," *Medical Record* 93 (February 16, 1918): 269–275.

33 Editorial, "Potential Criminals," *NYT,* October 11, 1913, p. 14.

34 Vineland visitors are cited in "Diary Notes" and "The Year's Calendar of Events" in annual reports.

35 Ibid. ERJ, "Diary Notes," *AR* (1917): 47.

36 ERJ, "Report of the Superintendent," *AR* (1912): 20. Unfortunately, these letters to Vineland have been lost; however, they existed as late as 1965. See McCaffrey, "Founders," pp. 242–243. "The Research Department was literally deluged with letters . . . ," McCaffrey writes. "The writer read hundreds of these letters that are preserved in the files of the Research Laboratory."

37 Mrs. Thos. Thompson to Children's Bureau, n.d. [received July 25, 1919]; Emma Lumberg to Thompson, July 26, 1919; Mrs. M. E. McDearman to *Mother's Magazine,* July 22, 1917 [letter forwarded to Children's Bureau]; Julia Lathrop to McDearman, July 31, 1917; Mrs. Wilson Runnette to Children's Bureau, n.d. [received June 4, 1919]; Lumberg to Runnette, June 12, 1919, CB. In letters to Thompson and Runnette, the Children's Bureau recommended contacting Johnstone, as well as several others; in answering McDearman, Lathrop recommended Beverly Farms, an Illinois institution.

38 Mabel L. Pollitzer to Julia Lathrop, November 1, 1917. "There are, of course, few institutions doing the kind [of?] work that is done in Vineland," Lundberg noted; Emma Lundberg to Mabel L. Pollitzer, November 9, 1917. Lathrop, "Memorandum: Address before Washington Alliance of Jewish Women, February 11, 1913," CB. On the role of women's organizations in such campaigns, see, for instance, Larson, *Sex, Race, and Science,* p. 73.

39 Dr. Dunning S. Wilson to Children's Bureau, August 26, 1915; Lathrop to Wilson, August 30, 1915; Christopher Carson Thurber to Julia Lathrop, June 14, 1915; Lathrop to Thurber, June 15, 1915; Rev. Paul Millholland to "Child Welfare Department," November 7, 1916; Lathrop to Millholland, February 4, 1917; Lathrop to HHG, January 4, 1917; CB.

40 Edgar Doll, "Three Phases of the Fifteenth International Congress," p. 43; and Doll, "Mental Tests at the Mental Hygiene Congress," *JEP* 4 (January 1913): 44–45.

41 "Here's Chance to Test Your Mental Development According to Famous Binet System," *Chicago Daily Tribune,* April 29, 1915, p. 5 (see illustration on p. 236).

42 For the literature on psychological professionalization, see Introduction, note 21.

43 Michael Sokal, "James McKeen Cattell and Mental Anthropometry: Nineteenth-Century Science and Reform and the Origins of Psychological Testing," in Sokal, ed., *Psychological Testing and American Society, 1890–1930* (New Brunswick, N.J.: Rutgers University Press, 1987), pp. 34–35. On the myriad types of psychological testing used in these decades, see Kurt Danziger, *Constructing the Subject: Historical Origins of Psychological Research* (Cambridge: Cambridge University Press, 1990); and Jill Morawski, ed., *The Rise of Experimentation in American Psychology* (New Haven, Conn.: Yale University Press, 1988).

44 Guy Whipple, *Manual of Mental and Physical Tests,* Part II: Complex Processes (Baltimore: Warwick & York, 1910; 2nd rev. ed., 1915), pp. 689–690; *JEP* 6 (September 1915): 456–457. Samuel Kohs, "The Binet–Simon Measuring Scale for Intelligence. An Annotated Bibliography," *JEP,* Part I, 5:4 (April 1914): 215–224; and Part II, 5:6 (June 1914): 335–346; Kohs, "An Annotated Bibliography of Recent Literature on the Binet-Simon Scale (1913–1917)," *JEP,* Part I, 8:7 (September 1917): 425–438; Part II, 8:8 (October 1917): 488–502; Part III, 8:9 (November 1917): 559–565; and Part IV, 8:10 (December 1917): 609–618.

45 "Conference on Binet-Simon Scale," *Transactions of the Fourth International Congress . . . ,* pp. 627–706. See also J. C. Bell et. al., "Communications and Discussions: Informal Conference on the Binet-Simon Scale: Some Suggestions and Recommendations," *JEP* 5 (1914): 95–100; and Lewis Terman, "A Report of the Buffalo Conference on the Binet-Simon Tests of Intelligence," *PS* 20 (December 1913): 549–554.

46 J. E. Wallace Wallin, "Current Misconceptions in Regard to the Functions of Binet Testing and of Amateur Psychological Testers," *Transactions of the Fourth International Congress . . . ,* pp. 687, 679, 681; Guy Whipple, "Editorials: Amateurism in Binet Testing Once More," *JEP* 4 (May 1913): 302. See also Whipple, "Communications and Discussions: The Fourth International Congress on School Hygiene," *JEP* 4 (October 1913): 476–478; and Edgar Doll, "Communications and Discussions: Inexpert Binet Examiners and Their Limitations," *JEP* 4 (December 1910): 607–609.

47 Edwin Starbuck to HHG, August 1, 1912; HHG to Starbuck, August 6, 1912. File M615, GP.

48 HHG, "The Reliability of the Binet-Simon Measuring Scale of Intelligence," *Transactions of the Fourth International Congress. . . ,* pp. 693–699; quotations pp. 693–694.

49 Ibid., p. 694.

50 Goddard learned about "general intelligence" from a 1911 article by Cyril Burt in the British journal *Child Study.* It described test performance as "the expression of a mental property neither merely specialized nor merely acquired, but something all-pervading, something inherited, something inborn"; see HHG, *FM,* pp. 556–557. On Spearman, see Raymond Fancher, *The Intelligence Men: Makers of the IQ Controversy* (New York: Norton, 1985), pp. 84–98. For testing in England, see Gillian Sutherland, *Ability, Merit and Measurement: Mental Testing and English Education, 1880–1940* (Oxford: Clarendon Press, 1984); and Nicolas Rose, *The Psychological Complex: Psychology, Politics and Society in England, 1869–1939* (London: Routledge & Kegan Paul,1985). Otherwise, Goddard apparently ignored Spearman's work, perhaps because it was not directly related to feeblemindedness.

51 Grace M. Fernald, "The Use of the Binet Scale with Delinquent Children," *Transactions of the Fourth International Congress . . . ,* pp. 670–677.

52 Ibid.

53 HHG and Helen Hill, "Delinquent Girls Tested by the Binet Scale," *TSB* 8 (June 1911): 55; HHG, "The Reliability of the Binet–Simon Measuring Scale," pp. 697–698.

54 Fernald, "Use of the Binet Scale," pp. 672–673.

55 Edmund Huey, "The Binet Scale for Measuring Intelligence and Retardation," *JEP* 1 (October 1910): 436. William Healy, *The Individual Delinquent* (Boston: Little, Brown, 1915), pp. 56–57.

56 See HHG, "Échelle Métrique de L'intelligence de Binet–Simon. Résultats Obtenus en Amérique a Vineland, N.J.," *L'Année Psychologique* 18 (1911): 288–326. HHG's copy states, "Article written at the request of Binet showing the results of the Binet–Simon Tests when given to 400 Feeble-minded in the Vineland Training School. Also the results from 2000 normal children in the public schools by Henry H. Goddard." File M31, GP.

57 Alfred Binet and Theodore Simon, *The Development of Intelligence in Children,* trans. Elizabeth Kite (Baltimore: Williams & Wilkins, 1916), pp. 316–329.

58 Ibid., pp. 318.

59 Ibid., pp. 318–321. Binet also believed that rich children were more likely to be in smaller classes, and thus to receive more individualized instruction.

60 HHG and Hill, "Delinquent Girls Tested," p. 52; Fernald, "Use of the Binet Scale," p. 673.

61 Isabel Lawrence, "A Study of the Binet Definition Tests," *PC* 5 (1911): 207–216. Lawrence recorded these answers to compare Binet–Simon scale evaluations with teachers' evaluations of students' abilities.

62 Failures on Binet's "Cutting Paper" test, recommended for age thirteen on the 1908 Binet scale and moved to "Adult" on Goddard's 1911 revision, are in HHG, "Two Thousand Normal Children Measured by the Binet Measuring Scale of Intelligence," *PS* 18 (June 1911): 253–254. Goddard tested eighteen psychologists, and six failed it.

63 While Binet's 1908 scale included tests for ages three through thirteen, his 1911 revision instead offered tests for ages three through ten, twelve, fifteen, and "Adults." See Binet and Simon, *Development of Intelligence,* p. 276. Goddard's 1911 Binet–Simon Revision offered tests for ages three through twelve, fifteen, and "Adult," thus adding tests for age eleven.

64 See, for example, Charles Scott Berry, "Some Limitations of the Binet–Simon Tests of Intelligence," *Transactions of the Fourth International Congress . . . ,* pp. 649–

654, which includes his study of University of Michigan seniors. Many others made similar complaints about tests for higher ages. "'Fifteen years' and 'Adult' have not proved reliable," Goddard stated in 1913. "One may try them for his own purpose, but there seems no way to score them, therefore we do not include them in [the] scale." HHG, "Standard Method for Giving the Binet Test," *TSB* 10 (April 1913): 30.

65 Mrs. E. Garfield-Gifford and HHG, "Defective Children in the Juvenile Court," *TSB* 8 (1912): 132–135. Fernald criticizes this study in "Use of the Binet Scale," pp. 674–675.

66 Lewis Terman, "Suggestions for Revising, Extending and Supplementing the Binet Intelligence Test; Part II: Psychological Principles Underlying the Binet–Simon Scale, and Some Practical Instructions for its Correct Use; Part III: The Significance of Intelligence Tests for Mental Hygiene," *JPA* 18: 1, 2, and 3 (1913–1914): 20–33, 93–104, 119–127. Lewis Terman, Grace Lyman, Dr. George Ordahl, Dr. Louise Ordahl, Neva Galbraith, and Wilford Talbert, "The Stanford Revision of the Binet–Simon Scale and Some Results from its Application to 1000 Non-selected Children," *JEP* 6 (November 1915): 551–562; and Lewis Terman, *The Measurement of Intelligence* (Boston: Houghton Mifflin, 1916).

67 Robert Yerkes and J. W. Bridges, "The Point Scale: A Method of Measuring Mental Capacity," *Boston Medical and Surgical Journal* 171 (1914): 857–865; Yerkes, James W. Bridges, and Rose S. Hardwick, *A Point Scale for Measuring Mental Ability* (Baltimore: Warwick & York, 1915).

68 Robert Yerkes and Helen Anderson, "The Importance of Social Status as Indicated by the Results of the Point-Scale Method of Measuring Mental Capacity," *JEP* 6 (March 1915): 137–150, quotations pp. 143, 149.

69 On Weschler's ideas, see Fancher, *Intelligence Men,* pp. 149–161.

70 J. E. Wallace Wallin, "Who is Feeble-Minded?" *Journal of Criminal Law and Criminology* (January 1916): 706–716, quotations pp. 706–707.

71 Ibid., pp. 707–715.

72 "Who's Loony? Why, Every One, Even the Mayor," *Chicago Daily Tribune* (December 29, 1915), p. 1; "Experts Assail Mentality Test Result as Joke," *Chicago Herald* (December 29, 1915), p. 1. Parts of this controversy are discussed, and the front page of the *Chicago Herald* reproduced, in Allan Chase, *The Legacy of Malthus: The Social Costs of the New Scientific Racism* (New York: Knopf, 1977), pp. 240–242. J. E. Wallace Wallin describes this incident in *The Odyssey of a Psychologist* (Wilmington, Del.: J. E. Wallace Wallin, 1955), pp. 113–116. Mary Campbell had been fired from her job as a psychological assistant in the Chicaco Municipal Court. Details of her dismissal, and her unsuccessful claim that the city owed her money, are in Harry Olson, Chief Justice, to Hon. P. W. Sullivan, July 2, 1915, Chicago Municipal Court Papers, Box 4, Folder 29, Chicago Historical Society.

73 Samuel Kohs, "'Who is Feeble-Minded?'" *Journal of Criminal Law and Criminology* 6 (1915–1916): 860–871.

74 Ibid. See also J. E. Wallace Wallin, "'Who is Feeble-Minded?': A Reply to Mr. Kohs," *Journal of Criminal Law and Criminology* 7 (1916–1917): 56–78; and Kohs, "'Who is Feeble-Minded?' A Rejoinder and a Rebuttal," *Journal of Criminal Law and Criminology* 7 (1916–1917): 219–228.

75 "Proceedings of the Twenty-Fourth Annual Meeting of the American Psychological Association," *PB* 13 (February 15, 1916): 49.

76 Ibid.

77 Robert Yerkes' letter to Carl Seashore, and Seashore's call for a symposium, are in "Communications and Discussions: Mentality Tests," *JEP* 7 (March 1916): 163–166.

78 Ibid. Replies from sixteen respondents, organized alphabetically, were published in three subsequent issues of this journal. See "Mentality Tests: A Symposium," *JEP* 7 (April 1916): 229–240; (May 1916): 278–286; (June 1916): 348–360. Replies are cited below by author's name.

79 Reply from Walter V. Bingham, 7 (April 1916): 230–231.

80 Reply from HHG, 7 (April 1916): 231–233.

81 Letter from Yerkes, "Communications and Discussions: Mentality Tests," pp. 163–164. On the psychiatric work done at the Boston Psychopathic Hospital, see Elizabeth Lunbeck, *The Psychiatric Persuasion: Knowledge, Gender, and Power in Modern America* (Princeton: Princeton University Press, 1994).

82 Reply from Terman, "Symposium," 7 (June 1916): 348–351.

83 Ibid., pp. 348–349.

84 Reply from Guy M. Whipple, 7 (June 1916): 357–360.

85 Reply from James R. Angell, 7 (April 1916): 229–230.

86 Ibid.

87 Goddard's correspondence with medical superintendents shows that he had clearly earned their respect. Despite later disagreements, Walter Fernald still argued that in explaining Binet's ideas, Goddard had done more for this field than anyone since Seguin. Fernald, "Thirty Years Progress in the Care of the Feeble-Minded," *JPA* 29 (June 1923/June 1924): 206–219.

88 A. J. Rosanoff, trans. and ed., *Manual of Psychiatry* by J. Rogues de Fursac, 3rd ed. (New York: John Wiley & Sons, 1911), pp. iii–iv; HHG, "The Educational Treatment of the Feeble-Minded," in William A. White and Smith Ely Jellife, eds., *The Modern Treatment of Nervous and Mental Diseases* (Philadelphia: Lea & Febiger, 1913), 1:143–194.

89 Walter Cornell's "Short Course" for medical inspectors ran September 5–15, 1911, and September 3–12, 1912. Both were advertised in the *TSB*. This course is described in McCaffrey, "Founders," pp. 257–258; she claims it was also given in 1913. In *Health and Medical Inspection of School Children* (Philadelphia: F. A. Davis, 1912), Cornell advocated a cautious approach toward Binet testing. He was criticized by Vineland personnel; see "Review of *Health and Medical Inspection of School Children*," *TSB* 9 (June 1912): 63–65.

90 A copy of Dr. Rosanoff's "Nassau County Survey. Special Modification of Binet-Simon Scale" is in the Children's Bureau Records. See Thomas Salmon to Emma O. Lundberg, September 11, 1916; Emma Lundberg to T. W. Salmon, September 12, 1916; A. J. Rosanoff to Emma Lundberg, September 15, 1916, CB.

91 G. M. W[hipple], "The Amateur and the Binet-Simon Tests," *JEP* 3 (January 1912): 118–119.

92 Dr. E. H. Mullan, *Mental Deficiency: Some of its Public Health Aspects, with Special Reference to Diagnosis*, Reprint No. 236, Public Health Reports, November 27, 1914 (rpt. Washington, D.C.: GPO, 1919), quotation p. 7.

93 Women psychologists who worked in medical settings include Leta Hollingworth, Augusta Bronner, Clara Harrison Town, Florence Mateer, and Grace Fernald. For a parallel discussion of psychiatry's hostility toward another group of professionalizing women, social workers, see Lunbeck, *Psychiatric Persuasion,* pp. 25–45.

94 Harry L. Hollingworth, *Leta Stetter Hollingworth: A Biography* (Lincoln: University of Nebraska Press, 1943), pp. 101–102. On Hollingworth, see also Rosenberg, *Beyond Separate Spheres,* pp. 84–86, 110–113; and Stephanie Shields, "Functionalism, Darwinism, and the Psychology of Women: A Study in Social Myth," *AP* 30 (July 1975): 739–754, esp. pp. 747–748.

95 "Court Bars Binet Test," *NYT* (July 19, 1916): 7; "Scorns the Aid of Science," *NYT* (July 20, 1916): 10.

96 Leta S. Hollingworth, "Existing Laws Which Authorize Psychologists to Perform Professional Services," *Journal of Criminal Law and Criminology* 13 (1922): 70–73. Wisconsin and New York passed similar laws in 1919, and South Dakota in 1921.

97 Lewis Terman, "Review of the Vineland Translation of Articles by Binet and Simon," *Journal of Delinquency* 1 (November 1916): 258.

98 Dr. William Burgess Cornell, "Psychology vs. Psychiatry in Diagnosing Feeble-Mindedness," *New York State Journal of Medicine* 17 (1917): 485–486.

99 Ibid. See also Lunbeck, *Psychiatric Persuasion,* pp. 25–45.

100 "Activities of Clinical Psychologists," *PB* 14 (June 15, 1917): 224–225.

101 Ibid.

102 Shepherd Ivory Franz, "Psychology and Psychiatry," *PB* 14 (June 15, 1917): 226–229.

103 Ibid.

104 Johnson, *Adventures,* pp. 416–417; Trent, *Inventing the Feeble Mind,* pp. 179–181, 206.

105 Johnson, *Adventures,* pp. 416–417.

106 Terman's letter is in Pastore, *The Nature–Nurture Controversy* (New York: King's Crown Press, 1949), pp. 94–95. "Arnold Gesell," in Edwin Boring et. al., eds., *A History of Psychology in Autobiography* (Worcester, Mass.: Clark University Press, 1952), 4:128–129.

Chapter 8

1 Julia Lathrop to HHG, May 4, 1914, Box 69 (1914–1920), CB. Lathrop also wrote to William Healy, Alexander Johnson, and W.H.C. Smith of the Beverly Farms institution. Wilson was responding to a letter from New Jersey educator Maximilian Groszmann, founder of the "National Association for the Study and Education of Exceptional Children," whom many contemporaries regarded as an opportunist. Although Groszmann never gained the president's endorsement, Wilson did consider his ideas about coordinating government work in this field.

2 Lathrop to HHG, May 4, 1914.

3 HHG to Lathrop, May 6, 1914, CB.

4 Ibid.

5 Children's Bureau studies included *Mental Defectives in the District of Columbia* (Washington, D.C.: GPO, 1915); *A Social Study of Mental Defectives in New Castle County, Delaware* (Washington, D.C.: GPO, 1917); and *Mental Defect in a Rural County* (Washington, D.C.: GPO, 1919).

6 The secondary literature on testing and immigration is extremely contentious; much of it responds to evidence in Leon Kamin, *The Science and Politics of I.Q.* (Potomac, Md.: Erlbaum, 1974), pp. 15–32. Most writers assess the influence (or lack of influence) of psychologists on the Immigration Restriction Act of 1924; Goddard's earlier studies of immigrants, however, are also considered. Kamin is challenged in Mark Snyderman and R. J. Herrnstein, "Intelligence Tests and the Immigration Act of 1924," *AP* 38 (September 1983): 986–995; for an overview of the criticisms, corrections, and responses this article generated, as well as the articles and commentaries that preceded it, see Steven Gelb, Garland Allen, Andrew Futterman and Barry Mehler, "Rewriting Mental Testing History: The View from the *American Psychologist*," *Sage Race Relations Abstracts* 11 (May 1986): 18–31. The best discussions of these issues are Franz Samelson, "Putting Psychology on the Map: Ideology and Intelligence Testing," in Allan R. Buss, ed., *Psychology in Its Social Context* (New York: Irvington, 1979), pp. 103–168; and Steven Gelb, "Henry H. Goddard and the Immigrants, 1910–1917: The Studies and Their Social Context," *JHBS* 22 (1986): 324–332.

7 On attitudes toward immigrants, see for instance John Higham, *Strangers in the Land: Patterns of American Nativism, 1860–1925* (New Brunswick, N.J.: Rutgers University Press, 1955); Barbara Solomon, *Ancestors and Immigrants: A Changing New England Tradition* (Cambridge, Mass.: Harvard University Press, 1956); Leonard Dinnerstein and David M. Reimers, *Ethnic Americans: A History of Immigration and Assimilation* (New York: Harper & Row, 1975); Alan Kraut, *The Huddled Masses: The Immigrant in American Society, 1880–1921* (Arlington Heights, Ill.: Harlan Davidson, 1982); and John Bodnar, *The Transplanted: A History of Immigrants in Urban America* (Bloomington: Indiana University Press, 1985).

8 Edward Ross, *The Old World in the New: The Significance of Past and Present Immigration to the American People* (New York: Century, 1914); Madison Grant, *The Passing of the Great Race* (New York: Scribner, 1916), pp. 89–90.

9 On the connections between eugenics, anti-immigration sentiments, and racist ideas, see Mark Haller, *Eugenics: Hereditarian Attitudes in American Thought* (New Brunswick, N.J.: Rutgers University Press, 1963), pp. 65, 144–159. Prescott Hall is quoted in Haller, p. 146.

10 HHG, "Feeble-Mindedness and Immigration," *TSB* 9 (October 1912): 91–94, quotation p. 91.

11 Of the hundreds of articles on Binet–Simon testing before 1917, a small number dealt with American racial differences. These include Alice Strong, "Three Hundred Fifty White and Colored Children Measured by the Binet–Simon Measuring Scale of Intelligence: A Comparative Study," *PS* 20 (1913): 485–515; Josiah Morse, "A Comparison of White and Colored Children Measured by the Binet Scale of Intelligence," *Transactions of the Fourth International Congress on School Hygiene* (Buffalo: Courier, 1914), 5:655–662; Morse, "A Comparison of White and Colored Children Measured by the Binet Scale of Intelligence," *Popular Science Monthly* 84 (1914): 75–79; and E. C. Rowe, "Five Hundred Forty-Seven White and Two Hundred Sixty-Eight Indian Children Tested by the Binet–Simon Tests," *PS* 21 (March 1914): 454–468. In his massive 1914 study, *FM,* Goddard cited none of these articles on race, nor did he cite racial issues as a social problem meriting the expertise of testers.

12 HHG, *FM,* pp. 74, 238. Of more than three hundred Vineland cases, Goddard specifically mentions immigration in ten, and suggests that children (or their parents) might have been detained by immigration inspectors in only four cases. See pp. 74, 122, 234, 238.

13 HHG, "Elementary Schools, Section E. – Ungraded Classes," in *Report on Educational Aspects of the Public School System of the City of New York . . .* Part II, Subdivision 1 (New York: City of New York, 1912), pp. 361–381. Elizabeth Farrell, "A Study of the School Inquiry Report on Ungraded Classes, Part 1" *PC* 8 (April 15, 1914): 40–44. (For this study, see Chapter 4.)

14 Lewis Terman, *Measurement of Intelligence* (Boston: Houghton Mifflin, 1916), pp. 91–92.

15 HHG, "The Feeble Minded Immigrant," *TSB* 9 (1912): 109–113, quotation p. 110.

16 On immigration restriction statutes, see Harlan Unrau, ed., *Ellis Island: Historic Resource Study,* 3 vols. (Washington, D.C.: U.S. Department of the Interior, 1984), 1:13–69. Dr. L. L. Williams, "The Medical Examination of Mentally Defective Aliens: Its Scope and Limitations," *American Journal of Insanity* 71 (October 1914): 257–268, quotation p. 258. See also Thomas Pitkin, *Keepers of the Gate: A History of Ellis Island* (New York: New York University Press, 1975).

17 Dr. George Stoner estimated the number examined to be 5,000 or more on busy days, and about 3,000 on ordinary days; see Stoner, "Insane and Mentally Defective Aliens Arriving at the Port of New York," *New York Medical Journal* 97 (May 10, 1913): 957. The number of Public Health Officers working as line inspectors apparently ranged from about six to twelve. In 1913, Goddard reported that a dozen physicians were examining about 29,000 cases a week. See Unrau, *Ellis Island,* 2:707; Elizabeth Yew, "Medical Inspection of Immigrants at Ellis Island, 1981–1924," *Bulletin of the New York Academy of Medicine* 56 1980): 488–510; Dr. E. H. Mullan, "Mental Examination of Immigrants: Administration and Line Inspection at Ellis Island," *Public Health Reports* 32 (May 18, 1917): 733–746; and HHG, "The Binet Tests in Relation to Immigration," *JPA* 18 (1913): 104–107. See also J. G. Wilson, "Some Remarks Concerning Diagnosis by Inspection," *New York Medical Journal* (July 6, 1911): 94–96; and C. P. Knight, "The Detection of the Mentally Defective Among Immigrants," *JAMA* 60 (January 11, 1913): 106–107. On the Public Health Service, see Alan Kraut, *Silent Travelers: Germs, Genes, and the "Immigrant Menace"* (Baltimore: Johns Hopkins University Press, 1994); and Fitzhugh Mullan, *Plagues and Politics: The Story of the United States Public Health Service* (New York: Basic Books, 1989).

18 HHG, "Feeble Minded Immigrant," p. 110. "HHG, "Feeble-Mindedness and Immigration," pp. 91–94, esp. Table II.

19 HHG, "Inquiry Relative to Feeble-Minded Immigrants," File M918, Vineland Research Laboratory Papers. A version of this questionnaire is reprinted in "Feeble-Mindedness and Immigration," p. 92.

20 HHG, "Feeble-Mindedness and Immigration," pp. 91–94. HHG, "Feeble Minded Immigrant," p. 109.

21 Ibid., pp. 109–113.

22 HHG to Robert Yerkes, November 5, 1912, Yerkes Papers.

23 HHG, "Feeble-Mindedness and Immigration," p. 113.

24 "Editorials: The Problem of the Feeble-Minded among Immigrants," *JAMA* 60 (January 18, 1913): 209–210. For the *JAMA*'s views, see "Editorials: Mental Disorders and Immigration," *JAMA* 58 (March 30, 1912): 938–939; "Editorials: Proposed Amendments to the Immigration Law," *JAMA* 58 (April 27, 1912): 1286–1287; and "Editorials: The Insanity Problem," *JAMA* 59 (August 17, 1912): 545.

25 E. K. Sprague, "Mental Examination of Immigrants," *Survey* 31 (January 17, 1914): 466–468.

26 E. H. Mullan, *Mental Hygiene,* Reprint No. 164, Public Health Reports (Washington, D.C.: GPO, 1914), p. 11; Williams, "Medical Examination," p. 263.

27 Bernard Glueck, "The Mentally Defective Immigrant," *New York Medical Journal* 93 (October 18, 1913): 760–766.

28 Williams, "Medical Examination," p. 265; Sprague, "Mental Examination," p. 468. Families of persons denied entrance often hired prominent attorneys and private physicians to challenge these diagnoses. Moreover, medical decisions were often overturned by nonmedical personnel; in 1912, for instance, 79 percent certified as unfit for all medical reasons were still allowed to land. See Yew, "Medical Inspection," pp. 498–500.

29 Goddard recognized these similarities. What his aides were doing was "really not more than is done by the physicians who stand there and at a glance, pick out the various physical defects and even insanity," he argued. HHG, "The Feeble Minded Immigrant," p. 112.

30 For an intriguing illustration of psychological influence on medical thinking, see E. H. Mullan, *Mentality of the Arriving Immigrant,* Public Health Bulletin No. 90 (Washington, D.C.: GPO, 1917; rpt. New York: Arno Press, 1970).

31 Mullan, "Mental Examination," p. 738; Knight, "Detection of the Mentally Defective," p. 107.

32 On this process, see Mullan, "Mental Examination,"; and U.S. Public Health Service, *Manual of the Mental Examinations of Aliens* (Washington, D.C.: GPO, 1918). Each certificate was signed by three different officers. See Sprague, "Mental Examination," p. 468. Kraut notes that doctors avoided making final decisions on exclusions in *Silent Travelers,* pp. 68–69. Yew notes the frequency of overturning medical certificates in "Medical Inspection," pp. 498–500.

33 Howard Knox, "A Scale, Based on the Work at Ellis Island, for Estimating Mental Defect," *JAMA* 62 (March 7, 1914): 741–747 (rpt. Chicago: American Medical Association, 1914). The "Steamship Picture Form Board" (later called the Ship Test by army psychologists) is illustrated in *Manual of the Mental Examination of Aliens.* Also used were the "Seguin Form Board, Vineland Model" and several "Construction Tests" (which involved arranging pieces within a frame) devised by William Healy. Other tests are in Mullan, *Mentality of the Arriving Immigrant,* Glueck, "The Mentally Defective Immigrant," and Knox, "Two New Tests for the Detection of Defectives," *New York Medical Journal* 93 (1913): 522–524.

34 Mullan, *Mentality of the Arriving Immigrant,* p. 20.

35 Howard Knox, "The Differentiation Between Moronism and Ignorance," *New York Medical Journal* 93 (September 20, 1913): 564–566; "Can't Bar All Defectives," *NYT,* October 14, 1913, p. 12. Many Binet questions are also described in *Manual of the Mental Examination of Aliens.*

36 Mullan, "Mental Examination," p. 746; Mullan, *Mental Deficiency: Some of Its Public Health Aspects, with Special Reference to Diagnosis,* Public Health Reports Reprint 236 (November 27, 1914; rpt. Washington, D.C.: GPO, 1919), p. 4.

37 Mullan, *Mental Deficiency,* p. 7. For other uses of testing by the Public Health Service, see also Dr. E. H. Mullan, *Mental Status of Rural School Children,* Public Health Reports 31:46 (November 17, 1916; rpt. Washington, D.C.: GPO, 1916).

38 Deportation rates for idiocy, imbecility, and feeblemindedness can be calculated from the statistics in Unrau, *Ellis Island,* 1:185, 200–206. This rate peaked in 1915 at 102 per 100,000 (although the number affected dropped from 1,077 to 335, reflecting the steep decline in immigration due to the war); it then dropped to 82 per 100,000 in 1916.

39 HHG, "Mental Tests and the Immigrant," *Journal of Delinquency* 2 (September 1917): 243–277.

40 Ibid., p. 244.

41 Ibid., pp. 260–261.

42 Ibid., p. 247. The question of setting boundaries between "normal," "borderline," and "feebleminded" test performance inspired a complex debate. Here Goddard was following Samuel Kohs, who set "normal" mentality at mental age XI^3 (eleven years plus three extra questions passed), "borderline" as between X^3 and XI^2, and feebleminded as below X^3. See Kohs, "The Practicability of the Binet Scale and the Question of the Borderline Case," pamphlet (Chicago: House of Correction Press, 1915).

43 HHG, "Mental Tests and the Immigrant," pp. 245–247. Kohs was probably Goddard's Yiddish-speaking tester.

44 Ibid., p. 247.

45 Ibid.

46 Ibid., p. 250–251.

47 Ibid., p. 251.

48 Ibid., pp. 249–250.

49 Ibid., p. 250.

50 *Manual of the Mental Examination of Aliens,* pp. 22–23.

51 Ibid., pp. 24–26.

52 HHG, "Mental Tests and the Immigrant," p. 249.

53 Ibid., pp. 266.

54 Ibid., pp. 266–268.

55 Ibid., p. 261–262; HHG, "The Binet Tests in Relation to Immigration," pp. 104–110; quotation p. 106.

56 HHG, "Mental Tests and the Immigrant," pp. 262–265.

57 Kite's field report is quoted in ibid., pp. 264–265.

58 HHG, "Mental Tests and the Immigrant," pp. 266–269.

59 Ibid., pp. 267–268.

60 Ibid., pp. 270–271.

61 Paul Popenoe, Editor, *American Breeders' Magazine,* to HHG, September 9, 1913; HHG to Popenoe, September 10, 1913, File M615, GP. Ironically, Popenoe was also an editor of the *Survey,* so he probably produced its summary of Goddard's article. See note 62 below.

62 "Two Immigrants Out of Five Feebleminded," *Survey* 38 (September 15, 1917): 528–529.

63 Letter "To the Editor" from Helen Winkler and Elinor Sachs, Council of Jewish Women, Department of Immigrant Aid; Response from "Editor"; both printed in "Testing Immigrants," *Survey* 39 (November 10, 1917): 152–153.

64 HHG, "Mental Tests and the Immigrant," p. 269.

65 Ibid.

66 James Weinstein, *The Corporate Ideal in the Liberal State: 1900–1918* (Boston: Beacon, 1968); Carol Signer Gruber, *Mars and Minerva: World War I and the Uses of the Higher Learning in America* (Baton Rouge: Louisiana State University Press, 1975).

67 The army testing project in World War I has received considerable attention. Valuable historical studies include Daniel Kevles, "Testing the Army's Intelligence: Psychologists and the Military in World War I," *Journal of American History* 55 (1968): 565–581; Joel Spring, "Psychologists and the War: The Meaning of Intelligence in the Alpha and Beta Tests," *History of Education Quarterly* 12 (1972): 3–15; Franz Samelson, "World War I Intelligence Testing and the Development of Psychology," *JHBS* 13 (1977): 274–282; Samelson, "Putting Psychology on the Map: Ideology and Intelligence Testing," in Allan R. Buss, ed., *Psychology in Its Social Context* (New York: Irvington, 1979), pp. 103–168; Richard von Mayrhauser, "The Manager, the Medic, and the Mediator: The Clash of Professional Psychological Styles and the Wartime Origins of Group Mental Testing," in Michael Sokal, ed., *Psychological Testing and American Society, 1890–1930* (New Brunswick, N.J.: Rutgers University Press, 1987), pp. 128–157; John Carson, "Army Alpha, Army Brass, and the Search for Army Intelligence," *Isis* 84 (1993): 278–309; Thomas Camfield, "Psychologists at War: The History of American Psychology and the First World War" (Ph.D. diss., University of Texas, 1969); and Richard von Mayrhauser, "The Triumph of Utility: The Forgotten Clash of American Psychologies in World War I" (Ph.D. diss., University of Chicago, 1986). My own summary of the army testing project uses insights from this literature, while also paying special attention to the ways that the army experience addressed conflicts of the previous decade – e.g., experts against amateurs, medical against educational paradigms, Yerkes' Point Scale against Terman's Stanford–Binet scale, etc. – and to the ways that Goddard's role was transformed.

68 HHG, "The Place of Intelligence in Modern Warfare," *United States Naval Medical Bulletin* 2 (1917): 283–289.

69 Ibid.

70 Robert Yerkes to the Council of the APA, April 6, 1917; rpr. in Robert Yerkes, ed., *Psychological Examining in the United States Army,* Memoirs of the National Academy of Sciences (Washington, D.C.: GPO, 1921), [hereafter *Psychological Examining*], 15:7–8. APA committees are in Yerkes, "Psychology and National Service," *PB* 14 (July 1917): 259–263; their work is analyzed in Camfield, "Psychologists at War," pp. 102–116.

71 Robert Yerkes to HHG, May 4, 1917, File N-1, National Research Council Records, National Academy of Sciences.

72 On Scott's committee, see Camfield, "Psychologists at War," pp. 218–230; and von Mayrhauser, "The Triumph of Utility." Von Mayrhauser has written extensively on the significant (and largely ignored) role played by Scott's committee. See "The Manager, the Medic, and the Mediator"; "Making Intelligence Functional: Walter Dill Scott and Applied Psychological Testing in World War I," *JHBS* 25 (1989): 60–72; and "The Practical Language of American Intellect," *History of the Human Sciences* 4 (1991): 371–393.

73 Robert Yerkes to George E. Hale [NRC Chairman], May 22, 1917, quoted in Camfield, "Psychologists at War," p. 128. Yerkes' efforts to secure funds are cited in Joseph Byers, *The Village of Happiness: The Story of the Training School* (Vineland, N.J.: Smith, 1934), p. 88.

74 Byers' letters to the Secretary of War and Secretary of the Navy, written on April 10, 1917, as well as his correspondence with Yerkes, are reprinted in Byers, *Village,* pp. 86–90. Yerkes to Byers, May 4, 1917, reprinted on pp. 87–88.

75 Ibid., p. 88. Byers cites the amount offered as up to $800; Yerkes cites $700 in *Psychological Examining,* p. 9.

76 Yerkes includes his original "Plan for the Psychological Examination of Recruits to Eliminate the Mentally Unfit," April 30, 1917, in "Psychology in Relation to the War," *Psychological Review* 25 (March 1918): 94–97, quotations p. 95.

77 Robert Yerkes to H. L. Hollingworth, May 22, 1917, quoted in Camfield, "Psychologists at War," p. 144. On the tense relationship between psychiatry and psychology and the work of the Joint Committee, see Camfield, pp. 143–150.

78 Yerkes reports that by the second day, his committee had decided to adopt Terman's strategy of developing group tests for all recruits. *Psychological Examining,* p. 299.

79 On the history of multiple-choice tests, see Franz Samelson, "Was Early Mental Testing: (a) Racist Inspired, (b) Objective Science, (c) A Technology for Democracy, (d) The Origin of the Multiple-Choice Exams, (e) None of the Above? (Mark the RIGHT Answer)," in Sokal, ed., *Psychological Testing,* pp. 113–127. On this committee's decision to adopt multiple-choice testing and stencil scoring, see *Psychological Examining,* pp. 299–300.

80 Copies of all Army A, Army Alpha, Army Beta, Individual Tests, and Performance Tests are in Yerkes, ed., *Psychological Examining.* Otis made Binet's test of "disarranged sentences" a stencil-scored test by asking subjects to rearrange a sentence and then to mark it as true or false (example: "Morning the rises every sun" – true). See *Psychological Examining,* p. 204. For the origins of each subtest, see *Psychological Examining,* p. 311.

81 The "Information Test" was borrowed in part from the Bureau of Salesmanship Research Tests developed at Carnegie Institute of Technology and in part from an information test devised by Wells. See *Psychological Examining,* p. 311. Questions asked in various versions of this subtest can be found in *Psychological Examining,* pp. 206, 213, 227, and 234.

82 The "Practical Judgment Test" used several Binet questions in multiple-choice form, as well as new ones. See *Psychological Examining,* pp. 208, 215, 222, and 229.

83 On letter grades, see *Psychological Examining,* pp. 421–424. The tension between the Point Scale and the Stanford–Binet was even more obvious in deciding how to retest individuals who did poorly on the Alpha. For this purpose, the committee recommended retesting "by means of the Point Scale or Stanford–Binet scale according to availability of materials and preference of the examiner" (p. 167). Thus, 23.3 percent were retested by the Point Scale, 47.9 by the Stanford–Binet, and 28.8 percent by the Performance Scale (p. 100).

84 The private trial was paid for by the Mental Hygiene War Work Committee. See Yerkes, ed., *Psychological Examining,* p. 10. On comparing Army A results with officers' assessments, see *Psychological Examining,* pp. 327–345.

85 Yerkes to Byers, August 13, 1917, reprinted in Byers, *Village,* pp. 89–90.

86 Yerkes, "Psychology and National Service," p. 261. On the rivalry between physicians and psychologists, see Yerkes, ed., *Psychological Examining,* pp. 97–99.

87 A summary schedule of monthly progress is in Yerkes, ed., *Psychological Examining,* pp. 91–95; the total tested is reported on p. 99.

88 All of the data would be published in Yerkes, ed., *Psychological Examining,* with parts written by Yerkes and parts by Terman. Army results were also published in C. S. Yoakum and Robert Yerkes, *Army Mental Tests* (New York: Henry Holt, 1920).

89 On strategies used to gauge literacy, see *Psychological Examining,* pp. 347–355.

90 Ibid., p. 100.

91 On the correlation with schooling, see *Psychological Examining,* p. 779. Terman wrote this section. Here, too, different standards were used, for some reported only correlations between schooling and Alpha scores, while others included Beta scores.

92 *Psychological Examinining,* pp. 785–791.

93 Ibid., pp. 791, 789.

94 Among officers, 31.4 percent were college graduates; 28.6 percent had from one to three years of college; and 14.6 percent had postgraduate degrees. Thus, 74.6 percent of officers had at least one year of college; the comparable figure for native-born white recruits was 5.4 percent. Among southern blacks, 19.4 percent reported no years of schooling. See *Psychological Examining,* Chart 302, p. 758, and the discussion on p. 760.

95 For careful analyses of the army's responses to testers, see Camfield, "Psychologists at War"; Kevles, "Testing the Army's Intelligence"; Carson, "Army Alpha, Army Brass"; Samelson, "Putting Psychology on the Map," and von Mayrhauser, "The Triumph of Utility."

96 Terman and Yerkes continued a very close personal and professional correspondence for the remainder of their lives, as evidenced in both the Terman and Yerkes Papers.

97 *Psychological Examining,* p. 124. Instructions for all the subtests can be found on pp. 123–128. For an excellent analysis of the militarization of these instructions, see Carson, "Army Alpha, Army Brass," pp. 299–300.

98 On Binet's decision to eliminate reading tests in 1911 (which Goddard followed), see Alfred Binet and Theodore Simon, *The Development of Intelligence in Children,* trans. Elizabeth Kite (Baltimore: Williams & Wilkins, 1916), p. 275. Yerkes' earlier concern with environmental factors is evident in Yerkes and Helen Anderson, "The Importance of Social Status as Indicated by the Results of the Point-Scale Method of Measuring Mental Capacity," *JEP* 6 (March 1915): 137–150. A good analysis of cultural biases in the Army Beta is in Allan Chase, *The Legacy of Malthus: The Social Costs of the New Scientific Racism* (New York: Knopf, 1977), pp. 247–248.

99 "The main purpose of the individual examination is to secure a more accurate measurement of the mental ability of those who have made D− in alpha or beta, or both," stated the Examiner's Guide. Yerkes, ed., *Psychological Examining,* p. 167. On men retested, see pp. 791–798.

100 *Psychological Examining,* p. 103.

101 Yerkes and Cattell are quoted in Samelson, "Putting Psychology on the Map," p. 106.

102 Byers, *Village,* p. 89.

103 HHG to Major Lewis Terman, November 25, 1918, File M33, GP.

104 HHG, "Anniversary Address," in Edgar A. Doll, ed., *Twenty-Five Years* (Vineland, N.J.: The Training School, 1932), p. 58.

105 HHG, *Psychology of the Normal and Subnormal* (New York: Dodd, Mead, 1919).

106 Ibid., pp. 3–51.

107 Ibid., pp. 56–57, 52. Goddard, following Terman, argued that the average intelligence level was about mental age sixteen, and the upper intelligence limit about mental age twenty.

108 Ibid., p. 148.

109 Ibid., p. 250.

110 Ibid., p. 234.

111 Ibid., p. 236.

112 Ibid., pp. 239–241. [Edwin Markham], "The Book of the Month: The Kallikak Family," *Hearst's Magazine* 23 (February 1913): 329–331. For a similar interpretation of Markham's poem, see Dr. C. P. Wertenbaker, "Eugenics and the Public Health," *New York Medical Journal* 93 (September 27, 1913): 602–608, esp. pp. 604–605.

113 HHG, *Psychology of the Normal and Subnormal,* p. 246.

114 Ibid., p. 237.

115 Charles Osgood to HHG, January 9, 1919, File M33, GP. Goddard's Vanuxem lectures were published as *Human Efficiency and Levels of Intelligence* (Princeton: Princeton University Press, 1920).

116 Ibid., pp. 1, vi, 34–94.

117 Ibid., p. 95–96.

118 Ibid., p. 99.

119 Ibid., pp. 100–101.

120 Ibid., pp. 101–102.

121 Ibid., pp. 127–128, 104.

122 Ibid., pp. 98–99.

123 On the debates that emerged in the 1920s, see Chapter 9.

Chapter 9

1 HHG, "Report of the Research Department," _TSB_ 15 (December 1918): 123–125.

2 Ibid. The comparison to the Mayo brothers was by Hastings Hart of the Sage Foundation.

3 HHG, "Pro" and "Con" List, File M33 AA3; HHG to H. Riddle, February 21, 1918, File M33.1, GP. "Ohio Gets Expert at Big Pay to Better Children," newspaper clipping, May 14 [1918], File M44, GP, states Goddard's salary as second only to the governor's.

4 Hoyt Landon Warner, _Progressivism in Ohio, 1897–1917_ (Columbus: Ohio State University Press, 1964), pp. 405– 406, 419– 420; Thomas Haines, "The Ohio Plan for the Study of Delinquency," _Popular Science Monthly_ 86 (June 1915): 576–580; and Hamilton Cravens, "Applied Science and Public Policy: The Ohio Bureau of Juvenile Research and the Problem of Juvenile Delinquency, 1913–1930," in Michael Sokal, ed., _Psychological Testing and American Society, 1890–1930_ (New Brunswick, N.J.: Rutgers University Press, 1987), pp. 158–194. "Ohio Gets Expert."

5 HHG, _Juvenile Delinquency_ (New York: Dodd, Mead, 1921), pp. 49–56. Since, as usual, Goddard's numbers do not add up correctly, I am using Florence Mateer's figures, which cite 3,578 cases examined, of which 236 spent some time in residence. Mateer, "Department of Clinical Psychology," in _Bureau of Juvenile Research: Review of the Work, 1918–1920_ (Mansfield: Ohio State Reformatory, 1921), p. 22.

6 William Healy, _The Individual Delinquent_ (Boston: Little, Brown, 1915); _Pathological Lying, Accusation, and Swindling_ (Boston: Little, Brown, 1915); _Mental Conflicts and Misconduct_ (Boston: Little, Brown, 1917). HHG, "The Sub-Normal Mind Versus the Abnormal," _Journal of Abnormal Psychology_ 16 (April 1921): 47–54.

7 HHG, _Juvenile Delinquency,_ pp. 27– 48. HHG, "The Bureau," in _Bureau of Juvenile Research: Review of the Work, 1918–1920,_ pp. 1–2.

8 On the health of delinquents, see HHG, _Juvenile Delinquency,_ pp. 57–66, 103–108, and Gertrude Transeau, "The Medical Department," _Bureau of Juvenile Research: Review of the Work, 1918–1920,_ pp. 3–5. Both considered congenital syphilis the most serious medical condition afflicting these delinquents. For Goddard's techniques for modifying behavior, see Florence Mateer, "Department of Clinical Psychology," pp. 35–38, as well as his text.

9 Mateer, "Department of Clinical Psychology," pp. 5–50. Violet H. Foster and HHG, "The Ohio Literacy Test," _PS_ 31 (1924): 340–358.

10 Ibid., p. 31. HHG, *Juvenile Delinquency,* pp. 31, 45–47.

11 Ibid., pp. 62–63.

12 Ibid., pp. 64–66.

13 Ibid., pp. iv, 119–120.

14 HHG to Mr. Burk, May 17, 1920, File M35.2, Misc. 2, GP. Letters of resignation from Transeau and ten other employees, dated April 4, 1921, and Mateer's resignation letter of April 17, 1921, are in File M31 BJR, GP, which also contains House Resolutions, internal memos, correspondence, and newspaper clippings pertaining to this scandal and investigation.

15 J. E. Wallace Wallin, "Shall We Continue to Train Psychologists for Second-String Jobs?" *PC* 18 (1929): 242–245. Thomas Verner Moore, "A Century of Psychology in Its Relationship to American Psychiatry," in G. K. Hall et. al., *One Hundred Years of American Psychiatry* (New York: Columbia University Press, 1944), pp. 443–477.

16 Clipping, "Bureau Endangered," *Cleveland Plain Dealer,* May 22, 1921, in File M31 BJR, GP. "Ohio Institute for Public Efficiency, Bulletin No. 16," May 20, 1921, File M31 BJR, GP. Goddard's new salary was actually below those of medical superintendents, for their contracts usually included "maintenance of the director and his family," while his did not. See also "Bureau of Juvenile Research, Budget for 1919–1921," File M33 AA3, GP.

17 HHG to George Arps, June 24, 1922, File M36, GP.

18 HHG, *Two Souls in One Body?* (New York: Dodd, Mead, 1927).

19 Ibid., p. 23. Additional newspaper clippings are in File M44, GP. For a discussion of this case within the context of broader questions raised by "multiple personality" diagnoses, see Ian Hacking, "Two Souls in One Body," *Critical Inquiry* 17 (Summer 1991): 838–867; and *Rewriting the Soul: Multiple Personality and the Sciences of Memory* (Princeton: Princeton University Press, 1995).

20 HHG, "A Case of Dual Personality," *Journal of Abnormal and Social Psychology* 21 (July 1926): 170–191. HHG, *Two Souls,* pp. v–ix, 46–53, 222–237.

21 HHG, *Two Souls,* p. 55; HHG, "Case of Dual Personality," p. 185. Hacking analyzes the evidence for incest, and Goddard's avoidance of this topic, in "Two Souls."

22 HHG, *Two Souls,* pp. 151–52. S. Weir Mitchell is quoted on pp. 213–219. Goddard's claim to be "curing" this case brought a strong protest from Dr. William Pritchard, superintendent of Columbus State Hospital, who noted that Bernice had been brought to his institution in a straightjacket. Goddard, however, believed that Bernice had indeed improved, but had temporarily lapsed into a violent phase when relatives refused to take her home. See W. H. Pritchard to HHG, December 30, 1926; HHG to Pritchard, January 17, 1927 (unsent letter); and HHG to Wm. H. Pritchard, January 29, 1927, File M33 AA2, GP.

23 HHG to Will Monroe, April 27, 1928, File M36, GP.

24 The Kallikak story was dissseminated through various media, including a five-page pamphlet (see illustration on p. 230). Goddard received many letters thanking him for speaking engagements, or for court appearances. See, for example, Fred M. Hunter, Superintendent of Oakland Public Schools, to HHG, September 21, 1923, File M31 Corr.; and Dr. Edmund Baehr to HHG, February 2, 1926, which thanks him for help in a case in which the state failed to win "a death verdict for the slayer of a policeman," File M33 AA3, GP.

25 Clipping, "'Bad Boy' Will Be Unknown in School of Future, Psychologist Addressing St. Paul Teachers Says," unidentified newspaper dated "Sept. 17, 1923" by HHG. Clipping, "Kids Are Like Blueberries," *Omaha World-Herald,* July 12, 1932. File M44, GP.

26 HHG, *School Training of Gifted Children* (Yonkers-on-Hudson, N.Y.: World Book, 1928), pp. v–viii.

27 Ibid., pp. v, 1–13, 76, pp. 58–63, 90–102.

28 Ibid., p. 63.

29 Ibid., pp. 81, 21–23; HHG, "Some Fundamental Errors in Educational Practice," *Phi Delta Kappan* 8:6 (June 1925), pp. 1–8.

30 HHG, *School Training of Gifted Children,* p. 53; HHG, "Fundamental Errors," pp. 4–7.

31 HHG to Children's Bureau, January 22, 1930, Box 353 (1929–1932), CB. See also "Special Education and the Gifted Child," *Educational Research Bulletin* 4 (April 1, 1925): 133–139; "The Gifted Child and His Education," *NEA Journal* 19 (November 1930): 275–276; "The Gifted Child," *Journal of Educational Sociology* 6 (1933): 354–361; and "The Psychology of the Status Quo of Exceptional Children," *Journal of Exceptional Children* 5 (April 1939): 181–183.

32 Robert Yerkes, ed., *Psychological Examining in the United States Army,* Memoirs of the National Academy of Sciences, vol. 15 (Washington, D.C.: GPO, 1921).

33 These quotations, as well as others illustrating the influence of the army findings on contemporary cultural debates, are in Roland Marchand, *Advertising the American Dream: Making Way for Modernity, 1920–1940* (Berkeley: University of California Press, 1985), p. 67. *Time* magazine refused to dilute its content for "morons."

34 Albert Wiggam, "The New Decalogue of Science," *Century Magazine* 103 (March 1922): 644. William Allen White, "What Is the Matter with America," *Collier's* 70 (July 1, 1922): 3–4, 18. White was famous for an earlier article, "What Is the Matter with Kansas."

35 Charles Gould, *America: A Family Matter* (New York: Scribner, 1922), pp. 18, 164, 3; Lothrop Stoddard, *The Revolt Against Civilization* (New York: Scribner, 1922), pp. 56–74.

36 The Lippmann–Terman debate is reprinted in N.J. Block and Gerald Dworkin, eds., *The IQ Controversy: Critical Readings* (New York: Pantheon, 1976), pp. 5–44. Citations below are to original articles.

37 Walter Lippmann, "The Mental Age of Americans," *New Republic* 32 (October 25, 1922): 213–215.

38 Lippmann, "The Mystery of the 'A' Men," *New Republic* 32 (November 1, 1922): 246–248. See also Lippmann, "The Reliability of Intelligence Tests," *New Republic* 32 (November 8, 1922): 275–277.

39 Lippmann, "The Abuse of the Tests," *New Republic* 32 (November 15, 1922): 297–298.

40 Lippmann, "Tests of Hereditary Intelligence," *New Republic* 32 (November 22, 1922): 328–330.

41 Lippmann, "A Future for the Tests," *New Republic* 33 (November 29, 1922): 9–11.

42 Lewis Terman, "The Great Conspiracy, or The Impulse Imperious of Intelligence Testers, Psychoanalyzed and Exposed by Mr. Lippmann," *New Republic* 33 (December 27, 1922): 116–120.

43 Ibid.

44 Lippmann, "The Great Confusion: A Reply to Mr. Terman," *New Republic* 33 (January 3, 1923): 145–146. See also "Correspondence," *New Republic* 33 (January 17, 1923): 201–202, which contains a brief letter by Terman, an answer by Lippmann, and three other letters to the editor, all offended by Terman's sarcastic tone; and Lippmann, "A Defense of Education," *Century* 106 (May 1923): 95–103.

45 William McDougall, *Is America Safe for Democracy?* (New York: Scribner, 1922). Carl Brigham, *A Study of American Intelligence* (Princeton: Princeton University Press, 1923), p. 182. Robert Yerkes, "Foreword," to Brigham, pp. v–viii. The role of testers in helping pass the Immigration Act of 1924 has generated much controversy. Leon Kamin stresses this role in *The Science and Politics of I.Q.* (Potomac, Md.: Erlbaum, 1974), pp. 15–32. Franz Samelson revises this view by arguing that although psychologists hoped to affect policy decisions, politicians hardly needed their help. See "Putting Psychology on the Map: Ideology and Intelligence Testing," in Allan R. Buss, ed., *Psychology in Its Social Context* (New York: Irvington, 1979), pp. 103–168. Mark Snyderman and R. J. Herrnstein challenged Kamin's evidence in "Intelligence Tests and the Immigration Act of 1924," *AP* 38 (September 1983): 986–995; responses challenging their evidence are summarized in Steven Gelb, Garland Allen, Andrew Futterman and Barry Mehler, "Rewriting Mental Testing History: The View from the *American Psychologist*," *Sage Race Relations Abstracts* 11 (May 1986): 18–31.

46 Brigham, *Study of American Intelligence,* pp. 191–192, attributes the higher scores of northern blacks to three factors: greater educational opportunity (which, he concedes, "does affect, to some extent, scores"); greater admixture of "white blood"; and social factors "which draw the more intelligent negro to the North."

47 Horace M. Bond, "What the Army 'Intelligence' Tests Measured," *Opportunity: A Journal of Negro Life* 2 (July 1924): 197–202. Charles S. Johnson, "Mental Measurements of Negro Groups," *Opportunity: A Journal of Negro Life* 1 (February 1923): 21–25. For black reactions to the testing controversy, see William B. Thomas, "Black Intellectuals' Critique of Early Mental Testing: A Little Known Saga of the 1920s," *American Journal of Education* 90 (1982): 258–292; Thomas, "Black Intellectuals on IQ Tests," in Russell Jacoby and Naomi Glauberman, eds., *The Bell Curve Debate: History, Documents, Opinions* (New York: Times Books, 1995), pp. 510–541; and Wayne Urban, "The Black Scholar and Intelligence Testing: The Case of Horace Mann Bond," *JHBS* 25 (1989): 323–334. See also Franz Samelson, "From 'Race Psychology' to 'Studies in Prejudice': Some Observations on the Thematic Reversal in Social Psychology," *JHBS* 14 (1978): 265–278.

48 Franz Boas, "Fallacies of Racial Inferiority," *Current History* 25 (February 1927): 681. On theories of "culture," see George Stocking, Jr., *Race, Culture, and Evolution: Essays in the History of Anthropology* (New York: Free Press, 1968); Hamilton Cravens, *The Triumph of Evolution: American Scientists and the Heredity–Environment Controversy, 1900–1941* (Philadelphia: University of Pennsylvania Press, 1978), pp. 89–156; Carl Degler, *In Search of Human Nature: The Decline and Revival of Darwinism in American Social Thought* (New York: Oxford, 1991), pp. 57–211; and Elazar

Barkan, *The Retreat of Scientific Racism* (Cambridge: Cambridge University Press, 1992), pp. 13–134.

49 William Bagley's essays are collected in *Determinism in Education* (Baltimore: Warwick & York, 1925), quotations pp. 12, 123. Citations here are from "Democracy and the IQ" (1922), and "The Army Tests and the Pro-Nordic Propaganda" (1923).

50 Ibid., pp. 12, 41, 131.

51 Lewis Terman, "Introduction" to "Nature and Nurture, Part I: Their Influence Upon Intelligence," *Twenty-Seventh Yearbook of the National Society for the Study of Education* (Bloomington, Ill.: Public School Publishing, 1928), pp. 1–7. On Terman's debates with Lippmann and Bagley, see Henry Minton, *Lewis M. Terman: Pioneer in Psychological Testing* (New York: New York University Press, 1988), pp. 100–109, 150–154.

52 L. W. Crafts, "Bibliography of Feeble-Mindedness in Relation to Juvenile Delinquency," *Journal of Delinquency* 1 (September 1916): 195–208; Arthur Fink, *Causes of Crime: Biological Theories in the United States, 1800–1915* (Philadelphia: University of Pennsylvania Press, 1938), pp. 211–239.

53 Carl Murchison, *Criminal Intelligence* (Worcester, Mass.: Clark University Press, 1926), p. 28.

54 Ibid., pp. 32, 289–291.

55 In New York, for instance, despite constant publicity emphasizing the need to institutionalize at least 40,000 more persons, funding was provided only for 1,660 more beds during the 1910–1920 decade. See Stanley Davies, *Social Control of the Feebleminded* (New York: National Committee for Mental Hygiene, 1923), pp. 68–70.

56 Charles Bernstein, "Colony and Extra-institutional Care for the Feebleminded," *Mental Hygiene* 4 (January 1920): 1–28. On Bernstein's "world test," see Davies, *Social Control,* pp. 108–170. See also James W. Trent, Jr., *Inventing the Feeble Mind: A History of Mental Retardation in the United States* (Berkeley: University of California Press, 1994), pp. 207–215; and Philip M. Ferguson, *Abandoned to Their Fate: Social Policy and Practice toward Severely Retarded People in America, 1820–1920* (Philadelphia: Temple University Press, 1994), pp. 105–127.

57 Walter Fernald, "After-care Study of the Patients Discharged from Waverly for a Period of Twenty-five Years," *Ungraded* 5 (November, 1919): 25–31; Davies, *Social Control,* pp. 85–107.

58 Walter Fernald, "Thirty Years Progress in the Care of the Feeble-Minded," *JPA* 29 (June 1923-June 1924): 206–219.

59 Ibid. On the "Myth of the Menace of the Feebleminded," see Mark Haller, *Eugenics: Hereditarian Attitudes in American Thought* (New Brunswick, N.J.: Rutgers University Press, 1963), pp. 95–110.

60 Davies, *Social Control.*

61 Ibid., pp. 18–19.

62 Ibid., p. 201.

63 H. S. Jennings, *Prometheus, or Biology and the Advancement of Man* (London: Kegan Paul, 1925), pp. 24–26. Daniel Kevles considers the criticisms of Jennings and others in *In the Name of Eugenics: Genetics and the Uses of Human Heredity* (New York: Knopf, 1985), pp. 114–137. See also Garland Allen, *Thomas Hunt Morgan: The*

Man and His Science (Princeton: Princeton University Press, 1978); Allen, "Genetics, Eugenics and Society: Internalists and Externalists in Contemporary History of Science," *Social Studies of Science* 6 (1976): 105–122; Kenneth Ludmerer, *Genetics and American Society: A Historical Appraisal* (Baltimore: Johns Hopkins University Press, 1972); William Provine, *The Origins of Theoretical Population Genetics* (Chicago: University of Chicago Press, 1971); Robert Kohler, *Lords of the Fly: Drosophila Genetics and the Experimental Life* (Chicago: University of Chicago Press, 1994); and Diane Paul, *Controlling Human Heredity: 1865 to the Present* (Atlantic Highlands, N.J.: Humanities Press, 1995).

64 Kevles, *Name of Eugenics,* pp. 121–122; Raymond Pearl, "The Biology of Superiority," *American Mercury* 12 (November 1927): 257–266; quotations p. 260.

65 Abraham Myerson, *The Inheritance of Mental Diseases* (New York: Williams & Wilkins, 1925). For Myerson's sharp attacks on eugenic "pseudoscience," see, for example, his review of Albert Wiggam's *New Decalogue of Science* in *Mental Hygiene* 8 (1924): 598–602.

66 Ibid., pp. 64, 77–80.

67 HHG to Elizabeth Kite, December 7, 1928, File M35.1, GP. Goddard's argument made little sense, for in the Kallikak study all known identities were disguised by pseudonyms; hence the phrase "nameless feebleminded girl" would not be a way to indicate protecting an identity.

68 Kite to HHG, December 13, 1928, File M35.1, GP.

69 Ibid. See also HHG to Kite, January 5, 1929, File M35.1, GP.

70 On popularizing eugenics, see Kevles, *Name of Eugenics,* pp. 57–112.

71 *Buck v. Priddy* (1924) reached the Supreme Court as *Buck v. Bell,* 274 U.S. 200 (1927). J. David Smith, *Minds Made Feeble: The Myth and Legacy of the Kallikaks* (Rockville, Md.: Aspen Systems Corporation, 1985), pp. 139–154; see also Allan Chase, *The Legacy of Malthus: The Social Costs of the New Scientific Racism* (New York: Knopf, 1977), pp. 302–318; Philip Reilly, *The Surgical Solution: A History of Involuntary Sterilization in the United States* (Baltimore: Johns Hopkins University Press, 1991), pp. 67–68, 86–88, 152–156; and Stephen Jay Gould, "Carrie Buck's Daughter," *Natural History* 93 (1984): 14–18.

72 HHG, "Anniversary Address," in Edgar Doll, ed., *Twenty-Five Years* (Vineland, N.J.: The Training School, 1932), p. 59. On legal battles over sterilization, see Reilly, *Surgical Solution,* and Edward Larson, *Sex, Race, and Science: Eugenics in the Deep South* (Baltimore: Johns Hopkins University Press, 1995).

73 Smith, *Minds Made Feeble,* pp. 139–150; Reilly, *Surgical Solution,* pp. 66–68, 86–87.

74 Oliver Wendell Holmes, *Buck v. Bell,* p. 207.

75 Carl Brigham's recantation is in "Intelligence Tests of Immigrant Groups," *Psychological Review* 37 (1930): 158–165. As early as 1920, Goddard began to change his views of what the army tests meant. "We have evidently been guilty of bad logic," he declared, for it was "an absurdity" to call everyone testing twelve years or less feebleminded. At fault was "the concept of who is feeble-minded." HHG, "The Subnormal Mind versus the Abnormal," p. 48.

76 HHG, "Who Is a Moron?" *Scientific Monthly* 24 (January 1927): 41–46.

77 Ibid.

78 HHG, "Feeblemindedness: A Question of Definition," *JPA* 33 (1928): 219–227.

79 Ibid.

80 Membership List, American Eugenics Society's Advisory Council of 1929, File M36, GP; HHG to Dr. Krenberger, April 8, 1929, is written on stationery listing HHG as a member of Ohio's Committee on the Legalization of Sterilization. File M36, GP.

81 Goddard's text made no comments about the Jewish children in Cleveland's gifted classes; however, his statistics show 45 Jewish children, a number second only to the "Americans" (121) and far larger than the next ethnic group, "Russians" (5). Moreover, notes by a school home visitor include miscellaneous comments such as "Ordinary Jewish family." See HHG, *School Training of Gifted Children,* pp. 129–141. The "Japanese genius" and his artwork are described on pp. 208–215. F. E. Stafford, Chairman, Interracial Committee on Moral Conduct, Department of Public Instruction, Territory of Hawaii, to HHG, May 11, 1927; pamphlet, "Standard of Moral Conduct and Social Ethics, Honolulu, Hawaii, U.S.A.," File M42; HHG, "Idle Thoughts: The Rising Tide of Character," File M37, GP.

82 HHG, "Fundamental Errors," p. 2; HHG to Edwin Holt, December 19, 1931, File M35. HHG's essay, entitled "The Child's Inheritance and What Can Be Done With It," is marked "Sent to Eugenics 2-14-1930 But not used. Returned by request." File M35.1, GP.

83 Roberta Bole to HHG, September 30, 1929, File M35.1 Misc. l; Bole to HHG, February 7, 1931, File M36; Professor Griboedoff, Leningrad Scientific-Medical Pedological Society, to HHG, November 1928, File M35; Notes to "As Luck Would Have It," GP.

84 HHG, "Anniversary Address," p. 59 (emphasis in original).

85 Lev Flournoy, "'Out There': A Psychologist Looks Into the Uncharted Future," *Columbus Citizen,* August 1, 1934, clipping in File M44, GP.

86 HHG, "Anniversary Address," pp. 53–69.

87 Ibid.

88 The first German edition of *Die Familie Kallikak* was published in 1914, and the second in November 1933; it was also reprinted in Karl Wilker, trans., "Die Familie Kallikak," *Friedrich Mann's Pedagogisches Magazin,* No. 1393 (1934). Smith, *Minds Made Feeble,* pp. 161–163. See also Stefan Kühl, *The Nazi Connection: Eugenics, American Racism, and National Socialism* (New York: Oxford University Press, 1994), pp. 40–42.

89 Although antisemitic remarks are common in social science publications of this decade, this is not true of Goddard's books. In *Minds Made Feeble,* pp. 191–193, Smith stresses a 1928 letter to Arnold Gesell. In explaining his attendance at a Battle Creek eugenics conference, Goddard states: "You see Kellogg entertains the whole association and there is just enough Scotch or Jew in me that I could not miss a free dinner!" he wrote. Later he adds: "We should start west at once but again that Jewish blood (I think it is Jewish – look at the nose!) makes me wait until the summer excursion tickets are on sale – May 15th." Goddard, Smith adds, "would have had great difficulty in seeing how the stereotypes expressed in his letter would be one of the seeds that led to the Holocaust." Clearly, Goddard accepted the popular stereotypes of Jews (and Scots). However, any analysis of his attitudes toward Jews should also consider his relationship with Samuel Kohs and his efforts to rescue Selina Krenberger – episodes not mentioned in Smith's book. (The Krenberger episode is discussed later in this chapter.)

90 On insect metaphors, see Edmund P. Russell III, "'Speaking of Annihilation': Mobilizing for War Against Human and Insect Enemies, 1914–1945," *Journal of American History* 84 (March 1996): 1505–1529. On German eugenics, see, for instance, Sheila Weiss, *Race, Hygiene and National Efficiency: The Eugenics of Wilhelm Schallmayer* (Berkeley: University of California Press, 1987); Robert Proctor, *Racial Hygiene: Medicine Under the Nazis* (Cambridge, Mass.: Harvard University Press, 1988); Paul Weindling, *Health, Race, and German Politics Between National Unification and Nazism, 1870–1945* (Cambridge: Cambridge University Press, 1989); and Henry Friedlander, *The Origins of Nazi Genocide: From Euthanasia to the Final Solution* (Chapel Hill: University of North Carolina Press, 1995).

91 Amram Scheinfeld, *You and Heredity* (New York: Frederick A. Stokes, 1939), pp. 361–362.

92 Knight Dunlap, "Antidotes for Superstitions Concerning Human Heredity," *Scientific Monthly* 51 (September 1940): 221–225.

93 HHG, "In Defense of the Kallikak Study," *Science* 95 (June 5, 1942): 574–576.

94 Ibid., p. 575. HHG is referring to Ernest Bayles and R. Will Burnett, *Biology for Better Living* (New York: Silver Burdett, 1942).

95 HHG, "In Defense," pp. 574–575.

96 Lewis Terman to HHG, August 3, 1942; Carl Seashore to HHG, June 15, 1942; Helen Hill to HHG, July 8, 1942; CBD to HHG, July 23, 1942, File M33.1, GP.

97 Scheinfeld, "The Kallikaks After Thirty Years," *Journal of Heredity* 35 (September 1944): 259–264.

98 Ibid. See also Scheinfeld, *The New "You and Heredity"* (Philadelphia: Lippincott, 1950), pp. 525–530.

99 Salomon Krenberger was editor of *Eos,* a small journal dedicated to the study of handicapped children, which translated several of Goddard's articles. Goddard's meeting with Freud is cited in Samuel Kohs to Edwin Schanfarber, September 28, 1938, File M31, GP.

100 Samuel Fels to HHG, August 10, 1938. Kohs to Schanfarber, September 28, 1938; Samuel Luchs of Schanfarber and Shanfarber, Attorneys, to HHG, October 3, 1938. Kohs was national field director of the National Coordinating Committee for Aid to Refugees and Emigrants Coming from Germany. Edwin Schanfarber was president of the Columbus Jewish Welfare Federation. Apparently, Krenberger's affidavits were from persons she did not know.

101 Selina Krenberger to HHG, March 10, 1939, GP. Delays were caused by the closing of embassies, regulations regarding the exchange of currency, and other bureaucratic obstacles, many designed purposely to discourage or impede emigration.

102 Krenberger to HHG [copy of letter sent to HHG's lawyer], October 26, 1939, File M37, GP. There are no further letters. According to Viennese records, Selina Krenberger died in a small Jewish hospital in 1941 from heart, liver, and lung problems, as well as jaundice, at the age of forty-eight. Mrs. H. Weiss, Israelitische Kultus-gemeinde Wien, Department of Records, to Leila Zenderland, May 21, 1996.

103 HHG, "A Partial Reply to Mrs. X's criticism of the Forgotten Gospel," File M31.1, GP. "Mrs. X" is probably Mrs. Bole.

104 Ibid.

105 Undated obituary, "Mrs. Goddard, Wife of O.S.U. Professor, Dies," File M32 Ephemeras; HHG to Mrs. Seashore, January 27, 1952, File M33.1, GP.

106 George Arps to Edgar Doll, October 23, 1936, File M227, Doll Papers. HHG's unwilling retirement is cited in H. E. Burtt and S. L. Pressey, "Henry Herbert Goddard: 1866–1957," *American Journal of Psychology* 70 (December 1957): 656–657. On Goddard's reclusive retirement years, see Marie Skodak Crissey to John Popplestone, File M35.2, GP.

107 Samuel Kohs to Edgar Doll, January 1, 1941, File M233; Edgar Doll to Martha Hall, March 23, 1939, and January 19, 1940, File M242; and Helen Hill to Edgar Doll, January 10, 1941, File M233, Doll Papers.

108 Cephas Guillet, *The Forgotten Gospel* (Dobbs Ferry, N.Y.: Cephas Guillet, 1940); Guillet, "Plan for a Peaceful World," File M31.1 Guillet Corr., GP. HHG, "Partial Reply to Mrs. X," [sent to Guillet]; Cephas Guillet to HHG, December 12, 1944, File M31.1, GP.

109 HHG, "Partial Reply to Mrs. X."

110 Ibid. HHG to Joseph Byers, June 17, 1949, File M35.2 Misc. 1, GP.

111 Ibid.

112 HHG to Albert Einstein, May 6, 1947; Einstein to HHG, October 8, 1947; File M31; William Donovan to HHG, January 24, 1951; HHG to Donovan, February 2, 1951, File M32, GP.

113 HHG, "A Suggested Definition of Intelligence," *American Journal of Mental Deficiency* 50 (1945): 245–250; HHG, "What Is Intelligence?" *Journal of Social Psychology* 24 (August 1946): 51–69.

114 HHG, Christmas letter [1945], Fischer Papers.

115 Helen Reeves to HHG, April 4, 1946, File M37, GP.

116 George Crane to HHG, May 8, 1945, File M36, GP.

117 Ibid.

118 HHG to George Crane, May 20, 1946, File M36, GP.

119 HHG to Dr. Ordway Tead, Harper and Brothers, November 20, 1943; Tead to HHG, January 13, 1944; "An Outside Reader's Opinion of the Manuscript," File M31, GP.

120 George Crane to HHG, May 26, 1946, File M36, GP. HHG, *How to Rear Children in the Atomic Age* (Mellott, Ind.: Hopkins Syndicate, 1948), p. vi (also cited as *Our Children in the Atomic Age,* which appears on its title page, apparently as a printing error). Crane's book is described in an advertising order form, File M36, GP. By 1949, HHG's book had sold 1,039 copies; see "Sales Report as of October 31, 1949," File M36, GP.

121 This circular of September 10, 1946, signed by C. E. Miller, is cited in HHG to George Crane, May 5, 1947, File M36, GP.

122 Ibid.

123 George Crane, "A Tribute," in HHG, *How to Rear Children,* p. i.

124 HHG, *How to Rear Children,* pp. 10–11.

125 George McClelland to HHG, June 7, 1946, File M31; HHG to McClelland, July 3, 1946, File M31; Program from "University of Pennsylvania Convocation of the University Council in Honor of the Founding of the Psychological Clinic," September 5, 1946, in Notebook #6, File M31; "Citation accompanying the honorary degree of Doctor of Science conferred on September 5, 1946," File M35, GP.

126 "Dr. Goddard Moves to California; Weighs Title for Autobiography," unidentified clipping from Columbus newspaper, Fischer Papers.

127 Robert Fischer to HHG, October 17, 1950, File M33, GP. HHG to Fischer, October 18, 1948; HHG to Fischer, February 15, 1948, Fischer Papers.

128 Pieces for an autobiography are scattered throughout the Goddard Papers. See, for example, "Memorandum of Summers of HHG," File M33, GP. Fischer to HHG, February 10, 1948, Fischer Papers.

129 HHG to Robert Fischer, February 15, 1948, and March 2, 1948, Fischer Papers.

130 HHG to Robert Fischer, October 23, 1950, File M33, GP.

131 Nicholas Pastore to Lewis Terman, February 28, 1949, Terman Papers. HHG's quotations suggest that he received the same letter.

132 Pastore's original version is in the Fischer Papers. I cite the published version, which contains some very minor stylistic changes; see *The Nature–Nurture Controversy* (New York: King's Crown Press, 1949), pp. 77–84.

133 HHG to Nicholas Pastore, April 3, 1948, File M32 Ephemeras, GP.

134 HHG to Robert Fischer, April 5, 1948, Fischer Papers.

135 Fischer to HHG, April 9, 1948, Fischer Papers.

136 HHG to Fischer, April 15, 1948, Fischer Papers.

137 Pastore, *Nature–Nurture;* Goodwin Watson, "Foreword," pp. vii–ix.

138 Terman to Pastore, March 4, 1948; Terman to Goodwin Watson, March 15, 1948; Watson to Terman, March 18, 1948; Terman Papers.

139 Terman to Pastore, March 4, 1948. On Terman's liberal politics, see Minton, *Lewis Terman,* pp. 234–242.

140 Pastore, *Nature–Nurture,* p. 95; parts of Terman's letter are reprinted on pp. 94–95.

141 HHG, Christmas card to "Dear Friends" [no year], Fischer Papers; HHG, Christmas card "Addendum" [c. 1951], File 35.2 Misc. 1, GP.

142 HHG to "Dear Friends (or 'would be' Friends)," March 6, 1952, File M37, GP.

143 HHG, Christmas card "Addendum."

144 Ibid. HHG to Mrs. Seashore, January 27, 1952, File M33.1, GP.

145 "Dr. Henry Goddard, Psychologist, Dies; Author of 'The Kallikak Family' Was 90," *NYT,* June 22, 1957, p. 15.

146 Alice Nash to J. E. Wallace Wallin, August 6, 1957, Wallin Papers.

Epilogue

1 Alfred Binet and Theodore Simon, *The Development of Intelligence in Children,* trans. Elizabeth Kite (Baltimore: Williams & Wilkins, 1916), pp. 262–273.

2 Ibid., p. 262.

3 H. L. Mencken, *The American Language* (New York: Knopf, 1936, rev. 1963), pp. 207–208. In *Supplement I* (New York: Knopf, 1945), pp. 377–379, Mencken included a letter from Goddard on this word's origin. B. A. Botkin, ed., *Treasury of American Folklore* (New York: Crown, 1944), pp. 461–464; Levette Jay Davidson, "Moron Stories," *Southern Folklore Quarterly* 7 (June 1943): 101–104; and Ernest Baughman, "'Little Moron' Stories," *Hoosier Folklore Bulletin* 2 (1943): 17–18.

4 Henry E. Garrett and Hubert Bonner, *General Psychology,* 2nd rev. ed. (New York: American Book Company, 1961). On Garrett's ideas, see Richard Kluger, *Simple Justice: The History of 'Brown v. Board of Education' and Black America's Struggle for Equality* (New York: Knopf, 1976), pp. 482–484, 502–506; Allan Chase, *The Legacy of Malthus: The Social Costs of the New Scientific Racism* (New York: Knopf, 1977), pp. 154–155, 450–456, and Stephan Chorover, *From Genesis to Genocide: The Meaning of Human Nature and the Power of Behavior Control* (Cambridge, Mass.: MIT Press, 1980), pp. 47–48. Garrett's pamphlet, *Breeding Down* (Richmond: Patrick Henry Press, n.d.), is reprinted in Clarence Karier, ed., *Shaping the American Educational State: 1900 to the Present* (New York: Free Press, 1975), pp. 419–428.

5 George Yanok and Stanley Ralph Ross, "TV or Not TV," script for *The Kallikaks,* August 1, 1977; Vernon Scott, "Ebsen's Bonnie is a Kallikak," undated UPI clipping [Bonnie Ebsen, daughter of *Beverly Hillbillies* star Buddy Ebsen, played one of the Kallikaks]. Five episodes ran as a 1977 summer replacement. According to one reviewer, it portrayed "a would-be sharpie from the West Virginia coal country who now owns a filling station in Nowhere, Calif., along with his wife – apparently a mental deficient, a Daisy Mae daughter and a pop-off teen son." Cecil Smith, "Homely Hokum of The Kallikaks," *Los Angeles Times,* August 3, 1977, Part IV, p. 18. At least some viewers loved the show. One chastised Smith for overlooking "the significance of the name. Those antediluvians among us who studied sociology when defective heredity was believed to be responsible for the perpetration of 'undesirable characteristics' remember that 'Kallikak' was a fictitious name given to an actual family. . . ," she wrote. "I assumed the family name in the series was intentional," Smith replied, "and did, at least tongue-in-cheek, refer to the famous study." Smith, "Fans Defend the Kallikaks," *Los Angeles Times,* August 17, 1977, Part IV, p. 19. R. Chast, "The Jukes and Kallikaks Today," *The New Yorker* 63 (June 8, 1987): 25.

6 Goddard's writings are analyzed, for example, in Mark Haller, *Eugenics: Hereditarian Attitudes in American Thought* (New Brunswick, N.J.: Rutgers University Press, 1963); Hamilton Cravens, *The Triumph of Evolution: American Scientists and the*

Heredity-Environment Controversy, 1900–1941 (Philadelphia: University of Pennsylvania Press, 1978); and Daniel Kevles, *In the Name of Eugenics: Genetics and the Uses of Human Heredity* (New York: Knopf, 1985).

7 The Jensen controversy led the *Harvard Educational Review* to republish his essay and several critical responses in *Environment, Heredity, and Intelligence* (Cambridge, Mass.: Harvard Educational Review, 1969). See also Raymond Fancher, *The Intelligence Men: Makers of the IQ Controversy* (New York: Norton, 1985), pp. 197–201; and N.J. Block and Gerald Dworkin, eds., *The IQ Controversy: Critical Readings* (New York: Pantheon, 1976). Richard Herrnstein and Charles Murray, *The Bell Curve: Intelligence and Class Structure in American Life* (New York: Free Press, 1994); Russell Jacoby and Naomi Glauberman, eds., *The Bell Curve Debate: History, Documents, Opinions* (New York: Times Books, 1995); and Steven Fraser, ed., *The Bell Curve Wars: Race, Intelligence, and the Future of America* (New York: Basic Books, 1995).

8 Leon Kamin, *The Science and Politics of I.Q.* (Potomac, Md.: Lawrence Erlbaum, 1974). On the involvement of both Kamin and Arthur Jensen in this controversy, see Fancher, *Intelligence Men,* pp. 201–223. On Burt, see L. S. Hearnshaw, *Cyril Burt, Psychologist* (Ithaca, N.Y.: Cornell University Press, 1979). Recent efforts to defend Burt include Robert B. Joynson, *The Burt Affair* (London: Routledge, 1989), and Ronald Fletcher, *Science, Ideology, and the Media: The Cyril Burt Scandal* (New Brunswick, N.J.: Transaction, 1991). For a sharply critical analysis of the problems with these defenses, see Franz Samelson, "Rescuing the Reputation of Sir Cyril [Burt]," *JHBS* 28 (1992): 221–233.

9 Stephen Jay Gould, *The Mismeasure of Man* (New York: Norton, 1981), pp. 171–174. This charge was repeated in Fancher, *Intelligence Men,* p. 114 (and later retracted – see note 11 below); and J. David Smith, *Minds Made Feeble: The Myth and Legacy of the Kallikaks* (Rockville, Md.: Aspen Systems Corporation, 1985), pp. 83–84.

10 HHG to CBD, July 4, 1910, Davenport Papers (all emphases in original). For a fuller discussion, see Leila Zenderland, "On Interpreting Photographs, Faces, and the Past," *AP* 43 (September 1988): 743–744.

11 Raymond E. Fancher, "Henry Goddard and the Kallikak Family Photographs: 'Conscious Skulduggery' or 'Whig History'?" *AP* 42 (June 1987): 585–590. Fancher's theory is supported by Michael J. Kral, "More on Goddard and the Kallikak Family Photographs," *AP* 43 (September 1988): 745–746; and Sigrid S. Glenn and Janet Ellis, "Do the Kallikaks Look 'Menacing' or 'Retarded'?" *AP* 43 (September 1988): 742–743 (which challenged Gould's contention that the doctored photos actually make their subjects look "menacing" or "retarded"), as well as Zenderland, "On Interpreting Photographs." For a statement supporting Gould's theory, see J. David Smith, "Fancher on Gould, Goddard, and Historical Interpretation: A Reply," *AP* 43 (September 1988): 744–745.

12 HHG, "In Defense of the Kallikak Study," *Science* 95 (June 5, 1942): 574–576.

13 See, for example, Seymour Sarason and John Doris, *Psychological Problems in Mental Deficiency* (New York: Harper & Row, 1969). "We must strenuously resist assuming, implicitly or explicitly, that our practices are the result of cold logic and scientific facts unmediated by the knotty problems of values," these authors emphasize (p. 274). See also David Hothersall, *History of Psychology* (New York: Random

House, 1984), the only text exploring Goddard's work with the gifted in addition to his earlier writings.

14 Philip J. Pauly, "How Did the Effects of Alcohol on Reproduction Become Scientifically Uninteresting?" *Journal of the History of Biology* 29 (1996): 1–28. *Alcohol and Birth Defects: The Fetal Alcohol Syndrome and Related Disorders,* U.S. Department of Health and Human Services (Washington, D.C.: GPO, 1987). For a more personal exploration of this subject, see Michael Dorris, *The Broken Cord* (New York: Harper & Row, 1989).

15 Jane R. Mercer, *Labeling the Mentally Retarded* (Berkeley: University of California Press, 1973).

16 *Penry v. Lynaugh, Director, Texas Department of Corrections,,* 492 U.S. 302. Shelley Clarke, "A Reasoned Moral Response: Rethinking Texas's Capital Sentencing Statute after *Penry v. Lynaugh*," *Texas Law Review* 69 (December 1990): 407–471.

17 For views emphasizing motives, see, for example, Gould, *Mismeasure of Man;* Chase, *Legacy of Malthus;* and Smith, *Minds Made Feeble.*

18 Healy is quoted in Florence Mateer to HHG, April 24, 1918, File M33 AA2; Dr. George Arps to Dr. Charles Berry, August 4, 1930, File M35; Lois Hilgeman to HHG, no date, File M31, GP.

19 Walter L. Connors to HHG, October 13, 1947, File M35.2, GP.

20 Marion Hibler to HHG, October 30, 1922, File M36; Mollie Woods to HHG, September 17, 1946, File M37; Martin L. Reymert to HHG, December 13, 1946, File M37; Samuel Kohs to HHG, January 8, 1946, File M42; Samuel Kohs to Edwin J. Schanfarber, September 28, 1938, File M31, GP. "Deborah Kallikak" is quoted in H. T. Reeves, "The Later Years of a Noted Mental Defective," *JPA* 43 (1938): 194.

21 Edgar Doll, "H. H. Goddard and the Hereditary Moron," *Science,* n. s. 126 (August 23, 1957): 343–344.

22 Barbara Kalbfell, "Dr. Marie Skodak Crissey Interview, October 31 and November 5, 1979," p. 16, AHAP. On Skodak Crissey and the Iowa Station, see Hamilton Cravens, *Before Head Start: The Iowa Station and America's Children* (Chapel Hill: University of North Carolina Press, 1993), pp. 191–212.

23 Criticisms of paternalism are too ubiquitous to cite. On recent debates analyzing the concept of "maternalism," see, for example, Lynn Weiner et. al., "Maternalism as a Paradigm: Defining the Issues," *Journal of Women's History* 5 (Fall 1993): 96–115, as well as the literature cited in Chapter 7, note 2.

24 David J. Rothman, "Introduction" to Willard Gaylin, Ira Glasser, Steven Marcus, and Rothman, *Doing Good: The Limits of Benevolence* (New York: Pantheon, 1978), p. x.

25 Explorations of "social control" have influenced the historical study of education, crime, poverty, political reform, and social welfare policies. See Stanley Cohen and Andrew Scull, eds., *Social Control and the State* (New York: St. Martin's Press, 1983); and Jack P. Gibbs, ed., *Social Control: Views from the Social Sciences* (Beverly Hills, Calif.: Sage, 1982). Especially influential in exploring the changing nature of political controls are Michel Foucault's works; see *Madness and Civilization: A History of Insanity in the Age of Reason* (New York: Pantheon, 1965); and *Discipline and Punish: The Birth of the Prison,* trans. Alan Sheridan (New York: Vintage, 1979).

26 For an intriguing analysis of this trend toward dichotomization in the history of institutions, see Gerald Grob, "Rediscovering Asylums: The Unhistorical History of the Mental Hospital," in Morris J. Vogel and Charles Rosenberg, eds., *The Therapeutic Revolution* (Philadelphia: University of Pennsylvania Press, 1979), pp. 135–157; and "The History of the Asylum Revisited: Personal Reflections," in Mark S. Micale and Roy Porter, eds., *Discovering the History of Psychiatry* (New York: Oxford University Press, 1994).

27 Mark Adams, ed., *The Wellborn Science: Eugenics in Germany, France, Brazil, and Russia* (New York: Oxford, 1990). A similar comparison could be made concerning intelligence testing in France, England, Germany, the Netherlands, and other countries.

28 Between 1995 and 1997, the Vineland Training School moved all of its 196 former residents into 29 small group homes located throughout southern New Jersey. On this transition, see the school's current publication, *LifeStyles* 1:1 (1995) and 1:2 (1996).

29 On Davenport's career, see Kevles, *Name of Eugenics,* pp. 41–56; on Laughlin, see Smith, *Minds Made Feeble,* pp. 137–158. Alexander Johnson, *Adventures in Social Welfare* (Fort Wayne, Ind.: Alexander Johnson, 1923).

30 Elizabeth Kite, "Something about the Author" from "Autobiography: A 'Conversion Story,'" Box 1, Kite Papers. Kite produced *Beaumarchais and the War of American Independence,* 2 vols. (Boston: Badger, 1918); *L'Enfant and Washington* (Baltimore: Johns Hopkins University Press, 1929); *Brigadier-General Louis Lebeque Duportail* (Baltimore: Johns Hopkins University Press, 1933); and *Lafayette and His Companions on the "Victoire"* (Philadelphia: Catholic Historical Society, 1934). On Jones, see Elizabeth Gray Vining, *Friend of Life: The Biography of Rufus M. Jones* (Philadelphia: Lippincott, 1958).

31 HHG to Robert Fischer, October 23, 1950, File M33, GP.

32 Ibid.

33 Joseph Byers, *The Village of Happiness: The Story of the Training School* (Vineland, N.J.: Smith, 1934), p. 73.

Major Manuscript Collections Consulted

*AMERICAN PHILOSOPHICAL SOCIETY, PHILADELPHIA, PENNSYL-
VANIA*
American Eugenics Society Papers
Charles Benedict Davenport Papers

*ARCHIVES OF THE HISTORY OF AMERICAN PSYCHOLOGY,
UNIVERSITY OF AKRON*
Edgar Doll Papers
Robert Fischer Papers
Henry Herbert Goddard Papers
Vineland Research Laboratory Papers
J. E. Wallace Wallin Papers

CHICAGO HISTORICAL SOCIETY
Chicago Municipal Court Papers

CLARK UNIVERSITY ARCHIVES
G. Stanley Hall Papers

COLLEGE OF PHYSICIANS OF PHILADELPHIA
Francis C. Wood Medical History Collection

HAVERFORD COLLEGE, HAVERFORD, PENNSYLVANIA
Quaker Collection

JOHNS HOPKINS UNIVERSITY MEDICAL SCHOOL
Adolf Meyer Papers

LIBRARY OF CONGRESS, WASHINGTON, D.C.
Arnold Gesell Papers

NATIONAL ACADEMY OF SCIENCES, WASHINGTON, D.C.
National Research Council Records

NATIONAL ARCHIVES, WASHINGTON, D.C.
Children's Bureau Records, Department of Labor

RUTGERS UNIVERSITY ARCHIVES
Elizabeth Kite Papers
Edward Ransom Johnstone Papers

STANFORD UNIVERSITY ARCHIVES
Lewis Madison Terman Papers

TEACHERS COLLEGE, COLUMBIA UNIVERSITY
New York City Board of Education Records

VASSALBORO HISTORICAL SOCIETY, EAST VASSALBORO, MAINE

YALE UNIVERSITY
Robert M. Yerkes Papers

Publications by
Henry Herbert Goddard

Listed below are the main publications by Henry Herbert Goddard. Not included are the reports Goddard produced annually for the Vineland Training School's *Annual Report;* commentaries by Goddard recorded in the proceedings of meetings or conferences; and foreign translations of Goddard's articles.

"Are Drugs Unnecessary to the Cure of Disease?" *The Hypnotic Magazine* (March 1897):155–158.

"The Effects of Mind on Body as Evidenced by Faith Cures." *American Journal of Psychology* 10 (1899): 431–502.

"A New Brain Microtome." *Journal of Comparative Neurology* 10 (May 1900): 209–213."

Child Study for Pennsylvania Teachers," in "Some Report of What Was Done in the Sections" of the "Round Table Conferences at State Association." *PSJ* 49:3 (September 1900): 127–130.

"Child Study: Special Circular from Child Study Department." *PSJ* 49:5 (November 1900): 227–229.

"Negative Ideals." In *Studies in Education,* edited by Earl Barnes. Philadelphia: Earl Barnes, 1902, 2:392–398.

"Child Study" in "Some Report of What Was Done in the Sections" of "Department Conferences at State Association." *PSJ* 54:3 (September 1905): 120–125.

"Ideals of a Group of German Children." *PS* 13 (1906): 208–220.

"Psychological Work Among the Feeble-Minded." *JPA* 12:1 (September 1907): 18–30.

"The Research Work." *TSB* Supplement 1, No. 46 (December 1907): 1–9.

"A Side Light on the Development of the Number Concept." *TSB* Supplement 1, No. 46 (December 1907): 20–25.

"A Group of Feeble-Minded Children with Special Regard to their Number Concept." *TSB* Supplement 2, No. 49 (March 1908): 1–16.

With Alice Morrison. "Teaching Numbers to Backward Children." *TSB* 5:2 (April 1908): 4–7.

"Two Months Among the European Institutions." *TSB* 5:5 (July 1908): 11–16.

"The Fool Class." *TSB* 5:7 (September 1908): 7–8.

"Impressions of European Institutions and Special Classes." *JPA* 13:1 (September 1908): 18–28.

"The Grading of Backward Children." *TSB* 5:9 (November 1908): 12–14.

"New Apparatus for Testing Hearing." *TSB* 5:9 (November 1908): 5.

"The Binet and Simon Tests of Intellectual Capacity." *TSB* 5:10 (December 1908): 3–9.

"Summer School." *TSB* 5:11 (January 1909): 10–11.

"Will the Backward Child Outgrow Its Backwardness?" *TSB* 5:11 (January 1909): 1 3.

"Bibliography of Mental Deficiency." *TSB* 6: 2 and 3 (April and May 1909): 11–19.

"A Growth Curve for Feeble-Minded Children: Height and Weight." *JPA* (September 1909): 9–13.

"A Measuring Scale for Intelligence," *TSB* 6:11 (January 1910): 146–155.

"Research in School Hygiene: Experiences in an Institution for the Feebleminded." *PS* 17 (March 1910): 51–53.

"The Story of Abbie." *TSB* 7:1 (March 1910): 182–185.

"The Story of Albert." *TSB* 7:2 (April 1910): 193–198.

"Suggestions for a Prognostical Classification of Mental Defectives." *JPA* 14: 1– 4 (September and December 1909/March and June 1910): 48–54.

"What Can the Public School Do for Sub-Normal Children?" *TSB* 7:5 (September 1910): 242–248.

"An American Institution for Feeble-Minded Children." *Education* (October 1910).

"The Story of Charlie." *TSB* 7:6 (October 1910): 262–265.

"The Institution for Mentally Defective Children an Unusual Opportunity for Scientific Research." *TSB* 7:7 (November 1910): 275–278, and 7:8 (December 1910): 293–296.

"Four Hundred Feeble-Minded Children Classified by the Binet Method." *JPA* 15:1 and 2 (September and December 1910): 17–30.

"Heredity of Feeble-Mindedness." *American Breeders' Magazine* 1 (Third Quarter, 1910):165–178. Also abridged in *Proceedings of the American Philosophical Society* 51 (July 1912): 173–177.

"The Application of Educational Psychology to the Problems of the Special Class." *JEP* 1 (1910): 521–531.

"Research in School Hygiene, in the Light of Experiences in an Institution for the Feeble Minded. Abstract." In *Lectures and Addresses Delivered Before the Departments of Psychology and Pedagogy in Celebration of the Twentieth Anniversary of the Opening of Clark University.* Worcester, Mass.: Clark University, 1910, pp. 51–53.

"What Can the Public Schools Do for Subnormal Children?" *NEA* (1910): 912–919.

"Two Thousand Children Tested by the Binet Scale." *TSB* 7:9 (January 1911): 310–312.

"The Elimination of Feeble-Mindedness." *Annals of the American Academy of Political and Social Science* 37 (March 1911): 505–516.

With Helen F. Hill. "Feeble-Mindedness and Criminality." *TSB* 8:1 (March 1911): 3–6.

With Helen F. Hill. "Delinquent Girls Tested by the Binet Scale." *TSB* 8:4 (June 1911): 50–56.

"The Menace of the Feeble-Minded." *Pediatrics* 83 (June 1911): 350–359.

"A Revision of the Binet Scale." *TSB* 8:4 (June 1911): 56–62.

"Two Thousand Normal Children Measured by the Binet Measuring Scale." *PS* 18 (June 1911): 232–259.

"Two Thousand Children Tested by the Binet Scale." *Journal of Education* 74 (July 20, 1911): 99.

"Causes of Backwardness and Mental Deficiency in Children." *Journal of Education* 74 (July 27, 1911): 125.

"The Story of Camilla." *TSB* 8:5 (September 1911): 70–76. Also in *American Baby* (November 1911).

"The Story of Logan." *TSB* 8:6 (October 1911): 84–88.

"Wanted: A Child to Adopt." *Survey* 27 (October 14, 1911): 1003–1006.

"The Story of Harry." *TSB* 8:7 (November 1911): 97–100.

"Experience Versus Training." *TSB* 8:8 (December 1911): 119–121.

"The Juvenile Court Studying Children." *TSB* 8:8 (December 1911): 121–122.

"Suggestions for a Special Class Room." *TSB* 8:8 (December 1911): 114–116.

"The Bearing of Heredity Upon Educational Problems." *JEP* 2 (1911): 491–497.

"Defectives," and "Defectives, Schools for." In *A Cyclopedia of Education.* Edited by Paul Monroe. New York: Macmillan, 1911, pp. 275–280.

"Échelle Métrique de L'intelligence de Binet-Simon. Résultats Obtenus en Amérique a Vineland, N.J.," *L'Année Psychologique* 18 (1911): 288–326.

With Charles Benedict Davenport, H. H. Laughlin, David Weeks, E. R. Johnstone. *The Study of Human Heredity.* Cold Spring Harbor, N.Y.: Eugenics Record Office, Bulletin No. 2, 1911.

"The Treatment of the Mental Defective Who Is Also Delinquent." *NCCC* (1911): 64–65.

"Two Thousand Children Tested by the Binet Measuring Scale for Intelligence." *NEA* (1911):870–878.

With Mrs. E. Garfield Gifford. "Defective Children in the Juvenile Court." *TSB* 8:9 (January 1912): 132–135.

"The Significance of Feeble Mindedness." *Life and Health* (January 1912): 26–29.

"The Story of David." *TSB* 8:9 (January 1912): 129–131.

"Estimated Number of Feeble Minded Persons in State Reformatories and Industrial Schools." *TSB* 9:1 (March 1912): 8–10.

"The Basis for State Policy: Social Investigation and Prevention." *Survey* 27 (March 2, 1912):1852–1856.

"Phonetic Keys and the Scientific Attitude." *JEP* 3 (April 1912): 219–221.

"How Shall We Educate Mental Defectives?" *TSB* 9:3 (May 1912): 42–45, and 9:4 (June 1912): 56–61.

"The Form Board as a Measure of Intellectual Development in Children." *TSB* 9:4 (June 1912): 49–52.

"Importance of Field Workers Studying Heredity in Charity Work." *TSB* 9:4 (June 1912): 52–54.

"Joint Meeting on the Feeble-Minded and Epileptic." *Survey* 28 (June 15, 1912): 446–447.

"Sterilization and Segregation." *Bulletin of the American Academy of Medicine* 13 (August 1912): 210–219. Reprinted in *The Child* (September 1912) and as *Publications of the Russell Sage Foundation, Pamphlet No. 12.* New York: Department of Child Helping, 1913.

"Extension of the Research Work." *TSB* 9 (September 1912): 68–70.

"The Responsibility of Children in the Juvenile Court." *Journal of Criminal Law and Criminology* 3 (September 1912): 365–375.

"Feeble-Mindedness and Immigration." *TSB* 9:6 (October 1912): 91–94.

"Hygiene of the Backward Child." *American Physical Education Review* 17 (October 1912): 537–539. Also in *TSB* 9:7 and 8 (November and December 1912): 114–116.

"The Feeble Minded Immigrant." *TSB* 9:7 and 8 (November and December 1912): 109–113.

"The Story of Billikens." *TSB* 9:7 and 8 (November and December 1912): 101–104.

"Feeble-Mindedness and Crime." *Proceedings of the American Prison Association* (1912): 353–357.

"The Height and Weight of Feeble-Minded Children in American Institutions." *Journal of Nervous and Mental Disorders* 39 (1912): 217–235.

"Heredity of Feeble-Mindedness." *Proceedings of the American Philosophical Society* 51 (1912): 173–177.

The Kallikak Family: A Study in the Heredity of Feeble-Mindedness. New York: Macmillan, 1912.

"Ungraded Classes." In *Report on Educational Aspects of the Public School System of the City of New York to the Committee on School Inquiry of the Board of Estimate and Apportionment.* Edited by Paul Hanus. Part II, Subdivision I, Elementary Schools, Section E, pp. 361–381. New York: City of New York, 1911–1912.

"Infant Mortality in Relation to the Hereditary Effects of Mental Deficiency." *Virginia Medical Semi-Monthly* 17 (1912–1913): 534–537.

"The Binet Tests and the Inexperienced Teacher." *TSB* 10:1 (March 1913): 9–11.

"Feeble-Mindedness a Source of Prostitution." *Vigilance* (April 1913): 3–11.

"Standard Method for Giving the Binet Test." *TSB* 10:2 (April 1913): 23–30.

With Drs. Charles L. Dana, William N. Berkeley, and Walter S. Cornell. "Functions of the Pineal Gland with Report of Feeding Experiments." *Medical Record* 83 (May 10, 1913): 835–847.

"Heredity in Relation to Efficiency." *American Physical Education Review* 18 (May 1913): 304–308.

"The Hereditary Factor in Feeble-Mindedness." *The Institution Quarterly* (June 1913): 9–11.

"A City School for Defectives." *TSB* 10:6 (October 1913): 91.

"Who is Mentally Defective – How Many Are There and How Can They Be Detected?" *New England Medical Gazette* 48 (October 1913): 541–546.

"Kallikak Family." *Hearst's Magazine* 23 (Fall 1913): 329–331.

"Teachability of the Feeble-Minded." *JPA* 18 (1913): 54–60.

"The Binet Tests in Relation to Immigration." *JPA* 18 (1913): 105–107.

"The Diagnosis of Feeble Mindedness." *Illinois Medical Journal* 24 (1913): 141–143.

"The Educational Treatment of the Feeble-Minded." In *The Modern Treatment of Nervous and Mental Diseases.* Edited by William A. White and Smith Ely Jelliffe. Philadelphia: Lea & Febiger, 1913, 1:143–194.

"The Prevention of Feeble-Mindedness." *Transactions of the Fifteenth International Congress on Hygiene and Demography* (Washington, D.C.: GPO, 1913), 3:463–468.

"Three Annual Testings of 400 Feeble-Minded Children and 500 Normal Children." *PB* 10 (1913): 75–77.

"Mental Defectives in Boston Schools." *TSB* 11:1 (March 1914): 9–10.

"Relation of Feeble-Mindedness to Crime." *Bulletin of the American Academy of Medicine* 15 (April 1914): 105–112.

"The Research Department: What It Is, What It Is Doing, What It Hopes to Do." *Vineland Training School Publication No. 1* (May 1914).

"The Binet Measuring Scale of Intelligence. What it is, and How it is to be Used." *TSB* 11:6 (October 1914): 86–91.

"Brief Report on Two Cases of Criminal Imbecility; with Discussion." *JPA* 19 (1914): 31–35.

Feeble-Mindedness: Its Causes and Consequences. New York: Macmillan, 1914.

"The Reliability of the Binet–Simon Measuring Scale of Intelligence." In *Transactions of the Fourth International Congress on School Hygiene* (Buffalo: Courier, 1914), 5:693–699.

School Training of Defective Children. Yonkers-on-Hudson, N.Y.: World Book, 1914.

"Who is Mentally Defective – How Many Are There – and How Can They Be Detected?" *Transactions of the Fourth International Congress on School Hygiene* (Buffalo: Courier, 1914), 5:621–626.

"The Adaptation Board as a Measure of Intelligence." *TSB* 11:10 (February 1915): 182–188.

"The Hygienic Value of Grading a School According to the Intelligence of the Pupils." *Proceedings of the Eighth Congress of the American School Hygiene Association* 5 (June 1915): 157–162.

The Criminal Imbecile: An Analysis of Three Remarkable Murder Cases. New York: Macmillan, 1915.

"The Possibilities of Research as Applied to the Prevention of Feeble-Mindedness." *NCCC* (1915): 307–312.

"The Size of the Special Class." *Child-Study* 8 (1915): 124–125. Also in *TSB* 12 (June 1915): 106–107.

"Tests of Intelligence." *Reference Handbook of the Medical Sciences* (1915): 607–613.

"Syphilis in Parents as a Cause of Feeble-Mindedness in Children." *New York State Journal of Medicine* 16 (March 1916): 129–133.

"A Course of Study for Teachers of Mental Defectives." *School and Society* 3 (April 1, 1916): 497–502.

"Defectives in the Schools." *Teaching* 2 (April 1, 1916): 5–18.

"Mentality Tests: A Symposium." *JEP* 7 (April 1916): 231–233.

"Schools and Classes for Exceptional Children." *JEP* 7 (May 1916): 287–294.

"Alcoholism and Feeble-Mindedness." *Interstate Medical Journal* 23:6 (1916): 1–4.

"Introduction." In Alfred Binet and Theodore Simon, *The Development of Intelligence in Children.* Baltimore: Williams & Wilkins, 1916, pp. 5–8.

"The Menace of Mental Deficiency from the Standpoint of Heredity," *Massachusetts Society for Mental Hygiene,* No. 15 (1916): 1–9. Also abridged in *Boston Medical and Surgical Journal* 175 (1916): 269–271.

"Mental Deficiency." *Reference Handbook of the Medical Sciences* (1916): 379–385.

"Tests of Intelligence." *Reference Handbook of the Medical Sciences* (1916): 607–613.

"The Vineland Spirometer." *TSB* 13:10 (February 1917): 234–236.

"Syphilis as an Etiological Factor in Mongolian Idiocy." *JAMA* 68 (April 7, 1917): 1057.

"The Vineland Experience with Pineal Gland Extract." *JAMA* 68 (May 5, 1917): 1340. Also in *TSB* 14:4 (June 1917): 70–72.

"The Possibilities of Mental Hygiene in Cases of Arrested Mental Development." *Proceedings of the Tenth Congress of the American School Hygiene Association* 7 (June 1917): 264–269. Also in *TSB* 15:5 (September 1918): 67–72.

"Mental Tests and the Immigrant." *Journal of Delinquency* 2 (September 1917): 243–277.

"The Criminal Instincts of the Feeble-Minded," *Journal of Delinquency* 2 (November 1917): 352–355.

"Eugenics from the Professional Standpoint." *Journal of the Medical Society of New Jersey* 14 (1917): 62–65.

"The Mental Level of a Group of Immigrants." *PB* 14 (1917): 68–69.

"The Place of Intelligence in Modern Warfare." *United States Naval Medical Bulletin* 2:3 (1917): 1–9.

"The Tilting Board and Rotation Table." *Journal of Experimental Psychology* 2 (1917): 313–314.

"A Neglected Factor." *Journal of Education* 87 (March 14, 1918): 288–289.

"Social Efficiency." *Independent and Weekly Review* 96 (December 21, 1918): 404, 409.

"Report of the Research Department." *TSB* 15:8 (December 1918): 123–125.

"What Physicians Can Do to Help the Bureau of Juvenile Research." *Ohio State Medical Journal* 15 (March 1919): 144–150.

"The Clearing House for Juvenile Courts." *National Conference of Social Work* (1919): 67–69.

With Arnold Gesell and J. E. Wallace Wallin. "The Field of Clinical Psychology as an Applied Science: A Symposium." *Journal of Applied Psychology* (1919): 81–95.

Psychology of the Normal and Subnormal. New York: Dodd, Mead, 1919.

"In the Light of Recent Developments, What Should Be Our Policy in Dealing with the Delinquents, Juvenile and Adult?" *Journal of Criminal Law* 11 (November 1920): 426–432.

"The Defective Delinquent." *Indiana Bulletin of Charities and Corrections,* No. 123 (1920): 349–357.

"Feeble-Mindedness and Delinquency." *JPA* 25 (1920): 168–176.

Human Efficiency and Levels of Intelligence. Princeton: Princeton University Press, 1920.

"Psychopathology and Delinquency." *Proceedings of the American Prison Association* (1920): 405–412.

"The Problem of the Psychopathic Child." *American Journal of Insanity* 77 (April 1921): 511–516.

The Bureau of Juvenile Research: Review of the Work, 1918–20. Columbus: Ohio Board of Administration, 1921.

Juvenile Delinquency. New York: Dodd, Mead, 1921.

"The Sub-normal Mind Versus the Abnormal." *Journal of Abnormal Psychology* 16 (1921): 47–54.

"Relation of Congenital Syphilis to Juvenile Delinquency." *Ohio State Medical Journal* 18 (March 1922): 193–195.

"A Scientific Program of Child Welfare." *Annals of the American Academy of Political and Social Science* 105 (January 1923): 256–266.

"Bridging the Gap Between Our Knowledge of Child Well-being and Our Care of the Young." In *The Child: His Nature and His Needs.* Edited by M. V. O'Shea. New York: Children's Foundation, 1924.

With Violet H. Foster. "The Ohio Literacy Test." *PS* 31 (1924): 340–358.

"Special Education and the Gifted Child." *Educational Research Bulletin* 4 (April 1, 1925): 133–139.

"Some Fundamental Errors in Educational Practice." *Phi Delta Kappan* 8 (June 1925): 1–8.

"A Case of Dual Personality." *Journal of Abnormal and Social Psychology* 21 (July 1926): 170–191.

Two Souls in One Body? New York: Dodd, Mead, 1927.

"Who Is a Moron?" *Scientific Monthly* 27 (1927): 24, 41–46.

"Feeblemindedness: A Question of Definition." *JPA* 33 (1928): 219–227.

School Training of Gifted Children. Yonkers-on-Hudson, N.Y.: World Book, 1928.

"Levels of Intelligence and the Prediction of Delinquency." *Journal of Juvenile Research* 13 (October 1929): 262–265.

"The Gifted Child and His Education." *NEA Journal* 19 (November 1930): 275–276.

"Anniversary Address." In *Twenty-Five Years: A Memorial Volume in Commemoration of the Twenty-Fifth Anniversary of the Vineland Laboratory.* Edited by Edgar A. Doll. Vineland, N.J.: The Training School, 1932, pp. 53–69.

"In the Beginning." *Understanding the Child* 3:2 (April 1933): 2–6.

"The Gifted Child." *Journal of Educational Sociology* 6 (1933): 354–361.

"The Psychology of the Status Quo of Exceptional Children." *Journal of Exceptional Children* 5 (April 1939): 181–183.

"In Defense of the Kallikak Study." *Science* 95 (1942): 574–576.

"Dr. Meta Anderson Post." *TSB* 40:1 (March 1943): 1–4.

"In the Beginning." *TSB* 40:8 (December 1943): 154–161.

"A Suggested Definition of Intelligence." *American Journal of Mental Deficiency* 50 (1945): 245–250.

"What Is Intelligence?" *Journal of Social Psychology* 24 (August 1946): 51–69.

"A Suggested Definition of Intelligence." *TSB* 43:10 (February 1947): 185–193.

"Scientific Interests." *TSB* 44:3 (May 1947): 41–43.

How to Rear Children in the Atomic Age. Mellott, Ind.: Hopkins Syndicate, 1948.

Index